AUTOMATIC CONTROL ENGINEERING

McGraw-Hill Series in Mechanical Engineering

Jack P. Holman, *Southern Methodist University*
  Consulting Editor

Anderson: *Modern Compressible Flow: With Historical Perspective*
Eckert and Drake: *Analysis of Heat and Mass Transfer*
Hinze: *Turbulence*
Hutton: *Applied Mechanical Vibrations*
Juvinall: *Engineering Considerations of Stress, Strain, and Strength*
Kane and Levinson: *Dynamics: Theory and Applications*
Kays and Crawford: *Convective Heat and Mass Transfer*
Martin: *Kinematics and Dynamics of Machines*
Phelan: *Dynamics of Machinery*
Phelan: *Fundamentals of Mechanical Design*
Pierce: *Acoustics: An Introduction to Its Physical Principles and Applications*
Raven: *Automatic Control Engineering*
Rosenberg and Karnopp: *Introduction to Physical System Dynamics*
Schlichting: *Boundary-Layer Theory*
Shames: *Mechanics of Fluids*
Shigley: *Kinematic Analysis of Mechanisms*
Shigley-Mitchell: *Mechanical Engineering Design*
Shigley and Uicker: *Theory of Machines and Mechanisms*
Stoecker and Jones: *Refrigeration and Air Conditioning*
Vanderplaats: *Numerical Optimization Techniques for Engineering Design:*
  With Applications

AUTOMATIC
CONTROL
ENGINEERING

Fourth Edition

Francis H. Raven

Professor of Mechanical Engineering
University of Notre Dame

McGraw-Hill Book Company

New York   St. Louis   San Francisco   Auckland   Bogotá   Hamburg
Johannesburg   London   Madrid   Mexico   Milan   Montreal   New Delhi
Panama   Paris   São Paulo   Singapore   Sydney   Tokyo   Toronto

This book was set in Times Roman by Santype International Limited.
The editor was Anne Duffy; the cover was designed by Scott Chelius;
the production supervisor was Diane Renda.
Project supervision was done by Santype International Limited.
R. R. Donnelley & Sons Company was printer and binder.

AUTOMATIC CONTROL ENGINEERING

1234567890 DOCDOC 8943210987

ISBN 0-07-051233-7

Library of Congress Cataloging-in-Publication Data

Raven, Francis H. (Francis Harvey).
  Automatic control engineering.

  (McGraw-Hill series in mechanical engineering)
  Includes bibliographical references and index.
  1. Automatic control.   I. Title.   II. Series.
TJ213.R38   1987   629.8'3   86-10705
ISBN 0-07-051233-7

To Therese

CONTENTS

PREFACE

In recent years, automatic control systems have been rapidly increasing in importance in all fields of engineering. The applications of control systems cover a very wide range, from the design of precision control devices such as delicate electronic equipment to the design of massive equipment such as that used for the manufacture of steel or other industrial processes. Microprocessors have added a new dimension to the capability of control systems. New applications for automatic controls are continually being discovered.

This text is the outgrowth of notes developed by the author to teach control engineering at the University of Notre Dame. The author has endeavored to give the principles a thorough presentation and yet make them clear and easy to understand. It is assumed that the reader has the general maturity and background of a third- or fourth-year engineering student, but no previous training in control engineering.

Although the principles of feedback control systems are presented in a manner which is appropriate to the interests of mechanical engineers, this text has also been successfully used to teach students in other fields of engineering. In addition, the author has taught night courses for practicing engineers. In the light of their enthusiastic comments, it is felt that this book will be of much value to the engineer in industry who did not have the opportunity to take such a course while in college.

The study of control engineering is begun by showing how typical control systems may be represented by block diagrams. This is accomplished by first demonstrating how to represent each component or part of a system as a simple block diagram, then explaining how these individual diagrams may be connected to form the overall block diagram, just as the actual components are connected to form the complete control system. Because actual control systems frequently contain nonlinear components, considerable emphasis is given to such components. This material is presented in the first three chapters. In the fourth chapter, it is shown that much important information concerning the basic or inherent operating characteristics of a system may be obtained from knowledge of the steady-state behavior.

This introduction to control theory differs from the usual "black-box" approach, in which the block diagram for a system is given outright. The "black-box" approach permits the introduction of Laplace transforms and other methods of system analysis at an earlier stage. However, it has been the author's experience that students attain a deeper understanding of the various techniques used in system analysis if they are first familiarized with the physical significance of feedback controls.

In Chapter 5, it is shown how the linear differential equations which describe the operation of control systems may be solved algebraically by the use of Laplace transforms. It is seen that this method reveals directly much interesting information about systems. General characteristics of transient behavior are described in Chapter 6. It is pointed out that the transient response is governed by the location of the roots of the characteristic equation. It is also shown how to program a digital computer for investigating the performance of control systems. The application of the root-locus method to the design of control systems is the topic of Chapter 7. The use of the analog computer to simulate control systems is explained in Chapter 8.

State-space methods are presented in Chapter 9. It is shown how the concepts of classical control theory are combined with state-space concepts to yield the "modern" control theory. By seeing how "modern" control theory builds on classical methods, the reader attains both a better understanding of modern methods and a deeper appreciation of the classical formulation. Digital computers and microprocessors are becoming increasingly important as elements of control systems. As is explained in Chapter 10, most of the methods (both classical and modern) used in the design of continuous-data systems may be extended to digital-control systems. The term digital-control system comprises a broad class of systems including computer-controlled systems, discrete-data systems, sampled-data systems, and time sharing systems. The use of frequency-response techniques to evaluate dynamic performance is explained in Chapters 11 and 12.

McGraw-Hill and the author wish to express their appreciation for the many fine suggestions made by teachers who have used the previous editions. These suggestions have been of great value in the preparation of this revised edition. Special gratitude is owed to Professor Haim Baruh, Rutgers University; Professor Celal Batur, University of Akron; Professor Ronald Bottaccini, University of Wisconsin; Professor Frank D'Souza, Illinois Institute of Technology; Professor Robert G. Leonard, Virginia Polytechnic Institute and State University; Professor Robert A. Medrow, University of Missouri, Rolla; Professor Ioannis O. Pendelidis, University of Maryland; Professor David J. Purdy, Texas A&M University; Professor Robert A. Smoak, Tennessee Technological University; and Professor Gerald Whitehouse, Louisiana State University, who reviewed the entire manuscript. Students have also been very helpful in their comments. Particular recognition is due to Betty Raven.

The author also wishes to express his gratitude for the continued encouragement of his colleagues at the University of Notre Dame, especially to Professor Albin A. Szewczyk, Chairman of the Department of Aerospace and Mechanical Engineering and to Roger A. Schmitz, Keating-Crawford Professor, and Dean, College of Engineering. Thanks are also due to Ms. Susan Jensen for her typing of the numerous revisions of the notes from which this textbook has been developed.

The author's wife, Therese, has faithfully worked with him throughout the development of this text. She has made innumerable suggestions and has been a constant source of encouragement.

Francis H. Raven

INTRODUCTION TO AUTOMATIC CONTROLS

1.1 HISTORICAL DEVELOPMENT

Early people had to rely upon their own brute strength or that of beasts of burden to supply energy for doing work. By the use of simple mechanical devices such as wheels and levers, they accomplished such feats as the building of high pyramids and Roman highways and aqueducts. They first supplemented their energy and that of beasts by utilizing power from natural sources, such as the wind for powering sailing vessels and windmills, and waterfalls for turning water wheels. The invention of the steam engine was a milestone in human progress because it provided people with useful power that could be harnessed at will. Since then, people have devised many different means for obtaining abundant and convenient sources of energy. Engineering effort is primarily concerned with the practical applications of using power to serve human purposes. That is, the engineer designs and develops machines and equipment by which people can utilize power.

Early machines and equipment had controls which were predominantly of a manual nature, and the adjustments had to be reset frequently in order to maintain the desired output or performance. The design of newer equipment with greater usefulness and capabilities is bringing about an ever-increasing growth in the development of control equipment. The reason is twofold. First, automatic controls relieve people of many monotonous activities so that they can devote their abilities to other endeavors. Second, modern complex controls can perform functions which are beyond the physical abilities of people to duplicate. For example, an elaborate automatic control system operates the engine of a modern jet airplane with only a minimum amount of the pilot's attention so that the pilot is free to maneuver and fly the airplane.

It is interesting to note that, as the applications and uses for controls have increased, so also have the demands upon the performance of these systems increased. There is no doubt that a major concern of the engineer today, and even more so in the future, is, and will be, the design and development of automatic control systems.

1.2 FEEDBACK CONTROL SYSTEMS

Various forms of transportation are illustrated in Fig. 1.1. All of the basic concepts of feedback control systems are contained in each of these means of transportation. For the basic form of transportation, walking, the desired speed at which the walker wishes to go is the reference input. When taking a leisurely stroll through a park, the desired speed is slow. When in a hurry, the desired speed is fast. The actual speed is the controlled variable (i.e., the quantity being controlled). The part of a system which compares the reference input (desired value) with the controlled variable (actual value) in order to measure the error is called the comparator. The brain serves as the comparator for the walker. Typically, the error signal goes to a power-amplifying device (muscles in the legs) which actuates the system to be controlled so as to reduce the error to zero. Thus, the actual speed is the same as the desired speed.

The driver of a car compares the actual speed with the desired speed. Again the brain serves as the comparator. In response to the error signal, the driver changes the position of the accelerator pedal. The engine serves as the power-amplifying device. That is, a little motion of the accelerator pedal causes a substantial change in the power produced by the engine.

Figure 1.1 Various forms of transportation.

For cruise control, a magnet is attached to the drive shaft which rotates at some constant times the speed of the wheels. A magnetic pick-up fixed to the body of the car generates a pulse every time the drive shaft makes one complete rotation. The rate at which pulses are being generated is a measure of the actual speed of the car. At the instant the cruise control is set, the rate at which pulses are being generated is stored so as to serve as the reference input (i.e., desired speed). A vacuum-operated device is attached to the accelerator linkage. When the actual speed is greater than the reference speed, the rate at which pulses are being generated is greater than the reference value. This error signal goes to the vacuum device which serves as a power amplifier to change the position of the accelerator linkage so as to decrease the actual speed.

The most difficult problem which the Wright brothers encountered in the process of making the airplane into a useful machine was the control problem. After much experimentation, they developed an ingenious pulley system to coordinate the motion of the aileron and the rudder so as to produce a smooth turning effort. Note that the pilot (Orville or Wilbur) would compare the actual path of the plane with the desired path, and then in response to the error signal he would actuate the pulley system. Aerodynamic forces acting on the wings and rudder would serve as the power-amplifying device to change the trajectory of the plane. Although modern airplanes have much more sophisticated means for controlling the position of the ailerons and rudder, the principles are still the same.

The controlling of temperature is a typical example of a feedback control system. The position of the temperature dial sets the desired temperature (i.e., the reference input). The actual temperature of the system is the controlled variable (i.e., the quantity which is being controlled). The thermostat, or comparator, compares the actual temperature with the desired temperature in order to measure the error. This error signal is the actuating signal, which is then sent to the heating units in order to correct the temperature. For example, if the actual temperature is less than the desired temperature, the actuating signal causes the heating elements to supply more heat. If there is no error, the actuating signal does not change the amount of heat which is being supplied. When the actual temperature is greater than the desired value, then the actuating signal calls for a decrease in the amount of heat.

For a system to be classified as a feedback control system, it is necessary that the controlled variable be fed back and compared with the reference input. In addition, the resulting error signal must actuate the control elements (power-amplifying device) to change the output so as to minimize the error. A feedback control system is also called a closed-loop system. Any system which incorporates a thermostat to control temperature is a feedback, or closed-loop, system. Well-known examples are electric frying pans, irons, refrigerators, and household furnaces with thermostatic control.

For speed control systems, the device which subtracts the feedback signal from the reference input (i.e., the comparator) is oftentimes a centrifugal governor. The amount of compression of a spring sets a force which is a measure of the desired speed. The centrifugal force of the flyweights is a measure of the actual speed. The difference in these forces is a measure of the error. The governor serves the same purpose that the thermostat does for temperature controls. That is, the governor compares the actual speed with the desired value and measures the error. This error signal then goes to a power amplifier such as a hydraulic servomotor which controls the position of a flow valve which in turn determines the rate of fuel flow to the engine. The same basic concepts apply to all types of feedback control systems, whether the controlled variable be temperature, speed, pressure, flow, position, force, torque, or any other physical quantity.

In an open-loop system there is no comparison of the controlled variable with the desired input. Each setting of the input determines a fixed operating position for the control elements. For example, for a given input temperature setting, the heating units are positioned to supply heat at a fixed rate. Note that there is no comparator, or thermostat, which measures the error and resets the heating units. The disadvantage of such a system is illustrated by the fact that if a fixed rate of heat is supplied to a house, the inside temperature will vary appreciably with changes in the outside temperature. Thus, for a given set input to an open-loop system, there may be a big variation of the controlled variable depending on the ambient temperature.

In this example, the ambient temperature is an external disturbance. By an external disturbance is meant something external to the system which acts to change or disturb the controlled variable. A major advantage of employing feedback control is that, because of the comparator, the actuating signal continually changes so that the controlled variable tends to become equal to the reference input regardless of the external disturbance. Another consideration is that with feedback one can generally use relatively inexpensive components and yet obtain better control than is possible with very expensive components in an open-loop system. The primary effort of this text will be devoted to feedback control systems.

1.3 SYSTEM REPRESENTATION

The mathematical relationships of control systems are usually represented by block diagrams. These diagrams have the advantage of indicating more realistically the actual processes which are taking place, as opposed to a purely abstract mathematical representation. In addition, it is easy to form the overall block diagram for an entire system by merely combining the block diagrams for each component or part of the system.

A comparator subtracts the feedback signal from the reference input r. For the case in which the controlled variable c is fed back directly (i.e., for unity-feedback systems), the signal coming from the comparator is $r - c$, which is equal to the actuating signal e. The mathematical relationship for this operation is

$$e = r - c \qquad (1.1)$$

Figure 1.2 Block diagram of a comparator.

Figure 1.3 Block diagram of the control elements.

As is illustrated in Fig. 1.2, a circle is the symbol which is used to indicate a summing operation. The arrowheads pointing toward the circle indicate input quantities, while the arrowhead leading away signifies the output. The sign at each input arrowhead indicates whether the quantity is to be added or subtracted.

The portion of a system between the actuating signal e and the controlled variable c is called the control elements. The relationship between the actuating signal e, which enters the control elements, and the controlled variable c, which is the output of the control, is expressed by the equation

$$c = G(D)e \tag{1.2}$$

where $G(D)$ represents the operation of the control elements. In Chaps. 2 and 3, it is shown how the actual values of $G(D)$ for specific control systems are obtained. The block-diagram representation for the preceding equation is shown in Fig. 1.3. A box is the symbol for multiplication. In this case, the input quantity e is multiplied by the function in the box $G(D)$ to obtain the output c. With circles indicating summing points and with boxes, or blocks, indicating multiplication, any linear mathematical expression may be represented by block-diagram notation.

The complete block diagram for an elementary unity-feedback control system is obtained by combining Figs. 1.2 and 1.3 to yield Fig. 1.4. This diagram shows the controlled variable c being fed back to the summing point, where it is compared with the reference input r. This diagram pictorially shows why a feedback control system is also called a closed-loop system.

When the controlled variable is fed back to the comparator, it is usually necessary to convert the form of the controlled variable to a form that is suitable for the comparator. For example, in a temperature control system the controlled temperature is generally converted to a proportional force or position for use in the comparator. This conversion is accomplished by feedback elements $H(D)$. The block-diagram representation for this more general case of a feedback control

Figure 1.4 Block diagram of a unity-feedback control system.

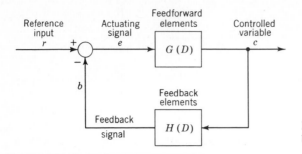

Figure 1.5 Block diagram of a feedback control system.

system is shown in Fig. 1.5. The signal which is fed back is

$$b = H(D)c \qquad (1.3)$$

The elements represented by $H(D)$ are called the feedback elements because they are located in the feedback portion of the control. The control elements represented by $G(D)$ are the feedforward elements because of their location in the feedforward portion of the loop. The actuating signal e is now $r - b$. This actuating signal e is a measure or indication of the error.

The term "feedback control system" is a general term which applies to any system in which the controlled variable is measured and fed back to be compared with the reference input. The terms "servomechanism" and "regulator" are distinguished as follows. A servomechanism is a particular type of feedback control system in which the controlled variable is a mechanical position (e.g., the angular position of a shaft). A regulator is distinguished as a feedback control system in which the reference input, although adjustable, is held fixed, or constant, for long periods of time (e.g., most temperature controllers).

1.4 MODERN CONTROL SYSTEMS

Commercial aviation as we know it today would not be possible without modern control systems. In a blind landing (i.e., an instrument landing), the pilot gets an error signal which lets him know the actual position of the plane relative to the ideal landing path. He then maneuvers the plane so as to keep it on the ideal landing path. Some modern airplanes have a completely automated landing system which automatically flies the vehicle along the ideal path of the landing beam.

The inertial guidance system for a space vehicle contains a stable platform which provides the reference orientation. Three gyroscopes are mounted on the stable platform so that their axes are mutually perpendicular. One axis is in an east-west direction and another axis in a north-south direction. The third axis is always directed along a line from the vehicle to the center of the earth. An accelerometer is mounted along each of the axes. A computer automatically integrates each of these accelerations to yield the corresponding components of velocity. Integration of these velocities yields the position of the vehicle relative to the earth. This actual position is then compared to the ideal position. When the error is sufficiently great, the astronaut fires the rockets so as to get back " on course."

The engine control system for a modern high-performance jet engine incorporates many subsystems which perform automatically to ensure the safe, efficient operation of the engine. For example, under certain conditions a jet engine can stall. The control system senses these conditions and automatically prevents the engine from running at a condition that would cause stall. At higher altitudes the air is considerably thinner. For the same amount of fuel flow, the air-fuel ratio would decrease considerably with altitude. The engine control system automatically compensates for changes in altitude. The control system also controls the nozzle jet area (this effects thrust), the position of the inlet guide vanes (this directs the air into the compressor for the best angle of attack), etc.

When the comparator is a piece of hardware such as a thermostat or a governor, the equations that describe the operation of the hardware are part of the control loop. Such controllers are referred to as analog-type controllers. The use of microprocessors is becoming increasingly popular in control systems. The microprocessor is in effect a digital computer and is programmed to yield the relationship that the designer desires. The microprocessor thus provides a flexibility that is not possible with analog-type controllers. With a microprocessor, a design change is accomplished by a simple change in the software, whereas with an analog controller a design change necessitates a hardware change. Because the microprocessor requires digital signals and the system to be controlled requires analog signals, the use of a microprocessor necessitates both an analog-to-digital converter and a digital-to-analog converter.

The microprocessor of a robot is programmed such that the robot can perform a variety of operations. Encoders are used to determine the actual position of a robot. The microprocessor compares the desired position with the actual position and sends out a signal to the motors controlling the robot so as to make the actual position the same as the desired position. The repeatability and accuracy of robots is extremely high. Robots come in a wide variety of sizes ranging from large industrial robots to small robots used for assembly of delicate electronic equipment. They have found extensive use in hazardous environments such as painting and welding in the automotive industry. Robots were sent into high radiation areas after the accident at the Three Mile nuclear power plant to help with the decontamination. Recent advances in robot technology have primarily been made possible by recent advances in control methods.

Manufacturing is being revolutionized by computer-controlled machines. The desired path of the cutter is preprogrammed into the computer (microprocessor). The actual position of the cutter is determined by an encoder. The desired position is compared to the actual position to determine the error. The actuating signal causes the cutting tool to move so as to reduce the error. Computer-controlled machines can manufacture complex shapes much more accurately than was heretofore possible. A program can be written to manufacture an entire piece. With a library of programs, it is simply a matter of selecting the program to produce the desired piece.

TWO

REPRESENTATION OF CONTROL COMPONENTS

To investigate the performance of control systems, it is necessary to obtain the mathematical relationship $G(D)$ relating the controlled variable c and the actuating signal e of the feedforward elements. This is accomplished by first obtaining the mathematical representation for each component between the actuating signal and the controlled variable and then expressing each of these equations as a block diagram. The combination of the block diagrams for each component yields the desired representation for $G(D)$. The value of $H(D)$ is obtained by applying the same technique to the components in the feedback portion of the control.

The quantity $G(D)$ could be obtained by writing the mathematical equation describing the operation of each component between e and c and then combining these individual equations algebraically to obtain the overall relationship between e and c. However, for all but the simplest systems, this procedure proves cumbersome because of the interaction between the various components in a typical control system. In addition, the block-diagram method gives one a better understanding of the system because of its visual representation.

The obtaining of block diagrams for typical elements used in control devices is illustrated in this chapter. In the next chapter, it is shown how these individual diagrams are combined to form entire control systems.

2.1 OPERATIONAL NOTATION

In writing equations for control systems, it is convenient to use the operational notation

$$D^n = \frac{d^n}{dt^n} \qquad n = 1, 2, 3, \ldots \qquad (2.1)$$

The operator D is a symbol which indicates differentiation with respect to time. For example, if x and y are functions of time, then

$$D(x + y) = \frac{d}{dt}(x + y) = \frac{dx}{dt} + \frac{dy}{dt} = Dx + Dy$$

This shows that the operator D obeys the distributive law; that is,

$$D(x + y) = Dx + Dy$$

Similarly, the commutative law

$$Dx + Dy = Dy + Dx$$

and the associative law

$$Dx + (Dy + Dz) = (Dx + Dy) + Dz$$

hold. It may also be shown that if a and b are constants, then

$$(D + a)(D + b)y = (D + a)\left(\frac{dy}{dt} + by\right)$$

$$= \frac{d}{dt}\left(\frac{dy}{dt} + by\right) + a\left(\frac{dy}{dt} + by\right)$$

$$= \frac{d^2y}{dt^2} + (a + b)\frac{dy}{dt} + aby$$

$$= [D^2 + (a + b)D + ab]y$$

Thus,

$$(D + a)(D + b)y = (D + b)(D + a)y$$

Any operator such as D^n which has the properties

$$D^n(x + y) = D^n x + D^n y \qquad D^n(ax) = aD^n x$$

is called a linear operator.

Consider the differential equation

$$\frac{d}{dt}x(t) + ax(t) = f(t)$$

In operational notation, this differential equation is

$$(D + a)x(t) = f(t) \qquad (2.2)$$

To solve the differential equation, first multiply through both sides by e^{at}. That is,

$$e^{at}\left[\frac{d}{dt}x(t) + ax(t)\right] = \frac{d}{dt}[e^{at}x(t)] = e^{at}f(t)$$

Integration yields

$$e^{at}x(t) = \int f(t)e^{at}\, dt + C$$

where C is the constant of integration. Solving the preceding for the function $x(t)$ which satisfies the differential equation gives

$$x(t) = e^{-at}\left[\int f(t)e^{at}\, dt + C\right] \tag{2.3}$$

To check that this function $x(t)$ does satisfy the differential equation, note that

$$Dx(t) = \frac{d}{dt}x(t) = -ae^{-at}\left[\int f(t)e^{at}\, dt + C\right] + e^{-at}[f(t)e^{at}]$$

and that

$$ax(t) = ae^{-at}\left[\int f(t)e^{at}\, dt + C\right]$$

Adding verifies that $Dx(t) + ax(t) = f(t)$. Cross-multiplying the operator in Eq. (2.2) yields the form

$$x(t) = \frac{1}{D + a}f(t) \tag{2.4}$$

Because Eqs. (2.2) and (2.4) are equivalent forms of the same differential equation, $x(t)$ as given by Eq. (2.3) must satisfy Eq. (2.4) as well as Eq. (2.2). That is,

$$\frac{1}{D + a}f(t) = e^{-at}\left[\int f(t)e^{at}\, dt + C\right] \tag{2.5}$$

It has previously been shown that the derivative d/dt of the right side plus a times the right side yields $f(t)$. Thus, operating on both sides of Eq. (2.5) by $D + a$ yields

$$(D + a)\frac{1}{D + a}f(t) = f(t) \tag{2.6}$$

Since the right side is $f(t)$, it follows that the operators on the left side may be cancelled. In general, it may be shown that

$$(D + a)^n \frac{1}{(D + a)^m}f(t) = (D + a)^{n-m}f(t) \tag{2.7}$$

The meaning of the reciprocal of D is obtained by noting that the integral of the derivative of a function is

$$y(t) = \int [Df(t)] \, dt = \int f'(t) \, dt = f(t) + C \tag{2.8}$$

where C is the constant of integration. The derivative of both sides with respect to time is

$$Dy(t) = Df(t)$$

Solving for $y(t)$ gives

$$y(t) = \frac{1}{D} [Df(t)] \tag{2.9}$$

Comparison of Eqs. (2.8) and (2.9) shows that

$$\frac{1}{D} [Df(t)] = \int [Df(t)] \, dt$$

Therefore, the symbol $1/D$ indicates integration.

The constant of integration is determined by evaluating Eq. (2.8) at some convenient initial time t_0. Thus, it is found that

$$C = y(t_0) - f(t_0)$$

Substitution of this result back into Eq. (2.8) shows that

$$y(t) = \int f'(t) \, dt = f(t) - f(t_0) + y(t_0) \tag{2.10}$$

The cancellation of operators in Eq. (2.9) would yield the erroneous result $y(t) = f(t)$. The algebraic cancellation of operators in Eq. (2.9) does not take into account the constant of integration that arises from the integration indicated by the $1/D$ term. Operators cannot be canceled when integration is the last operation to be performed unless there is no constant of integration, as is the case when all initial conditions are zero or $y(t_0) = f(t_0)$. The form of Eq. (2.9) that results after cancellation of operators is $y(t) = f(t)$. Note that when $y(t_0) = f(t_0)$, the initial conditions are such as to satisfy this form that results after cancellation.

In general, it may be shown that the cancellation

$$y(t) = \frac{1}{D^m} D^n f(t) = D^{n-m} f(t)$$

is valid if the initial conditions are such as to satisfy the expression that results after cancellation. For $n > m$, this form is $y(t_0) = D^{n-m} f(t_0)$. For $n = m$, this form is $y(t_0) = f(t_0)$. For $m > n$, this form is $D^{m-n} y(t_0) = f(t_0)$. The resulting form is automatically satisfied when all the initial conditions are zero.

The process of differentiating an integral may be expressed in the form

$$y(t) = D^n \frac{1}{D^m} f(t) = D^{n-m} f(t) \tag{2.11}$$

Because no initial-condition terms arise from the differentiation process, one may always cancel operators when differentiating an integral.

Illustrative example 2.1 For the operational relationship

$$y(t) = \frac{1}{D} D^2 f(t)$$

determine the initial conditions required for cancellation of operators to be valid. For $f(t)$ use the function $f(t) = t^2 + 3t + 5$.

SOLUTION The form that results after cancellation is

$$y(t) = \frac{1}{D} D^2 f(t) = Df(t) = f'(t) = 2t + 3$$

Evaluation at $t = 0$ yields the necessary initial condition

$$y(0) = f'(0) = (2t + 3)\Big|_{t=0} = 3$$

This result is substantiated by performing the indicated operations. That is,

$$y(t) = \frac{1}{D} D^2 f(t) = \frac{1}{D} f''(t) = f'(t) + C$$

Evaluation at $t = 0$ shows that $C = y(0) - f'(0)$. In order that $y(t) = f'(t)$, it is necessary that $y(0) = f'(0) = (2t + 3)|_{t=0} = 3$, so that $C = 0$.

Illustrative example 2.2 For the operational relationship

$$y(t) = \frac{1}{D^2} Df(t)$$

determine the initial conditions necessary for cancellation of operators to be valid. For $f(t)$ use the function $f(t) = 3t^2 + 2t + 1$.

SOLUTION The form that results after cancellation is

$$y(t) = \frac{1}{D^2} Df(t) = \frac{1}{D} f(t)$$

or

$$Dy(t) = f(t)$$

Evaluation at $t = 0$ yields

$$y'(0) = f(0) = (3t^2 + 2t + 1)\Big|_{t=0} = 1$$

To substantiate this result by performing the indicated operations, note that

$$y(t) = \frac{1}{D^2} Df(t) = \frac{1}{D} \int f'(t) \, dt = \frac{1}{D} [f(t) + C]$$

Hence

$$y'(t) = f(t) + C$$

Evaluation at time $t = 0$ shows that $C = y'(0) - f(0)$. In order that $y'(t) = f(t)$, it is necessary that $y'(0) = f(0) = 1$, so that $C = 0$.

The operator $1/(D + a)$ has properties similar to those of $1/D$. To show this, consider the differential form

$$y(t) = \frac{1}{D + a} (D + a) f(t) = \frac{1}{D + a} f'(t) + \frac{1}{D + a} af(t)$$

Application of Eq. (2.5) shows that

$$y(t) = e^{-at} \left[\int f'(t) e^{at} \, dt + C_1 \right] + ae^{-at} \left[\int f(t) e^{at} \, dt + C_2 \right]$$

Integration by parts [$\int u \, dv = uv - \int v \, du$] in which $u = e^{at}$, $du = ae^{at}$, $v = f(t)$, and $dv = f'(t) \, dt$ gives

$$y(t) = e^{-at} \left[f(t) e^{at} - a \int f(t) e^{at} \, dt + C_1 \right] + ae^{-at} \left[\int f(t) e^{at} \, dt + C_2 \right]$$

$$= f(t) + Ce^{-at}$$

where $C = C_1 + aC_2$. Evaluation at some convenient initial time t_0 gives $C = [y(t_0) - f(t_0)]e^{at_0}$. Substitution of this result back into the preceding expression shows that

$$y(t) = f(t) + [y(t_0) - f(t_0)]e^{-a(t - t_0)}$$

As indicated by Eq. (2.5), the term $1/(D + a)$ indicates integration. As is the case with the operator $1/D$, operators cannot be canceled when integration is the last operation to be performed unless there is no constant of integration. The constant of integration vanishes when the initial conditions are such as to satisfy the form that results after cancellation. For the preceding example, this resulting form is $y(t) = f(t)$. When $y(t_0) = f(t_0)$, then $C = [y(t_0) - f(t_0)]e^{-a(t - t_0)} = 0$.

In general it may be shown that the cancellation

$$y(t) = \frac{1}{(D + a)^m} (D + a)^n f(t) = (D + a)^{n - m} f(t) \tag{2.12}$$

is valid if the initial conditions are such as to satisfy the expression that results after cancellation.

In summary, the operators D and $D + a$ have similar properties. As indicated by Eqs. (2.7) and (2.11), cancellation of operators is always valid when differentiation is the last operation to be performed. As indicated by Eqs. (2.10) and (2.12), when integration is the last operation to be performed, cancellation is valid if the initial conditions are such as to satisfy the expression that results after cancellation. This resulting expression is automatically satisfied when all the initial conditions are zero.

2.2 MECHANICAL COMPONENTS

The load-deflection characteristics for a mechanical spring are shown in Fig. 2.1a. The spring force F_s required to deflect a spring a distance X from its free length is given by the equation

$$F_s = KX \qquad (2.13)$$

where K, the spring rate, is a constant which is equal to the slope of the plot of the load F_s versus deflection X. The input to a spring is usually the force F_s, and the output is the deflection X, so that the block-diagram representation for Eq. (2.13) is as shown in Fig. 2.1b.

For a viscous damper as illustrated in Fig. 2.2a, the force F_d required to move one end of the dashpot at a velocity V relative to the other end is equal to the product of the damping coefficient B and the velocity. That is,

$$F_d = BV = B \frac{dX}{dt}$$

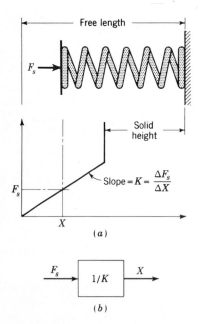

(a)

(b)

Figure 2.1 Spring characteristics.

(a)

(b)

Figure 2.2 Viscous damper.

Figure 2.3 Acceleration of a mass.

The substitution of the operator symbol $D = d/dt$ into the preceding expression yields

$$F_d = BDX \tag{2.14}$$

With the force F_d as the input and the displacement X as the output, the block-diagram representation for Eq. (2.14) is shown in Fig. 2.2b.

By Newton's second law of motion, it follows that the summation of the external forces $\sum F_e$ acting on a mass is equal to the product of the mass and the acceleration:

$$\sum F_e = MA = M \frac{d^2X}{dt^2} = MD^2X$$

The displacement X is given by the equation

$$X = \frac{1}{MD^2} \sum F_e \tag{2.15}$$

This is represented diagrammatically in Fig. 2.3.

For the mass-spring-damper combination shown in Fig. 2.4a, the spring force and damper force oppose, or resist, the motion caused by the applied load F. The summation of the forces acting on the mass is

$$\sum F_e = F + Mg - F_s - F_d = MD^2X$$

or $$F = (MD^2 + BD + K)X - Mg \tag{2.16}$$

This is the equation for the total forces acting on the system. For control work, it is usually more convenient to make measurements with respect to some reference operating point. A lowercase letter is used to designate the variation or change in

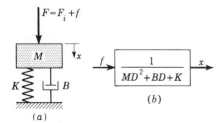

(a)

(b)

Figure 2.4 Series mass-spring-damper combination.

displacement x from the reference position X_i, so that $x = X - X_i$. The force equation at the reference operating condition is

$$F_i = KX_i - Mg$$

where F_i is the force required to maintain the mass in equilibrium at the reference position X_i. Subtracting the equation for the forces at the reference position from that for the total forces gives the equation of operation for changes about the reference position. That is,

$$F - F_i = (MD^2 + BD)X + K(X - X_i)$$

If lowercase letters are used to represent changes about the reference position, then the change in force is $f = F - F_i$ and similarly the change in displacement is $x = X - X_i$. Because X_i is a constant, $DX = D(X_i + x) = Dx$ and $D^2X = D^2x$. As would be expected, velocity and acceleration are independent of the reference position from which displacement is measured. The preceding equation of operation with respect to the reference position may thus be written in the form

$$f = (MD^2 + BD + K)x \tag{2.17}$$

This result could have been obtained directly by summing the change in forces from the reference operating condition. It is to be noted that forces which remain constant, such as the gravitational force Mg, have no change and thus do not appear.

Although x and f are measured from the reference operating point, Eq. (2.17) is a general equation describing the dynamic behavior of the system. It is not necessary that the system be initially at this reference operating point or that the system be initially at rest. As is later explained, it is usually much easier to obtain the equation of operation with respect to some convenient reference point rather than using total values. When total values are desired, it is an easy matter to add the reference value to the variation. The block diagram for Eq. (2.17) is shown in Fig. 2.4b.

Rotational Mechanical Components

A torsional spring is characterized by the equation

$$T_s = K_s \theta \tag{2.18}$$

where T_s = torque tending to twist spring
$\quad K_s$ = torsional spring rate
$\quad \theta$ = angular displacement of spring

A well-known example of a torsional spring is a shaft as is shown in Fig. 2.5. The right end of the shaft is displaced an angle θ with respect to the left end because of the twisting torque T_s. For a straight shaft, the torsional spring rate is

$$K_s = \frac{\pi d^4 G}{32L}$$

Figure 2.5 Shaft acting as a torsional spring.

where G = modulus of elasticity in shear
  d = diameter of shaft
  L = length of shaft

The torque T_d required to overcome viscous friction of a rotating member is

$$T_d = B_v \omega = B_v \frac{d\theta}{dt} = B_v D\theta \qquad (2.19)$$

where B_v = coefficient of viscous friction
  ω = angular velocity

A disk rotating in a viscous medium and supported by a shaft is shown in Fig. 2.6a. The applied torque tending to rotate the disk is T. The shaft torque and viscous friction oppose the motion, so that

$$\sum T_e = T - T_s - T_d = J\alpha = JD^2\theta \qquad (2.20)$$

where $\sum T_e$ is the summation of external torques acting on the disk, J is the mass moment of inertia of the disk, and α is the angular acceleration. The substitution of T_s from Eq. (2.18) and T_d from Eq. (2.19) into Eq. (2.20) yields

$$T = (JD^2 + B_v D + K_s)\theta \qquad (2.21)$$

The block-diagram representation for this system is shown in Fig. 2.6b.

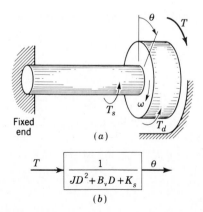

(a)

$$T \longrightarrow \boxed{\dfrac{1}{JD^2 + B_v D + K_s}} \longrightarrow \theta$$

(b)

Figure 2.6 Torsional inertia-spring-damper combination.

2.3 ELECTRICAL COMPONENTS

The resistor, inductor, and capacitor are the three basic components of electric circuits. The equation for the voltage drop E_R across a resistor is

$$E_R = RI \qquad (2.22)$$

where R is the resistance in ohms and I is the current flowing through the resistor in amperes.

For an inductor, the voltage drop E_L is given by the equation

$$E_L = L \frac{dI}{dt} = LDI \qquad (2.23)$$

where L is the inductance in henrys.

Similarly, the voltage drop E_C across a capacitor is

$$E_C = \frac{1}{C} \int I \, dt = \frac{1}{CD} I \qquad (2.24)$$

where C is the capacitance in farads.

The diagrammatic representations of Eqs. (2.22) to (2.24) are shown in Fig. 2.7.

For the series RLC circuit shown in Fig. 2.8a, the total voltage drop E is the sum of that across the inductor E_L, that across the resistor E_R, and that across the capacitor E_C:

$$E = E_L + E_R + E_C = \left(LD + R + \frac{1}{CD} \right) I \qquad (2.25)$$

The charge Q is the time integral of the current, that is, $Q = (1/D)I$ or $I = DQ$. By noting that $LDI = LD^2Q$, $RI = RDQ$, and $(1/CD)I = (1/C)Q$, then Eq. (2.25) becomes

$$E = \left(LD^2 + RD + \frac{1}{C} \right) Q \qquad (2.26)$$

The overall block-diagram representation for this RLC circuit is shown in Fig. 2.8b.

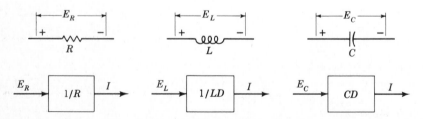

Figure 2.7 Representation of a resistor, inductor, and capacitor.

(a)

(b)

Figure 2.8 RLC series circuit.

2.4 SERIES AND PARALLEL LAWS

Elements are usually connected in either a series or a parallel arrangement. Arriving at the equation for such systems can be greatly simplified by the use of the laws for series and the laws for parallel combinations.

Series Electric Circuits

A general series circuit is shown in Fig. 2.9a. In a series circuit, the total voltage drop E is the sum of the individual voltage drops across each element, and the same current I flows through each element. The equation for the summation of the voltage drops is

$$E = \left(L_1 D + L_2 D + R_1 + R_2 + \frac{1}{C_1 D} + \frac{1}{C_2 D} \right) I \qquad (2.27)$$

The equation relating the voltage E and current I for any electric circuit can be expressed in the form

$$E = ZI$$

where Z is the impedance for the circuit, E is the voltage drop across the circuit, and I is the total current flowing through the circuit. For the case of Eq. (2.27) the impedence Z is

$$Z = L_1 D + L_2 D + R_1 + R_2 + \frac{1}{C_1 D} + \frac{1}{C_2 D} \qquad (2.28)$$

For elements in series, the total impedance is the sum of the individual impedance of each element. The block-diagram representation is shown in Fig. 2.9b.

(a)

(b)

Figure 2.9 General series circuit.

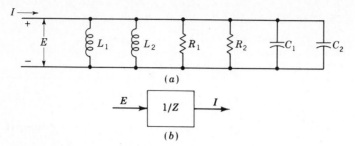

Figure 2.10 General parallel circuit.

Parallel Electric Circuits

A general combination of electric elements in parallel is shown in Fig. 2.10a. The distinguishing features of a parallel arrangement are that the voltage drop E across each element is the same, and the total current I flowing into the system is the sum of the current flowing through each element. Thus,

$$I = \frac{E}{L_1 D} + \frac{E}{L_2 D} + \frac{E}{R_1} + \frac{E}{R_2} + \frac{E}{1/C_1 D} + \frac{E}{1/C_2 D} \qquad (2.29)$$

or $$E = \frac{1}{1/L_1 D + 1/L_2 D + 1/R_1 + 1/R_2 + C_1 D + C_2 D} I = ZI$$

The impedance Z is

$$Z = \frac{1}{1/L_1 D + 1/L_2 D + 1/R_1 + 1/R_2 + C_1 D + C_2 D} \qquad (2.30)$$

For elements in parallel, the total impedance is equal to one divided by the sum of the reciprocal of the impedance of each element. The block-diagram representation is shown in Fig. 2.10b.

Illustrative example 2.3 For the circuit shown in Fig. 2.11, let it be desired to determine the equation relating the output voltage E_2 to the input voltage E_1.

SOLUTION The parallel combination of R_1 and C_1 is in series with R_2, so that the total impedance Z is

$$Z = Z_1 + R_2 = \frac{1}{1/R_1 + C_1 D} + R_2 = \frac{R_1}{1 + R_1 C_1 D} + R_2 \qquad (2.31)$$

Figure 2.11 Electric circuit.

The voltage E_1 is given by the equation

$$E_1 = ZI = \frac{R_1 + R_2 + R_1 R_2 C_1 D}{1 + R_1 C_1 D} I \qquad (2.32)$$

Similarly, E_2 is

$$E_2 = R_2 I \qquad (2.33)$$

The substitution of I from Eq. (2.32) into Eq. (2.33) yields the desired answer:

$$E_2 = \frac{R_2(1 + R_1 C_1 D)}{R_1 + R_2 + R_1 R_2 C_1 D} E_1 \qquad (2.34)$$

Series Mechanical Elements

A series arrangement of mechanical elements is shown in Fig. 2.12a. In general, it is better to use the equivalent "grounded-chair" representation for a mass, as shown in Fig. 2.12b, rather than the more common representation of Fig. 2.12a. The fact that the mass is in series with the other elements is more readily seen from Fig. 2.12b than from Fig. 2.12a. In determining inertia force, the acceleration of a mass is always taken with respect to ground. Thus, providing the grounded chair to indicate motion relative to ground is a more justifiable representation than Fig. 2.12a, which shows better the actual physical arrangement of the elements in the system. For series mechanical elements, the force f is equal to the summation of the forces acting on each individual component, and each element undergoes the same displacement. Thus,

$$f = (K_1 + K_2 + B_1 D + B_2 D + M D^2)x = Zx \qquad (2.35)$$

where x and f are measured from a convenient reference operating point. The equivalent impedance is

$$Z = K_1 + K_2 + B_1 D + B_2 D + M D^2 \qquad (2.36)$$

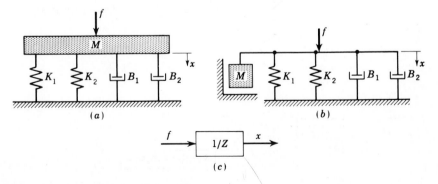

Figure 2.12 Mechanical elements in series.

For mechanical elements in series, the total impedance is the sum of the individual impedance of each element. The block-diagram representation for this system is shown in Fig. 2.12c.

Parallel Mechanical Elements

A parallel combination of mechanical elements is shown in Fig. 2.13a. For parallel elements, the same force f is transmitted through each element. In addition, the total deflection x is seen to be the sum of the individual deflection of each element. Thus,

$$x = \frac{f}{K_1} + \frac{f}{K_2} + \frac{f}{B_1 D} + \frac{f}{B_2 D} \tag{2.37}$$

or

$$f = \frac{1}{1/K_1 + 1/K_2 + 1/B_1 D + 1/B_2 D} x = Zx$$

The impedance is

$$Z = \frac{1}{1/K_1 + 1/K_2 + 1/B_1 D + 1/B_2 D} \tag{2.38}$$

For mechanical elements in parallel, the total impedance is equal to one divided by the sum of the reciprocal of the individual impedance of each element. The block-diagram representation is shown in Fig. 2.13b.

   A necessary condition for parallel elements is that the same force be transmitted through each element. Springs and dampers satisfy this condition because the force is the same on both sides. However, this is not the case for a mass such as that shown in Fig. 2.14a because the difference in forces acting on both sides of a mass is utilized in acceleration. Thus, a mass located between other elements cannot be in parallel with them. A mass can be in parallel only if it is the last element, as shown in Fig. 2.15. For this system, the displacement x is

$$x = (x - y) + (y - z) + z = \left(\frac{1}{K} + \frac{1}{BD} + \frac{1}{MD^2} \right) f$$

(a)

(b)

Figure 2.13 Mechanical elements in parallel.

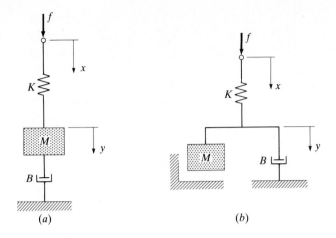

Figure 2.14 Mechanical circuit: (a) schematic representation, (b) grounded-chair representation.

Figure 2.15 Parallel mass-spring-damper combination.

Thus,

$$f = \frac{1}{1/K + 1/BD + 1/MD^2}\, x \qquad (2.39)$$

Parallel and series laws for rotational mechanical components may also be developed by extending the preceding techniques.

Illustrative example 2.4 For the mass-spring-damper combination shown in Fig. 2.14a, determine the equation relating f and x, the equation relating f and y, and the equation relating x and y.

SOLUTION The first step is to draw the equivalent grounded-chair system, in which the motion of the mass with respect to ground is clearly indicated as shown in Fig. 2.14b. The spring K is in parallel with the series combination of M and B. Application of the laws for parallel and series elements gives

$$f = \frac{1}{1/K + 1/Z}\, x \qquad (2.40)$$

The impedance Z of the series combination of M and B is

$$Z = MD^2 + BD$$

Thus,

$$f = \frac{1}{1/K + 1/(MD^2 + BD)} \, x = \frac{K(MD^2 + BD)}{MD^2 + BD + K} \, x \qquad (2.41)$$

The force f is transmitted through the spring K and acts upon the series combination of M and B. Thus, the equation of motion for this part of the system which relates f and y is

$$f = (MD^2 + BD)y \qquad (2.42)$$

The desired relationship between x and y is obtained by eliminating f from Eqs. (2.41) and (2.42). That is,

$$y = \frac{K}{MD^2 + BD + K} \, x \qquad (2.43)$$

Grounded-Chair Representation

The general procedure for constructing the grounded-chair representation is as follows:

1. Draw coordinates such that the coordinate at which the force acts is at the top and ground is at the bottom.
2. Insert each element in its correct orientation with respect to these coordinates.

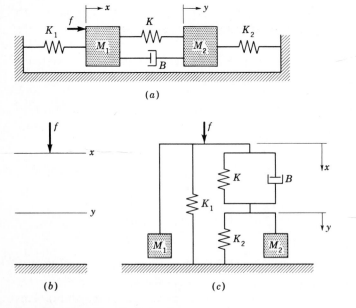

(a)

(b)   (c)

Figure 2.16 General mechanical system.

For the system shown in Fig. 2.16a, the coordinates are x, y, and ground. Step 1 is carried out by drawing the coordinates as shown in Fig. 2.16b. To do step 2 we must note that for the spring K_1 and mass M_1 the coordinates are x and ground, for the spring K and damper B the coordinates are x and y, and finally for the mass M_2 and spring K_2 the coordinates are y and ground. Inserting these elements between the proper coordinates as shown in Fig. 2.16c completes the grounded-chair representation. The series and parallel combinations are now recognized directly from this grounded-chair representation. The equation for the force f is

$$f = \left[M_1 D^2 + K_1 + \frac{1}{1/(K + BD) + 1/(K_2 + M_2 D^2)} \right] x$$

Degrees of Freedom

By degrees of freedom is meant the number of coordinates required to specify the position of all the elements in a mechanical system. Thus, the system shown in Fig. 2.14 has two degrees of freedom, while that shown in Fig. 2.15 has three degrees of freedom.

Rather than using the series and parallel laws, an alternate method of determining the equation of operation of mechanical systems is to write the force balance at each coordinate. For example, the force equation at the x coordinate of Fig. 2.14b is

$$K(x - y) = f \tag{2.44}$$

The compression of the spring K is $(x - y)$. As x increases so does the spring force; but as y increases, the spring force decreases. The force balance at the y coordinate is

$$K(x - y) - BDy = MD^2 y \tag{2.45}$$

The two preceding equations may be written in the form

$$
\begin{aligned}
Kx & \quad -Ky & = f \\
Kx & \quad -(MD^2 + BD + K)y & = 0
\end{aligned}
\tag{2.46}
$$

These two equations may be solved simultaneously to yield any of the desired relationships between x, y, and f. For example, solving for y gives

$$y = \frac{\begin{vmatrix} K & f \\ K & 0 \end{vmatrix}}{\begin{vmatrix} K & -K \\ K & -(MD^2 + BD + K) \end{vmatrix}} = \frac{1}{MD^2 + BD} f \tag{2.47}$$

Suppose that it is desired to consider y positive for upward motion rather than for downward motion (downward motion is shown in Fig. 2.14). This reversal would change the sign of each y term in Eqs. (2.44) and (2.45), which would change the sign of y in the resultant expression given by Eq. (2.47). In effect, reversing the positive sense of a coordinate merely changes its sign.

In applying the series or parallel laws to mechanical elements, care must be exercised to take the positive sense of motion of each coordinate (this is indicated by the arrow at each coordinate) in the same direction as that of the applied force. It is a simple matter to later change the positive sense of a coordinate by merely changing the sign of the corresponding term in the derived equation.

2.5 ANALOGIES

The equation of operation for the series mechanical system of Fig. 2.4a is given by Eq. (2.17), and the equation for the series electric circuit of Fig. 2.8a is given by Eq. (2.26). That is,

$$f = (MD^2 + BD + K)x \tag{2.17}$$

$$E = \left(LD^2 + RD + \frac{1}{C}\right)Q \tag{2.26}$$

Comparison of corresponding terms in Eqs. (2.17) and (2.26) shows that the differential equation of operation for each system has the same form. The terms which occupy corresponding positions are called analogous quantities. This particular analog is referred to as the direct analog. The analogous quantities for a direct analog are shown in Table 2.1.

The total force acting on a group of mechanical elements in series is equal to the sum of the forces exerted on each element. Similarly, the total voltage drop across a group of electrical elements in series is equal to the sum of the voltage drops across each element. Thus, in constructing a direct analog, series mechanical elements are replaced by analogous series electrical elements.

For parallel mechanical elements, the force acting on each element is the same, and for parallel electrical elements the voltage drop across each element is the same. Thus, in a direct analog, parallel mechanical elements should be replaced by equivalent electrical elements in parallel. A direct analog is also called a force-voltage analog in that force and voltage are analogous quantities.

Table 2.1 Analogous quantities in a direct (force-voltage) analog

Translational mechanical system	Force	Velocity	Displacement	Mass	Viscous damping coefficient	Spring constant
	f	$\dot{x} = Dx$	x	M	B	K
Electrical system	Voltage	Current	Charge	Inductance	Resistance	Reciprocal of capacitance
	E	$I = DQ$	Q	L	R	$\dfrac{1}{C}$

Table 2.2 Analogous quantities in an inverse (force-current) analog

Translational mechanical system	Force	Velocity	Spring constant	Damping coefficient	Mass
	f	$\dot{x}$	K	B	M
Electrical system	Current	Voltage	Reciprocal of inductance	Reciprocal of resistance	Capacitance
	I	E	$\dfrac{1}{L}$	$\dfrac{1}{R}$	C

The other type of analog is the inverse analog. To construct an inverse analog, it should first be noted that the total current flowing through a group of electrical elements in parallel is the sum of the currents in each element. This is analogous to the fact that the total force acting on a group of mechanical elements in series is the sum of the forces acting on each element. Thus, to construct an inverse analog, series mechanical elements must be replaced by parallel electrical elements. Similarly, in an inverse analog, it may be shown that parallel mechanical elements should be replaced by series electrical elements. Thus, the arrangement of series and parallel elements is inverted in constructing an inverse analog (i.e., series elements are replaced by parallel elements, and vice versa). An inverse analog is also called a force-current analog in that force and current are analogous quantities.

Analogous quantities for an inverse analog may be determined by comparing the equation of operation for the parallel mechanical system of Fig. 2.15 with that for the series electrical system of Fig. 2.8a. The equation of operation for the parallel mechanical system of Fig. 2.15 is given by Eq. (2.39). Multiplication of both sides of Eq. (2.39) by D gives

$$\dot{x} = \left(\frac{D}{K} + \frac{1}{B} + \frac{1}{MD} \right) f \tag{2.48}$$

The operation of the series electric circuit of Fig. 2.8a is described by Eq. (2.25), which has the same form as Eq. (2.48). That is,

$$E = \left(LD + R + \frac{1}{CD} \right) I \tag{2.25}$$

Comparison of corresponding terms in Eqs. (2.25) and (2.48) yields the analogous quantities for an inverse analog that are shown in Table 2.2.

Illustrative example 2.5 Let it be desired to determine the electrical analog for the mechanical system of Fig. 2.14 by using (a) the direct analog and (b) the inverse analog.

SOLUTION (a) The grounded-chair representation for the mechanical system of Fig. 2.14 is shown in Fig. 2.17a. The direct analog is shown in Fig. 2.17b.

(a)

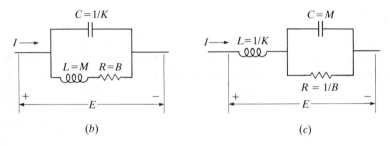

(b)   (c)

Figure 2.17 Analogs: (a) mechanical circuit, (b) direct analog, (c) inverse analog.

Note that the capacitor C is in parallel with the series combination of L and R, just as the spring K of Fig. 2.17a is in parallel with the series combination of M and B. The equation of operation for the electric circuit of Fig. 2.17b is

$$E = \frac{1}{CD + 1/(LD + R)} I = \frac{1}{C + 1/(LD^2 + RD)} \frac{I}{D}$$

$$= \frac{1}{C + 1/(LD^2 + RD)} Q = \frac{(1/C)(LD^2 + RD)}{LD^2 + RD + 1/C} Q \qquad (2.41a)$$

The equation of operation for the mechanical system of Fig. 2.17a is given by Eq. (2.41):

$$f = \frac{1}{1/K + 1/(MD^2 + BD)} x = \frac{K(MD^2 + BD)}{MD^2 + BD + K} x \qquad (2.41)$$

Comparison of corresponding terms in Eqs. (2.41) and (2.41a) verifies the analogous quantities given in Table 2.1.

(b) The inverse analog is shown in Fig. 2.17c. It is to be noted that the inductor L is in series with the parallel combination of C and R, whereas in Fig. 2.17a the spring K is in parallel with the series combination of M and B.

The equation of operation for the electric circuit of Fig. 2.17c is

$$E = \left(LD + \frac{1}{CD + 1/R}\right)I$$

or

$$I = \frac{1}{LD + 1/(CD + 1/R)}E = \frac{1}{L + 1/[CD^2 + (1/R)D]}\frac{E}{D}$$

$$= \frac{(1/L)[CD^2 + (1/R)D]}{CD^2 + (1/R)D + 1/L}\frac{E}{D} \tag{2.41b}$$

Comparison of corresponding terms in Eqs. (2.41) and (2.41b) verifies the analogous quantities given in Table 2.2. In the inverse analog, velocity is analogous to voltage, and thus displacement is analogous to the integral of voltage ($x = E/D$).

Comparison of corresponding terms in Eqs. (2.17) and (2.21) shows that a direct (torque-force) analog may be developed in which series translational mechanical elements are replaced by series rotational elements. Similarly, parallel translational elements should be replaced by parallel rotational elements.

2.6 SCALE FACTORS

Because of the convenience of working with electrical equipment, ordinarily an electrical analog is constructed for some other system. Scale factors ensure that the values of the voltages and currents in the electrical analog will be reasonable. For example, in a direct analog, analogous quantities are force-voltage and velocity-current; hence,

$$f = k_E E \tag{2.49}$$

$$\frac{dx}{dt} = k_I I \tag{2.50}$$

Thus, if the maximum value of f is 1000 lb and the maximum desired voltage is 100 V, then

$$k_E = \frac{f}{E} = 10 \text{ lb/V}$$

In this case, 1 V corresponds to 10 lb of force.

To determine the other analogous relationships, first write the force equations for a mass, a damper, and a spring. That is,

$$f = M \frac{d}{dt}\left(\frac{dx}{dt}\right) \tag{2.51}$$

$$f = B \frac{dx}{dt} \tag{2.52}$$

$$f = Kx \quad \text{or} \quad \frac{df}{dt} = K \frac{dx}{dt} \tag{2.53}$$

The voltage equations for an inductor, resistor, and capacitor are

$$E = L \frac{d}{dt} I \qquad E = RI \qquad \frac{dE}{dt} = \frac{1}{C} I$$

The preceding expressions may be written in the form

$$k_E E = \frac{k_E}{k_I} L \frac{d}{dt} k_I I \tag{2.54}$$

$$k_E E = \frac{k_E}{k_I} R k_I I \tag{2.55}$$

$$\frac{d}{dt} k_E E = \frac{k_E}{k_I} \frac{1}{C} k_I I \tag{2.56}$$

From a comparison of Eqs. (2.51) and (2.54), it follows that to have $f = k_E E$ and $dx/dt = k_I I$, then $M = (k_E/k_I)L$ or

$$L = kM \tag{2.57}$$

where $k = k_I/k_E$.

Comparison of Eqs. (2.52) and (2.55) shows that

$$R = kB \tag{2.58}$$

Comparison of Eqs. (2.53) and (2.56) yields

$$C = \frac{1}{kK} \tag{2.59}$$

For the case in which $k = 1$, the preceding analogous relationships are the same as those given in Table 2.1.

Time Scale

For systems which are extremely fast acting, it may be desired to slow down the solution on the electric analog. Similarly, for extremely slow systems it may be desired to speed up the solution. By letting t represent actual time for a phenomenon to occur in the actual system and τ represent the time for the corresponding phenomenon to take place in the analog, then

$$\tau = at \tag{2.60}$$

where t = actual time
τ = time in analog

If a phenomenon takes 1 s to complete in actual time t and if $a = 10$, then $\tau = at = 10$, or the solution has been slowed down on the analog by a factor of 10. Hence, for $a > 1$ the solution is slowed down and for $a < 1$ the solution is speeded up.

When a time scale change is effected, events in the electric analog occur in analog time τ; hence, writing Eqs. (2.54), (2.55), and (2.56) in terms of τ gives

$$k_E E = \frac{L}{k} \frac{d}{d\tau} k_I I$$

$$k_E E = \frac{R}{k} k_I I$$

$$\frac{d}{d\tau} k_E E = \frac{1}{kC} k_I I$$

Differentiation of Eq. (2.60) shows that $d\tau/dt = a$; hence

$$\frac{dI}{d\tau} = \frac{dt}{d\tau} \frac{dI}{dt} = \frac{1}{a} \frac{dI}{dt} \quad \text{and} \quad \frac{dE}{d\tau} = \frac{1}{a} \frac{dE}{dt}$$

Expressing the equations for the electrical elements in terms of time t gives

$$k_E E = \frac{L}{ka} \frac{d}{dt} k_I I \tag{2.61}$$

$$k_E E = \frac{R}{k} k_I I \tag{2.62}$$

$$\frac{d}{dt} k_E E = \frac{a}{kC} k_I I \tag{2.63}$$

From a comparison of Eqs. (2.51) and (2.61), it follows that to have $f = k_E E$ and $dx/dt = k_I I$, then $M = L/ka$ or

$$L = akM \tag{2.64}$$

Comparison of Eqs. (2.52) and (2.62) shows that

$$R = kB \tag{2.65}$$

Comparison of Eqs. (2.53) and (2.63) shows that

$$C = \frac{a}{k} \frac{1}{K} \tag{2.66}$$

For the case in which $a = 1$ and $k = 1$ the preceding results are identical to those given in Table 2.1.

The corresponding relationships for an inverse analog are developed in a similar manner. In this analog, force and current are analogous quantities, as are voltage and velocity; hence, the desired scale factors are

$$f = k_I I \tag{2.67}$$

$$\frac{dx}{dt} = k_E E \tag{2.68}$$

Equations (2.61), (2.62), and (2.63) are modified forms of the basic relationships for an inductor, resistor, and capacitor respectively. Solving Eqs. (2.61), (2.62), and (2.63) for $k_I I$ gives

$$\frac{d}{dt} k_I I = \frac{ka}{L} k_E E \tag{2.69}$$

$$k_I I = \frac{k}{R} k_E E \tag{2.70}$$

$$k_I I = \frac{kC}{a} \frac{d}{dt} k_E E \tag{2.71}$$

Comparison of Eqs. (2.53) and (2.69) reveals that to have $f = k_I I$ and $dx/dt = k_E E$, then $K = ka/L$ or

$$L = ak \frac{1}{K} \tag{2.72}$$

Similarly, comparing Eq. (2.52) with Eq. (2.70) and comparing Eq. (2.51) with Eq. (2.71) shows that

$$R = k \frac{1}{B} \tag{2.73}$$

$$C = \frac{a}{k} M \tag{2.74}$$

For the case in which $a = 1$ and $k = 1$, these relationships become identical to those given in Table 2.2.

Units

In the British gravitational system, force is measured in pounds (lb_f), length in feet (ft), time in seconds (s), and mass in slugs. From Newton's equation of motion ($f = ma$), it follows that the unit relationship is

$$lb_f = slug \cdot ft/s^2$$

For a direct analog, the units of k_E are lb_f/V and the units of k_I are $(ft/s)/A = ft/(s \cdot A)$. Thus, $k = k_I/k_E$ has units of $ft \cdot V/(lb_f \cdot A \cdot s)$. From Eq. (2.64), the resulting units for $L = akM$ are

$$L = \frac{ft \cdot V}{lb_f \cdot A \cdot s} \frac{lb_f \cdot s^2}{ft} = \frac{V \cdot s}{A}$$

Substitution of units into the relationship $E = L(dI/dt)$ shows that inductance L has units of $V \cdot s/A$, which are called henrys. Similarly, from Eq. (2.65) it is found that resistance R has units of V/A, which are ohms, and from Eq. (2.66) it is found that capacitance C has units of $A \cdot s/V$, which are farads. For the inverse analog, k_I has units of lb_f/A and k_E has units of $(ft/s)/V = ft/(V \cdot s)$. The resulting units for $k = k_I/k_E$ are $lb_f \cdot V \cdot s/(ft \cdot A)$. From Eq. (2.72), the resulting units for $L = ak/K$ are

$$L = \frac{lb_f \cdot V \cdot s/(ft \cdot A)}{lb_f/ft} = \frac{V \cdot s}{A}$$

These are the units for henrys. Similarly, Eqs. (2.73) and (2.74) yield the proper units for R and C respectively.

In the SI system, force is measured in newtons (N), mass in kilograms (kg), length in meters (m), and time in seconds (s). The relationship among the units in the SI system is

$$N = \frac{kg \cdot m}{s^2}$$

The use of SI units in the preceding analysis yields the same results as those obtained with the British gravitational system. The only difference is that ft is replaced by m, lb_f is replaced by N, and slug $= lb_f \cdot s^2/ft$ is replaced by $kg = N \cdot s^2/m$.

2.7 THERMAL SYSTEMS

For small temperature differences, the rate of heat transferred into a body is proportional to the temperature difference across the body:

$$Q = hA(T_1 - T) = \frac{T_1 - T}{R_T} \tag{2.75}$$

where Q = rate of heat flow
$\quad h$ = coefficient of heat transfer of the surface of the body
$\quad A$ = surface area
$\quad T$ = temperature of the body
$\quad T_1$ = temperature of the surrounding medium

and the symbol $R_T = 1/hA$ means equivalent thermal resistance and will soon be shown to be analogous to the electrical resistance R.

The rate of change of temperature of the body dT/dt is related to the rate of heat transfer into the body by the expression

$$Q = Mc \frac{dT}{dt} = C_T DT \tag{2.76}$$

where c = average specific heat of the body
$\quad M$ = mass
$\quad C_T = Mc$ is the equivalent thermal capacitance

The equation for a resistor and a capacitor may be written in the form

$$I = \frac{1}{R} E \tag{2.77}$$

and

$$I = CDE \tag{2.78}$$

Comparison of Eq. (2.77) with Eq. (2.75) and of Eq. (2.78) with Eq. (2.76) shows the following quantities to be analogous:

$$\begin{array}{cc} T \sim E & Q \sim I \\ R_T \sim R & C_T \sim C \end{array} \tag{2.79}$$

This is the direct, or temperature-voltage, analog.

In Fig. 2.18a is shown an insulated container. The liquid is stirred so that the temperature T of the liquid is constant throughout. The equation for the heat transfer into the liquid is given by Eq. (2.75). Because this equation has the same form as the resistor equation, it is represented as a thermal resistor R_T in the equivalent thermal circuit of Fig. 2.18b. Similarly, the equation for the heat stored in the liquid is given by Eq. (2.76). Because this equation has the same form as the capacitor equation, it is represented as a thermal reservoir in Fig. 2.18b. The equation of operation for this thermal circuit is

$$Q = \frac{T_1 - T}{R_T} = C_T DT$$

Solving for T gives

$$T = \frac{T_1}{1 + (R_T C_T)D} \tag{2.80}$$

(a)   (b)   (c)

(d)

Figure 2.18 Thermal system.

The electric circuit which is the direct analog of the thermal circuit is shown in Fig. 2.18c. The equation of operation for the electric circuit is

$$I = \frac{E_1 - E}{R} = CDE$$

Solving for E gives

$$E = \frac{E_1}{1 + (RC)D} \tag{2.81}$$

In the direct analog, series thermal elements are replaced by analogous electrical elements in series. Similarly, parallel thermal elements are replaced by analogous electrical elements in parallel. For thermal systems, there are only two elements.

The inverse, or temperature-current, analog is developed by first writing the equation for a resistor and an inductor in the form

$$E = RI \tag{2.82}$$

$$E = LDI \tag{2.83}$$

Comparison with Eqs. (2.75) and (2.76) reveals the following analogous relationships:

$$T \sim I \qquad Q \sim E$$

$$R_T \sim \frac{1}{R} \qquad C_T \sim L \tag{2.84}$$

The inverse analog of Fig. 2.18b is constructed by replacing the thermal resistor R_T and thermal capacitor C_T, which are in series, by an electrical resistor R and inductor L in parallel as shown in Fig. 2.18d. Because the thermal elements R_T and C_T have the same rate of heat flow Q, each electrical element R and L must have the same voltage E. The temperature of the thermal reservoir is T, and thus the inductor current is I. Similarly, because the temperature difference across the thermal resistor is $T_1 - T$, the electric resistor current is $I_1 - I$. The equation of operation for the inverse analog (Fig. 2.18d) is

$$E = R(I_1 - I) = LDI$$

Solving for I gives

$$I = \frac{I_1}{1 + (L/R)D} \tag{2.85}$$

For any circuit, there are but two possible analogs: the inverse analog in which the series-parallel arrangement is inverted, and the direct analog in which the series-parallel arrangement is retained.

2.8 FLUID SYSTEMS

In working with fluid systems, it is necessary to distinguish if the fluid is incompressible or compressible. For incompressible fluids, it suffices to work with the volume rate of flow. For compressible fluids, it is necessary to work with the mass rate of flow.

Incompressible Fluids

When the pressure difference across a flow restriction is small, the volume rate of flow Q is proportional to the pressure drop $P_1 - P$ across the restriction:

$$Q = \frac{P_1 - P}{R_F} \tag{2.86}$$

where R_F is the equivalent fluid resistance.

The rate of flow into a tank, such as that shown in Fig. 2.19a, is equal to the cross-sectional area A of the tank times the rate of change of height. Thus,

$$Q = ADH = \frac{A}{\rho} DP = C_F DP \tag{2.87}$$

where $P = \rho H$, in which ρ is the density of the fluid
$\quad H = $ the head
$\quad C_F = A/\rho = $ equivalent fluid capacitance

The equation of operation for the fluid system of Fig. 2.19a is

$$Q = \frac{P_1 - P}{R_F} = C_F DP$$

Solving for P gives

$$P = \frac{P_1}{1 + (R_F C_F)D} \tag{2.88}$$

Replacing P by ρH yields the equation for the head H. In Fig. 2.19b is shown the fluid circuit representation for this system.

(a) $\qquad\qquad\qquad\qquad$ (b)

Figure 2.19 Fluid system.

Comparison of Eqs. (2.86) and (2.87) with Eqs. (2.77) and (2.78) shows the following quantities to be analogous:

$$P \sim E \qquad Q \sim I$$
$$R_F \sim R \qquad C_F \sim C \qquad (2.89)$$

These are analogous quantities for the direct, or pressure-voltage, analog. The electric circuit which is the direct analog for the fluid system of Fig. 2.19*b* is the same as that shown in Fig. 2.18*c*.

Comparison of Eqs. (2.86) and (2.87) with Eqs. (2.82) and (2.83) yields the analogous relationships for the inverse, or pressure-current, analog. Thus,

$$P \sim I \qquad Q \sim E$$
$$R_F \sim \frac{1}{R} \qquad C_F \sim L \qquad (2.90)$$

The inverse analog for the fluid system of Fig. 2.19*b* is the same as that shown in Fig. 2.18*d*.

Compressible Fluids

For small pressure differences, the mass rate of flow M through a restriction is proportional to the pressure difference $P_1 - P$:

$$M = \frac{P_1 - P}{R_F} \qquad (2.91)$$

where R_F is the equivalent fluid resistance. In Fig. 2.20*a* is shown a tank of constant volume V. The equation of state for the fluid in the tank is

$$PV = WRT$$

The flow into such a tank is usually isothermal. Thus, differentiating both sides of the equation of state with respect to time and solving for $M = dW/dt$ gives

$$M = \frac{dW}{dt} = \frac{V}{RT} \frac{d}{dt} P = \frac{V}{RT} DP = C_F DP \qquad (2.92)$$

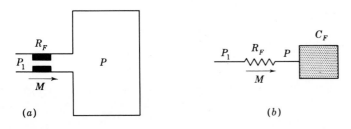

(a) (b)

Figure 2.20 Fluid system.

where $C_F = V/RT$ is the equivalent fluid capacitance. In Fig. 2.20b is shown the fluid circuit representation for this system. The equation of operation is

$$P = \frac{P_1}{1 + (R_F C_F)D} \tag{2.93}$$

The analogous relationships for the direct, or pressure-voltage, analog are the same as those given by Eq. (2.89) except that Q is replaced by M. The electric circuit which is the direct analog is the same as Fig. 2.18c.

Similarly, the analogous relationships for the inverse, or pressure-current, analog are the same as those given by Eq. (2.90) with Q replaced by M, and the electric circuit which is the inverse analog is the same as Fig. 2.18d.

A major use of analogs is that often it is easier to study experimentally one type of system rather than another. For example, it may be easier to change a resistor rather than a coefficient of viscous friction. Whenever possible, it is best to work with the system directly rather than to consider the operation of an analogous system. This eliminates the chance of error in construction of the analogy. Also, when carried far enough analogies usually break down because things which are physically possible for one component may be impossible for the analogous component.

PROBLEMS

2.1 For $f(t) = 4t$ and $y(0) = 0$, perform the indicated operations to evaluate $y(t)$ for each of the following cases:

(a) $y(t) = \dfrac{1}{D} f(t)$

(b) $y(t) = \dfrac{1}{D + 2} f(t)$

2.2 Same as Prob. 2.1 except $f(t) = 4 \sin 2t$.

2.3 For $f(t) = e^{-t}$ and $y(0) = 0$, perform the indicated operations to obtain $y(t)$ when

(a) $y(t) = \dfrac{1}{D} f(t)$

(b) $y(t) = \dfrac{1}{D} Df(t)$

(c) $y(t) = \dfrac{1}{D - 1} f(t)$

(d) $y(t) = \dfrac{1}{D - 1} (D - 1) f(t)$

2.4 For $f(t) = \sin t$ and $y(0) = 0$, perform the indicated operations to obtain $y(t)$ when

(a) $y(t) = \dfrac{1}{D} f(t)$

(b) $y(t) = \dfrac{1}{D} Df(t)$

(c) $y(t) = \dfrac{1}{D + 1} f(t)$

(d) $y(t) = \dfrac{1}{D + 1} (D + 1) f(t)$

2.5 For each of the following operational relationships, determine the required initial conditions such that cancellation of operators is valid. For $f(t)$ use the function $f(t) = e^{2t}$.

(a) $y(t) = \dfrac{1}{D} Df(t)$

(b) $y(t) = \dfrac{1}{D} D^2 f(t)$

(c) $y(t) = \dfrac{1}{D + 1} (D + 1) f(t)$

(d) $y(t) = \dfrac{1}{D + 1} (D + 1)^2 f(t)$

2.6 For each of the mechanical systems shown in Fig. P2.6,
  (a) Determine the equation which relates f and x.
  (b) Determine the equation which relates f and y.
  (c) Determine the equation which relates x and y.

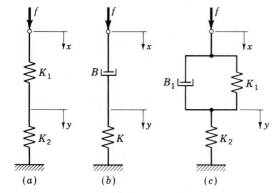

(a)   (b)   (c)   **Figure P2.6**

2.7 For each of the electrical networks shown in Fig. P2.7,
  (a) Determine the equation which relates E_1 and I.
  (b) Determine the equation which relates E_1 and E_2.

(a)   (b)

Figure P2.7

2.8 A schematic diagram of an accelerometer for measuring the linear acceleration d^2x/dt^2 is shown in Fig. P2.8. Determine the operational form for the differential equation which relates y (the change in the position of the mass relative to the frame) to the acceleration D^2x of the frame.

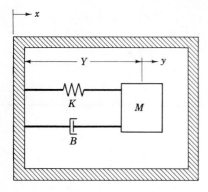

Figure P2.8

2.9 A viscous damper (shock absorber) is shown in Fig. P2.9. The rate of flow of fluid q is proportional to the pressure drop $q = C(p_1 - p_2)$. The force transmitted across the damper is $f = A(p_1 - p_2)$ where A is the cross-sectional area. The relative velocity of the piston relative to the cylinder is $(\dot{x} - \dot{y})$. The rate of change of volume is $q = A(\dot{x} - \dot{y})$. Show that the equation of operation of this viscous damper may be expressed in the form $f = B(\dot{x} - \dot{y})$. Determine the equation for the coefficient of viscous damping B.

Figure P2.9

2.10 For each of the mechanical systems shown in Fig. P2.10, construct the grounded-chair representation and then determine the equation which relates f and x.

(a)   (b)   Figure P2.10

2.11 For each of the mechanical systems shown in Fig. P2.11, construct the equivalent grounded-chair representation and then determine the equation which relates f and x.

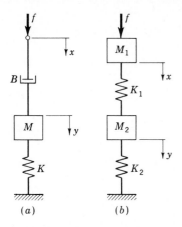

(a) (b) **Figure P2.11**

2.12 For the mechanical system shown in Fig. P2.12, construct the electric circuit which is the

Figure P2.12

(a) Direct analog. Determine the equation relating E and I of this electrical system, and compare this equation with that relating f and x of Fig. P2.12.

(b) Inverse analog. Determine the equation relating E and I of this electrical system, and compare this equation with that relating x and f of Fig. P2.12.

2.13 For the electrical system shown in Fig. P2.13 construct the mechanical circuit which is the

(a) Direct analog. Determine the equation relating f and x of this mechanical system, and compare this equation with that relating E and I of Fig. P2.13.

(b) Inverse analog. Determine the equation relating f and x of this mechanical system, and compare this equation with that relating E and I of Fig. P2.13.

Figure P2.13

2.14 For the mechanical system shown in Fig. P2.14,
  (*a*) Determine the differential equation relating f and x.
  (*b*) Construct the direct (force-voltage) analog.
  (*c*) Construct the inverse (force-current) analog.

Figure P2.14

2.15 For the mechanical system shown in Fig. P2.15,
  (*a*) Construct the grounded-chair representation.
  (*b*) Obtain the overall equation relating f and x.
  (*c*) Construct the electric circuit which is the direct analog.
  (*d*) Indicate what current in the analog corresponds to the velocity of the mass M.

Figure P2.15

2.16 Same as Prob. 2.15 except construct the inverse analog, and then indicate what voltage in this analog corresponds to the velocity of the mass M.

2.17 For the mechanical system shown in Fig. P2.17,
  (*a*) Draw the grounded-chair representation.
  (*b*) Determine the equation relating f and x.
  (*c*) Draw the electric circuit which is the direct analog.
  (*d*) On the analog show where you would place a voltmeter to obtain a voltage proportional to the force across the damper B_2.

Figure P2.17

2.18 Same as Prob. 2.17 except construct the inverse analog, and then indicate where you would place an ammeter to obtain a current proportional to the force across the damper B_2.

2.19 The mass-spring-damper system shown in Fig. P2.19 represents the suspension for an automobile in which B is the shock absorber, K is the spring, and M is the mass of the car.

(a) Construct the grounded-chair representation and then determine the equation relating f and x and the equation relating y and x.

(b) Construct the electric circuit which is the direct analog and then show where you would place an ammeter to obtain a current proportional to the velocity $\dot{y}$ of the mass.

(c) Construct the electric circuit which is the inverse analog and then show where you would place a voltmeter to determine a voltage proportional to the velocity $\dot{y}$ of the mass.

Figure P2.19

2.20 For the mechanical system shown in Fig. P2.14 determine the numerical values for the electrical elements of the direct analog when

(a) $M = 0.5$ slugs, $B = 10$ lb$_f \cdot$ s/ft, $K = 50$ lb$_f$/ft, $k_E = 2$ lb$_f$/V, and $k_I = 200$ ft/(s $\cdot$ A).

(b) $M = 0.5$ kg, $B = 10$ N $\cdot$ s/m, $K = 50$ N/m, $k_E = 2$ N/V, and $k_I = 200$ m/(s $\cdot$ A).

2.21 For the mechanical system shown in Fig. P2.14 determine numerical values for the electrical elements of the inverse analog when

(a) $M = 0.5$ slugs, $B = 10$ lb$_f \cdot$ s/ft, $K = 50$ lb$_f$/ft, $k_E = 2$ ft/(s $\cdot$ V), and $k_I = 200$ lb$_f$/A.

(b) $M = 0.5$ kg, $B = 10$ N $\cdot$ s/m, $K = 50$ N/m, $k_E = 2$ m/(s $\cdot$ V), and $k_I = 200$ N/A.

2.22 The parameters for a mechanical system are $M = 10$ slugs, $B = 5000$ lb$_f \cdot$ s/ft, and $K = 2500$ lb$_f$/ft. It is estimated that $f_{max} = 500$ lb$_f$ and $\dot{x}_{max} = 5$ ft/s. Determine what size resistor, inductor, and capacitor to use in the direct analog such that $E_{max} = 100$ V and $I_{max} = 10$ A.

After the analog has been made, it is found that $E_{max} = 50$ V and $I_{max} = 4$ A. What were the actual values of f_{max} and $\dot{x}_{max}$?

What size resistor, inductor, and capacitor should be used to speed up the solution of this problem by a factor of 10?

2.23 Same as Prob. 2.22 except for an inverse rather than direct analog.

2.24 The differential equation for a series RLC circuit is

$$E = \left(LD + R + \frac{1}{CD} \right) I$$

(a) Construct the mechanical circuit which is the direct (force-voltage) analog for the series RLC circuit. Determine the equation for the force f as a function of velocity $\dot{x}$. (Label mass as M_a, spring K_a, and damper B_a.)

(b) Construct the mechanical circuit which is the inverse analog for the series RLC circuit. Determine the equation for velocity $\dot{x}$ as a function of force f. (Label mass as M_b, spring K_b, and damper B_b.)

(c) Compare the results of parts (a) and (b) above, and then suggest analogous quantities in constructing the dual of a mechanical circuit. (Duals are two different mechanical circuits whose differential equations have the same form.)

2.25 For the lever shown in Fig. P2.25 the variation in the applied force is f and the variation in spring position is x. The horizontal line represents the reference position of the lever.

(a) Determine the equation relating f and x.

(b) Determine the relationship between t and θ (where $t = fL_f$ is the variation in applied torque and $x = L_2\,\theta$).

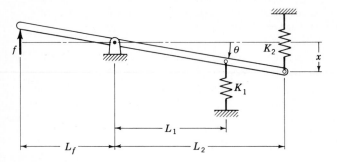

Figure P2.25

2.26 The lever system shown in Fig. P2.26 is drawn in its reference position. The variation in spring position is designated by x and the variation in applied force is designated by f (f and x are zero at the reference position).

(a) Determine the equation relating f and x.

(b) Determine the relationship between t and θ where $t = fL_f$ and $x = L\theta$.

Figure P2.26

2.27 Oftentimes mechanical systems have gearing. Such systems may be represented by equivalent systems without gearing. Figure P2.27a shows a geared system and Fig. P2.27b shows an equivalent system without gearing. The gear ratio $n = \omega_1/\omega_2$ is the ratio of the speed of the driving shaft ω_1 to that of the driven shaft ω_2. In order for the systems to be equivalent, the kinetic energy of the system without gearing $J_e\omega_1^2/2$ must be the same as that of the system with gearing $J\omega_2^2/2$. Thus, show that $J_e = J/n^2$.

(a)   (b)

Figure P2.27

2.28 As shown in Fig. P2.28a, the torque T is transmitted through gearing to the load, which is viscous friction. The resisting torque due to the viscous friction is $B\omega_2$, where B is the coefficient of viscous friction. The rate at which power is being dissipated is equal to the product of the resisting torque and the angular velocity $P = (B\omega_2)\omega_2 = B\omega_2^2$. The system without gearing is shown in Fig. P2.28b. In order for the systems to be equivalent, the rate of power dissipation of the system without gearing $B_e\omega_1^2$ must be the same as that for the system with gearing. Thus, show that $B_e = B/n^2$.

(a)   (b)

Figure P2.28

2.29 As shown in Fig. P2.29a, the torque T is transmitted to the load, which is a torsional spring. The torque applied to the torsional spring is $K\theta_2$, where K is the torsional spring rate and θ is the angle of twist. The potential energy stored in the spring is $K\theta_2^2/2$. The corresponding system without gearing is shown in Fig. P2.29b. In order for the systems to be equivalent, the potential energy stored in the system without gearing $K_e\theta_1^2/2$ must be the same as that for the system with gearing. Thus, show that $K_e = K/n^2$.

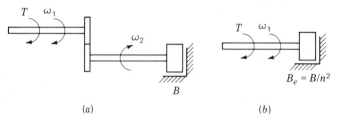

(a)   (b)

Figure P2.29

2.30 For the manometer shown in Fig. P2.30, determine the equation of motion relating the pressure P at the open end to the position x. The length of the measuring column is L, the cross-sectional area is A, and the density is ρ.

Figure P2.30

2.31 For the thermometer shown in Fig. P2.31, the temperature of the surrounding medium is T_1. The temperature of the glass enclosure is T_2 and the temperature of the fluid in the thermometer is T. The rate of heat flow from the surrounding medium to the glass is $Q_1 = (T_1 - T_2)/R_{T_1}$. The rate of heat flow from the glass to the fluid is $Q_2 = (T_2 - T)/R_{T_2}$.

The rate of temperature change of the glass is $DT_2 = (Q_1 - Q_2)/C_{T_1}$ and the rate of change of temperature of the fluid is $DT = Q_2/C_{T_2}$.

Construct the thermal circuit representation for this system and then determine the equation for the temperature T of the fluid as a function of the surrounding temperature T_1.

T

T_2

T_1

Figure P2.31

2.32 In Fig. P2.32 is shown an oven which is supplied with heat from an electric source. The rate of heat supplied is proportional to the voltage, $Q_s = KE_s$. The rate at which heat is lost through the walls is $Q = (T - T_1)/R_T$. The rate of change of temperature of the oven is

$$DT = \frac{Q_s - Q}{C_T}$$

Construct the thermal circuit representation for this system and then determine the equation for the temperature T of the oven as a function of the applied voltage E_s and the surrounding temperature T_1.

E_s

T T_1

Figure P2.32

2.33 In Fig. P2.33 is shown a fluid system of two tanks in series.

Construct the fluid circuit representation for this system. Determine the equation for the pressure P (head $H = P/\rho$) as a function of the inlet pressure P_1(P_2 should not appear in this equation).

H_2

H

P_1 P_2 P 0

R_{F_1} C_{F_1} R_{F_2} C_{F_2} R_{F_3} **Figure P2.33**

2.34 In Fig. P2.34 is shown a tank in which flow is supplied at a rate Q_s.

Construct the fluid circuit representation for this system. Determine the equation for the pressure P (head $H = P/\rho$) as a function of P_1 and Q_s.

Figure P2.34

2.35 In Fig. P2.35 is shown a tank with two inlets. The fluid is compressible. Determine the equation for the pressure P as a function of P_1 and P_2.

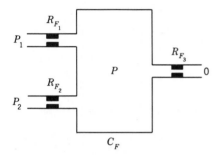

Figure P2.35

2.36 In Fig. P2.36 is shown two tanks in series. For a compressible fluid, determine the equation for the pressure P as a function of the inlet pressure P_1 (P_2 should not appear in this equation).

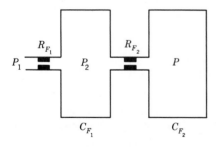

Figure P2.36

THREE

REPRESENTATION OF CONTROL SYSTEMS

In this chapter it is shown how to obtain the overall block-diagram representation for some typical control systems. In brief, the method employed is to obtain the block diagram for each component or process and then "hook up," or connect, the corresponding inputs and outputs for each diagram to obtain the one overall representation for the system. The techniques which are presented in later chapters for determining the operating characteristics of control systems are based on a knowledge of the overall block-diagram representation for the system.

3.1 LINEARIZATION OF NONLINEAR FUNCTIONS

The most powerful methods of system analysis have been developed for linear control systems. For a linear control system all the relationships between the variables are linear differential equations, usually with constant coefficients. The reason that differential equations rather than algebraic equations are obtained is that in feedback control systems the variables are functions of time. For example, in controlling temperature, the actuating signal causes a change in heat flow, but time is required for this added heat to bring the temperature to its desired value. In speed control systems, the actuating signal causes a change in power of the prime mover, but time is required for the engine to accelerate or decelerate to its desired speed. Similarly, in pressure control systems, it takes time to bring the pressure in a chamber to some desired value.

Actual control systems usually contain some nonlinear elements. Such elements would in turn yield nonlinear differential equations for the system. In the following it is shown how the equations for nonlinear elements may be linearized. Thus, the resulting differential equation of operation for the system becomes linear. A linear equation in the n variables $x_1, x_2, \ldots, x_n$ is one which has the form

$$Y = Y_i + c_1 x_1 + c_2 x_2 + \cdots + c_n x_n$$

where $c_1, c_2, \ldots, c_n$ and Y_i are constants. Note that each of the n variables is to the first power and no products of these variables appear. A nonlinear equation results if one or more of the variables is to some power other than the first power (for example, x_n^3) or if the equation contains products of the variables (for example, $x_2 x_5$).

A plot of the nonlinear relationship

$$Y = X^2 \tag{3.1}$$

is shown in Fig. 3.1. In the vicinity of the point (X_i, Y_i), the function is closely approximated by the tangent. For example, consider the point (X, Y) on the curve of the nonlinear function. The abscissa X is displaced a distance x from X_i. This abscissa X intersects the nonlinear function a vertical distance $y + \varepsilon$ from Y_i, and it intersects the tangent a distance y from Y_i. The equation for Y is

$$Y = Y_i + y + \varepsilon \approx Y_i + y \tag{3.2}$$

Lowercase letters indicate the variation of the capital-letter parameters from the reference point. From Fig. 3.1, it is seen that the slope of the tangent line is

$$\frac{y}{x} = \frac{dY}{dX}\bigg|_i = \text{slope at point } (X_i, Y_i)$$

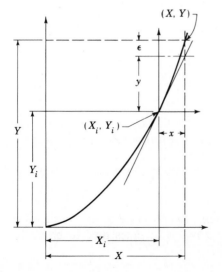

Figure 3.1 Graph of function $Y = X^2$.

The symbol $|_i$ means that the derivative is to be evaluated at the reference point. Thus

$$y = \frac{dY}{dX}\bigg|_i x = \frac{d}{dX}(X^2)\bigg|_i x = 2X_i x \tag{3.3}$$

Substitution of y from Eq. (3.3) into Eq. (3.2) yields the following linear approximation for Y:

$$Y \approx Y_i + 2X_i x \tag{3.4}$$

Illustrative example 3.1 Effect a linear approximation for the equation $Y = X^2$ for values of X in the neighborhood of 10, and find the error when using this approximation for $X = 11$.

SOLUTION The reference values are $X_i = 10$ and $Y_i = X_i^2 = 100$. The variation from the reference value is $x = X - X_i = 11 - 10 = 1$. Substitution of these values into Eq. (3.4) yields

$$Y \approx Y_i + 2X_i x = 100 + (2)(10)(1) = 120$$

The exact value is $Y = X^2 = 121$; thus the error is 1 part in 121, or less than 1 percent.

A general procedure for obtaining a linear approximation is to use the expression derived in calculus for approximating the variation ΔY for a function $Y = Y(X_1, X_2, \ldots, X_n)$ of n independent variables. That is,

$$\Delta Y = \frac{\partial Y}{\partial X_1}\bigg|_i \Delta X_1 + \frac{\partial Y}{\partial X_2}\bigg|_i \Delta X_2 + \cdots + \frac{\partial Y}{\partial X_n}\bigg|_i \Delta X_n$$

Using lowercase letters to represent variations from the reference values, then

$$y = \Delta Y = Y - Y_i$$
$$x_1 = \Delta X_1 = X_1 - X_{1i}$$
$$x_2 = \Delta X_2 = X_2 - X_{2i}$$
$$\cdots\cdots\cdots\cdots\cdots\cdots$$
$$x_n = \Delta X_n = X_n - X_{ni}$$

Thus, the general expression for obtaining a linear approximation for a nonlinear function is

$$y = C_1 x_1 + C_2 x_2 + \cdots + C_n x_n \tag{3.5}$$

where

$$C_1 = \frac{\partial Y}{\partial X_1}\bigg|_i \qquad C_2 = \frac{\partial Y}{\partial X_2}\bigg|_i \qquad \text{etc.}$$

Evaluation of these partial derivatives at the reference condition yields constants. As is illustrated by Eq. (3.5), in a linear equation each variable term $x_1, x_2, \ldots, x_n$ is of the first power, and the contribution of each term is added independently.

The application of Eq. (3.5) to the nonlinear equation $Y = X^2$ is effected as follows. The independent variable is X, which corresponds to X_1 in the general equation. Thus,

$$C_1 = \left.\frac{\partial Y}{\partial X_1}\right|_i = \left.\frac{\partial Y}{\partial X}\right|_i = \left.\frac{\partial}{\partial X} X^2\right|_i = 2X_i$$

The variation is

$$y = C_1 x = 2X_i x$$

This is the result obtained by the preceding geometric interpretation and given by Eq. (3.3).

Illustrative example 3.2 For the triangle shown in Fig. 3.2, the length of the base is X, the length of the adjacent side is Y, and the included angle is θ. The altitude h is $Y \sin \theta$. The equation for the area of a triangle (one-half the altitude times the base) may thus be written in the form

$$A = \tfrac{1}{2}XY \sin \theta$$

Determine the linear approximation for the area A. The reference values are $X_i = 10$, $Y_i = 16$, and $\theta_i = 30°$. Determine the approximate area when $X = 9$, $Y = 18$, and $\theta = 33°$.

SOLUTION The reference value of the area is

$$A_i = \tfrac{1}{2}X_i \, Y_i \sin \theta_i = \tfrac{1}{2}(10)(16)(0.5) = 40$$

Figure 3.2 Triangle.

The change a in the area is

$$a = \frac{\partial A}{\partial X}\bigg|_i x + \frac{\partial A}{\partial Y}\bigg|_i y + \frac{\partial A}{\partial \theta}\bigg|_i \Delta\theta$$

$$= (\tfrac{1}{2}Y \sin \theta)_i\, x + (\tfrac{1}{2}X \sin \theta)_i\, y + (\tfrac{1}{2}XY \cos \theta)_i\, \Delta\theta$$

It is convenient to express linear approximations in nondimensional form. This is accomplished by expressing the partial derivatives in terms of the reference value of the function being approximated. For the preceding, the reference value is $A_i = (\tfrac{1}{2}XY \sin \theta)_i$. Thus,

$$a = (\tfrac{1}{2}XY \sin \theta)_i\, \frac{x}{X_i} + (\tfrac{1}{2}XY \sin \theta)_i\, \frac{y}{Y_i} + (\tfrac{1}{2}XY \sin \theta)_i\, \frac{\Delta\theta}{\tan \theta_i}$$

$$= \left(\frac{x}{X_i} + \frac{y}{Y_i} + \frac{\Delta\theta}{\tan \theta_i}\right) A_i$$

The variations are $x = X - X_i = 9 - 10 = -1$, $y = Y - Y_i = 18 - 16 = 2$, $\Delta\theta = \theta - \theta_i = (33 - 30)\pi/180 = \pi/60$ radians. Note that each of the ratios x/X_i, y/Y_i, and $\Delta\theta/\tan \theta_i$ are nondimensional; thus the change in the angle $\Delta\theta$ must be measured in radians:

$$a = \left(-\frac{1}{10} + \frac{2}{16} + \frac{3\sqrt{3}}{60}\right) 40 = 4.464$$

The approximation for the total area A is

$$A = A_i + a = 40.0 + 4.464 = 44.464$$

The need for linearizing nonlinear relationships is frequently encountered in control engineering. For example, most mechanical speed control systems incorporate a flyball governor for sensing the speed error. This is a centrifugal device, so that a force is obtained which is proportional to the square of the speed. In the design of hydraulic equipment in which the working medium is an incompressible fluid, one encounters the nonlinear equations which govern such fluid flow. Similarly, the working medium for pneumatic equipment is air, whose flow is described by nonlinear relationships.

Illustrative example 3.3 Effect the linear approximation for P in the equation of state $PV = WRT$. The reference conditions are $P_i = 100$ lb$_f$/ft^2, $V_i = 100$ ft^3, $W_i = 10/53.3$ lb$_m$, and $T_i = 1000°$R. Determine the percent error in using this approximation for P when $V = 110$ ft^3, $T = 1200°$R, and W remains the same. The constant R is 53.3 ft $\cdot$ lb$_f$/(lb$_m \cdot °$R).

SOLUTION From the equation of state and the fact that W remains constant, it is seen that P is a function of the independent variables T and V, or $P = P(T, V)$. Application of Eq. (3.5) to obtain the variation p of the pressure from its reference value yields

$$p = \frac{\partial P}{\partial T}\bigg|_i t + \frac{\partial P}{\partial V}\bigg|_i v \qquad (3.6)$$

The partial derivatives are evaluated from the equation of state as follows:

$$\frac{\partial P}{\partial T}\bigg|_i = \frac{\partial}{\partial T}\left(\frac{WRT}{V}\right)\bigg|_i = \frac{WR}{V}\bigg|_i = \frac{(10)(53.3)}{(53.3)(100)} = 0.10$$

$$\frac{\partial P}{\partial V}\bigg|_i = \frac{\partial}{\partial V}\left(\frac{WRT}{V}\right)\bigg|_i = \frac{-WRT}{V^2}\bigg|_i = -\frac{(10)(53.3)(1000)}{(53.3)(100)^2} = -1.0$$

The linearized approximation for P is

$$P \approx P_i + p = P_i + 0.1t - v$$

From the given information, it follows that

$$v = V - V_i = 110 - 100 = 10$$

$$t = T - T_i = 1200 - 1000 = 200$$

Thus,

$$P \approx 100 + (0.1)(200) - (10) = 110 \ \text{lb}_f/\text{ft}^2$$

To express the preceding approximation in nondimensional form, note that the reference value of the function being approximated is $P_i = (WRT/V)|_i$. Thus,

$$\frac{\partial P}{\partial T}\bigg|_i = \frac{WR}{V}\bigg|_i = \left(\frac{WRT}{V}\right)\bigg|_i \frac{1}{T_i} = \frac{P_i}{T_i}$$

$$\frac{\partial P}{\partial V}\bigg|_i = -\frac{WRT}{V^2}\bigg|_i = -\left(\frac{WRT}{V}\right)\bigg|_i \frac{1}{V_i} = \frac{-P_i}{V_i}$$

Substitution of these results into Eq. (3.6) gives

$$p = \left(\frac{t}{T_i} - \frac{v}{V_i}\right)P_i \qquad (3.7)$$

Note that t/T_i and v/V_i are nondimensional ratios. Evaluation of the resulting linear approximation gives

$$P = P_i + p = 100 + 100\left(\frac{200}{1000} - \frac{10}{100}\right) = 100 + (20 - 10) = 110 \ \text{lb}_f/\text{ft}^2$$

This checks the previously attained result. The exact value of P is

$$P = \frac{WRT}{V} = \frac{(10)(53.3)(1200)}{(53.3)(110)} = 109.1$$

Therefore, the percent error is

$$\frac{(110 - 109.1)100}{109.1} = 0.82\%$$

In the SI system, pressure has units of newtons per square meter (N/m²), volume has units of cubic meters (m³), mass has units of kilograms (kg), and temperature has units of kelvins (K). When using SI units, the value of R in the equation of state ($PV = WRT$) is $R = 287$ N·m/(kg·K).

Illustrative example 3.4 The equation for the flow of an incompressible fluid through a restriction is

$$Q = C_c A \sqrt{\frac{2g_c}{\rho}} (P_1 - P_2) \qquad (3.8)$$

where Q = rate of flow, ft³/s
C_c = coefficient of discharge (dimensionless)
A = area of restriction, ft²
$P_1 - P_2$ = pressure drop across restriction, lb$_f$/ft²
ρ = density of fluid, lb$_m$/ft³
g_c = mass conversion constant = 32.2 lb$_m$/slug

Determine the linear approximation for the variation q. Letting $g_c = (32.2)(12) = 386$ converts the length units in Eq. (3.8) from feet to inches. Thus, the flow units are cubic inches per second (in³/s), the area units are square inches (in²), the pressure units are pounds-force per square inch (lb$_f$/in²), and the density units are pounds-mass per cubic inch (lb$_m$/in³).

SOLUTION Because Q is a function of the area A and pressure drop $P_1 - P_2$, then

$$Q = Q[A, (P_1 - P_2)]$$

The variation q is

$$q = \left.\frac{\partial Q}{\partial A}\right|_i a + \left.\frac{\partial Q}{\partial (P_1 - P_2)}\right|_i (p_1 - p_2)$$

The partial derivatives are

$$\left.\frac{\partial Q}{\partial A}\right|_i = C_c \sqrt{\frac{2g_c}{\rho}} (P_1 - P_2)\bigg|_i = \frac{Q}{A}\bigg|_i$$

and

$$\left.\frac{\partial Q}{\partial (P_1 - P_2)}\right|_i = \frac{C_c A}{2} \sqrt{\frac{2g_c}{\rho}} \frac{1}{P_1 - P_2}\bigg|_i = \frac{1}{2} \frac{Q}{P_1 - P_2}\bigg|_i$$

Thus,

$$q = Qi\left[\frac{a}{A_i} + \frac{1}{2} \frac{p_1 - p_2}{(P_1 - P_2)_i}\right] \qquad (3.9)$$

When using SI units, the g_c term does not appear in Eq. (3.8); thus $Q = C_c A \sqrt{(P_1 - P_2)/\rho}$. Substitution of units into this expression shows that the units of Q are $m^2 \sqrt{(N/m^2)/(kg/m^3)} = m^2 \sqrt{N \cdot m/kg} = m^2 \sqrt{N \cdot m/(N \cdot s^2/m)} = m^2 \sqrt{m^2/s^2} = m^3/s$.

For fluid flow, it is customary to express mass in units of pounds-mass (lb_m) rather than slugs. The mass of a body in slugs is equal to its mass in pounds-mass divided by 32.2. The units relationship is slug $= lb_m/g_c = lb_m/32.2$, where $g_c = 32.2 \, lb_m/slug$ is the mass conversion constant. The constant $g_c = 32.2 \, lb_m/slug$ is not to be confused with the local value of g, which is the acceleration of a freely falling body (ft/s^2).

In the flow equation, the kinetic energy term ($mv^2/2$) has units of $(slug)(ft/s)^2 = (lb_f \cdot s^2/ft)(ft/s)^2 = ft \cdot lb_f$. When mass is expressed in units of pounds-mass, the kinetic energy term is ($mv^2/2g_c$). Note that dividing m by g_c converts mass from pounds-mass to slugs.

Illustrative example 3.5 For sonic flow of air through a restriction, the mass rate of flow is

$$M = \frac{0.53}{\sqrt{T}} AP \qquad (3.10)$$

where M = mass rate of flow, lb_m/s
T = the inlet temperature, °R
A = area of restriction, ft^2
P = inlet pressure, lb_f/ft^2

Determine the linearized approximation for the variation m when the inlet temperature T is constant. It is interesting to note that because the length unit cancels in Eq. (3.10), any unit of length may be used. When using inches, the area is in square inches (in^2) and the pressure is in pounds-force per square inch (lb_f/in^2).

SOLUTION Because the temperature is constant, the mass rate of flow is a function of A and P. Thus,

$$m = \frac{\partial M}{\partial A}\bigg|_i a + \frac{\partial M}{\partial P}\bigg|_i p$$

The partial derivatives are

$$\frac{\partial M}{\partial A}\bigg|_i = \frac{0.53}{\sqrt{T}} P \bigg|_i = \frac{M}{A}\bigg|_i$$

and

$$\frac{\partial M}{\partial P}\bigg|_i = \frac{0.53}{\sqrt{T}} A \bigg|_i = \frac{M}{P}\bigg|_i$$

Thus,

$$m = \left(\frac{a}{A_i} + \frac{p}{P_i} \right) M_i \qquad (3.11)$$

When using SI units, Eq. (3.10) becomes $M = 0.04AP/\sqrt{T}$. The units for mass rate of flow M are kilograms per second (kg/s), the units for area A are square meters (m^2), the units of pressure P are newtons per square meter (N/m^2), and the units for temperature T are kelvins (K).

Illustrative example 3.6 For subsonic flow of air through a restriction, the mass rate of flow is

$$M = \frac{1.05}{\sqrt{T}} A\sqrt{(P_1 - P_2)P_2} \qquad (3.12)$$

where M = mass rate of flow, lb$_m$/s
$\quad T$ = inlet temperature, °R
$\quad A$ = area of restriction, ft^2
$\quad P_1$ = inlet pressure, lb$_f$/ft^2
$\quad P_2$ = outlet pressure, lb$_f$/ft^2

Determine the linearized approximation for the variation m when the inlet temperature is constant. As was the case for Eq. (3.10), any unit of length may be used in Eq. (3.12).

SOLUTION The mass rate of flow is a function of A, $P_1 - P_2$, and P_2. Thus

$$m = \frac{\partial M}{\partial A}\bigg|_i a + \frac{\partial M}{\partial(P_1 - P_2)}\bigg|_i (p_1 - p_2) + \frac{\partial M}{\partial P_2}\bigg|_i p_2$$

Evaluating the partial derivatives yields

$$m = \left[\frac{a}{A_i} + \frac{p_1 - p_2}{2(P_1 - P_2)_i} + \frac{p_2}{2(P_2)_i} \right] M_i \qquad (3.13)$$

The same result is obtained if P_1 and P_2 are regarded as the variables rather than $P_1 - P_2$ and P_2.

When using SI units, Eq. (3.12) becomes $M = 0.08A\sqrt{(P_1 - P_2)P_2/T}$. The units are kilograms, meters, newtons, seconds and kelvin.

Geometric Interpretation of Error Introduced by a Linear Approximation

From the linear approximation for the area of a rectangle, it is possible to represent geometrically the error which is introduced. In Fig. 3.3 is shown a rectangle in which the reference length is L_i and the width W_i. The area of a rectangle is

Figure 3.3 Geometric representation of error.

a function of the length L and the width W, so that the variation in the area from its reference size is obtained as follows:

$$A = A(L, W)$$

$$a = \frac{\partial A}{\partial L}\bigg|_i l + \frac{\partial A}{\partial W}\bigg|_i w$$

The preceding partial derivatives are evaluated from the equation $A = LW$:

$$\frac{\partial A}{\partial L}\bigg|_i = W_i \quad \text{and} \quad \frac{\partial A}{\partial W}\bigg|_i = L_i$$

Thus the linearized expression for the area A is

$$A \approx A_i + a = A_i + W_i l + L_i w \tag{3.14}$$

Each term in the preceding expression is represented by an area in Fig. 3.3. The difference between this approximation and the actual area LW is the small shaded portion lw.

Illustrative example 3.7 Determine the linearized representation for Eq. (2.16), that is,

$$F = (MD^2 + BD + K)X - Mg \tag{2.16}$$

SOLUTION Because F is a function of X, it follows that

$$f = \frac{\partial F}{\partial X}\bigg|_i x$$

The operator D is not a function of X. Thus, the operator is regarded as a constant in evaluating the preceding partial derivative. That is,

$$\frac{\partial F}{\partial X}\bigg|_i = MD^2 + BD + K$$

The resulting linearized equation is

$$f = (MD^2 + BD + K)x \tag{2.17}$$

This verifies the linearized form given by Eq. (2.17). As illustrated by this example, for equations which are already linear the substitution of lowercase letters for the capital-letter variable terms and dropping out the constant terms (for example, $-Mg$) yields directly the linear equation of operation about the reference point. Constant terms drop out because the partial derivative of a constant is zero.

3.2 LINEARIZATION OF OPERATING CURVES

In the preceding section, it was shown how equations which are nonlinear could be linearized. For many components encountered in control systems, the operating characteristics are given in the form of general operating curves rather than equations. For example, in Fig. 3.4 is shown a family of curves of constant values of Z (that is, $Z = 15$, 20, and 25). For $X = 1000$ and $Z = 20$, the corresponding value of Y is 60. This point is indicated by A. The general functional relationship is

$$Y = Y(X, Z)$$

Linearization about a reference point of operation gives

$$y = \frac{\partial Y}{\partial X}\bigg|_i x + \frac{\partial Y}{\partial Z}\bigg|_i z = \frac{\partial Y}{\partial X}\bigg|_Z x + \frac{\partial Y}{\partial Z}\bigg|_X z \tag{3.15}$$

In obtaining the partial derivative of Y with respect to X, all variables in Y are maintained constant except X. Because $Y = Y(X, Z)$, the notation $\partial Y / \partial X$ means that in differentiating Y with respect to X, Z is regarded as constant. The term $\partial Y / \partial X|_i = \partial Y / \partial X|_Z$ is the change of Y per change in X with Z held constant at its reference value. Similarly, $\partial Y / \partial Z|_i = \partial Y / \partial Z|_X$ is the change in Y per change in Z with X fixed at its reference value.

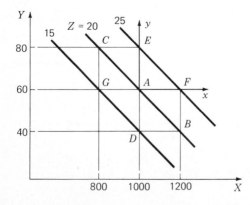

Figure 3.4 General operating curves.

Thus, the term $\partial Y/\partial X|_Z$ is evaluated by dividing changes ΔY by corresponding changes ΔX in which Z is maintained constant at its reference value. For $Z_i = 20$, the value of Y at point B is 40 and the value at point C is 80. Similarly, the value of X at point B is 1200 and the value at point C is 800. Thus

$$\left.\frac{\partial Y}{\partial X}\right|_Z = \left.\frac{\Delta Y}{\Delta X}\right|_Z = \frac{Y_B - Y_C}{X_B - X_C} = \frac{40 - 80}{1200 - 800} = -\frac{40}{400} = -0.10$$

This partial derivative is the slope of the curve $Z_i = 20$ at the reference operating point.

The term $\partial Y/\partial Z|_X$ is obtained by dividing changes ΔY by corresponding changes ΔZ in which X is maintained constant at the reference value. For $X_i = 1000$, the value of Y at point D is 40 and the value of Z is 15. Similarly, at point E the value of Y is 80 and the value of Z is 25. Thus

$$\left.\frac{\partial Y}{\partial Z}\right|_X = \left.\frac{\Delta Y}{\Delta Z}\right|_X = \frac{Y_D - Y_E}{Z_D - Z_E} = \frac{40 - 80}{15 - 25} = \frac{-40}{-10} = 4$$

This partial derivative, which is obtained from a vertical (X_i = constant) extrapolation of the curves, determines the vertical spacing. That is,

$$\Delta Y = 4\Delta Z$$

As Z changes by 5 units, the vertical distance ΔY between lines of constant Z is $\Delta Y = 4(5) = 20$ units. Substitution of the preceding results into Eq. (3.15) yields

$$y = -0.1x + 4z \tag{3.16}$$

The origin of the (x, y) system is at point A, which is the reference point. Note that $X = X_i + x = 1000 + x$ and $Y = Y_i + y = 60 + y$.

At point B, $x = X - X_i = 1200 - 1000 = 200$ and $z = Z - Z_i = 20 - 20 = 0$. Thus, $y = -0.1(200) + 0 = -20$. The value of Y at point B is $Y = Y_i + y = 60 - 20 = 40$. At point E, $x = 1000 - 1000 = 0$ and $z = 25 - 20 = 5$. Thus, $y = 0 + 4(5) = 20$ and $Y = Y_i + y = 60 + 20 = 80$.

The preceding relationship may be solved for z. That is,

$$z = \frac{0.1x + y}{4} = 0.025x + 0.25y \tag{3.17}$$

The curves shown in Fig. 3.4 may be interpreted as expressing any one of the three variables (X, Y, or Z) as a function of the other two. Consider now that they represent Z as a function of X and Y. That is,

$$Z = Z(X, Y)$$

Linearization gives

$$z = \left.\frac{\partial Z}{\partial X}\right|_Y x + \left.\frac{\partial Z}{\partial Y}\right|_X y \tag{3.18}$$

The term $\partial Z/\partial X|_Y$ is obtained by dividing changes in Z by corresponding changes in X in which Y is maintained constant. For $Y_i = 60$, the value of Z at point F is 25 and the value of X is 1200. Similarly, at point G the value of Z is 15 and the value of X is 800. Thus

$$\frac{\partial Z}{\partial X}\bigg|_Y = \frac{\Delta Z}{\Delta X}\bigg|_Y = \frac{Z_F - Z_G}{X_F - X_G} = \frac{25 - 15}{1200 - 800} = 0.025$$

This partial derivative, which is obtained from a horizontal $(Y_i = \text{constant})$ extrapolation of the curves, determines the horizontal spacing. That is,

$$\Delta X = \frac{1}{0.025} \Delta Z = 40 \, \Delta Z$$

As Z changes by 5 units, the horizontal spacing ΔX between lines of constant Z is $\Delta X = 40(5) = 200$ units. The term $\partial Z/\partial Y|_X$ is obtained by dividing changes in Z by corresponding changes in Y with X fixed. Thus

$$\frac{\partial Z}{\partial Y}\bigg|_X = \frac{\Delta Z}{\Delta Y}\bigg|_X = \frac{15 - 25}{40 - 80} = \frac{-10}{-40} = 0.25$$

Note that $\partial Z/\partial Y|_X = \frac{1}{4}$ is the reciprocal of $\partial Y/\partial Z|_X = 4$, which was previously evaluated. Substitution of the results for $\partial Z/\partial X|_Y$ and $\partial Z/\partial Y|_X$ into Eq. (3.18) verifies the result given by Eq. (3.17).

The functional relationship between X, Y, and Z represented in Fig. 3.4 may be expressed in the implicit form

$$G(X, Y, Z) = 0$$

From calculus, it is known that for an implicit function of n variables, the product of the n partial derivatives is $(-1)^n$. For this function of the three variables, X, Y, and Z, this relationship is

$$\frac{\partial X}{\partial Y}\bigg|_Z \frac{\partial Y}{\partial Z}\bigg|_X \frac{\partial Z}{\partial X}\bigg|_Y = (-1)^3 = -1 \tag{3.19}$$

Substitution of the values $\partial X/\partial Y|_Z = 1/(-0.1) = -10$, $\partial Y/\partial Z|_X = 4$, and $\partial Z/\partial X|_Y = 0.025$ verifies that $(-10)(4)(0.025) = -1$. In effect, if any two of the three quantities—horizontal spacing $\partial X/\partial Z|_Y$, vertical spacing $\partial Y/\partial Z|_X$, or slope $\partial Y/\partial X|_Z$—are known, then the third may be determined in accordance with Eq. (3.19).

A typical family of operating curves for an engine is shown in Fig. 3.5. Usually such curves are determined experimentally, and it would be quite tedious and difficult to express them as equations. The linearized equation for the operation of the engine about some reference operating point is obtained as follows. From Fig. 3.5, it is seen that the speed N is a function of the rate of fuel flow Q and the engine torque T; thus

$$N = N(Q, T)$$

Figure 3.5 Characteristic curves for an engine.

Linearization gives

$$n = \frac{\partial N}{\partial Q}\bigg|_i q + \frac{\partial N}{\partial T}\bigg|_i t$$

The term $\partial N/\partial Q\,|_i$ is the change in speed per change in fuel flow with all other parameters held constant (in this case with T constant). This partial derivative is equal to the reciprocal of the slope of the line of constant torque evaluated at the reference point. That is,

$$\frac{\partial N}{\partial Q}\bigg|_i = \frac{2400 - 1600}{32 - 20} = 66.7$$

The partial derivative $\partial N/\partial T\,|_i$ is the change in speed per change in torque with Q held constant. This is evaluated from a horizontal interpolation of the characteristic operating curves as

$$\frac{\partial N}{\partial T}\bigg|_i = \frac{2730 - 1530}{80 - 160} = -15$$

The minus sign indicates that for a constant Q the speed decreases as the torque increases. Thus, for operation in the vicinity of the point $N_i = 2000$, $Q_i = 26$, and $T_i = 120$, the linearized approximation for N is

$$N \approx N_i + n = 2000 + 66.7q - 15t$$

The main difference in working with characteristic operating curves for a component rather than equations is that the partial derivatives are evaluated from a physical interpretation of the curves rather than mathematically from the equations.

3.3 HYDRAULIC SYSTEMS

A schematic diagram of a hydraulic amplifier is shown in Fig. 3.6a. Such amplifiers are very rapid acting and are capable of producing very large forces. The position of the valve is designated by x and the position of the large piston which moves the load is y. This type of valve is called a balanced valve because the pressure forces acting on it are all balanced so that it requires little force to change its position. When the valve is moved upward, the supply pressure is connected to the upper side of the piston to admit oil. The fluid in the lower side of the piston is connected to the drain to permit return flow to the pump where it is recirculated in the system. For the reverse process, the valve is moved downward so that the supply pressure is connected to the bottom side of the big piston. The upper side of this piston is connected to the upper drain to permit return flow to the pump.

When the mass of the load is negligible, the pressure drop across the valve remains constant. For this case the rate of flow to the piston is proportional to the area uncovered by the valve, which is seen to be proportional to the distance x. Thus,

$$q = Cx$$

where q is the rate of flow through the valve to the piston chamber. This rate of flow q into the piston chamber is equal to the rate of change of volume of the chamber which is equal to the piston velocity Dy times the cross-sectional area A of the piston:

$$q = ADy$$

Equating the preceding expressions for q and solving for y gives

$$y = \frac{C}{AD} x = \frac{C}{A} \frac{1}{D} x \qquad (3.20)$$

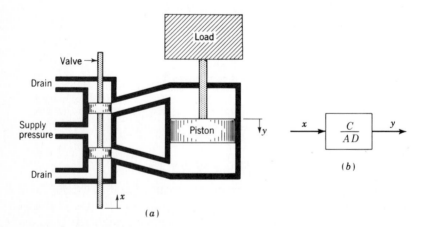

(b)

(a)

Figure 3.6 Hydraulic amplifier.

The $1/D$ term indicates that this hydraulic valve and piston combination in effect integrates hydraulically. That is, the position y is proportional to the integral of the position x. The block-diagram representation is shown in Fig. 3.6b.

The effect of the load M upon the operation of this hydraulic amplifier may be ascertained from the operating curves shown in Fig. 3.7. The rate of flow Q to the cylinder is a function of the valve position X and the pressure drop P across the power piston. That is,

$$Q = Q(X, P)$$

Linearization gives

$$q = \frac{\partial Q}{\partial X}\bigg|_P x + \frac{\partial Q}{\partial P}\bigg|_X p = C_1 x - C_2 p \tag{3.21}$$

where $C_1 = \partial Q/\partial X\,|_P$ and $C_2 = -\partial Q/\partial P\,|_X$. The term $\partial Q/\partial X\,|_P$ is obtained from a vertical interpolation of the curves in which P is maintained constant at the reference value. Because Q increases as X increases, $\partial Q/\partial X\,|_P$ is a positive constant. The term $\partial Q/\partial P\,|_X$ is the slope of the curve at the reference value of X. Because Q decreases as P increases (negative slope), the term $\partial Q/\partial P\,|_X$ is negative. The constant $C_2 = -\partial Q/\partial P\,|_X$ is a positive constant. In analyzing control systems, it is desirable that all constants be positive. This enables the control engineer to determine directly from the equations of operation and the resulting block diagrams whether the output increases or decreases when the input undergoes a certain change.

The force transmitted to the load by the power piston is equal to the product of the pressure drop across the piston and the cross-sectional area A of the piston. Thus,

$$pA = M\,\frac{d^2 y}{dt^2} = MD^2 y \tag{3.22}$$

where M is the mass of the load and $d^2 y/dt^2$ is the acceleration. Substitution of p from Eq. (3.22) into Eq. (3.21) gives

$$q = C_1 x - \frac{C_2 M}{A} D^2 y$$

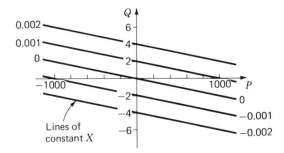

Figure 3.7 Operating curves for a hydraulic amplifier.

The rate of flow q into the cylinder is equal to the rate of change of volume, which is the cross-sectional area A times the velocity dy/dt. Thus

$$q = A \frac{dy}{dt} = ADy$$

Equating the preceding expressions for q and solving for the output position y as a function of the input position x gives

$$y = \frac{(C_1/A)x}{D(1 + \tau D)} \tag{3.23}$$

where $\tau = C_2 M/A^2$. Note that when the load is neglected ($M = 0$), then $\tau = C_2 M/A^2 = 0$ and Eq. (3.23) reduces to Eq. (3.20). Usually C_2 is quite small, so that $\tau \approx 0$. In general the approximation given by Eq. (3.20) is quite good.

Hydraulic Servomotor

A hydraulic servomotor is shown in Fig. 3.8. A linkage called a walking beam connects the input position x, the valve position e, and the piston position y. The centerline of the lever when the servomotor is in its reference position is indicated in Fig. 3.8. The variations in x, e, and y from their reference positions are also indicated. When e is zero, the valve is "line on line" and no flow can go to or from the big piston.

The operation of this servomotor may be visualized as follows. When the input x is changed from the reference position, the walking beam first pivots about the connection at y because the large forces acting on the piston hold it in

Figure 3.8 Hydraulic servomotor.

place temporarily. This position of the walking beam is shown by the dashed line in Fig. 3.8. Because of the corresponding movement of e, the valve now admits fluid to the big piston to move it in the direction which makes e zero. The final position of the walking beam, in which e is again zero and the piston has moved a distance y, is indicated in Fig. 3.8. For steady-state operation ($e = 0$), the relationship between the input x and the output y is

$$\frac{y}{b} = \frac{x}{a} \quad \text{or} \quad y = \frac{b}{a} x$$

The overall block diagram which describes the dynamic as well as the steady-state operation of this servomotor is obtained as follows. In Fig. 3.9a is shown the walking-beam linkage. For small variations about the reference position,

$$e = \left. \frac{\partial E}{\partial X} \right|_i x + \left. \frac{\partial E}{\partial Y} \right|_i y \tag{3.24}$$

The value of $\partial E / \partial X |_i$ is obtained by finding the ratio of the change in E for a change in X with all other parameters held constant at the reference position. Figure 3.9b illustrates the linkage with Y fixed in the reference position. From similar triangles,

$$\left. \frac{\partial E}{\partial X} \right|_i = \lim_{\substack{\Delta E \to 0 \\ \Delta X \to 0}} \frac{\Delta E}{\Delta X} = \frac{b}{a+b}$$

Similarly, from Fig. 3.9c in which X is fixed in the reference position,

$$\left. \frac{\partial E}{\partial Y} \right|_i = \lim_{\substack{\Delta E \to 0 \\ \Delta Y \to 0}} \frac{\Delta E}{\Delta Y} = \frac{-a}{a+b}$$

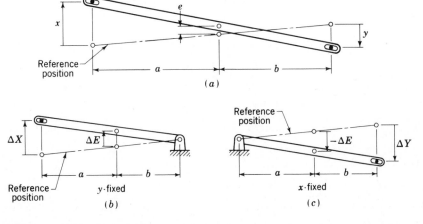

Figure 3.9 Walking-beam linkage.

The minus sign arises because e decreases as y increases. The substitution of the preceding results into Eq. (3.24) yields the following expression for the walking-beam linkage:

$$e = \frac{b}{a + b} x - \frac{a}{a + b} y \qquad (3.25)$$

This equation shows that the walking-beam linkage is actually a comparator or summing point. The preceding result could have been obtained directly by a closer examination of Fig. 3.9a. It is apparent that the motion of e is the sum of the contribution due to changing x with y fixed, that is, $[b/(a + b)]x$, and that due to changing y with x fixed, that is, $[-a/(a + b)]y$.

For the case in which $a = b$,

$$e = \frac{x - y}{2} \qquad (3.26)$$

The block-diagram representation for Eq. (3.26) is shown in Fig. 3.10a.

The equation for the valve-and-piston combination is given by Eq. (3.20), but with x replaced by e. Thus,

$$y = \frac{C}{AD} e \qquad (3.27)$$

The block-diagram representation for the preceding expression is shown in Fig. 3.10b. Combining Fig. 3.10a and b yields the overall block diagram for the servomotor as shown in Fig. 3.10c.

The overall relationship between the input x and the output y is obtained as follows from the block diagram of Fig. 3.10c:

$$(x - y) \frac{C}{2AD} = y$$

or $$(1 + \tau D)y = x \qquad (3.28)$$

where $$\tau = \frac{2A}{C}$$

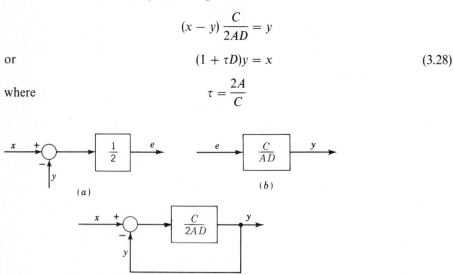

Figure 3.10 Block diagram for (a) walking-beam linkage, (b) valve and piston, (c) servomotor.

Equation (3.28) is the differential equation relating x and y. For steady-state operation, both x and y are constant. The quantity $Dy = dy/dt$ is zero when y is constant, and thus for steady-state operation Eq. (3.28) becomes

$$y = x$$

First-Order Systems

To determine the transient response of y for a given change in x, it is necessary to solve Eq. (3.28), which is a first-order linear differential equation with constant coefficients. Many elements encountered in control systems are described by a first-order differential equation of the form

$$(1 + \tau D)y = Kx \tag{3.29}$$

where K is a constant. For $K = 1$, this equation reduces to Eq. (3.28). To solve Eq. (3.29), first rewrite it in the form

$$\frac{dy}{dt} + \frac{1}{\tau} y = \frac{K}{\tau} x$$

Multiplying through by the integrating factor $e^{t/\tau}$ gives

$$e^{t/\tau} \frac{dy}{dt} + \frac{e^{t/\tau}}{\tau} y = \frac{d}{dt} (e^{t/\tau}y) = \frac{Ke^{t/\tau}}{\tau} x$$

Integration shows that

$$e^{t/\tau}y = \frac{K}{\tau} \int xe^{t/\tau} \, dt$$

Thus,

$$y = \frac{Ke^{-t/\tau}}{\tau} \int xe^{t/\tau} \, dt$$

For a given input x, the preceding may be solved for the resulting response y. Consider the case in which the input x changes instantaneously from its initial value $x = 0$ at time $t = 0$ to a new or final value $x = 1$ for $t > 0$. A unit step change, as is illustrated graphically in Fig. 3.11a, has occurred. Solving for y when $x = 1$ gives

$$y = \frac{Ke^{-t/\tau}}{\tau} \int e^{t/\tau} \, dt = \frac{Ke^{-t/\tau}}{\tau} (\tau e^{t/\tau} + C)$$

The constant of integration C is evaluated from the initial condition $y(0) = 0$ when $t = 0$. Thus,

$$0 = \frac{K}{\tau} (\tau + C)$$

or

$$C = -\tau$$

Figure 3.11 Response of first-order system to a unit step function.

The resulting solution is

$$y = K(1 - e^{-t/\tau}) \tag{3.30}$$

A graph of the response $y(t)$ is shown in Fig. 3.11b. A characteristic feature of such an exponential response curve is that when $t = \tau$, then $y(t)$ has undergone 63.2 percent of its total change. This is proved by letting $t = \tau$ in Eq. (3.30); thus,

$$y(\tau) = K(1 - e^{-1}) = K(1 - 0.368) = 0.632K$$

When $t = 2\tau$, $y(t)$ has undergone 86.5 percent of its total change, and when $t = 3\tau$, $y(t)$ has undergone 95 percent of its total change. The final steady-state value of $y(t)$ is K. Because y is constant for the steady state, then $Dy = 0$ in Eq. (3.29). The resulting equation for steady-state operation is

$$y = Kx$$

Thus a one-unit change in x results in K-units change in y. The parameter K is referred to as the steady-state gain. Another feature of an exponential response, as is shown in Fig. 3.11b, is that the tangent to the curve at $t = 0$ intersects the final value at time $t = \tau$. This is proved by differentiating Eq. (3.30) to obtain the slope of the response curve and then evaluating this result at time $t = 0$. Thus,

$$\left.\frac{dy}{dt}\right|_{t=0} = \left.\frac{K}{\tau} e^{-t/\tau}\right|_{t=0} = \frac{K}{\tau}$$

As shown in Fig. 3.11b, the slope of the tangent line at $t = 0$ is K/τ. The term τ is called the time constant and is a measure of the speed of response. When τ is small, the system approaches its new operating condition very fast, and when τ is large, more time is required for the change to occur.

Jet Pipe Amplifier

A schematic representation of a jet pipe amplifier is shown in Fig. 3.12a. The jet pipe pivots about a fixed pivot point. The position of the jet pipe at the centerline of the springs is indicated by e and the position at the nozzle end is indicated by

Figure 3.12 Jet pipe amplifier: (*a*) schematic diagram, (*b*) initial block diagram, (*c*) final block diagram, (*d*) position x determined by cam.

w. The pipe is supplied with a high-pressure hydraulic fluid such that a steady stream is continually flowing out the nozzle end. The rate of flow q to the power piston is $q = Cw$, where C is a constant depending on the particular jet pipe and the supply pressure.

The position x is the input position and y, which is the position of the power piston, is the output position. The compression of spring K_1 is $(x - e)$ and the compression of spring K_2 is $(e + y)$; hence $K_1(x - e) = K_2(e + y)$ or $K_1 x - K_2 y = (K_1 + K_2)e$. For a typical jet pipe, the distance from the fixed pivot to the end where w is measured is twice the distance from the fixed pivot to the center of the springs where e is measured; thus $w = 2e$. The rate of change of volume in the power piston is $q = A\,dy/dt = ADy$. Combining the preceding equations yields the block-diagram representation shown in Fig. 3.12b. Multiplying the feedforward elements together yields the block diagram shown in Fig. 3.12c. The differential equation of operation for the jet pipe amplifier is obtained from Fig. 3.12c as follows:

$$y = \frac{K_1/K_2}{1 + (A/2C)[(K_1 + K_2)/K_2]D}\ x$$

This has the form of Eq. (3.29), that is,

$$y = \frac{K}{1 + \tau D}\ x \tag{3.29}$$

Thus, the steady-state gain is K_1/K_2 and the time constant is $A(K_1 + K_2)/2CK_2$. Usually C is very large, in which case the time constant is very small. In many applications the position x is determined by a cam pressing against the bottom of the spring seat as shown in Fig. 3.12d. For each angular position θ of the cam there is one position x. For steady-state operation, $y = (K_1/K_2)x$. A typical application is that in which the load is a metering valve. Each position y of the metering valve results in a different amount of fluid flow through the valve; hence a very accurate flow-metering device results.

The jet pipe amplifier shown in Fig. 3.12a has force feedback. That is, the spring K_2 provides a force proportional to the output position y. The spring K_1 provides a force proportional to the desired position. The difference causes a movement of the jet pipe. If the desired input x is connected to one end of a walking-beam linkage, the output y to the other end, and the jet pipe position e to the center, then a jet pipe amplifier with position feedback results.

3.4 PNEUMATIC SYSTEMS

A flapper valve, as shown in Fig. 3.13a, is one in which small changes in the position X of the flapper cause large variations in the controlled pressure P_2 in the chamber. When the flapper is closed off the pressure P_2 in the chamber is equal to the supply pressure P_1. If the flapper is opened wide, the chamber pressure approaches the ambient pressure P_a. If a spring-loaded bellows which is free

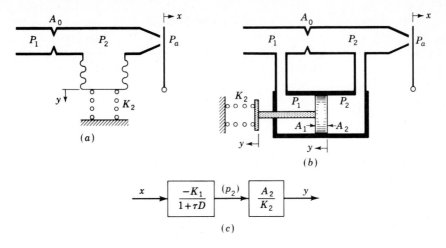

Figure 3.13 Pneumatic amplifier.

to expand as the chamber pressure increases is attached, this becomes a position control device as well as a pressure controller. An equivalent system which uses a piston rather than a bellows is shown in Fig. 3.13b.

Although flapper valves have numerous applications in hydraulic systems, they are more extensively used in pneumatic controls. A major reason for this is that spool valves do not lend themselves to pneumatic applications because of the excessive leakage of air that is the result of the very low viscosity of air. When the flapper valves of Fig. 3.13 are supplied with air, pneumatic amplifiers result.

The procedure used to obtain the equation of operation for a pneumatic amplifier is similar to that used for a hydraulic amplifier, with the exception that because of compressibility effects, the mass rate of flow must be considered rather than the volume rate. For a constant supply pressure P_1 and fixed inlet orifice, the mass rate of flow into the chamber M_{in} is a function of the chamber pressure P_2 only. Linearization gives

$$m_{in} = \frac{\partial M_{in}}{\partial P_2}\bigg|_i p_2 = -C_1 p_2$$

The minus sign indicates that as p_2 increases, m_{in} decreases. For a constant ambient pressure P_a, the mass rate of flow out from the chamber M_o is a function of X and P_2. Thus,

$$m_o = \frac{\partial M_o}{\partial X}\bigg|_i x + \frac{\partial M_o}{\partial P_2}\bigg|_i p_2 = C_2 x + C_3 p_2$$

When the pressure ratio across an orifice is less than 0.528, the flow is sonic, and when it is greater than 0.528, the flow is subsonic. For sonic flow the mass rate of flow is given by Eq. (3.10) and for subsonic flow it is given by Eq. (3.12).

The change in mass w of air in the chamber is the integral of $m_{in} - m_o$. That is,

$$w = \frac{m_{in} - m_o}{D} = \frac{-C_1 p_2 - (C_2 x + C_3 p_2)}{D}$$

Multiplying through by D shows that

$$-C_1 p_2 - C_2 x - C_3 p_2 = Dw \tag{3.31}$$

From the equation of state, the total mass W of air in the chamber is

$$W = \frac{P_2 V_2}{RT_2}$$

where V_2 is the volume of the chamber and T_2 is the stagnation temperature of air in the chamber. For the usual case of adiabatic flow, T_2 is equal to the stagnation temperature of the supply, which is constant. Linearization yields, for the change in mass w of air in the chamber,

$$w = \frac{\partial W}{\partial V_2}\bigg|_i v_2 + \frac{\partial W}{\partial P_2}\bigg|_i p_2 = C_4 v_2 + C_5 p_2 \tag{3.32}$$

where

$$\frac{\partial W}{\partial V_2}\bigg|_i = \frac{P_2}{RT_2}\bigg|_i = \frac{W}{V_2}\bigg|_i = C_4$$

and

$$\frac{\partial W}{\partial P_2}\bigg|_i = \frac{V_2}{RT_2}\bigg|_i = \frac{W}{P_2}\bigg|_i = C_5$$

The term $C_4 v_2$ represents the change in mass required to account for the change in volume, and the term $C_5 p_2$ represents the change in mass required to account for the change in pressure. This is the compressibility effect. The change in volume of the chamber is equal to the area A_2 times the change in position:

$$v_2 = A_2 y$$

From Fig. 3.13a, the summation of forces acting on the bellows is

$$P_2 A_2 = K_2 Y$$

Similarly, from Fig. 3.13b, the summation of forces acting on the piston is

$$P_2 A_2 - P_1 A_1 = K_2 Y$$

Because $P_1 A_1$ is constant, linearization of either of the preceding gives the same result. That is,

$$y = \frac{A_2}{K_2} p_2$$

Substituting w from Eq. (3.32) into Eq. (3.31) and then using the preceding expressions to eliminate v_2 and y yields the result

$$p_2 = -\frac{K_1}{1 + \tau D} x \tag{3.33}$$

where $K_1 = C_2/(C_1 + C_3)$ and $\tau = (C_5 + A_2^2 C_4/K_2)/(C_1 + C_3)$. The minus sign indicates that as x increases, p_2 decreases. Because $y = (A_2/K_2)p_2$, it follows that

$$y = \frac{A_2}{K_2}\frac{-K_1}{1 + \tau D} x$$

The block-diagram representation is shown in Fig. 3.13c. For most flapper valves the time constant τ may generally be regarded as negligible.

A nondimensional family of curves (Fig. A.1) for determining the equilibrium chamber pressure P_2 is developed in Appendix A. Lines of constant area ratio A_2/A_1 are plotted, where A_1 is the area of the first orifice times its coefficient of discharge and A_2 is the area of the second orifice times its coefficient of discharge. The abscissa is the overall pressure ratio P_1/P_3, where P_1 is the supply pressure and P_3 is the discharge pressure, which is usually the ambient pressure P_a. The ordinate P_2/P_1 is the ratio of the chamber pressure to the supply pressure. If the area ratio A_2/A_1 and the overall pressure ratio P_1/P_3 are known, then the ratio P_2/P_1 may be determined. The desired chamber pressure is equal to this ratio (P_2/P_1) times the supply pressure P_1. A typical plot of P_2 versus X is shown in Fig. 3.14. Note that absolute pressure is used for pneumatic amplifiers, whereas gauge pressure is used for hydraulic amplifiers.

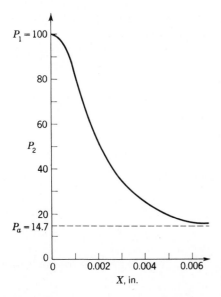

Figure 3.14 Equilibrium curve of P_2 versus X for a pneumatic amplifier.

For steady-state operation, P_2 is constant so $\tau DP_2 = 0$ in Eq. (3.33). Thus, the equation for steady-state operation is

$$p_2 = -K_1 x \tag{3.34}$$

The constant $-K_1 = p_2/x$ is the slope of Fig. 3.14 at the reference point of operation. As shown in Fig. 3.14, this slope remains quite constant over a considerable portion of the operating range. Note that it is not necessary to evaluate C_1, C_2, and C_3 individually in order to attain $K_1 = C_2/(C_1 + C_3)$.

3.5 DC MOTORS

A major reason for the use of dc machines in electromechanical control systems is the ease with which speed can be controlled. The polarity of the applied voltage determines the direction of rotation. Also, dc machines are capable of providing large power amplifications.

The field and armature windings of dc motors may be shunt-connected, series-connected, compounded, or separately excited. The motors used in control systems are generally separately excited. There are two types of separate excitation: field control with fixed armature current and armature control with fixed field.

Field Control

A separately excited motor in which the armature current I_a is maintained constant is shown in Fig. 3.15a. The constant current I_a may be supplied by a dc generator or from an ac line. The latter method requires the use of transformers and rectifiers to obtain the proper rectification. The voltage E_f applied to the field is obtained from the output of an amplifier in a low-power application or from a dc generator when greater power is needed. In the field circuit, the resistance of the windings is R_f and the inductance is designated by L_f. The torque T

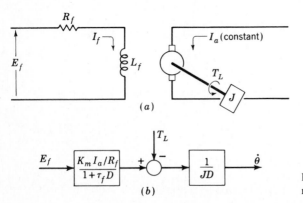

Figure 3.15 Field-controlled dc motor.

developed by a motor is proportional to the product of the armature current I_a and the magnetic flux ϕ of the field:

$$T = K_1 \phi I_a \qquad (3.35)$$

where K_1 is a constant for any motor and depends upon the total number of armature conductors, the number of field poles, etc.

A typical curve of flux ϕ versus field current I_f is shown in Fig. 3.16. When the field current I_f becomes great enough to cause the iron to saturate, the flux ϕ no longer increases linearly with the current. Motors used in control systems usually operate over the linear portion of this curve, in which case

$$\phi = K_2 I_f \qquad (3.36)$$

where K_2 is the slope of the linear portion of the curve as shown in Fig. 3.16. The substitution of the preceding result into Eq. (3.35) yields

$$T = K_1 K_2 I_a I_f = K_m I_a I_f \qquad (3.37)$$

where $K_m = K_1 K_2$.

If the moment of inertia of the armature is J, the coefficient of viscous friction B_v, and the load torque T_L, then from a summation of torques acting on the armature it follows that

$$T = (B_v D + J D^2)\theta + T_L \qquad (3.38)$$

where θ is the angular position of the armature, or motor shaft.

The equation for the field current I_f is obtained from the equivalent field circuit of Fig. 3.15a:

$$I_f = \frac{E_f}{R_f + L_f D} = \frac{E_f}{R_f(1 + \tau_f D)} \qquad (3.39)$$

where $\tau_f = L_f/R_f$ is the time constant of the field circuit.

Substituting T from Eq. (3.38) and I_f from Eq. (3.39) into Eq. (3.37) and solving for θ gives

$$\theta = \frac{1}{D(B_v + JD)} \left[\frac{(K_m I_a/R_f) E_f}{1 + \tau_f D} - T_L \right] \qquad (3.40)$$

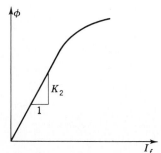

Figure 3.16 Plot of flux versus field current.

Multiplication of the preceding expression by D gives the angular velocity $\dot{\theta} = D\theta$:

$$\dot{\theta} = \frac{1}{B_v + JD} \left[\frac{(K_m I_a/R_f)E_f}{1 + \tau_f D} - T_L \right]$$ (3.41)

Generally, the damping B_v is negligible, so that the block-diagram representation for the speed of this field-controlled dc motor is as shown in Fig. 3.15b.

Armature Control

A dc motor with armature control is one in which the speed is controlled by the armature voltage E_a. An armature-controlled motor in which the field current I_f is kept constant is shown in Fig. 3.17a. The armature voltage E_a is usually supplied by a generator, which in turn may be supplied by an amplifier. The voltage E_c is the counter emf induced by the rotation of the armature windings in the magnetic field. The counter emf is proportional to the product of the armature speed $\dot{\theta}$ and the field strength ϕ. That is,

$$E_c = K_3 \phi \dot{\theta}$$ (3.42)

where K_3 is a constant for any particular motor. The substitution of ϕ from Eq. (3.36) into the preceding gives

$$E_c = K_2 K_3 I_f \dot{\theta} = K_c I_f \dot{\theta}$$ (3.43)

(a)

(b)

Figure 3.17 Armature-controlled dc motor.

where $K_c = K_2 K_3$. The circuit equation for the armature portion of Fig. 3.17a is

$$E_a - K_c I_f \dot{\theta} = R_a I_a + L_a D I_a = R_a(1 + \tau_a D)I_a \qquad (3.44)$$

where $\tau_a = L_a/R_a$ and the term $K_c I_f \dot{\theta}$ is the counter emf developed in the armature. The torque developed by the motor is given by the equation

$$T = K_m I_f I_a \qquad (3.45)$$

and the torque balance for the output shaft is

$$T = (B_v D + J D^2)\theta + T_L \qquad (3.46)$$

The block-diagram representation for this armature-controlled motor is obtained by combining the block-diagram representations for Eqs. (3.44), (3.45), and (3.46), as shown in Fig. 3.17b. The counter emf is responsible for the feedback.

A complete generator and armature-controlled motor combination is shown in Fig. 3.18. This is in effect a Ward-Leonard system, or motor-generator set. The voltage E supplied to the generator may be quite small, as in the case of that coming from an amplifier. The resistance of the field of the generator is R_{fg}, and the inductance is L_{fg}. The armature of the generator is driven at a constant speed by a prime mover. The output voltage of the generator E_g goes directly to the armature of the motor, so that $E_g = E_a$.

The circuit equation for the generator field is

$$E = (R_{fg} + L_{fg} D)I_{fg} \qquad (3.47)$$

The voltage induced in the armature is the generated voltage $E_g = E_a$, which is

$$E_a = K_{cg} \dot{\theta}_g I_{fg} = K_c I_{fg} \qquad (3.48)$$

where $\dot{\theta}_g$ is the angular velocity of the prime mover, which is constant, so that $K_{cg} \dot{\theta}_g = K_c$. The substitution of I_{fg} from Eq. (3.47) into Eq. (3.48) yields

$$E_a = \frac{K_c E}{R_{fg} + L_{fg} D} = \frac{(K_c/R_{fg})E}{1 + \tau_{fg} D} \qquad (3.49)$$

where $\tau_{fg} = L_{fg}/R_{fg}$ is the time constant of the generator field.

The overall block diagram relating the input voltage E and the velocity $\dot{\theta}$ of this armature-controlled motor-generator system is obtained by connecting the block-diagram representation for the output E_a from Eq. (3.49) to that for E_a in Fig. 3.17b.

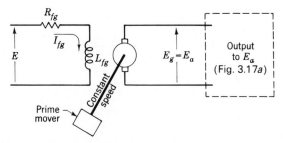

Figure 3.18 Complete generator and armature-controlled motor combination.

DC Tachometer

A dc tachometer is a generator in which the magnetic flux is usually supplied by a permanent magnet. Because the flux is maintained constant, the equation of operation for a tachometer is obtained from Eq. (3.42) as follows:

$$E_c = E_g = K_3 \phi \dot{\theta} = K_g \dot{\theta} \tag{3.50}$$

where $K_g = K_3 \phi$ is a constant and E_g is the generated voltage. Thus a tachometer is seen to supply a voltage E_g which is proportional to the speed at which it is driven.

Remote-Control Positional Servomechanism

A remote-control positional servomechanism is shown in Fig. 3.19a. The wiper arm of the input potentiometer is positioned by the desired input position θ_r, so that the voltage E_r is proportional to θ_r (that is, $E_r = K_r \theta_r$). Similarly, the controlled shaft position θ_c determines the position of the wiper arm for the other potentiometer, so that $E_c = K_c \theta_c$. The error signal $E_e = E_r - E_c$ is amplified by the amplifier and the resultant voltage is applied to the field of a field-controlled motor, so that $E_f = K_1 E_e$. The operational representation of the differential equation for the motor is given by Eq. (3.40). The overall block diagram for this system is shown in Fig. 3.19b.

The motor must be located at the output shaft, while the input potentiometer is usually situated in any convenient location. A major advantage in using such electrical equipment for position control systems is the ease of connecting the input and output by means of wires.

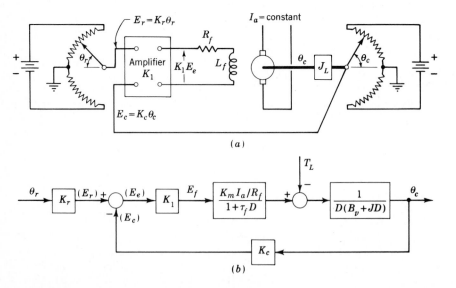

Figure 3.19 Remote-control positional servomechanism.

The preceding position controller may be converted to a speed control system by connecting the output shaft to a tachometer rather than to a potentiometer. In this case, the voltage signal E_c coming from the tachometer is proportional to the speed $\dot{\theta}_c$ (that is, $E_c = K_c\dot{\theta}_c$). Similarly, each wiper position of the input potentiometer corresponds to a desired speed setting $\dot{\theta}_r$, rather than position θ_r, so that the reference voltage is $E_r = K_r\dot{\theta}_r$.

3.6 AC MOTORS

An ac two-phase motor is used for simple low-power applications. One of the phases is supplied with a fixed ac voltage which acts as the reference voltage. The other phase is connected to the controlled voltage. A schematic representation of a two-phase motor is shown in Fig. 3.20a. Because the reference voltage E_R is constant, the speed depends upon the control voltage E. The direction of rotation is reversed by changing the polarity of the control voltage.

As is shown in Fig. 3.20a, the reference and control windings are displaced by $90°$ in the stator of the motor. Thus, although the voltage applied to each winding has the same frequency, there is a $90°$ phase shift of one with respect to the other.

Typical performance curves relating the developed motor torque T and the angular velocity $\dot{\theta}$ for constant values of control current I are shown in Fig. 3.20b. The equation describing the operation of a two-phase motor about some equilibrium point of operation is derived as follows. From Fig. 3.20b it is to be noticed that the speed is a function of T and I:

$$\dot{\theta} = F(T, I)$$

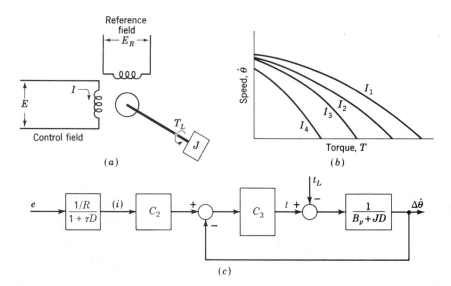

Figure 3.20 (a) Two-phase ac motor, (b) typical performance curves, (c) block-diagram representation.

Linearization about the reference point of operation gives

$$\Delta\dot\theta = \left.\frac{\partial\dot\theta}{\partial T}\right|_I t + \left.\frac{\partial\dot\theta}{\partial I}\right|_T i = -C_1 t + C_2 i \qquad (3.51)$$

where $\Delta\dot\theta$ is the change in speed from the reference value, $C_1 = -\partial\dot\theta/\partial T\,|_I$ and $C_2 = \partial\dot\theta/\partial I\,|_T$. The term $\partial\dot\theta/\partial T\,|_I$ is the slope of the operating curves at the reference operating condition, which is negative. The minus sign converts C_1 to a positive constant. For a given torque, the speed increases as the current increases, so that $C_2 = \partial\dot\theta/\partial I\,|_T$ is positive. Solving Eq. (3.51) for t gives

$$t = \frac{1}{C_1}\,[C_2 i - \Delta\dot\theta] = C_3(C_2 i - \Delta\dot\theta) \qquad (3.52)$$

where $C_3 = 1/C_1 = -\partial T/\partial\dot\theta\,|_I$. The torque balance for the armature is

$$T - T_L = (B_v + JD)\dot\theta$$

For small departures,

$$t - t_L = (B_v + JD)\Delta\dot\theta \qquad (3.53)$$

The equation of operation for the control windings is

$$E = (R + LD)I = R(1 + \tau D)I$$

where $\tau = L/R$. For small departures, the preceding becomes

$$e = R(1 + \tau D)i \qquad (3.54)$$

The block-diagram representation for this two-phase ac motor is obtained from Eqs. (3.52), (3.53), and (3.54), as shown in Fig. 3.20c.

3.7 BLOCK-DIAGRAM ALGEBRA

It is often desirable to rearrange the form of a block diagram. In Fig. 3.21 is shown a number of rearrangements which are commonly employed. It is to be noticed that in all cases the rearrangement does not affect the overall relationship between the input elements (i.e., elements with arrowheads pointing into the diagram) and the output elements (i.e., elements with arrowheads pointing away from the diagram). There are many possible rearrangements for systems. However, it is usually desirable to make the ultimate form of the block diagram the same as that shown in Fig. 3.22. The reason for this is that the methods to be presented later for evaluating the performance of systems are based on systems which are represented in this general form.

Figure 3.21 Equivalent block diagrams: (*a*) combining interconnected summing points, (*b*) moving a summing point behind an element, (*c*) moving a summing point ahead of an element, (*d*) moving a take-off point behind an element, (*e*) moving a take-off point ahead of an element, (*f*) eliminating a minor feedback loop.

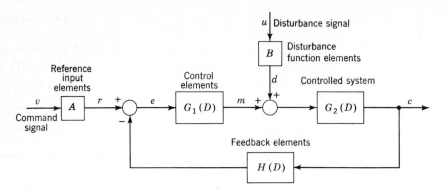

Figure 3.22 General block-diagram representation for a control system.

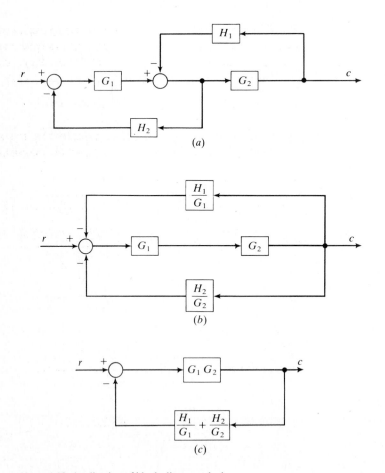

Figure 3.23 Application of block-diagram algebra.

Illustrative example 3.8 Use block-diagram algebra to put the block diagram shown in Fig. 3.23a into the general form shown in Fig. 3.22.

SOLUTION The first step is to move the point where the upper feedback path (element H_1) enters the feedforward loop from the right side of element G_1 to the left side. This is accomplished by the application of the technique shown in Fig. 3.21b. As shown in Fig. 3.23b, the resulting upper feedback element is H_1/G_1. The next step is to move the point where the lower feedback path (element H_2) leaves the feedforward path from the left side of element G_2 to the right side. This is accomplished by application of the technique shown in Fig. 3.21e. The resulting lower feedback element is H_2/G_2. The upper and lower feedback paths of Fig. 3.23b may now be combined into a single feedback path as is illustrated by the block diagram shown in Fig. 3.23c, which has the same form as Fig. 3.22.

3.8 SPEED CONTROL SYSTEMS

All engines derive their power from some source of energy, i.e., fuel. The power developed by an engine is controlled by controlling the rate of flow of fuel to the engine. When the power developed by an engine exceeds the power required (load on the engine), the excess power accelerates the engine, thus resulting in an increased speed. Conversely, when the load is greater than the power developed, a decrease in speed results. A speed governor automatically controls the speed of an engine.

In Fig. 3.24a is shown a speed governor. The flyweights are pin-connected to a rotating ballhead. This ballhead is connected by gears to the engine so that the speed of rotation of the flyweights is equal to a constant times the engine speed. Because of the bearing located between the toes of the flyweights and the spring, the spring does not rotate. In Fig. 3.24b is shown a schematic representation of the governor shown in Fig. 3.24a.

Figure 3.25 shows a typical speed control system for gas turbines, steam turbines, or diesel engines. The position of the throttle lever sets the desired speed of the engine. The speed control is drawn in some reference operating position so that the values of all the lowercase parameters are zero. The positive direction of motion of these parameters is indicated by the arrowhead on each.

If the speed of the engine were to drop below its reference value, then the centrifugal force of the flyweights would decrease, thus decreasing the force exerted on the bottom of the spring. This causes x to move downward, which in turn moves e downward. Fluid then flows to the bottom of the big piston to increase y and thus open wider the flow control valve. When more fuel is supplied, the speed of the engine increases until equilibrium is again reached. For steam turbines, the flow control valve controls the flow of steam rather than fuel, as is the case with gas turbines and diesels.

(a)

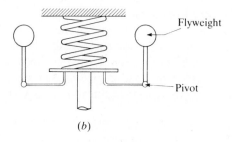

(b)

Figure 3.24 (a) Flyball governor, (b) schematic representation.

Suppose that the throttle lever is moved to a higher speed setting, which in turn causes z to move downward. This in turn causes x to move downward. As just discussed, moving x downward opens the fuel flow valve, which increases the speed.

The overall block-diagram representation for this system is obtained as follows. The position Z of the top of the spring is a function of the desired speed only [that is, $Z = Z(N_{in})$]. Thus, the variation of the top of the spring z from its reference position is

$$z = \left.\frac{\partial Z}{\partial N_{in}}\right|_i n_{in} = C_2 n_{in} \tag{3.55}$$

where $n_{in} = N_{in} - N_i$ is the change in desired speed and $C_2 = \left.\partial Z/\partial N_{in}\right|_i$ is the change in position Z per change in desired speed N_{in}.

Figure 3.25 Speed control system.

In Fig. 3.26 is shown the free-body diagram of the right flyweight of Fig. 3.24b. The centrifugal force F_c acting on the flyweight is

$$F_c = MR\omega^2$$

where M = mass of the flyweight
$R = R_i + r$ = distance from center of rotation to center of mass of the flyweight
ω = angular velocity of the flyweight

Usually, the governor is geared directly to the output shaft such that ω is equal to the gear ratio times the output speed; i.e.,

$$\omega = C_g N_o$$

Figure 3.26 Free-body diagram of flyweight.

where C_g is the gear ratio. Substitution of this value for ω into the preceding expression for F_c gives

$$F_c = C_g^2 MRN_o^2 \tag{3.56}$$

Because there are two flyweights, half of the spring force $F_s/2$ acts on the toe of each flyweight. The perpendicular distance from the pivot to the line of action of the force $F_s/2$ acting on the toe is a. The centrifugal force F_c acts radially outward through the center of mass of the flyweight. The perpendicular distance from the pivot to the line of action of F_c is b. Taking the moment of forces about the pivot point shows that

$$\frac{aF_s}{2} = bF_c$$

or
$$F_s = 2\frac{b}{a}F_c = 2\frac{b}{a}C_g^2 MRN_o^2 = C_f C_r RN_o^2 \tag{3.57}$$

where $C_f = 2C_g^2 M$
$C_r = b/a$

The two independent variables are R and N_o; linearization gives

$$f_s = C_3 r + C_4 n_o \tag{3.58}$$

where

$$C_3 = \left.\frac{\partial F_s}{\partial R}\right|_i = C_f C_r N_i^2$$

and
$$C_4 = \left.\frac{\partial F_s}{\partial N_o}\right|_i = 2C_f C_r R_i N_i$$

The compression of the spring from its reference length is $z - x$. Thus, the variation in force exerted by the spring is

$$f_s = K_s(z - x) \tag{3.59}$$

where K_s is the spring constant. Setting Eqs. (3.58) and (3.59) equal,

$$K_s(z - x) = C_3 r + C_4 n_o$$

The geometry of Fig. 3.25 shows that the motions of r and x are related so that $r = -(b/a)x = -C_r x$. The reason for the minus sign is that as r increases, x decreases. Eliminating r from the preceding equation yields

$$K_s z - K_s x = -C_r C_3 x + C_4 n_o$$

or
$$x = \frac{K_s z - C_4 n_o}{K_s - C_r C_3} \tag{3.60}$$

The block-diagram representation for Eqs. (3.55) and (3.60) is shown in Fig. 3.27, which is the comparator for the speed control system.

The operation of the servomotor was discussed in Sec. 3.3 and the block diagram was given in Fig. 3.10c.

The flow through the flow control valve is a function of the position Y [that is, $Q = Q(Y)$]. Linearization gives

$$q = \frac{\partial Q}{\partial Y}\bigg|_i y = C_5 y \tag{3.61}$$

where C_5 is the slope of the curve of Q versus Y evaluated at the reference position.

The speed N_o of an engine is a function of the fuel flow Q supplied to the engine and the engine torque T:

$$N_o = N_o(Q, T)$$

The linearized form of this expression is

$$n_o = \frac{\partial N_o}{\partial Q}\bigg|_T q + \frac{\partial N_o}{\partial T}\bigg|_Q t = C_6 q - C_7 t \tag{3.62}$$

where $C_6 = \partial N_o/\partial Q|_T$ is the change in speed per change in fuel flow with T maintained at the reference value and $C_7 = -\partial N_o/\partial T|_Q$ is the change in speed

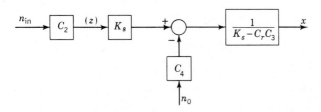

Figure 3.27 Block diagram for comparator.

per change in torque with Q maintained constant at the reference value. For a constant T, the speed N_o increases as Q increases, so that $\partial N_o/\partial Q \, |_T$ is a positive number. For a given Q, the speed decreases as the torque increases, so that $\partial N_o/\partial T \, |_Q$ is a negative number. The minus sign $(C_7 = -\partial N_o/\partial T \, |_Q)$ converts C_7 to a positive constant. For convenience in analyzing control systems, it is desirable that all constants be positive numbers. The values of the preceding partial derivatives are obtained from the curve of operating characteristics of the particular engine under consideration. Typical operating characteristics for an engine are shown in Fig. 3.5.

The torque t produced by the engine minus the load torque t_L is the net torque available for acceleration. That is,

$$t - t_L = J\alpha = JDn_o \tag{3.63}$$

where $\alpha = dn_o/dt = Dn_o$ is the angular acceleration of the engine and J is the rotational inertia of the engine. Substituting t from Eq. (3.63) into Eq. (3.62) gives

$$n_o = C_6 q - C_7 t_L - C_7 JDn_o$$

or

$$n_o = \frac{C_6 q - C_7 t_L}{1 + C_7 JD} = \frac{C_6}{1 + \tau_2 D} (q - C_8 t_L) \tag{3.64}$$

where

$$\tau_2 = C_7 J \quad \text{and} \quad C_8 = \frac{C_7}{C_6} = \frac{-\partial N_o/\partial T \, |_i}{\partial N_o/\partial Q \, |_i} = \frac{\partial Q}{\partial T}\bigg|_i$$

By writing $N_o = N_o(Q, T)$ in the implicit form $G(Q, T, N_o) = 0$, then the product of the partial derivatives is

$$\frac{\partial Q}{\partial T} \frac{\partial T}{\partial N_o} \frac{\partial N_o}{\partial Q} = -1 \quad \text{or} \quad \frac{-\partial N_o/\partial T}{\partial N_o/\partial Q} = \frac{\partial Q}{\partial T} \tag{3.65}$$

The block-diagram representation for Eqs. (3.61) and (3.64) is shown in Fig. 3.28.

The block-diagram representation for this speed control system is obtained by combining Figs. 3.27, 3.10c, and 3.28, as is shown in Fig. 3.29, in which $\tau_1 = 2A/C$.

Letting $K_1 = C_5/(K_s - C_r C_3)$ and eliminating the minor feedback loop, Fig. 3.29 may be represented as shown in Fig. 3.30. The operational form of the differential equation relating the output n_o to the input n_{in} and external disturbance t_L for the speed control system represented by the block diagram of

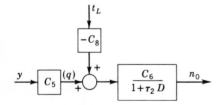

Figure 3.28 Block diagram for engine.

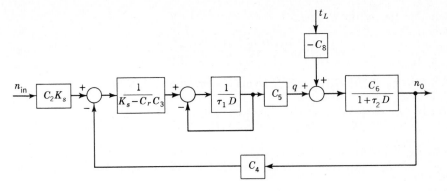

Figure 3.29 Combined block diagram.

Fig. 3.30 is obtained as follows. Subtract the feedback signal $C_4 n_o$ from the reference input $C_2 K_s n_{in}$ and perform the mathematical operations indicated by the feedforward portion of the block diagram until the output n_o is obtained. That is,

$$\left[(C_2 K_s n_{in} - C_4 n_o) \frac{K_1}{1 + \tau_1 D} - C_8 t_L\right] \frac{C_6}{1 + \tau_2 D} = n_o \tag{3.66}$$

Solving for n_o yields

$$n_o = \frac{C_2 C_6 K_1 K_s n_{in} - C_6 C_8 (1 + \tau_1 D) t_L}{(1 + \tau_1 D)(1 + \tau_2 D) + C_4 C_6 K_1} \tag{3.67}$$

In determining the dynamic behavior of a system, one is interested in the variation of the system parameters from some reference condition. This is the type of information that is available from the block-diagram representation shown in Fig. 3.30. If absolute values are desired, it is an easy matter to convert from n_{in} to N_{in} or from n_o to N_o by merely adding the reference value.

In Fig. 3.22 is shown the block-diagram representation for a typical feedback control system. The overall equation of operation is

$$\{[r(t) - H_1(D)c(t)]G_1(D) + d(t)\}G_2(D) = c(t)$$

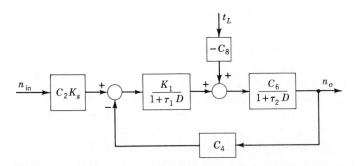

Figure 3.30 Overall block diagram for speed control system.

Solving for the output $c(t)$ gives

$$c(t) = \frac{G_1(D)G_2(D)r(t) + G_2(D)d(t)}{1 + G_1(D)G_2(D)H(D)} \tag{3.68}$$

The operators $G_1(D)$, $G_2(D)$, and $H(D)$ may be written in the form

$$G_1(D) = \frac{N_{G_1}}{D_{G_1}} \qquad G_2(D) = \frac{N_{G_2}}{D_{G_2}} \qquad H(D) = \frac{N_H}{D_H} \tag{3.69}$$

where N_{G_1} is the numerator of G_1 and D_{G_1} is the denominator of G_1, etc. Substitution of $G_1(D)$, $G_2(D)$, and $H(D)$ from Eq. (3.69) into Eq. (3.68) yields the general form

$$c(t) = \frac{N_{G_1}N_{G_2}D_H r(t) + N_{G_2}D_H D_{G_1}d(t)}{N_{G_1}N_{G_2}N_H + D_{G_1}D_{G_2}D_H} \tag{3.70}$$

It is to be noted that the coefficient $r(t)$ is the product of the numerator terms $N_{G_1}N_{G_2}$ from $r(t)$ to the output and the denominator term D_H from the output back to $r(t)$. Similarly, the coefficient for $d(t)$ is the product of the numerator term N_{G_2} from $d(t)$ to the output and the denominator terms $D_H D_{G_1}$ from the output back to $d(t)$. The denominator for Eq. (3.70) is seen to be the product of all the numerator terms in the loop ($N_{G_1}N_{G_2}N_H$) plus the product of all the denominator terms ($D_{G_1}D_{G_2}D_H$). In general, the coefficient for a signal entering any place in a loop is equal to the product of the numerator terms from the point where the signal enters the loop to the output times the product of the denominator terms from the output back to the point where the signal enters the loop. The denominator of the differential equation is always the product of all the denominator terms in the loop plus the product of all the numerator terms.

The form given by Eq. (3.70) saves considerable time and effort in obtaining the differential equation of operation for a control system. For example, from Fig. 3.30, it is to be noted that $N_{G_1} = K_1$, $N_{G_2} = C_6$, $N_H = C_4$, $D_{G_1} = (1 + \tau_1 D)$, $D_{G_2} = (1 + \tau_2 D)$, $D_H = 1$, $r(t) = C_2 K_s n_{in}$, $d(t) = -C_8 t_L$, and $c(t) = n_o$. Substitution of these values into Eq. (3.70) yields directly Eq. (3.67).

3.9 GENERALIZED FEEDBACK CONTROL SYSTEM

A general representation for a feedback control system is shown in Fig. 3.31. It is to be noticed that the command signal, or desired input, does not usually go directly to the comparator but must be converted to a suitable input for this device. Similarly, the controlled variable, or output, in the general case must also be changed by the feedback elements $H(D)$ before it can be measured by the comparator. The actuating signal e is amplified by the control elements $G_1(D)$ before it enters the system $G_2(D)$ being controlled. An external disturbance, as shown in Fig. 3.31, is a disturbance which acts independently to affect the operation of the system. Although in Fig. 3.31 the external disturbance is shown

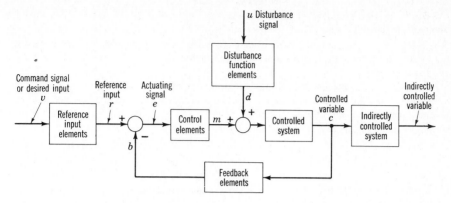

Figure 3.31 Generalized feedback control system.

entering the system between the control elements and the controlled system, in general, the external disturbance may enter the system at any point.

It is also to be noticed from this generalized representation of a control system that the controlled variable is not necessarily the quantity which it is desired to control. For example, a household thermostat controls the temperature of the air around the thermostat and depending upon the circulation of air in the house, the temperature of other areas may vary considerably. In addition, the idealized purpose of this control is to maintain the comfort of the persons of the household which depends upon humidity, their clothing, their amount of physical activity, etc. Thus, it is apparent that the controlled variable is not necessarily the ultimate quantity which it is desired to control.

PROBLEMS

3.1 Determine the linear approximation for each of the following functions:
(a) $Y = \sqrt{X}$
(b) $Y = 1/\sqrt{X}$
For $X_i = 100$, use this approximation to evaluate Y when $X = 98$.

3.2 Determine the linear approximation for each of the following functions:
(a) $Y = \sin X$
(b) $Y = \tan X$
For $X_i = 30°$, use this approximation to evaluate Y when $X = 33°$.

3.3 Determine the linear approximation for the equation $F = 32/(12 - X)$. For $X_i = 4$, what is the approximate value of F when $X = 3$?

3.4 Effect a linear approximation for $Y = X^{1/5}$. For $X_i = 32$, what is the approximate value of $\sqrt[5]{30}$?

3.5 Effect a linear approximation for $Y = X^{2/3}$. For $X_i = 1000$, what is the approximate value of $(997)^{2/3}$?

3.6 Effect a linear approximation for $Y = \log X$. For $X_i = 100$, what is the approximate value of $\log 101$?

3.7 Determine the linear approximation for the equation $Z = \sin X \cos Y$. For $X_i = 60°$ and $Y_i = 30°$, what is the approximate value of Z when $X = 63°$ and $Y = 28°$?

3.8 Effect a linear approximation for the function $Z = X^2 \log Y$. For $X_i = 5$ and $Y_i = 100$, what is the approximate value of Z when $X = 5.1$ and $Y = 98$?

3.9 The volume V of a sphere is $V = \frac{4}{3}\pi R^3$. Determine the equation for the linear approximation to V. If $R_i = 10$, what percent error results from using this approximation for V when $R = 11$? What is the percent error for $R = 9$?

3.10 The equation for the volume of a cylinder is $V = \pi R^2 H$, where R is the radius of the base and H is the altitude. Determine the equation for the variation v in the volume due to a variation r in the radius and a variation h in the altitude.

3.11 The period of oscillation of a simple pendulum is

$$T = 2\pi \sqrt{\frac{L}{g}}$$

For $L_i = 100$ in and $g = 32.2$ ft/s^2, determine the change in the period due to
  (a) An increase in L of 1 in.
  (b) A decrease in g of 0.1 ft/s^2.

3.12 Linearize the following equation:

$$V = \frac{D}{T}$$

where V is velocity, D is distance, and T is time.
  (a) Determine the linear approximation for V when $V_i = 60$ mi/h, $D_i = 1$ mi, and $T_i = 60$ s.
  (b) With the speedometer indicating 60 mi/h, it is observed that 62 s are required to travel between successive mile indicator markers on the thruway. Approximately what is the velocity of the vehicle?

3.13 To linearize $V = D/T$ of Prob. 3.12 by the perturbation method, replace V by $V_i + v$, D by $D_i + d$, and T by $T_i + t$. Neglect higher-order variations (differential quantities) and then subtract off the reference condition $V_i T_i = D_i$. Thus, verify the linearized form obtained in Prob. 3.12.

3.14 From the law of cosines, it follows that the length of the side Z of a triangle is

$$Z^2 = X^2 + Y^2 - 2XY \cos \theta$$

where θ is the included angle between sides X and Y. Designate variations as x, y, z, and $\Delta\theta$, and then proceed to determine the equation for the variation z. For $X_i = 50$, $Y_i = 80$, and $\theta_i = 60°$, what is the approximate value of Z when $X = 52$, $Y = 76$, and $\theta = 60°$?

3.15 A chord of length L subtended by the angle θ is illustrated in Fig. P3.15. The equation for the length is $L = 2R \sin \theta/2$ where R is the radius. Determine the equation for the variation l in the length

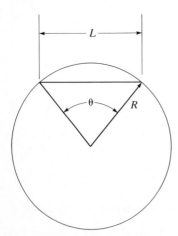

Figure P3.15

due to a change r in the radius and a change $\Delta\theta$ of the angle. For $R_i = 10$ and $\theta_i = 60°$, what is the approximate value of L when $R = 12$ and $\theta = 63°$?

3.16 The volume of a torus ("donut" shaped, Fig. P3.16) is $V = 2\pi^2 R r^2$.

(a) Determine the equation for the variation ΔV in the volume due to a change ΔR in R and a change Δr in r.

(b) The change in volume due to a change in Δr only is $S\,\Delta r$, where S is the surface area. Proceed to determine the equation for the surface area S.

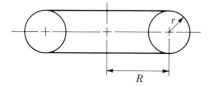

Figure P3.16

3.17 The error incurred in measuring the diameter D of a circle is d. Effect a linear approximation for the variation a of the area A of the circle. How accurately must the diameter be measured such that the area shall be correct to 1 percent?

3.18 The firing of a projectile on an inclined surface is illustrated in Fig. P3.18. The angle of inclination of the incline is β. The initial velocity of the projectile V is at an angle α from the horizontal. The equation for the range R is

$$R = \frac{V^2}{g\,\cos^2 \beta}\,[\sin(2\alpha - \beta) - \sin \beta]$$

Effect a linear approximation for the change in range r due to a variation v in the initial velocity and a variation $\Delta\alpha$ of the angle α. For $V_i = 100$ ft/s, $\alpha_i = 75°$, $\beta = 30°$, and $g = 32.2$ ft/s², what is the approximate value of R when $V = 110$ ft/s and $\alpha = 70°$?

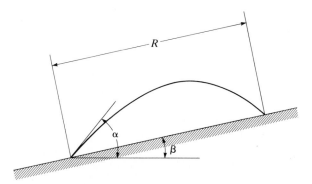

Figure P3.18

3.19 A certain motion is governed by the differential equation $\ddot{R} - R^{-3} + R^{-2} = 0$. Assuming that the variation of R from its reference value of unity remains small, linearize the differential equation. That is, determine the linear differential equation which describes the variation r of R.

3.20 The linearized equation of operation about a typical operating point for a hydraulic amplifier used in a machine tool is

$$f = 20\,000e - 100\dot{y}$$

where $\dot{y} = Dy =$ velocity of the power piston
$\quad f =$ change in force acting on the piston
$\quad e =$ position of the valve which controls the piston

The load is that due to the tool reactive force f_L and inertia force MD^2y; hence

$$f = f_L + MD^2y$$

The reference position x and controlled position y are connected by a walking-beam linkage such that

$$e = \frac{x - y}{2}$$

For $M = 1$, obtain the overall block-diagram representation for this system in which x is the reference input, y the controlled variable, and f_L the external disturbance.

3.21 Typical operating curves for a dc motor are shown in Fig. P3.21. These are curves of torque T versus operating speed N for constant values of voltage E applied to the motor. These curves are a plot of the function $N = N(T, E)$. Effect a linear approximation for N. Evaluate the partial derivatives in this approximation when T_i is 2 in · lb and E_i is 20 V.

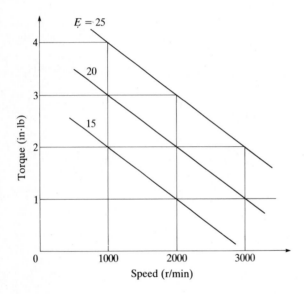

Figure P3.21

3.22 For the tank shown in Fig. P3.22, Q_{in} is constant. The flow out is a function of the head $Q_o = Q_o(H)$, where $H = H_i + h$. The equilibrium level is such that $Q_{in} = Q_o(H_i)$. The linear approximation for the flow out is $Q_o = Q_o(H_i) + (\partial Q_o/\partial H)_i h = Q_o(H_i) + Ch$, where $C = (\partial Q_o/\partial H)_i$. Show that the differential equation of operation for this tank is $dh/h = -(C/A)dt = -dt/\tau$, where $\tau = A/C$. Integrate both sides of this differential equation, and show that $h = h_0 e^{-t/\tau}$, where h_0 is the value of h at time $t = 0$. Note that as time approaches infinity, the level returns to its equilibrium value $h = 0$.

Figure P3.22

3.23 In Fig. P3.23 is shown a hydraulic servomotor which is similar to the power-amplifying device used in power-steering units. A movement in the x direction of the valve is seen to open passage 1 to the supply pressure which in turn causes the big piston to move to the right. Because the sleeve is directly connected to this piston, the sleeve also moves to the right to close off flow from the valve. Determine the block diagram relating the input position x to the output y. Identify the time constant.

Figure P3.23

3.24 In Fig. P3.24 is shown a modification of the hydraulic power amplifier discussed in Prob. 3.23. Determine the block-diagram representation for this device, in which x is the input and y is the output. Note that the position of the sleeve is $[a/(a + b)]y$.

Figure P3.24

3.25 For the hydraulic amplifier shown in Fig. P3.25, determine the block diagram for the walking-beam linkage and also the block diagrams relating e to y and y to w. Combine these diagrams to determine the overall block-diagram representation for the system.

Figure P3.25

3.26 Derive the equation relating p_2 and z for each of the pneumatic control elements shown in Fig. P3.26a and b. Explain the significance of each partial derivative which occurs in these equations.

(a)

(b)   Figure P3.26

3.27 In Fig. P3.27 is shown a flapper amplifier. Determine the overall block diagram which results when the controlled pressure P_2 is connected to each of the pneumatic control elements shown in Fig. P3.26. The position z is fed back to the walking-beam linkage as indicated. Note that for this overall block diagram the input is the error e and the output is the position z. (The time constant for the flapper amplifier may be considered negligible so that $p_2 = -K_1 x$.)

Figure P3.27

3.28 In Fig. P3.28 is shown a tank which is used as a mixer. Pure water flows in at a constant rate Q_w. A salt solution at a concentration C_s flows in at a constant rate Q_s. The mixed solution of concentration C_o leaves at the rate Q_o. Because the volume V of fluid in the tank is kept constant, the outflow Q_o is equal to $Q_w + Q_s$. The rate of accumulation of salt in the tank is $Q_s C_s - Q_o C_o = VDC_o$. The salt concentration C_s is controlled by a valve (not shown) such that $C_s = K\theta$.

Determine the differential equation relating the concentration C_o of the tank to the valve position θ in terms of the parameters Q_s, Q_w, and V. Identify the time constant.

Figure P3.28

3.29 A block of material at temperature T_0 is placed in an oven which is held at the constant temperature T_1. The rate of heat flow q into the material is given by $q = hA(T_1 - T)$, where h is the coefficient of heat transfer of the surface, A is the area of the surface, and T is the temperature of the block. The rate of change of temperature of the block is $dT/dt = q/Mc$, where M is the mass and c is the average specific heat. Determine the differential equation which describes the temperature of the block as a function of time. Identify the time constant.

3.30 Determine the differential equation relating y and x for the hydraulic system described by the block diagram shown in Fig. P3.30. Identify the time constant and steady-state gain.

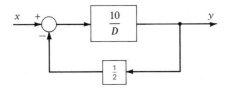

Figure P3.30

3.31 The speed torque curves for a dc motor are shown in Fig. P3.31. V is the applied voltage, N is the speed, and T is the output torque. Determine the linear approximation for the change in torque t due to a change in speed n and a change in voltage v. The motor drives an inertial load such that $t = J(d/dt)n$, where J is the mass moment of inertia. For $J = 0.1$, determine the differential equation relating the change in speed n to the change in voltage v. Determine the time constant τ and the steady-state gain.

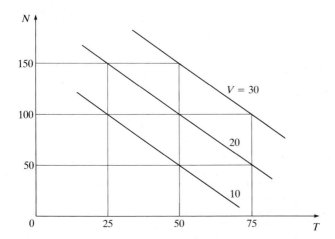

Figure P3.31

3.32 The characteristics of an engine are described by the family of curves shown in Fig. P3.32. Determine the linear approximation for the torque t delivered by the engine. When the torque t is greater than the load torque t_L, the excess $(t - t_L)$ is used to accelerate the engine so that $t - t_L = J\,dn/dt$, where J is the rotational inertia of the engine. For $J = 0.025$, determine the differential equation relating the change in speed n to the change in fuel flow q and the change in load torque t_L. What is the time constant τ?

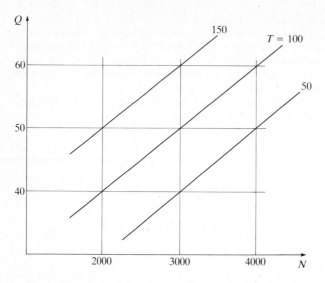

Figure P3.32

3.33 The tank shown in Fig. P3.33 is initially empty ($t < 0$). A constant rate of flow q_{in} is added for $t > 0$. The rate at which flow leaves the tank is $q_o = Ch$. The cross-sectional area of the tank is A. Determine the differential equation for the head h. Identify the time constant. What is the final steady-state value of the head?

Figure P3.33

3.34 The dynamics of an electronic amplifier can be approximated by the differential equation

$$E_o = \frac{K}{1 + aD} E_{in}$$

where E_o is the output voltage and E_{in} is the input voltage. Explain the physical significance of the parameters K and a. Describe the details of an experiment to determine the values for K and a. A power source, a switch, and a sensitive voltmeter are available.

3.35 In Fig. P3.35 is shown an electrical speed control system. The input potentiometer provides a reference input voltage E_r proportional to the desired speed N_{in}, that is, $E_r = K_r N_{in}$. A voltage signal E_c which is proportional to the controlled output speed N_o is provided by the tachometer ($E_c = K_c N_o$). The error $E_r - E_c$ is amplified by an electronic amplifier whose output is $E_f = K_a(E_r - E_c)$. The voltage E_f is applied to the field of a field-controlled dc motor. The torque exerted on the shaft by the motor (air-gap torque) is proportional to the field current, that is, $T = K_m I_a I_f$. Determine the overall block diagram for this speed control system for the case in which the load torque consists of an inertia $JD^2\theta = JDN_o$ and an external torque T_L.

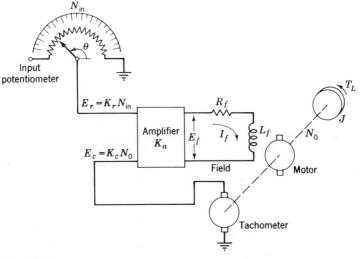

Figure P3.35

3.36 A field-controlled dc motor is shown in Fig. P3.36. The motor drives the load through a gearbox, so that $\omega_c = n\omega$, where n is the gear ratio, ω is the motor speed, and ω_c is the speed of the load (i.e., the controlled speed). The output shaft is connected to a tachometer, which produces a voltage proportional to the controlled speed ($E_c = K_c \omega_c$). An electronic amplifier is used to amplify the error signal by a factor K_a, that is, $E_f = K_a(E_r - E_c)$. Complete the overall block-diagram representation for this system.

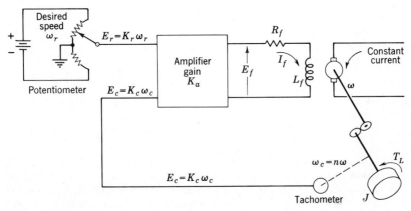

Figure P3.36

3.37 In Fig. P3.37 is shown a generator which is used as a voltage amplifier. The prime mover drives the generator at a constant speed. Determine the equation of operation for the amplification ratio E_2/E_1.

Figure P3.37

3.38 In Fig. P3.38 is shown a tension-regulating apparatus such as is used in the paper industry. To ensure uniform winding, it is necessary to maintain a constant tension F_c as the sheet is being wound on the wind-up roll. To increase the tension in the paper, the tension control lever is raised. This raises the torque control arm of the motor, which increases the torque T_m applied by the motor to the wind-up roll. The change in torque provided by the motor is $t_m = K_m e/(1 + \tau D)$. For the wind-up roll, it follows that $F_c = T_m/R$, where R is the radius of the wheel. Determine the overall block diagram relating a variation f_r of the reference or desired tension to a variation of the controlled tension f_c.

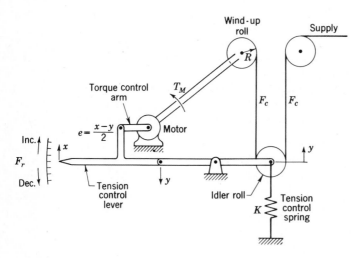

Figure P3.38

3.39 In Fig. P3.39 is shown a liquid-level controller. To raise the level of the fluid, the control lever is moved up (i.e., position z is raised). This raises the valve position $e = (z - h_o)/2$, which increases y, thereby admitting more flow Q_{in}. The flow Q_{in} is a function of the flow valve opening Y and the supply pressure P_s. The change in volume of liquid in the tank is the time integral $(q_{in} - q_o)/D$, which is equal to the cross-sectional area of the tank A_T times the change in level h_o. The flow out Q_o is seen to depend on the pressure head H_o. Determine the overall block diagram for this controller.

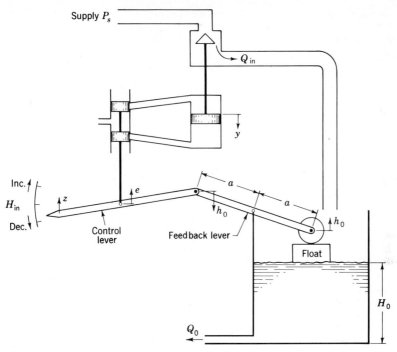

Figure P3.39

3.40 A system for controlling flow is shown in Fig. P3.40. Increasing the desired flow setting increases the compression on spring K_1, which causes x and position e of the balanced valve to move up. This in turn causes the flow valve to move down, which increases the flow. The amount of flow out is

Figure P3.40

measured by a venturi-type flowmeter, so that the pressure drop $P_1 - P_2$ is a function of Q_o. The diaphragm prevents leakage from the high pressure P_1 to the low pressure P_2, but it permits motion, just as a piston would. The effective area of the diaphragm is A_d. The flow Q_o is seen to be a function of the flow valve opening Y and the supply pressure P_s. Determine the overall block-diagram representation for this system.

3.41 The linearized equation of operation for the governor shown in Fig. 3.24 is given by Eq. (3.60). The governor shown in Fig. P3.41 is the same as that shown in Fig. 3.24 except $z = 0$. Thus, letting $z = 0$ in Eq. (3.60) yields the equation for the position x of the governor shown in Fig. P3.41, $x = -C_4 n_o/(K_s - C_r C_3)$. The rate of flow through the valve is $q = C_2 x$. The rate of change of volume in the power piston is $q = A\, dy/dt$. Because Z is fixed, N_{in} is constant and thus $n_{in} = 0$. The block diagram for this governor is shown in Fig. P3.41. Show that $G_1(D) = K/D$. Determine K.

Figure P3.41

3.42 The linearized equation of operation for the governor shown in Fig. 3.24 is given by Eq. (3.60). The governor shown in Fig. P3.42 is the same as that shown in Fig. 3.24 except $z = -y/2$. Thus, letting $z = -y/2$ in Eq. (3.60) yields the equation for the position x of the governor shown in Fig. P3.42, $x = -(K_s y/2 + C_4 n_o)/(K_s - C_r C_3)$. The rate of flow through the valve is $q = C_2 x$. The rate of change of volume in the power piston is $q = A\, dy/dt$. Because N_{in} is constant, then $n_{in} = 0$. The block diagram for this governor is shown in Fig. P3.42. Show that $G_1(D) = K/(1 + \tau D)$. Determine K and τ.

3.43 The linearized equation of operation for the governor shown in Fig. 3.24 is given by Eq. (3.60). The governor shown in Fig. P3.43 is the same as that shown in Fig. 3.24 except $z = 0$. Thus, letting $z = 0$ in Eq. (3.60) yields the equation for the position x of the governor shown in Fig. P3.43, $x = -C_4 n_o/(K_s - C_r C_3)$. Because of the sleeve around the valve, the opening is $(x - y)$. The rate of flow through the valve is $q = C_2(x - y)$. The rate of change of volume in the power piston is $q = A\, dy/dt$. Because Z is fixed, N_{in} is constant and thus $n_{in} = 0$. The block diagram for this governor is shown in Fig. P3.43. Show that $G_1(D) = K/(1 + \tau D)$. Determine K and τ.

$n_{in}=0$ $\xrightarrow{}$ ⊖ $\xrightarrow{}$ $\boxed{G_1(D)}$ $\xrightarrow{y}$

$-$ $\uparrow$ n_0

Increase flow

Decrease flow

Figure P3.42

$n_{in}=0$ $\xrightarrow{}$ ⊕ $\xrightarrow{}$ $\boxed{G_1(D)}$ $\xrightarrow{y}$

$+$

$-$ $\uparrow$ n_0

Increase flow

Decrease flow

Figure P3.43

3.44 In Fig. P3.44 is shown a speed-sensing device in which the position x is proportional to the angular velocity of the gear pump. The rate of flow q through the pump is proportional to the speed of rotation, $q = C_1 n$. Because of the fixed orifice, the pressure p is proportional to the rate of flow q, $p = C_2 q$. The force balance of the bellows-spring combination is $K_s x = Ap$, where K_s is the spring rate and A is the area of the bellows. Determine the equation relating the position x to the speed n.

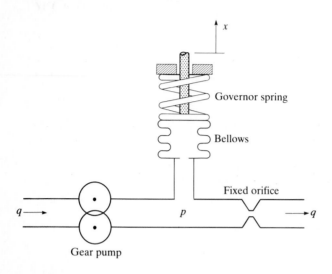

Figure P3.44

3.45 For the system shown in Fig. P3.45, $G_1(D) = K_1/(1 + \tau_1 D)$. Determine the differential equation of operation when $A = 1$, $B = -5$, $K_1 = 1$, $K_2 = 2$, $K_H = 0.5$, $\tau_1 = \frac{1}{6}$, and $\tau_2 = 1$.

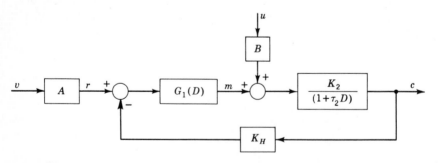

Figure P3.45

3.46 For the system shown in Fig. P3.45, $G_1(D) = K_1/D$. Determine the differential equation of operation when $A = 1$, $B = -1$, $K_1 = K_2 = 0.75$, $K_H = 1.0$, and $\tau_2 = 0.25$.

3.47 Same as Prob. 3.46 except $K_2 = 4$.

FOUR

STEADY-STATE OPERATION

By steady-state operation is meant the equilibrium state attained such that there is no change with respect to time of any of the system variables. The system remains at this equilibrium state of operation until it is excited by a change in the desired input or in the external disturbance. A transient condition is said to exist as long as any of the variables of the system is changing with time. In this chapter, it is shown that considerable information about the basic character of a system may be obtained from an analysis of its steady-state operation.

4.1 STEADY-STATE ANALYSIS

The general block-diagram representation for a feedback control system is shown in Fig. 4.1a. For steady-state operation, c, v, and u will have constant values, and therefore terms resulting from powers of D operating on these constant quantities will be zero. The equation describing the steady-state operation of a control system is obtained by letting $D = 0$ in the differential equation of operation for the system. Similarly, the block diagram that describes the steady-state operation of a system is obtained by letting $D = 0$ in the general block-diagram representation for the system. The block diagram which describes the steady-state operation of the system of Fig. 4.1a is shown in Fig. 4.1b, in which

$$K_{G_1} = [G_1(D)]_{D=0} \qquad K_{G_2} = [G_2(D)]_{D=0} \qquad K_H = [H(D)]_{D=0}$$

(a)

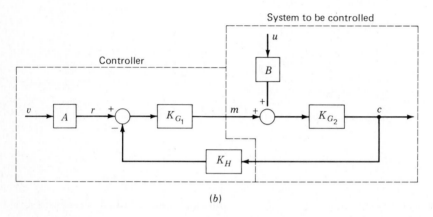

(b)

Figure 4.1 General block-diagram representation for a control system.

where K_{G_1} is obtained by letting $D = 0$ in the differential operator $G_1(D)$, etc. From Fig. 4.1b, the equation for steady-state operation is found to be

$$[(Av - K_H c)K_{G_1} + Bu]K_{G_2} = c$$

or

$$c = \frac{AK_{G_1}K_{G_2}}{1 + K_{G_1}K_{G_2}K_H}v + \frac{BK_{G_2}}{1 + K_{G_1}K_{G_2}K_H}u \qquad (4.1)$$

The constant A which appears in Eq. (4.1) is, in effect, the scale factor for the input dial. To have the coefficient of the v term equal to unity, A must be selected such that

$$\frac{AK_{G_1}K_{G_2}}{1 + K_{G_1}K_{G_2}K_H} = 1$$

or

$$A = \frac{1 + K_{G_1}K_{G_2}K_H}{K_{G_1}K_{G_2}} = \frac{1}{K_{G_1}K_{G_2}} + K_H \qquad (4.2)$$

When A is chosen in accordance with Eq. (4.2), the coefficient of the v term is unity, in which case Eq. (4.1) becomes

$$c = v + \frac{BK_{G_2}}{1 + K_{G_1}K_{G_2}K_H}\, u = v + \frac{B}{1/K_{G_2} + K_{G_1}K_H}\, u \qquad (4.3)$$

To have the controlled variable c equal to the command signal v (that is, $c = v$), it is necessary that the coefficient of the u term be zero. This coefficient is zero if either K_{G_1} or K_H is infinite. From Eq. (4.1) it follows that if K_H is infinite, then c must be zero regardless of the value of v or u. In effect, no control is possible when K_H is infinite. Also note from Eq. (4.2) that an infinite value of K_H would necessitate A being infinite, which is physically impossible. Thus only K_{G_1} can be made infinite. This is accomplished by having an integrator in the control elements to yield a $1/D$ term, which gives the effect of an infinite constant during steady-state operation. This type of system is called an integral control system.

The left portion of the control system enclosed by the dotted lines in Fig. 4.1b is the controller. For the controller it follows that

$$(Av - K_H c)K_{G_1} = m$$

or
$$c = -\frac{1}{K_{G_1}K_H}\, m + \frac{A}{K_H}\, v \qquad (4.4)$$

In Fig. 4.2a is shown a plot of the steady-state operating characteristics for a typical controller. Lines of constant values for the command signal V are plotted with the controlled variable C as the abscissa and the manipulated variable M as the ordinate. For $v = 0$, Eq. (4.4) shows that

$$-K_{G_1}K_H = \frac{m}{c}\bigg|_{v=0} = \frac{\Delta M}{\Delta C}\bigg|_{\Delta V=0} = \frac{\partial M}{\partial C}\bigg|_{V} \qquad (4.5)$$

where $v = \Delta V$ is the change in V from the reference value, $m = \Delta M$ is the change in M from the reference value, and $c = \Delta C$ is the change in C from the reference

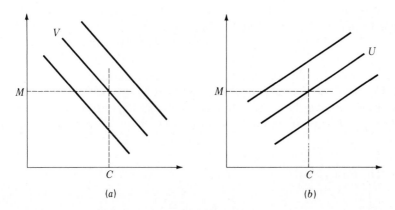

Figure 4.2 Steady-state operating characteristics for (a) controller and (b) system to be controlled.

value. For $v = 0$, then V must be constant. The term $\partial M/\partial C\,|_V$ is the slope of the controller curves shown in Fig. 4.2a.

For the case in which $c = 0$, Eq. (4.4) becomes

$$AK_{G_1} = \frac{m}{v}\bigg|_{c=0} = \frac{\Delta M}{\Delta V}\bigg|_{\Delta C = 0} = \frac{\partial M}{\partial V}\bigg|_C \tag{4.6}$$

For $c = 0$, then C must be constant. A line of constant C is a vertical line in Fig. 4.2a. The term $\partial M/\partial V\,|_C$ determines the vertical spacing between the lines of constant V.

Finally, for the case in which $m = 0$, Eq. (4.4) shows that

$$\frac{K_H}{A} = \frac{v}{c}\bigg|_{m=0} = \frac{\Delta V}{\Delta C}\bigg|_{\Delta M = 0} = \frac{\partial V}{\partial C}\bigg|_M \tag{4.7}$$

For $m = 0$, then M must be constant. A line of constant M is a horizontal line. The term $\partial V/\partial C\,|_M$ determines the horizontal spacing between lines of constant V.

The right portion of Fig. 4.1b enclosed by the dotted lines represents the system to be controlled. The equation for the steady-state operation of the system to be controlled is

$$(m + Bu)K_{G_2} = c$$

or
$$c = K_{G_2}m + BK_{G_2}u \tag{4.8}$$

From an analysis similar to that for the controller, it follows that

$$\frac{\partial M}{\partial C}\bigg|_U = \frac{1}{K_{G_2}} \qquad \frac{\partial M}{\partial U}\bigg|_C = -B \qquad \frac{\partial C}{\partial U}\bigg|_M = BK_{G_2} \tag{4.9}$$

Typical steady-state operating curves for the system to be controlled are shown in Fig. 4.2b. The first partial $\partial M/\partial C\,|_U$ is the slope, the second partial $\partial M/\partial U\,|_C$ determines the vertical spacing, and the last partial $\partial C/\partial U\,|_M$ determines the horizontal spacing. In summary, the individual constants in the block diagram of Fig. 4.1b may be obtained from the steady-state operating curves, and vice versa.

Illustrative example 4.1 The block diagram for a feedback control system is shown in Fig. 4.3. The reference operating point is $V_i = C_i = 100$, $M_i = 50$, and $U_i = 10$. Determine the steady-state constants for this system and then sketch the steady-state operating curves. Select A in accordance with Eq. (4.2).

SOLUTION The steady-state constants are $B = -5$, $K_H = 0.5$, and

$$K_{G_1} = \frac{1}{1 + \tau_1 D}\bigg|_{D=0} = 1 \qquad K_{G_2} = \frac{2}{1 + \tau_2 D}\bigg|_{D=0} = 2$$

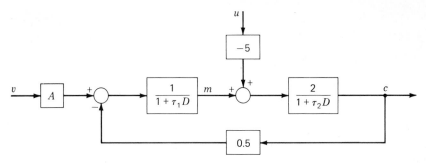

Figure 4.3 A feedback control system.

The value of A is

$$A = \frac{1}{K_{G_1}K_{G_2}} + K_H = \frac{1}{(1)(2)} + 0.5 = 1.0$$

The slope of the operating curves for the controller is

$$\left.\frac{\partial M}{\partial C}\right|_V = -K_{G_1}K_H = -(1.0)(0.5) = -0.5$$

The line of $V_i = 100$ may now be drawn through the reference point ($M_i = 50$, $C_i = 100$) with a slope of (-0.5), as shown in Fig. 4.4a. The horizontal spacing is

$$\left.\frac{\Delta V}{\Delta C}\right|_M = \frac{K_H}{A} = 0.5$$

For $\Delta C = 20$, then $\Delta V = 0.5\,\Delta C = (0.5)(20) = 10$. Note in Fig. 4.4$a$ that when M is constant (a horizontal line), the horizontal spacing is such that as V changes by 10 units, then C changes by 20 units. The vertical spacing is

$$\left.\frac{\Delta M}{\Delta V}\right|_C = AK_{G_1} = (1.0)(1.0) = 1.0$$

For $\Delta V = 10$, then $\Delta M = 1.0\,\Delta V = (1.0)(10) = 10.0$. Note in Fig. 4.4$a$ that when C is constant (a vertical line), the vertical spacing is such that as V changes by 10 units, then M changes by 10 units.

If any two of the three quantities (slope, horizontal spacing, or vertical spacing) are known, then the third quantity is determined. Thus, only two of these three quantities are independent. This fact is proved by noting that for the controller enclosed by the dashed box in Fig. 4.1b, the output is the manipulated variable M and the inputs are the command signal V and the controlled variable C. For a given controller the manipulated variable M is a function of V and C. That is,

$$M = M(V, C)$$

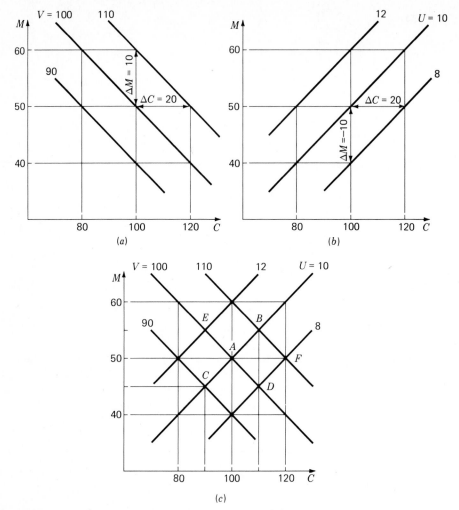

Figure 4.4 Steady-state characteristics for the system of Fig. 4.3.

The implicit form for this relationship is

$$G(M, V, C) = 0$$

For an implicit function of n variables, the product of the n partial derivatives is $(-1)^n$. For the implicit function G of the three variables M, V, and C, it follows that

$$\left.\frac{\partial M}{\partial V}\right|_C \left.\frac{\partial V}{\partial C}\right|_M \left.\frac{\partial C}{\partial M}\right|_V = -1 \tag{4.10}$$

Solving for the slope $\partial M/\partial C|_V$ gives

$$\left.\frac{\partial M}{\partial C}\right|_V = -\left.\frac{\partial M}{\partial V}\right|_C \left.\frac{\partial V}{\partial C}\right|_M = -(AK_{G_1})\left(\frac{K_H}{A}\right) = -K_{G_1}K_H$$

This illustrates the statement that if the vertical spacing $\partial M/\partial V \,|_C = AK_{G_1}$ and the horizontal spacing $\partial V/\partial C \,|_M = K_H/A$ are known, then the slope $\partial M/\partial C \,|_V = -K_{G_1}K_H$ is automatically determined.

The slope of the load lines for the system to be controlled is

$$\left.\frac{\partial M}{\partial C}\right|_U = \frac{1}{K_{G_2}} = \frac{1}{2} = 0.5$$

As shown in Fig. 4.4b, the load line for $U_i = 10$ is drawn through the reference point $(M_i = 50, C_i = 100)$ with a slope of (0.5). The horizontal spacing is

$$\left.\frac{\Delta C}{\Delta U}\right|_M = BK_{G_2} = (-5)(2) = -10$$

For $\Delta U = -2$, then $\Delta C = (-10) \Delta U = (-10)(-2) = 20$. Note in Fig. 4.4b that when M is constant (a horizontal line), the horizontal spacing between the load lines is such that as C changes by 20 units, then U changes by -2 units. The vertical spacing is

$$\left.\frac{\Delta M}{\Delta U}\right|_C = -B = -(-5) = 5$$

For $\Delta U = -2$, then $\Delta M = 5 \Delta U = 5(-2) = -10$. Thus, when C is constant (a vertical line), the vertical spacing between load lines is such that as M changes by -10 units, then U changes by -2 units.

As was the case for the lines of operation for the controller, the load lines can be completely determined if any two of the three quantities (slope, horizontal spacing, or vertical spacing) are known. As is shown in Fig. 4.1b, for the system to be controlled, the output is the controlled variable C and the inputs are the manipulated variable M and the disturbance U. Thus, the output C is a function of M and U. That is,

$$C = C(M, U)$$

The implicit form for this relationship is

$$G(C, M, U) = 0$$

The product of the partial derivatives is

$$\left.\frac{\partial C}{\partial M}\right|_U \left.\frac{\partial M}{\partial U}\right|_C \left.\frac{\partial U}{\partial C}\right|_M = -1 \qquad (4.11)$$

Solving for the slope $\partial M/\partial C \,|_U$ gives

$$\left.\frac{\partial M}{\partial C}\right|_U = -\left.\frac{\partial M}{\partial U}\right|_C \left.\frac{\partial U}{\partial C}\right|_M = -(-B)\frac{1}{BK_{G_2}} = \frac{1}{K_{G_2}}$$

Thus, by knowing the vertical spacing $\partial M/\partial U \,|_C = -B$ and the horizontal spacing $\partial U/\partial C \,|_M = 1/BK_{G_2}$, the slope $\partial M/\partial C \,|_U = 1/K_{G_2}$ can be determined.

Because Figs. 4.4a and b have the same coordinates (C, M), the two diagrams may be superimposed upon each other as shown in Fig. 4.4c. Substitution of the values of the constants into Eq. (4.1) yields the overall equation for steady-state operation

$$c = \frac{AK_{G_1}K_{G_2}}{1 + K_{G_1}K_{G_2}K_H} v + \frac{BK_{G_2}}{1 + K_{G_1}K_{G_2}K_H} u$$

$$= \frac{(1)(1)(2)}{1 + (1)(2)(0.5)} v + \frac{(-5)(2)}{1 + (1)(2)(0.5)} u = v - 5u \qquad (4.12)$$

The coefficient of the v term and the coefficient of the u term may be obtained directly from Fig. 4.4c without the need to evaluate all the steady-state constants. For the overall system shown in Fig. 4.1b, the output is the controlled variable C and the inputs are the command signal V and the disturbance U. Thus, for any given system, C is a function of V and U. That is,

$$C = C(V, U)$$

Linearization gives

$$c = \frac{\partial C}{\partial V}\bigg|_U v + \frac{\partial C}{\partial U}\bigg|_V u \qquad (4.13)$$

Comparison of this result with Eq. (4.1) shows that

$$\frac{\partial C}{\partial V}\bigg|_U = \frac{AK_{G_1}K_{G_2}}{1 + K_{G_1}K_{G_2}K_H} \qquad (4.14)$$

and

$$\frac{\partial C}{\partial U}\bigg|_V = \frac{BK_{G_2}}{1 + K_{G_1}K_{G_2}K_H} \qquad (4.15)$$

The term $\partial C/\partial V|_U$ is evaluated by dividing changes in C by corresponding changes in V with U fixed. In Fig. 4.4c, with U = 10, at point B the value of C is 110 and the value of V is 110. At point C, the value of C is 90 and the value of V is 90. Thus

$$\frac{\Delta C}{\Delta V}\bigg|_U = \frac{C_B - C_C}{V_B - V_C} = \frac{110 - 90}{110 - 90} = 1$$

where C_B is the value of C at point B, V_B is the value of V at point B, etc. The term $\partial C/\partial U|_V$ is evaluated by dividing changes in C by corresponding changes in U with V fixed. In Fig. 4.4c, with V = 100, at point D the value of C is 110 and the value of U is 8. At point E, the value of C is 90 and the value of U is 12. Thus

$$\frac{\Delta C}{\Delta U}\bigg|_V = \frac{110 - 90}{8 - 12} = \frac{20}{-4} = -5$$

The point A shown in Fig. 4.4c is the reference point of operation. In going from A to B the command signal changes from 100 to 110, so that v = 10. There is no change in the load, so that u = 0. Application of Eq. (4.12) gives

$c = v - 5u = 10$. Note that in going from point A to point B, the controlled variable changes from 100 to 110, so that $c = 10$. In going from point A to point F, the command signal changes from 100 to 110, so that $v = 10$. The load U changes from 10 to 8, so that $u = -2$. Application of Eq. (4.12) gives $c = v - 5u = 10 - 5(-2) = 20$. In going from A to F the controlled variable changes from 100 to 120, so that $c = 20$.

Because the input scale factor has been selected in accordance with Eq. (4.2), at the reference load $U_i = 10$ ($u = 0$) the controlled variable is equal to the command signal ($c = v$). Note in Fig. 4.4c that at point A, $V = C = 100$; at point B, $V = C = 110$; and at point C, $V = C = 90$.

4.2 EQUILIBRIUM

In Fig. 4.5a is shown a typical operating line for a controller in which V is the value of the command signal. When the value of the controlled variable is C_1, the value of the manipulated variable being supplied by the controller is M_1. In Fig. 4.5b is shown a typical load line for a system to be controlled in which U is the load. When the value of the manipulated variable being supplied to the system is M_1, the output from the system is C_1. As indicated in Fig. 4.5c by point A, the intersection of the line of operation of the command signal V for the controller and the load line U for the system determines the equilibrium point of operation for the system. That is, at point A the amount of the manipulated variable M_1 being supplied by the controller is the same as that required to maintain the system output at C_1. If the system output were C_2, then point B would be the point of operation for the controller and point C would be the point of operation for the system to be controlled. The amount of the manipulated variable required to maintain the system at point C is M_C. Because the controller is only supplying the amount M_B, the system output C decreases until equilibrium is attained at point A.

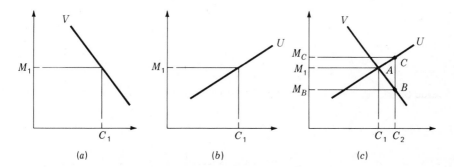

Figure 4.5 (a) Typical operating line for a controller, (b) typical load line for system to be controlled, and (c) superposition of Fig. 4.5a and b.

The characteristics shown in Fig. 4.6a illustrate the effect of changing the command signal from V_1 to V_2 with the load U maintained constant. Initially the system is at equilibrium at point A, the intersection of the V_1 controller curve and the U load curve. When the command signal is changed from V_1 to V_2, the new operating point for the controller is at point B, which is the intersection of the value of the output C_1 and the V_2 controller curve. The point of operation for the system to be controlled remains at point A, which is the intersection of the output C_1 and the U load curve. Because the controller is supplying more of the manipulated variable than is required to maintain the system to be controlled at point A $(M_B > M_1)$, the output C increases. The new equilibrium point of operation will be at point C, where the value of the manipulated variable supplied by the controller M_2 is the value required to maintain the system output at C_2. If the input dial is calibrated in such a way that the value of the command signal V_1 is equal to the output C_1, the value of the command signal V_2 is equal to the output C_2, etc., then the system is calibrated so that the output will always equal the input for a given loading condition. When this is so, Eq. (4.2) is automatically satisfied.

The characteristics shown in Fig. 4.6b illustrate the effect of changing the load from U_1 to U_2 with the command signal V maintained fixed. Initially the system is at equilibrium at point A, the intersection for the controller curve V and the load line U_1. When the load is changed to U_2, the new point of operation for the system to be controlled is at point B, which is the intersection of the value of the output C_1 and the load line U_2. The point of operation of the controller remains at point A, which is the intersection of the value of the output C_1 and the controller line V. Because the amount of manipulated variable M_1 being supplied by the controller is less than the amount M_B required to maintain the system to be controlled at point B, the output C decreases. The new equilibrium point of operation will be at point C, which is the intersection of the new load line U_2 and the controller line V. Note that with the command signal V maintained fixed, the output has changed from C_1 to C_2 due to the change in load.

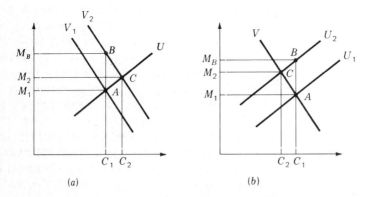

(a)   (b)

Figure 4.6 Effect of changing (a) the command signal and (b) the load.

The slope of the controller lines V is $-K_{G_1}K_H$. For an integral-type controller, K_{G_1} is infinite. The resulting controller lines are vertical. For this case there is no change in the output due to a change in the load. Replacing the control element $G_1(D) = 1/(1 + \tau_1 D)$ by the element $G_1(D) = 1/D$ would yield such a system. The resulting steady-state characteristics are shown in Fig. 4.7a. Note that for steady-state operation, the value of the controlled variable C is always equal to the command signal V regardless of the value of the disturbance U. Using the points B and C in Fig. 4.7a to evaluate the coefficient $\partial C/\partial V \mid_U$ and using the points D and E to evaluate the coefficient $\partial C/\partial U \mid_V$ gives

$$\left.\frac{\Delta C}{\Delta V}\right|_U = \frac{C_B - C_C}{V_B - V_C} = \frac{110 - 90}{110 - 90} = 1$$

$$\left.\frac{\Delta C}{\Delta U}\right|_V = \frac{C_D - C_E}{U_D - U_E} = \frac{100 - 100}{8 - 12} = 0$$

Substitution of these results into Eq. (4.13) yields the equation for steady-state operation

$$c = v$$

Thus for an integral-type control system, the controlled variable c is equal to the command signal v regardless of the disturbance.

For an open-loop system, there is no feedback path. When $K_H = 0$, the feedback path of Fig. 4.1b is disconnected so that an open-loop system results. For this case the slope of the controller lines $(-K_{G_1}K_H = 0)$ is zero. As is shown in Fig. 4.7b, for an open-loop system the lines of operation for the controller are

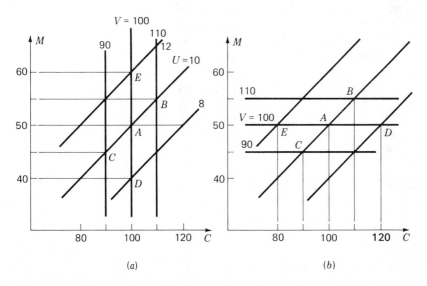

Figure 4.7 Steady-state characteristics for (a) an integral control system and (b) an open-loop system.

horizontal. Using the points B and C in Fig. 4.7b to evaluate the coefficient $\partial C/\partial V \mid_U$ and the points D and E to evaluate the coefficient $\partial C/\partial U \mid_V$ gives

$$\left.\frac{\Delta C}{\Delta V}\right|_U = \frac{110 - 90}{110 - 90} = 1 \qquad \left.\frac{\Delta C}{\Delta U}\right|_V = \frac{120 - 80}{8 - 12} = \frac{40}{-4} = -10$$

Substitution of these results into Eq. (4.13) yields the equation for steady-state operation

$$c = v - 10u$$

The slope of the controller lines $\partial M/\partial C \mid_V = -K_{G_1}K_H$ varies from zero for an open-loop system ($K_H = 0$) to infinity for an integral control system ($K_{G_1} = \infty$). A proportional control system is one for which the controller lines have a finite slope, as shown in Fig. 4.4c. Usually the slope of the control lines for a proportional controller is very steep, so that they are almost vertical. The steeper the slope, the less the variation in the controlled variable C due to a change in the disturbance U. An open-loop system has the greatest change in the controlled variable due to a disturbance, whereas an integral controller has no change in the controlled variable due to a disturbance.

Proportional control systems usually exhibit better transient characteristics than integral control systems. In addition, proportional control systems provide the operator with a "feel" as to what is going on. For example, a power-steering system feeds back some of the torque applied to the steering wheel so that the driver has a measure or feel for the turning effort being applied to the wheels. Satisfactory performance may usually be achieved by making the control lines sufficiently steep; this makes the coefficient of the u term sufficiently small so that variations in the external disturbance cause only slight errors. For integral control systems, the coefficient of the u term is zero. For proportional control systems, the coefficient of the u term is finite. The coefficient attains its maximum value for an open-loop system.

4.3 PROPORTIONAL CONTROL SYSTEMS

The differential equation relating the output n_o to the input n_{in} and external disturbance t_L for the speed control system represented by the block diagram of Fig. 3.30 is given by Eq. (3.67). Letting $D = 0$ in Eq. (3.67) yields for the equation describing the steady-state operation of the speed control system

$$n_o = \frac{C_2 C_6 K_1 K_s n_{in} - C_6 C_8 t_L}{1 + C_4 C_6 K_1} \qquad (4.16)$$

Letting $D = 0$ in the overall block diagram of Fig. 3.30 yields the block diagram for steady-state operation shown in Fig. 4.8a. Comparison with the general block diagram of Fig. 4.8b shows that $c = n_o$, $v = n_{in}$, $u = t_L$, $A = C_2 K_s$, $B = -C_8$, $K_{G_1} = K_1$, $K_{G_2} = C_6$, and $K_H = C_4$. Substitution of these results into Eq. (4.1) verifies Eq. (4.16).

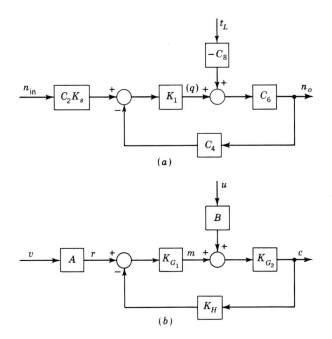

Figure 4.8 Block diagram for steady-state operation.

The speed control system shown in Fig. 3.30 is a proportional control system. Substitution of the corresponding values for this speed control system into Eq. (4.2) gives

$$A = C_2 K_s = \frac{1}{C_6 K_1} + C_4$$

or

$$C_2 = \frac{1}{K_s}\left(\frac{1}{C_6 K_1} + C_4\right) \tag{4.17}$$

The term $C_2 = \partial Z/\partial N_{in}\,|_i$ is the scale factor for the speed-setting dial.

Because some of the terms in Eq. (4.17) are partial derivatives evaluated at the reference operating condition, the value of the scale factor C_2 is seen to vary for different reference points. This would result in a nonlinear scale for the input speed dial. The use of a nonlinear scale may be avoided by connecting the throttle lever to a cam which in turn sets the desired position of the top of the spring, as is shown in Fig. 4.9. It is then a relatively easy matter to set up the speed control system so that Eq. (4.17) is satisfied for any reference condition. When this is so, Eq. (4.16) becomes

$$n_o = n_{in} - \frac{C_6 C_8}{1 + C_4 C_6 K_1} t_L \tag{4.18}$$

Equation (4.18) is the typical form of the steady-state relationship that exists between the input, output, and external disturbance for a proportional control

Figure 4.9 Cam to avoid nonlinear input scale.

system. When the load torque T_L is not equal to the reference value (that is, $t_L \neq 0$), then n_o is not equal to n_{in}. For example, suppose that this is the speed control system for the gas turbine of a jet airplane and that T_{L_i} is the torque required for the airplane in level flight. When the airplane is inclined to gain altitude, a greater load torque T_L is required than that for level flight (that is, $t_L > 0$). Thus, Eq. (4.18) shows that the output speed is slightly less than the desired value for this flight condition.

The physical reason for this can be seen by looking at the schematic diagram of Fig. 3.25 for the speed control system. For level flight, the system is set up so that $N_o = N_{in}$. When the airplane is gaining altitude, the load torque is increased. This increased torque results in a decreased speed, which in turn causes a lower position for x. Because of the lower position for x, there is a greater flow of fuel. To have the airplane continue to gain altitude, more flow is required than for level flight. To maintain this increased flow, the engine speed must be slightly less than for level flight.

The steady-state operating curves for this speed control system are shown in Fig. 4.10. For an airplane in level flight, the curve of fuel flow Q required to maintain various speeds N_o is indicated by $T_{L_i} = T_2$. The curve marked T_3 would correspond to operation of the airplane at a certain angle of inclination. Similarly, the curve T_1 would correspond to the airplane losing altitude at a certain angle of declination.

The operating line AB for the controller is obtained by fixing the speed setting at some value N_{in} and then plotting corresponding values of fuel flow Q

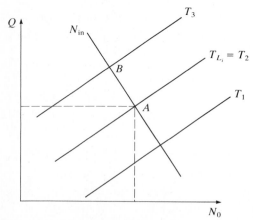

Figure 4.10 Steady-state operating curves for a speed control system.

coming from the controller, for various speeds of rotation N_o. Note from Fig. 3.25 that an increased speed N_o of the flyweights increases the centrifugal force, which causes x to move up. This in turn causes y to move down, and this decreases the flow Q supplied by the controller.

Steady-state operation exists at the intersection of the line of operation of the controller and the torque line for the given flight condition, because at this intersection just enough flow is being supplied to maintain the flight condition. For example, if the airplane is in level flight and the operating line for the desired speed setting is AB, then the intersection at point A of the line T_2 and the operating line AB is the steady-state operating point for the system. The speed-setting dial is calibrated by setting the value of speed N_{in} on the dial equal to the steady-state value of the output speed at the reference load T_2. For a given speed setting, such as that indicated by the line AB, if the load is increased to T_3 while the desired speed is unchanged, then the new operating point must be on the line of T_3 at point B. Because AB is not a vertical line, variations in the load are seen to cause variations in the output speed. A proportional controller is sometimes called a droop controller and the line AB is referred to as the droop line.

Illustrative example 4.2 A typical family of steady-state operating curves for a unity feedback ($K_H = 1$) speed control system is shown in Fig. 4.11. At the reference operating condition (point A), $N_{in} = N_o = 4000$, $Q_i = 1000$, and $T_i = 200$. Determine the steady-state constants and the equation for steady-state operation. With N_{in} held fixed at its reference value, what is the change in speed N_o when the load T changes from the reference value $T_i = 200$ to $T = 300$? By what factor should the slope of the controller lines be changed so as to reduce this change by a factor of 50?

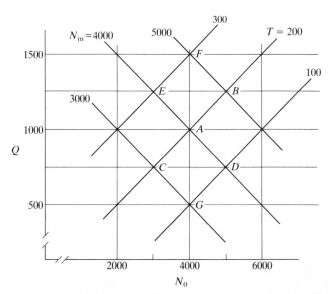

Figure 4.11 Typical family of steady-state operating curves for a speed control system.

SOLUTION The constants for the system to be controlled are

$$\frac{1}{K_{G_2}} = \frac{\partial M}{\partial C}\bigg|_U = \frac{\Delta Q}{\Delta N_o}\bigg|_T = \frac{Q_B - Q_C}{N_{oB} - N_{oC}} = \frac{1250 - 750}{5000 - 3000} = \frac{1}{4}$$

$$B = -\frac{\partial M}{\partial U}\bigg|_C = -\frac{\Delta Q}{\Delta T}\bigg|_{N_o} = -\frac{Q_F - Q_G}{T_F - T_G} = -\frac{1500 - 500}{300 - 100} = -5$$

where Q_B is the value of Q at point B, N_{oB} is the value of N_o at point B, etc.
The constants for the controller are

$$-K_{G_1}K_H = \frac{\partial M}{\partial C}\bigg|_V = \frac{\Delta Q}{\Delta N_o}\bigg|_{N_{in}} = \frac{Q_D - Q_E}{N_{oD} - N_{oE}} = \frac{750 - 1250}{5000 - 3000} = -\frac{1}{4}$$

$$AK_{G_1} = \frac{\partial M}{\partial V}\bigg|_C = \frac{\Delta Q}{\Delta N_{in}}\bigg|_{N_o} = \frac{Q_F - Q_G}{N_{inF} - N_{inG}} = \frac{1500 - 500}{5000 - 3000} = \frac{1}{2}$$

For $K_H = 1$, then $K_{G_1} = \frac{1}{4}$ and $A = 2$. The equation for steady-state operation is

$$n_o = \frac{AK_{G_1}K_{G_2}}{1 + K_{G_1}K_{G_2}K_H} n_{in} + \frac{BK_{G_2}}{1 + K_{G_1}K_{G_2}K_H} t$$

$$= \frac{(2)(1/4)(4)}{1 + (1/4)(4)(1)} n_{in} + \frac{(-5)(4)}{1 + (1/4)(4)(1)} t = n_{in} - 10t$$

This result may be verified by application of Eq. (4.13). That is,

$$c = \frac{\partial C}{\partial V}\bigg|_U v + \frac{\partial C}{\partial U}\bigg|_V u \tag{4.13}$$

Thus,

$$n_o = \frac{\partial N_o}{\partial N_{in}}\bigg|_T n_{in} + \frac{\partial N_o}{\partial T}\bigg|_{N_{in}} t = \frac{N_{oB} - N_{oC}}{N_{inB} - N_{inC}} n_{in} + \frac{N_{oD} - N_{oE}}{T_D - T_E} t$$

$$= \frac{5000 - 3000}{5000 - 3000} n_{in} + \frac{5000 - 3000}{100 - 300} t = n_{in} - 10t$$

With N_{in} fixed at the reference value, then $n_{in} = 0$. The change in speed n_o due to changing the load from the reference value $T_i = 200$ to $T = 300$ ($t = T - T_i = 100$) is

$$n_o = n_{in} - 10t = 0 - 10(100) = -1000$$

To decrease this change by a factor of 50, the coefficient of the t term must be changed from (-10) to $(-10/50) = (-1/5)$. Thus

$$\frac{BK_{G_2}}{1 + K_{G_1}K_{G_2}K_H} = \frac{(-5)(4)}{1 + 4K_{G_1}K_H} = -\frac{1}{5}$$

Solving for $K_{G_1}K_H$ gives

$$K_{G_1}K_H = \frac{99}{4}$$

To decrease the speed error by a factor of 50, the slope of the controller lines must be increased by a factor of 99 (i.e., from $-K_{G_1}K_H = -1/4$ to $-99/4$).

4.4 INTEGRAL CONTROL SYSTEMS

By eliminating the linkage between x and y of Fig. 3.25 and using the hydraulic integrator shown in Fig. 4.12, the proportional control system is converted to an integral control system. The block-diagram representation for the integrator is also shown in Fig. 4.12. The substitution of this diagram for that of the servomotor which it replaces in Fig. 3.28 yields the block-diagram representation shown in Fig. 4.13a.

The value of K_{G_1} is computed as follows:

$$K_{G_1} = \left[\frac{C_1 C_5}{(K_s - C_r C_3)A_1 D}\right]_{D=0} = \left(\frac{K_I}{D}\right)_{D=0} = \infty \qquad (4.19)$$

where $K_I = C_1 C_5/(K_s - C_r C_3)A_1$ is the constant associated with the integrating portion of the system, as is shown in Fig. 4.13b.

Figure 4.12 Integral control system.

Figure 4.13 Block diagram for integral control system.

Because K_{G_1} is infinite, e must be zero for steady-state operation. Thus subtracting the feedback signal from the reference input in Fig. 4.13b gives

$$C_2 K_s n_{in} - C_4 n_o = e = 0$$

or

$$n_o = \frac{C_2 K_s}{C_4} n_{in} \qquad (4.20)$$

The preceding expression shows that the speed is independent of the load torque for an integral control system. It is an easy matter to adjust the scale factor C_2 for the input speed dial so that $C_2 K_s/C_4 = 1$, in which case

$$n_o = n_{in} \qquad (4.21)$$

The operation of an integral control system may be visualized as follows. From Fig. 4.12, it can be seen that if x momentarily changes and then returns to its line-on-line position, the position of y has been changed permanently and so has the amount of flow going to the engine. Therefore, changing the amount of flow to account for a new operating torque does not change the steady-state position of x, which must be line on line. Because neither x nor the spring compression changes, the output speed must always be equal to the desired value in order that the flyweight force balances the spring force. (Note that, for the proportional control system, changing the fuel flow requires a permanent change in the position x.)

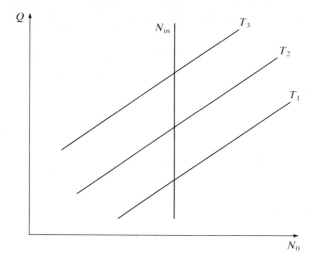

Figure 4.14 Steady-state operating curves for integral control system.

An integral control is easily recognized because there must be an integrating component yielding a $1/D$ term in the block diagram between the comparator and the point where the external disturbance enters the system. The line of operation of an integral controller is a vertical line. The operating characteristics of an integral control system are shown in Fig. 4.14.

An integral controller is also called a floating controller because of the floating action of the position y of the flow-setting valve. Two other terms used for an integral controller are reset controller and isochronous controller.

4.5 PROPORTIONAL PLUS INTEGRAL CONTROL SYSTEMS

From a consideration of steady-state operation only, integral control systems seem preferable to proportional systems. However, it is generally easier to achieve good transient behavior with a proportional system than with an integral system (techniques for determining the transient behavior of systems are presented in Chaps. 5 through 12). It is possible to combine the basic features of a proportional controller and an integral controller to form a proportional plus integral controller.

The action of a proportional plus integral controller in response to a change in the input or external disturbance is initially similar to that of a proportional controller, but as the new equilibrium point is reached, the control action becomes the same as that of an integral controller. (In effect, the slope of the controller line continually increases.)

A proportional plus integral controller combines the desirable transient characteristics of a proportional controller and the feature of no steady-state error of the integral controller.

A proportional plus integral controller is shown in Fig. 4.15. The proportional action is provided by unit 1, which is the same as that for the proportional controller shown in Fig. 3.25. The equation for the proportional action is

$$y_1 = \frac{1}{1 + \tau_1 D} x \tag{4.22}$$

The integral action is provided by unit 2, which is the same as that for the integral controller shown in Fig. 4.12. The equation for this integral action is

$$y_2 = \frac{C}{AD} x \tag{4.23}$$

The proportional and integral actions are added by a walking-beam linkage such that

$$y = \frac{y_1 + y_2}{2} \tag{4.24}$$

The substitution of y_1 and y_2 into the preceding expression gives

$$y = \frac{1}{2} \left(\frac{1}{1 + \tau_1 D} + \frac{C}{AD} \right) x \tag{4.25}$$

Figure 4.15 Proportional plus integral control system.

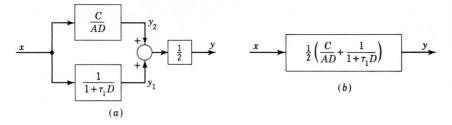

Figure 4.16 Block diagram for proportional plus integral action.

The individual block diagrams for Eqs. (4.22) to (4.24) are shown in Fig. 4.16a. The combined diagram is shown in Fig. 4.16b. The substitution of this combined diagram into its corresponding position between x and y of Fig. 4.13a yields the resulting representation for this proportional plus integral control system, as is shown in Fig. 4.17.

The value of K_{G_1} for this proportional plus integral control system is

$$K_{G_1} = \frac{C_5}{2(K_s - C_r C_3)} \left(\frac{1}{1 + \tau_1 D} + \frac{C}{AD} \right)_{D=0} = \infty \qquad (4.26)$$

The proportional plus integral actions are clearly evidenced by Eq. (4.26). Because K_{G_1} is infinite for steady-state operation, it follows that e is zero during steady-state operation. Thus, from Fig. 4.17,

$$C_2 K_s n_{in} - C_4 n_o = e = 0$$

or

$$n_o = \frac{C_2 K_s}{C_4} n_{in} = n_{in} \qquad (4.27)$$

Comparison of Eqs. (4.27) and (4.20) shows that the steady-state operation of a proportional plus integral control system is the same as that of an integral control system alone. A proportional plus integral control is sometimes referred to as a compensated isochronous control. To better understand the action of this control, suppose that the throttle lever is moved to increase the speed. This

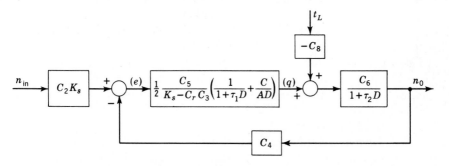

Figure 4.17 Overall block diagram for proportional plus integral control system.

causes the position x to move down, as does e'. The time constant τ_1 of the proportional unit is small, so that y_1 changes rapidly to increase the flow setting. The resulting motion of y_1 returns e' to its line-on-line position.

For the integrating unit, the quantity C/A is small, so that y_2 continues to move at a slower rate to provide corrective action. As the speed increases, the position x moves up. The integration unit continues to provide corrective action until x is returned to its line-on-line position (that is, $x = 0$). In summary, for proportional plus integral control, the initial effect is provided primarily by the proportional action and the final effect is provided by the integrator.

4.6 MODES OF CONTROL

In addition to proportional, integral, and proportional plus integral control, another mode of control is derivative, or rate, action. For a derivative controller, the steady-state expression for the control elements is

$$K_{G_1} = (KD)_{D=0} = 0 \tag{4.28}$$

The output of a derivative controller is proportional to the rate of change of error. For any constant value of the actuating signal e, the output of the control elements is zero. Thus, a steady state m exist in a derivative control system with any constant value of error signal. Because a derivative controller operates on the rate of change of error and not the error itself, the derivative mode of control is never used alone, but rather in combination with a proportional, or integral, or proportional plus integral controller. The advantage of using derivative action is that the derivative is a measure of how fast the signal is changing and thus tends to give the effect of anticipation. The addition of derivative action is limited primarily to systems which respond very slowly, such as large industrial processes.

The selection of the control elements $G_1(D)$ is seen to have a predominant effect upon the steady-state operation of a system. For more complex control systems, it becomes increasingly difficult, if not impossible, to distinguish the individual modes of control. However, regardless of the various modes that may be present, it is a relatively simple matter to determine whether K_{G_1} is finite or infinite. For an infinite value, the integral action predominates and there is no steady-state error due to variations in the external disturbance. For a finite value, the system behaves as a proportional control system.

A major problem in the design of control systems is the determination of the system parameters to obtain satisfactory transient performance. The transient behavior of a system is prescribed by the differential equation of operation for the system. In the next chapter, it is shown how such differential equations may be solved algebraically by the use of Laplace transforms. In Chap. 6, it is shown that the transient behavior is governed primarily by the roots of the characteristic equation for the system. Thus, the transient characteristics of a system may be ascertained directly from a knowledge of the roots of the characteristic equation.

PROBLEMS

4.1 For the control system shown in Fig. P4.1, determine the steady-state equation relation v, u, and c for each of the following cases:

(a) $G_1(D) = \dfrac{K_1}{1 + \tau_1 D}$ (b) $G_1(D) = \dfrac{K_1}{D}$ (c) $G_1(D) = \dfrac{K_1}{D} + \dfrac{K_1}{1 + \tau_1 D}$

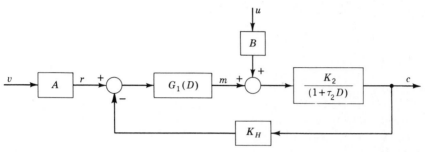

Figure P4.1

4.2 The steady-state operating curves for a proportional temperature control system are shown in Fig. P4.2.

(a) Determine the equation for steady-state operation about point A.

(b) If this were an open-loop rather than a closed-loop system, what would be the steady-state equation of operation?

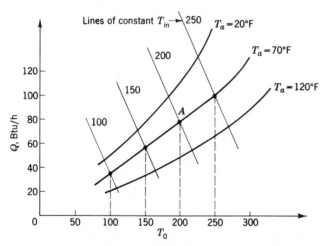

Figure P4.2

4.3 For the system shown in Fig. P4.3a, determine the differential equation of operation and the equation for steady-state operation. At the reference operating condition ($r = c = d = 0$), $R_i = C_i = 100$, $D_i = 40$, and $M_i = 60$. In Fig. P4.3b, sketch the load line ($D_i = 40$) and the controller line ($R_i = 100$). Also sketch the load line for $D = 60$ and the controller line for $R = 110$. Note that because $A = 1$, then $R = V$. Similarly, because $B = 1$, then $D = U$.

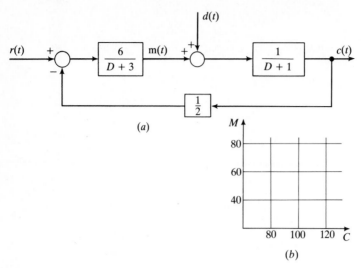

(a)

(b)

Figure P4.3

4.4 For the system shown in Fig. P4.4a, determine the differential equation of operation and the equation for steady-state operation. At the reference operating condition ($r = c = d = 0$), $R_i = C_i = 100$, $M_i = 250$, and $D_i = 150$. In Fig. P4.4b, sketch the load line ($D_i = 150$) and the controller line ($R_i = 100$). Also sketch the load line for $D = 200$ and the controller line for $R = 120$. Note that for $A = 1$, then $R = V$. Similarly, for $B = 1$, then $D = U$.

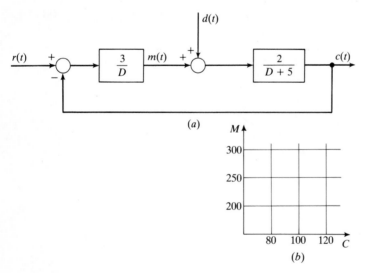

(a)

(b)

Figure P4.4

4.5 The steady-state operating curves for a unity feedback system ($K_H = 1$) are shown in Fig. P4.5. Construct the block diagram that describes the steady-state operation of this system.

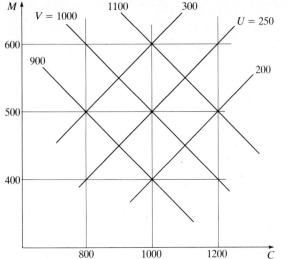

Figure P4.5

4.6 A controller is to be designed such that for no change in the command signal ($v = 0$), the output c will not change by more than one unit when the load u increases by 40 units. The steady-state operating curves for the system have a slope of 2. For $B = 0.3$, what is the required slope of the controller lines?

4.7 The steady-state operating curves for a system to be controlled are shown in Fig. P4.7. The reference operating condition is $V_i = C_i = 500$, $U_i = 100$, and $M_i = 200$. Determine the slope of the controller lines such that the controlled variable does not decrease by more than 4 percent of its reference value ($c = 0.04 \times 500 = 20$) when $U = 125$.

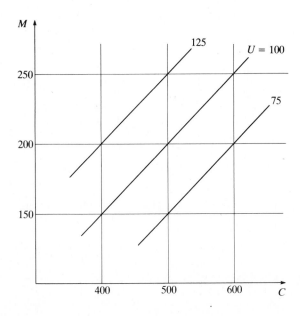

Figure P4.7

4.8 The block diagram for a feedback control system is shown in Fig. P4.8. At the reference operating point $V_i = C_i = 100$, $M_i = 25$, and $U_i = 50$. With V held fixed at its reference value ($V = V_i = 100$), calculate the change c in the controlled variable when U changes from its reference value $U_i = 50$ to $U = 60$. What is the new value of C?

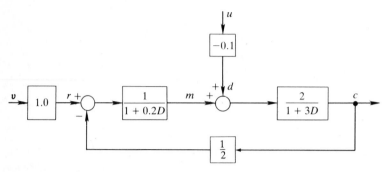

Figure P4.8

4.9 The block diagram for a control system is shown in Fig. P4.9. At the reference operating point $C_i = V_i = 500$, $M_i = 250$, and $U_i = 100$. Determine the value of A such that the coefficient of the v term in the equation for steady-state operation is unity. With the command signal V held constant at its reference value $V = V_i = 500$, determine the change c in the output (error) when the load changes from its reference value $U_i = 100$ to $U = 88$. Determine the required slope of the controller lines so as to reduce this error by a factor of 6. By what factor has the slope of the lines of constant V been changed?

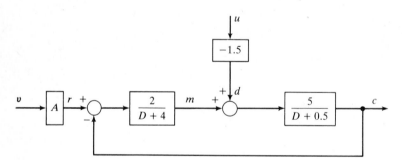

Figure P4.9

4.10 The block diagram of a feedback control system for operation about a reference operating point is shown in Fig. P4.10. The steady-state equation of operation is

$$c = \frac{\partial C}{\partial V}\Big|_U v + \frac{\partial C}{\partial U}\Big|_V u$$

(a) Determine the value of K_{G_1} such that $\partial C/\partial U\,|_V = 0.05$.
(b) Determine the value of A such that $\partial C/\partial V\,|_U = 1$.

Figure P4.10

(c) For the system to be controlled, what is the slope of the lines of constant U?

(d) For the system to be controlled, determine the horizontal spacing ΔC between the lines of constant U when $\Delta U = 10$ units.

4.11 For the system shown in Fig. P4.11,

(a) Determine K_{G_1} such that $\partial C/\partial U \mid_V = 0.1$.

(b) Determine A such that $\partial C/\partial V \mid_U = 1$.

Figure P4.11

4.12 The steady-state operating curves for a system to be controlled are shown in Fig. P4.12.

(a) Determine B and K_{G_2} at the reference operating point P.

(b) Determine the required slope of the controller lines $(-K_{G_1}K_H)$ such that the variation in C will not exceed 2 percent of its reference value $[c = 0.02(200) = 4 \text{ units}]$ when the external disturbance varies from its reference value to its maximum value $(u = U_{max} - U_i = 20 \text{ units})$.

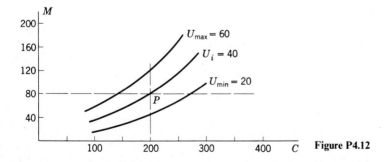

Figure P4.12

4.13 At the reference operating point $C_i = 100$, $M_i = 5$, and $U_i = 10$, the characteristics of a system to be controlled are $B = 0.5$ and $K_{G_2} = 25$. Determine the slope of the load lines U, and then

(a) Draw the nominal load line $U = 10$ for the system to be controlled, and also draw the $U = 6$ and the $U = 14$ load lines.

Note: Straight lines will suffice because we are working with linearized information about the reference operating point.

(b) Calculate the required slope of the controller line such that the output C will not change by more than 1 unit when the load is changed from its nominal value $U_i = 10$ to $U = 14$.

4.14 Same as Prob. 4.13 except $B = -0.5$.

4.15 The linearized equation of operation for an engine, $N = N(Q, T)$, about the reference point $N_i = 2000$ r/min, $T_i = 400$ ft · lb, and $Q_i = 40$ lb/min is

$$n = 50q - 10t$$

Determine $\Delta N/\Delta Q|_i$, $\Delta N/\Delta T|_i$, and $\Delta T/\Delta Q|_i$. Sketch the operating curve of the engine for $T_i = 400$ ft · lb and the curve for $T = 500$ ft · lb.

4.16 The block diagram for a feedback control system is shown in Fig. P4.16. At the reference operating condition $c = u = v = 0$.

(a) Determine the value of A such that $c = v$ for steady-state operation.

(b) For $u = \Delta U = 10$ units, what is the vertical distance ΔM between the iines of constant U?

(c) For $v = \Delta V = 10$ units, what is the horizontal distance ΔC between the lines of constant V?

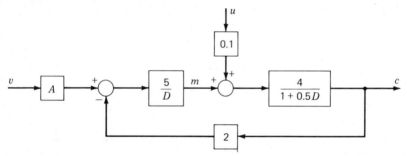

Figure P4.16

4.17 The steady-state operating curves for a unity-feedback system are shown in Fig. P4.17.

(a) Evaluate B, K_{G_2}, K_{G_1}, K_H, AK_{G_1}, $\partial C/\partial V|_U$, and $\partial C/\partial U|_V$.

(b) If the slope of the controller lines is increased by a factor of 10, what steady-state error results for $V = 400$ ($v = 0$) when U changes from 12 to 10 ($u = -2$)?

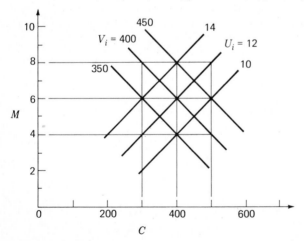

Figure P4.17

4.18 For the control system shown in Fig. P4.18, determine

(a) The slope of the controller lines and the slope of the load lines.

(b) The value of A such that $c = v$ for steady-state operation.

(c) The value of B such that the horizontal spacing between the load lines is $\Delta C = 0.5\ \Delta U$.

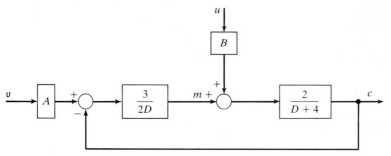

Figure P4.18

4.19 For the control system shown in Fig. P4.19, the reference operating point is $C_i = V_i = 4000$ r/min, $M_i = 2000$ lb/h, and $U_i = 200$ ft·lb.

(a) Determine A such that $c = v$ when $u = 0$.

(b) For $V = V_i = 4000$ r/min ($v = \Delta V = 0$), determine the speed error $c = \Delta C$ when $U = 150$ ft·lb ($u = \Delta U = U - U_i = -50$).

(c) Determine the required slope of the controller lines such that this speed error is reduced to 25 r/min.

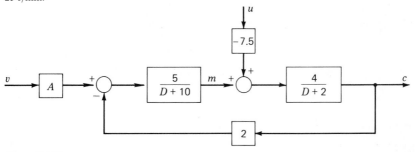

Figure P4.19

4.20 When the system to be controlled has an integrating element such as to make K_{G_2} infinite, the steady-state operating curves for the system become horizontal straight lines, as shown in Fig. P4.20. The steady-state operating curves for a remote-control positioning device are shown in Fig. P4.20.

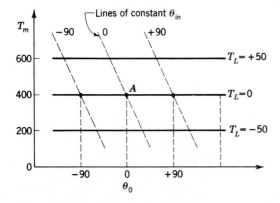

Figure P4.20

The controlled shaft position is designated by θ_o and the set position by θ_{in}. Variations are to be indicated by $\Delta\theta$. For this remote-control positioning system, the external disturbance is a load torque T_L, and the manipulated variable is a motor torque T_m.

(a) Determine the steady-state equation of operation relating $\Delta\theta_o$, $\Delta\theta_{in}$, and t_L for operation about point A.

(b) To decrease the error caused by variations in the external disturbance by a factor of 5, what should be the new slope of the controller lines?

4.21 A controller is to be designed for a system such that the output C will not change by more than one unit when the load U increases by 40 units. The steady-state operating curves for the system to be controlled have a slope of 2 units. What is the required slope of the controller lines (a) when $B = 0.3$ and (b) when $B = -0.3$?

4.22 In Fig. P4.22 is shown the steady-state block diagram for a system which is subjected to two external disturbances. Determine the steady-state equation relating c, v, u_1, and u_2. For which of the following cases will the steady-state operation be independent of u_1? For which cases will it be independent of variations in u_2?

Note: Steady-state operation is independent of K_{G_3}.

(a) $K_{G_1} = \infty$, K_{G_2} and K_H are finite.

(b) $K_{G_2} = \infty$, K_{G_1} and K_H are finite.

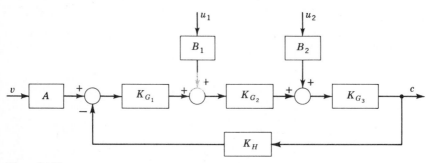

Figure P4.22

4.23 A reproducing shaper is shown in Fig. P4.23. The position y of the duplicating cutter is seen to follow the position x of the master cutter. Determine the mode of operation of the shaper. What modifications would be necessary to convert this to a proportional plus integral controller?

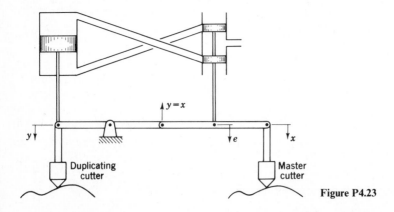

Figure P4.23

FIVE

LAPLACE TRANSFORMS

By transient response is meant the manner in which a system changes from some initial operating condition to some final condition. For example, in Fig. 5.1 it is to be seen that at some arbitrary time $t = 0$ the output is $y(0)$. The curve marked (*a*) represents the transient response of a system in which the output $y(t)$ slowly approaches its new operating condition. The curve marked (*b*) shows a system which successively overshoots and undershoots, but these oscillations gradually die out as the new operating condition is obtained.

A linear control system or component is one in which the operation is described by a linear differential equation, usually with constant coefficients. For a known input, classical methods could be used for determining the output. However, considerable time is saved by using the Laplace transform method of solving linear differential equations. In addition, as is explained in later chapters, the Laplace transform analysis is closely related to other methods for evaluating system performance.

A brief review of classical methods for solving differential equations with constant coefficients is first presented. This is done so that, when the Laplace transform method is explained, a clearer understanding of the similarities and the differences between the two methods results.

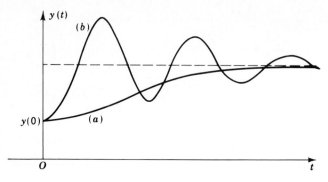

Figure 5.1 Transient response.

5.1 CLASSICAL METHODS

The transient response of a control system may be obtained by solving differential equations of the general form

$$(D^n + b_{n-1}D^{n-1} + \cdots + b_1 D + b_0)y(t)$$
$$= (a_m D^m + a_{m-1}D^{m-1} + \cdots + a_1 D + a_0)f(t)$$

If the coefficient b_n of the $D^n y(t)$ term is not unity, it is an easy matter to divide each term by b_n to yield the preceding form. Solving for $y(t)$ gives

$$y(t) = \frac{a_m D^m + a_{m-1}D^{m-1} + \cdots + a_1 D + a_0}{D^n + b_{n-1}D^{n-1} + \cdots + b_1 D + b_0} f(t) \qquad (5.1)$$

where $a_0, a_1, \ldots, a_m$ and $b_0, b_1, \ldots, b_{n-1}$ are constants. The term $f(t)$ represents the excitation to the system. This is called the forcing function because it forces or excites the system. The output $y(t)$ is called the response function because it responds to the forcing function $f(t)$. The denominator of Eq. (5.1) is the characteristic function of the differential equation. The equation which results from setting the characteristic function equal to zero is called the characteristic equation. The value of the exponent n, the highest power of D in the characteristic function, is the order of the differential equation. It is now shown how Eq. (5.1) which is an nth order differential equation may be written as the sum of n first-order differential equations. First rewrite Eq. (5.1) in the form

$$y(t) = \frac{L_m(D)}{L_n(D)} f(t) \qquad (5.2)$$

where $L_n(D) = D^n + b_{n-1}D^{n-1} + \cdots + b_1 D + b_0$
$L_m(D) = a_m D^m + a_{m-1}D^{m-1} + \cdots + a_1 D + a_0$

The characteristic function $L_n(D)$ may be factored into the form

$$L_n(D) = (D - r_1)(D - r_2) \cdots (D - r_n) \qquad (5.3)$$

where $r_1, r_2, \ldots, r_n$ are the roots of the characteristic equation $L_n(D) = 0$. In factoring the polynomial $L_n(D)$ as shown in Eq. (5.3), it is to be noted that D may be treated as an algebraic quantity. For example, consider the function

$$L_n(D) = D^2 + 3D + 2 = (D - r_1)(D - r_2)$$

Setting $L_n(D)$ equal to zero yields

$$D^2 + 3D + 2 = 0$$

The roots of the preceding equation are

$$r_{1,2} = \frac{-3 \pm \sqrt{9 - 8}}{2} = -1, -2$$

Therefore, $L_n(D) = [D - (-1)][D - (-2)] = (D + 1)(D + 2)$.

Because $r_1, r_2, \ldots, r_n$ are the values of D for which $L_n(D) = 0$, then $r_1, r_2, \ldots, r_n$ are also called the zeros of the function $L_n(D)$. Thus, the roots of the characteristic equation $L_n(D) = 0$ are the zeros of the characteristic function $L_n(D)$. The zeros of $L_n(D)$ are said to be distinct if each zero has a different value (that is, $r_1 \neq r_2 \neq r_3 \neq \cdots \neq r_n$). When two or more zeros are equal, the characteristic function is said to have repeated zeros. The case in which $L_n(D)$ has distinct zeros is considered first.

For distinct zeros, $L_m(D)/L_n(D)$ in Eq. (5.2) may be written in the form

$$\frac{L_m(D)}{L_n(D)} = \frac{K_1}{D - r_1} + \frac{K_2}{D - r_2} + \cdots + \frac{K_i}{D - r_i} + \cdots + \frac{K_n}{D - r_n} \tag{5.4}$$

The procedure for obtaining any constant K_i is as follows. First multiply both sides of Eq. (5.4) by $D - r_i$, that is,

$$(D - r_i)\frac{L_m(D)}{L_n(D)} = \frac{D - r_i}{D - r_1}K_1 + \frac{D - r_i}{D - r_2}K_2 + \cdots + K_i + \cdots + \frac{D - r_i}{D - r_n}K_n \tag{5.5}$$

The multiplication of the K_i term in Eq. (5.4) by $D - r_i$ is seen to cancel the denominator, thus leaving K_i alone as shown in Eq. (5.5). Letting $D = r_i$ in Eq. (5.5), each term of the right side becomes zero except for K_i, which remains. Thus

$$K_i = \lim_{D \to r_i} \left[(D - r_i)\frac{L_m(D)}{L_n(D)} \right] \tag{5.6}$$

Successive application of Eq. (5.6), in which $i = 1, 2, \ldots, n$, yields each of the constants $K_1, K_2, \ldots, K_n$ respectively in Eq. (5.4). As an example of the use of this partial-fraction–expansion technique, let it be desired to expand $L_m(D)/L_n(D)$, where

$$\frac{L_m(D)}{L_n(D)} = \frac{5D + 8}{D^2 + 3D + 2} = \frac{K_1}{D + 2} + \frac{K_2}{D + 1}$$

Application of Eq. (5.6) yields

$$K_1 = \lim_{D \to -2} \left[(D + 2) \frac{5D + 8}{(D + 2)(D + 1)} \right]$$

$$= \lim_{D \to -2} \frac{5D + 8}{D + 1} = \frac{-10 + 8}{-2 + 1} = 2$$

and

$$K_2 = \lim_{D \to -1} \frac{5D + 8}{D + 2} = \frac{-5 + 8}{-1 + 2} = 3$$

Thus,

$$\frac{L_m(D)}{L_n(D)} = \frac{2}{D + 2} + \frac{3}{D + 1}$$

The general form for expressing an nth-order differential equation as a sum of n first-order differential equations is obtained by substitution of $L_m(D)/L_n(D)$ from Eq. (5.4) into Eq. (5.2). That is,

$$y(t) = \frac{K_1}{D - r_1} f(t) + \cdots + \frac{K_i}{D - r_i} f(t) + \cdots + \frac{K_n}{D - r_n} f(t)$$

$$= \sum_{i=1}^{n} \frac{K_i}{D - r_i} f(t) = \sum_{i=1}^{n} K_i y_i(t) \qquad (5.7)$$

where

$$y_i(t) = \frac{1}{D - r_i} f(t)$$

To solve this differential equation, first write it in the form

$$\frac{d}{dt} y_i(t) - r_i y_i(t) = f(t) \qquad (5.8)$$

Next, multiply through by the integrating factor $e^{-r_i t}$:

$$e^{-r_i t} \frac{d}{dt} y_i(t) - r_i e^{-r_i t} y_i(t) = \frac{d}{dt} [e^{-r_i t} y_i(t)] = f(t) e^{-r_i t}$$

Note that the integrating factor makes it possible to write the left side as a perfect differential. Integration of the second and last terms on the right side shows that

$$e^{-r_i t} y_i(t) = \int f(t) e^{-r_i t} \, dt + c_i$$

Multiplying through by $e^{r_i t}$ yields, for the desired solution,

$$y_i(t) = e^{r_i t} \left[\int f(t) e^{-r_i t} \, dt + c_i \right]$$

or

$$y_i(t) = c_i e^{r_i t} + e^{r_i t} \int f(t) e^{-r_i t} \, dt \qquad (5.9)$$

Because the constant of integration c_i is displayed separately in Eq. (5.9), it suffices to evaluate the integral at time t only. Substitution of Eq. (5.9) into Eq. (5.7) yields the following general solution:

$$y(t) = \sum_{i=1}^{n} k_i e^{r_i t} + \sum_{i=1}^{n} K_i e^{r_i t} \int f(t) e^{-r_i t} \, dt \qquad (5.10)$$

where $k_i = c_i K_i$ is a constant. The first summation on the right side of Eq. (5.10) is the complementary solution, and the second summation is the particular solution. That is,

$$y_c(t) = \sum_{i=1}^{n} k_i e^{r_i t} \qquad (5.11)$$

and

$$y_p(t) = \sum_{i=1}^{n} K_i e^{r_i t} \int f(t) e^{-r_i t} \, dt \qquad (5.12)$$

where $y_c(t)$ is the complementary solution and $y_p(t)$ is the particular solution.

Illustrative example 5.1 Determine the solution of the following differential equation when all the initial conditions are zero and $f(t) = 4t$:

$$(D^2 + 3D + 2)y(t) = f(t)$$

or

$$y(t) = \frac{1}{(D+1)(D+2)} f(t) = \left(\frac{K_1}{D+1} + \frac{K_2}{D+2} \right) f(t)$$

$$= \left(\frac{1}{D+1} - \frac{1}{D+2} \right) f(t) \qquad (5.13)$$

where

$$K_1 = \lim_{D \to -1} \left[(D+1) \frac{1}{(D+1)(D+2)} \right] = 1$$

and

$$K_2 = \lim_{D \to -2} \left[(D+2) \frac{1}{(D+1)(D+2)} \right] = -1$$

SOLUTION The substitution of $r_1 = -1$ and $r_2 = -2$ into Eq. (5.11) yields, for the complementary solution,

$$y_c = k_1 e^{r_1 t} + k_2 e^{r_2 t} = k_1 e^{-t} + k_2 e^{-2t}$$

For $K_1 = 1$ and $K_2 = -1$, the particular solution is evaluated from Eq. (5.12) as follows:

$$y_p = K_1 e^{-t} \int (4t) e^t \, dt + K_2 e^{-2t} \int (4t) e^{2t} \, dt$$

$$= e^{-t}[4e^t(t-1)] - e^{-2t}[e^{2t}(2t-1)]$$

$$= 4(t-1) - (2t-1) = 2t - 3$$

The general solution is the sum of the complementary and the particular solutions. That is,

$$y(t) = k_1 e^{-t} + k_2 e^{-2t} + 2t - 3$$

The derivative $dy/dt = y'(t)$ is

$$y'(t) = -k_1 e^{-t} - 2k_2 e^{-2t} + 2$$

Because the initial conditions are zero, $y(0) = y'(0) = 0$. Evaluation of $y(t)$ and $y'(t)$ at time $t = 0$ gives

$$y(0) = k_1 + k_2 - 3 = 0$$

$$y'(0) = -k_1 - 2k_2 + 2 = 0$$

Solving these two equations simultaneously for k_1 and k_2 yields $k_1 = 4$ and $k_2 = -1$. Thus, the desired result is

$$y(t) = 4e^{-t} - e^{-2t} + 2t - 3 \tag{5.14}$$

Different initial conditions would yield different values for the constants k_1 and k_2.

Repeated Zeros

Suppose that the characteristic function $L_n(D)$ has a multiple or repeated zero r which occurs q times; that is,

$$L_n(D) = (D - r)^q (D - r_1)(D - r_2) \cdots (D - r_{n-q}) \tag{5.15}$$

For repeated zeros the partial-fraction expansion has the general form

$$y(t) = \frac{L_m(D)}{L_n(D)} f(t)$$

$$= \frac{C_q f(t)}{(D - r)^q} + \frac{C_{q-1} f(t)}{(D - r)^{q-1}} + \cdots + \frac{C_i f(t)}{(D - r)^i} + \cdots + \frac{C_1 f(t)}{D - r}$$

$$+ \frac{K_1 f(t)}{D - r_1} + \frac{K_2 f(t)}{D - r_2} + \cdots + \frac{K_{n-q} f(t)}{D - r_{n-q}} \tag{5.16}$$

The constants $K_1, K_2, \ldots, K_{n-q}$ are evaluated as before by application of Eq. (5.6); however, the constants $C_q, C_{q-1}, \ldots, C_1$, which arise from the partial-fraction expansion of the repeated zero, are evaluated as follows:

$$C_q = \lim_{D \to r} \left[(D - r)^q \frac{L_m(D)}{L_n(D)} \right]$$

$$C_{q-1} = \lim_{D \to r} \left\{ \frac{1}{1!} \frac{d}{dD} \left[(D - r)^q \frac{L_m(D)}{L_n(D)} \right] \right\} \tag{5.17}$$

$$C_{q-k} = \lim_{D \to r} \left\{ \frac{1}{k!} \frac{d^k}{dD^k} \left[(D - r)^q \frac{L_m(D)}{L_n(D)} \right] \right\}$$

Consider now the case of a root r which is repeated twice. That is,

$$y(t) = \frac{f(t)}{(D-r)^2} \tag{5.18}$$

This may be written as

$$(D-r)(D-r)y(t) = f(t)$$

To solve this differential equation, let

$$y_1 = (D-r)y(t) \tag{5.19}$$

Replacing $(D-r)y(t)$ with y_1 yields

$$(D-r)y_1 = f(t)$$

Multiplying through by the integrating factor e^{-rt} gives

$$e^{-rt}\frac{d}{dt}y_1 - re^{-rt}y_1 = \frac{d}{dt}(e^{-rt}y_1) = f(t)e^{-rt}$$

Integration of the second and last terms on the right side shows that

$$e^{-rt}y_1 = \int f(t)e^{-rt}\,dt + c_1$$

where c_1 is the constant of integration. Multiplying through by e^{rt} yields, for y_1,

$$y_1 = e^{rt}\int f(t)e^{-rt}\,dt + c_1 e^{rt} \tag{5.20}$$

Equating Eqs. (5.19) and (5.20) shows that

$$(D-r)y(t) = e^{rt}\int f(t)e^{-rt}\,dt + c_1 e^{rt}$$

Again, multiplying through by the integrating factor e^{-rt},

$$e^{-rt}\frac{d}{dt}y(t) - re^{-rt}y(t) = \frac{d}{dt}[e^{-rt}y(t)] = \int f(t)e^{-rt}\,dt + c_1$$

Integration yields

$$e^{-rt}y(t) = \iint f(t)e^{-rt}\,dt\,dt + c_1 t + c_0$$

where c_0 is another constant of integration. Multiplying through by e^{rt} yields for the desired result

$$y(t) = e^{rt}\iint f(t)e^{-rt}\,dt\,dt + (c_0 + c_1 t)e^{rt} \tag{5.21}$$

By proceeding in this manner, the response due to a root which is repeated i times may be obtained. Thus, the response due to the term $C_i f(t)/(D - r)^i$ in Eq. (5.16) is

$$y_i(t) = \frac{C_i f(t)}{(D - r)^i}$$

$$= (c_0 + c_1 t + \cdots + c_{i-1} t^{i-1}) e^{rt} + C_i e^{rt} \int \cdots \int f(t) e^{-rt} (dt)^i$$

$$i = 2, 3, \ldots, q \qquad (5.22)$$

where $c_0, c_1, \ldots, c_{i-1}$ are constants which must be evaluated from the initial conditions. The first term containing the c constants is the complementary solution, whereas the second term on the right side of the preceding expression is the particular solution. The response due to the distinct zeros in Eq. (5.16) may be evaluated by application of Eq. (5.10).

Illustrative example 5.2 Determine the solution of the following differential equation when all the initial conditions are zero and $f(t) = e^{-t}$:

$$(D + 1)^2 (D + 2) y(t) = f(t)$$

or
$$y(t) = \frac{1}{(D + 1)^2 (D + 2)} f(t)$$

$$= \left[\frac{C_2}{(D + 1)^2} + \frac{C_1}{D + 1} + \frac{K_1}{D + 2} \right] f(t) \qquad (5.23)$$

SOLUTION For this example q is 2, so that

$$C_2 = \lim_{D \to -1} \left[(D + 1)^2 \frac{1}{(D + 1)^2 (D + 2)} \right] = \left(\frac{1}{D + 2} \right)_{D = -1} = 1$$

$$C_1 = \lim_{D \to -1} \left(\frac{d}{dD} \frac{1}{D + 2} \right) = \left[\frac{-1}{(D + 2)^2} \right]_{D = -1} = -1$$

$$K_1 = \lim_{D \to -2} \left[(D + 2) \frac{1}{(D + 1)^2 (D + 2)} \right] = \left[\frac{1}{(D + 1)^2} \right]_{D = -2} = 1$$

Application of Eq. (5.22) in which $i = 2$ yields, for the response due to the first term on the right side of Eq. (5.23),

$$y = (c_0 + c_1 t) e^{-t} + C_2 e^{-t} \int\int e^{-t} e^t \, dt \, dt$$

$$= (c_0 + c_1 t) e^{-t} + 0.5 t^2 e^{-t}$$

The response due to the second term is

$$y = c_2 e^{-t} + C_1 e^{-t} \int e^{-t} e^t \, dt = c_2 e^{-t} - t e^{-t}$$

Application of Eq. (5.10) yields, for the response due to the last term,

$$y = k_3 e^{-2t} + K_1 e^{-2t} \int e^{-t} e^{2t} \, dt = k_3 e^{-2t} + e^{-t}$$

The general solution is the sum of the responses due to each of these three terms. That is,

$$y(t) = [(c_0 + c_2) + c_1 t]e^{-t} + k_3 e^{-2t} + (1 - t + 0.5t^2)e^{-t}$$

$$= (k_1 + k_2 t)e^{-t} + k_3 e^{-2t} + 0.5t^2 e^{-t}$$

where $k_1 = c_0 + c_2 + 1$ and $k_2 = c_1 - 1$. Successive differentiation yields

$$y'(t) = [-k_1 + k_2(1 - t)]e^{-t} - 2k_3 e^{-2t} + (t - 0.5t^2)e^{-t}$$

$$y''(t) = [k_1 - k_2(2 - t)]e^{-t} + 4k_3 e^{-2t} + (1 - 2t + 0.5t^2)e^{-t}$$

Because the initial conditions are zero, $y(0) = y'(0) = y''(0) = 0$. Thus, evaluation of $y(t)$, $y'(t)$, and $y''(t)$ at $t = 0$ gives

$$y(0) = k_1 + k_3 = 0$$

$$y'(0) = -k_1 + k_2 - 2k_3 = 0$$

$$y''(0) = k_1 - 2k_2 + 4k_3 + 1 = 0$$

Solving these three equations simultaneously yields $k_1 = 1$, $k_2 = -1$, and $k_3 = -1$. Thus the desired result is

$$y(t) = (1 - t + 0.5t^2)e^{-t} - e^{-2t} \tag{5.24}$$

Numerous techniques, such as the method of undetermined coefficients, variation of parameters, etc., have been developed for solving linear differential equations with constant coefficients. However, the method of Laplace transforms, which is next described, is best suited for solving the type of problems which are of interest to control engineers. In many ways, the Laplace transform method is similar to the preceding method of using the partial-fraction expansion to reduce an nth-order equation to the sum of n first-order equations. A major difference is that, in the Laplace transform method, the response due to each term in the partial-fraction expansion is determined directly from the transform table. Thus, there is no need to perform the integrations indicated by either Eq. (5.10) or Eq. (5.22). Because initial conditions are automatically incorporated into the Laplace transform, the resulting response expression yields directly the total solution (i.e., complementary plus particular solution). Thus, the constants arising from the initial conditions are automatically evaluated, so that the final desired result is obtained directly.

5.2 LAPLACE TRANSFORM METHOD

This method of solving differential equations is somewhat analogous to the process of multiplying or dividing using logarithms. In the well-known transformation of logarithms, numbers are transformed into powers of the base 10 or some other base. This process in effect makes it possible to multiply and divide by using the simpler operations of addition and subtraction. After obtaining the desired answer in logarithms, the transformation back to the real-number system is accomplished by finding antilogarithms.

In the method of Laplace transforms, transformation of the terms of the differential equation yields an algebraic equation in another variable s. Thereafter the solution of the differential equation is effected by simple algebraic manipulations in the s domain (the new variable is s rather than time t). To obtain the desired time solution, it is necessary to invert the transform of the solution from the s domain back to the time domain. Actually, for much control work, information obtained in the s domain suffices so that it may be unnecessary to invert back to the time domain.

The Laplace transformation $F(s)$ of a function of time $f(t)$ is[1]

$$F(s) = \mathcal{L}[f(t)] = \int_0^\infty f(t)e^{-st}\, dt \tag{5.25}$$

where $\mathcal{L}$ is the symbol for taking the Laplace transform. The symbol $\mathcal{L}$ is read "transform of," so that $\mathcal{L}[f(t)]$ means "transform of $f(t)$." For the integral on the right side of Eq. (5.25) the variable t vanishes after evaluation between the limits of integration. Thus, the resulting expression is a function of s only [that is, $F(s)$]. The mathematical derivation of the Laplace transform method including Eq. (5.25) is presented in Appendix B.

Transforming Functions from the Time Domain to the s Domain

Some input functions which are frequently used for investigating the characteristics of a control system are the step function, pulse function, impulse function, exponentially decaying function, and sinusoidal function.

Step function A graphical representation of a step function is shown in Fig. 5.2. A step function is designated by the symbol $hu(t)$, where h is the height and $u(t)$ is the symbol for a unit step function whose height is 1. Application of Eq. (5.25) gives

$$F(s) = \mathcal{L}[hu(t)] = \int_0^\infty he^{-st}\, dt = -\frac{he^{-st}}{s}\bigg|_0^\infty$$

$$= \frac{h(-e^{-s(\infty)} + e^{-s(0)})}{s} = \frac{h}{s} \tag{5.26}$$

Figure 5.2 Step function.

In evaluating a transform, the term s is regarded as any constant which makes $F(s)$ convergent. As illustrated by Eq. (5.26), if s is any positive constant $(s > 0)$, then $e^{-s(\infty)} = 0$ and $e^{-s(0)} = e^{-0} = 1$, so that the result follows. However, it should be noted that for negative values of s, then $e^{-s(\infty)} = e^{\infty} = \infty$, in which case $F(s)$ would be divergent. The operator s must be taken as any constant such that $F(s)$ is convergent. Although there is a range of values of s over which $F(s)$ is convergent, there is but one transform $F(s)$ corresponding to each time function $f(t)$. In Table 5.1 is shown a list of time functions $f(t)$ and their corresponding transforms $F(s)$. In solving problems by Laplace transforms, the term s acts as a dummy operator, and thus there is no need to know the range of values over which $F(s)$ exists.

The listing of transform pairs [i.e., corresponding values of $F(s)$ and $f(t)$] given in Table 5.1 is adequate for the solution of most problems which arise in control engineering. The derivation of most of these transform pairs is now explained.

Pulse function A pulse function is shown in Fig. 5.3. The height of the function is h and the width t_o, so that its area is ht_o. The Laplace transform is obtained by applying Eq. (5.25), in which $f(t) = h$ for $0 < t < t_o$ and $f(t) = 0$ for $t > t_o$.

$$F(s) = \int_0^{t_o} he^{-st}\, dt = h\left(\frac{-e^{-st}}{s}\right)_0^{t_o} = \frac{h}{s}(1 - e^{-t_o s}) \qquad (5.27)$$

Table 5.1 Laplace transform pairs

$f(t)$	$F(s)$	$f(t)$	$F(s)$
$u_1(t)$	1	$t^n e^{at}$	$\dfrac{n!}{(s-a)^{n+1}}$
$u(t)$	$\dfrac{1}{s}$	$\sin \omega t$	$\dfrac{\omega}{s^2 + \omega^2}$
t	$\dfrac{1}{s^2}$	$\cos \omega t$	$\dfrac{s}{s^2 + \omega^2}$
e^{at}	$\dfrac{1}{s-a}$	$e^{at} \sin \omega t$	$\dfrac{\omega}{(s-a)^2 + \omega^2}$
t^n	$\dfrac{n!}{s^{n+1}}$	$e^{at} \cos \omega t$	$\dfrac{s-a}{(s-a)^2 + \omega^2}$

Figure 5.3 Pulse function.

Impulse function A special case of a pulse function is an impulse function. If the height is designated as $h = k/t_o$, it follows that the area is always equal to k. Now, as the width t_o approaches zero, the height becomes infinite but the area remains equal to k. This limiting case of a pulse function is called an impulse. The symbol $ku_1(t)$ represents an impulse function whose area is k. Substitution of $h = k/t_o$ into Eq. (5.27) and taking the limit as t_o approaches zero gives the following transform for an impulse:

$$F(s) = \mathscr{L}[ku_1(t)] = \lim_{t_o \to 0} \left[\frac{k}{t_o s} (1 - e^{-t_o s}) \right] = \frac{0}{0}$$

Application of L'hopital rule for evaluating the preceding indeterminant gives

$$F(s) = \lim_{t_o \to 0} \frac{(d/dt_o)[k(1 - e^{-t_o s})]}{(d/dt_o)(t_o s)} = \lim_{t_o \to 0} \frac{kse^{-t_o s}}{s} = k \qquad (5.28)$$

The transform of an impulse function is thus seen to be equal to the area of the function. The impulse function whose area is unity, $u_1(t)$, is called a unit impulse. Much information about the transient behavior of a system may be obtained by determining the manner in which a system returns to its equilibrium state after the system has been excited by a momentary disturbance such as a pulse or an impulse.

Exponentially decaying function The function $f(t) = e^{-at}$ is shown in Fig. 5.4. Applying Eq. (5.25) gives the transform of this exponentially decaying function:

$$F(s) = \mathscr{L}(e^{-at}) = \int_0^\infty e^{-(a+s)t}\, dt = -\frac{e^{-(a+s)t}}{s+a}\bigg|_0^\infty = \frac{1}{s+a} \qquad (5.29)$$

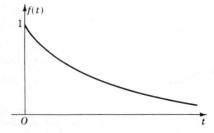

Figure 5.4 Exponentially decaying function.

Sinusoidal function A sinusoidal function is shown in Fig. 5.5. The equation for this sinusoidal is $f(t) = \sin \omega t$. Thus

$$F(s) = \mathcal{L}(\sin \omega t) = \int_0^\infty e^{-st} \sin \omega t \; dt \tag{5.30}$$

The preceding integration is simplified by making use of Euler's equations:

$$e^{j\theta} = \cos \theta + j \sin \theta$$
$$e^{-j\theta} = \cos \theta - j \sin \theta \tag{5.31}$$

Adding Euler's equations and dividing by 2 yields

$$\cos \theta = \frac{e^{j\theta} + e^{-j\theta}}{2} \tag{5.32}$$

Subtracting the second of Euler's equations from the first and dividing by $2j$ yields

$$\sin \theta = \frac{e^{j\theta} - e^{-j\theta}}{2j} \tag{5.33}$$

Substitution of the exponential form for $\sin \theta$ as given by Eq. (5.33) into Eq. (5.30) gives

$$F(s) = \int_0^\infty \frac{e^{j\omega t} - e^{-j\omega t}}{2j} e^{-st} \; dt$$

$$= \frac{1}{2j}\left[\frac{1}{s - j\omega} - \frac{1}{s + j\omega}\right] = \frac{\omega}{s^2 + \omega^2} \tag{5.34}$$

The response of a system to a sinusoidal forcing function forms the basis for appraising the performance of systems by frequency-response techniques, as discussed in Chaps. 11 and 12.

The validity of Euler's equations is proved by expanding $e^{j\theta}$, $\cos \theta$, and $\sin \theta$ by use of Maclaurin's series. Thus,

$$e^{j\theta} = 1 + j\theta + \frac{(j\theta)^2}{2!} + \frac{(j\theta)^3}{3!} + \frac{(j\theta)^4}{4!} + \frac{(j\theta)^5}{5!} + \cdots$$

$$= \left(1 - \frac{\theta^2}{2!} + \frac{\theta^4}{4!} - \cdots\right) + j\left(\theta - \frac{\theta^3}{3!} + \frac{\theta^5}{5!} - \cdots\right)$$

$$\cos \theta = 1 - \frac{\theta^2}{2!} + \frac{\theta^4}{4!} - \cdots$$

$$\sin \theta = \theta - \frac{\theta^3}{3!} + \frac{\theta^5}{5!} - \cdots$$

The results of Eq. (5.31) follow directly from the preceding expansions.

Figure 5.5 Sinusoidal function.

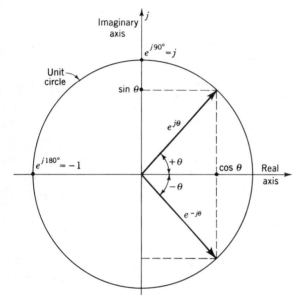

Figure 5.6 Physical significance of unit vector, $e^{j\theta}$.

Because the vector $e^{j\theta}$ is the vector sum of $\cos \theta + j \sin \theta$, the magnitude of $e^{j\theta}$ is always $\cos^2 \theta + \sin^2 \theta = 1$. The physical significance of the unit vector $e^{j\theta}$ is shown in Fig. 5.6. Here it is to be noticed that $e^{j\theta}$ is a unit vector which is rotated counterclockwise an angle θ from the real axis. For θ equal to 90°, it is seen from the unit circle that $e^{j90°} = j$. Squaring the preceding unit vector shows that $j^2 = e^{j90°} e^{j90°} = e^{j180°}$. From the unit circle, this vector is seen to be equal to -1. Thus, $j^2 = -1$ and $j = \sqrt{-1}$.

5.3 TRANSFORM PROPERTIES

Significant characteristics and theorems pertaining to Laplace transformations are developed in this section. A list of Laplace transform properties is given in Table 5.2. With the aid of these properties, the usefulness of the Laplace transformation method is greatly extended. In addition, these theorems help us to obtain a much better understanding of the method.

Table 5.2 Laplace transform properties

Time function	Laplace transform
$kf(t)$	$kF(s)$
$f_1(t) \pm f_2(t)$	$F_1(s) \pm F_2(s)$
$f'(t)$	$sF(s) - f(0)$
$f''(t)$	$s^2 F(s) - sf(0) - f'(0)$
$f^n(t)$	$s^n F(s) - s^{n-1}f(0) - \cdots - f^{n-1}(0)$
$f^{(-1)}(t)$	$\dfrac{F(s)}{s} + \dfrac{f^{(-1)}(0)}{s}$
$f^{(-n)}(t)$	$\dfrac{F(s)}{s^n} + \dfrac{f^{(-1)}(0)}{s^n} + \cdots + \dfrac{f^{(-n)}(0)}{s}$
$f(at)$	$\dfrac{1}{a} F\left(\dfrac{s}{a}\right)$
$e^{at}f(t)$	$F(s-a)$
$t^n f(t)$	$(-1)^n \dfrac{d^n}{ds^n} F(s)$
$f(\tau) = f(t - t_0)$	$e^{-t_0 s} F(s)$
$\displaystyle \int_0^t f(\lambda)g(t - \lambda)\, d\lambda$	$F(s)G(s)$

Real Translation

The function $f(\tau)$ shown in Fig. 5.7 begins at time $t = t_0$ rather than at $t = 0$. Note that $\tau = 0$ when $t = t_0$. Because $\tau = t - t_0$, then $f(\tau) = f(t - t_0)$. From Eq. (5.25), the Laplace transform for $f(\tau)$ is

$$\mathscr{L}[f(\tau)] = \int_0^\infty f(\tau)e^{-st}\, dt = \int_{t_0}^\infty f(\tau)e^{-st}\, dt \qquad (5.35)$$

where $f(\tau) = f(t - t_0)$ is zero for $t < t_0$. In making the change of variable from t to τ, note that $t = t_0 + \tau$, $dt = d\tau$, and the lower limit of integration $t = t_0$ corresponds to $\tau = 0$. Thus, the preceding integral becomes

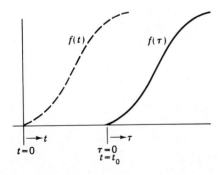

$f(t)$ $f(\tau)$

$t = 0$

$\tau = 0$
$t = t_0$

Figure 5.7 Delayed time function.

$$\mathcal{L}[f(\tau)] = \int_0^\infty f(\tau)e^{-s(t_o+\tau)} \, d\tau = e^{-t_o s} \int_0^\infty f(\tau)e^{-s\tau} \, d\tau \qquad (5.36)$$

The right side may be written as

$$\int_0^\infty f(\tau)e^{-s\tau} \, d\tau = \int_0^\infty f(t)e^{-st} \, dt = F(s)$$

where $F(s)$ would be the transform of the function if it were not delayed. Substitution of the preceding result into Eq. (5.36) gives

$$\mathcal{L}[f(\tau)] = \mathcal{L}[f(t - t_o)] = e^{-t_o s}F(s) \qquad (5.37)$$

Thus, the transform of a function which is delayed a time t_o is obtained by multiplying the transform of the function if it were not delayed [that is, $F(s)$] by $e^{-t_o s}$.

An application of Eq. (5.37) is immediately evident by noting that the pulse function shown in Fig. 5.3 may be regarded as a step function of height h which begins at $t = 0$ minus a step function of height h which begins at $t = t_o$. The transform for the first step function is h/s, while that for the delayed step function is $(h/s)e^{-t_o s}$. Subtracting the delayed step function from the first yields the following transform for the pulse:

$$\frac{h}{s}(1 - e^{-t_o s})$$

This is the result given by Eq. (5.27).

Transform of a Derivative

Any linear differential equation will of course have derivatives of various orders. The order of each derivative is the same as the exponent of the operator D in the operational representation of the term. The general expression for transforming derivatives is obtained as follows. The form for integration by parts is

$$\int u \, dv = uv - \int v \, du$$

Let $u = f(t)$ and $v = -e^{-st}/s$; then $du = (d/dt)[f(t)] \, dt$ and $dv = e^{-st} \, dt$. Integration between the limits of zero and infinity gives

$$\int_0^\infty f(t)e^{-st} \, dt = -f(t)\frac{e^{-st}}{s}\Big|_0^\infty + \frac{1}{s}\int_0^\infty \frac{d}{dt} f(t)e^{-st} \, dt \qquad (5.38)$$

The left side is $F(s)$. The upper limit of the first term on the right side is zero. Thus,

$$F(s) = \frac{f(0)}{s} + \frac{1}{s}\int_0^\infty \frac{d[f(t)]}{dt} e^{-st} \, dt$$

$$= \frac{f(0)}{s} + \frac{1}{s} \mathcal{L}\left[\frac{df(t)}{dt}\right] \qquad (5.39)$$

where $f(0)$ is the initial value of $f(t)$. Solving Eq. (5.39) for the transform of the derivative gives

$$\mathscr{L}\left[\frac{df(t)}{dt}\right] = \mathscr{L}[f'(t)] = sF(s) - f(0) \qquad (5.40)$$

By the extension of the preceding techniques to higher-order derivatives, the following equations for transforms of higher-order derivatives are obtained:

$$\mathscr{L}[f''(t)] = s^2 F(s) - sf(0) - f'(0)$$

$$\mathscr{L}[f'''(t)] = s^3 F(s) - s^2 f(0) - sf'(0) - f''(0) \qquad (5.41)$$

$$\mathscr{L}[f^n(t)] = s^n F(s) - s^{n-1}f(0) - \cdots - f^{n-1}(0)$$

where $f'(t) = df(t)/dt, f''(t) = d^2 f(t)/dt^2, \ldots, f^n(t) = d^n f(t)/dt^n$, and $f'(0)$ is the initial value of $f'(t)$, etc. The initial conditions $f(0), f'(0), f''(0), \ldots$ associated with a particular differential equation must, of course, be given.

An interesting result is obtained when Eq. (5.40) is applied to the function shown in Fig. 5.8a. The initial value of this function is $f(0)$. Because of the step change of height h_c, the value of the function for $t > 0$ is $h = h_c + f(0)$. The transform of this function is

$$F(s) = \mathscr{L}[f(t)] = \mathscr{L}\{[h_c + f(0)]u(t)\} = \frac{h_c + f(0)}{s}$$

Application of Eq. (5.40) to obtain the transform of the derivative of this function gives

$$\mathscr{L}\left[\frac{d}{dt}f(t)\right] = sF(s) - f(0) = \frac{s[h_c + f(0)]}{s} - f(0) = h_c \qquad (5.42)$$

The preceding transform is the same as that obtained for an impulse function. Thus, the derivative of a step change is an impulse function whose area is equal to the change in height h_c of the step.

This result may also be verified geometrically. In Fig. 5.8b is shown the function $f(t)$ whose initial value is $f(0)$. The function increases linearly from its initial value $f(0)$ to the value $h = h_c + f(0)$ at time t_o. For $t > t_o$, the function maintains the value h. Note that as t_o approaches zero, this function approaches the step change shown in Fig. 5.8a. The derivative $f'(t)$ is shown in Fig. 5.8c. For $0 < t < t_o$, the slope of $f(t)$ in Fig. 5.8b is h_c/t_o. For $t > t_o$, the slope is zero. The derivative $f'(t)$ is a pulse function of height h_c/t_o and width t_o. As t_o approaches zero, the derivative approaches an impulse of area h_c, where h_c is equal to the change in height of the step.

When a function $f(t)$ has a derivative which is a constant times the original function, then the transform $F(s)$ may be obtained by application of the equation for the transform of the derivative of the function. For example, for $f(t) = \sin \omega t$, it follows that $f'(t) = \omega \cos \omega t$ and $f''(t) = -\omega^2 \sin \omega t = -\omega^2 f(t)$. Thus, application of the general expression for the transform of the second derivative gives

$$\mathscr{L}[f''(t)] = \mathscr{L}[-\omega^2 f(t)] = -\omega^2 F(s) = s^2 F(s) - sf(0) - f'(0)$$

(a)

(b)

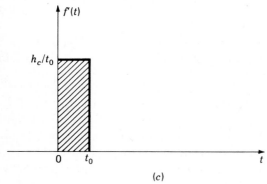

(c)

Figure 5.8 Derivative of step change h_c is impulse of area h_c.

Substitution of $f(0) = [\sin \omega t]_{t=0} = 0$ and $f'(0) = [\omega \cos \omega t]_{t=0} = \omega$ into the preceding and solving for $F(s) = \mathscr{L}[f(t)] = \mathscr{L}[\sin \omega t]$ gives

$$F(s) = \mathscr{L}[\sin \omega t] = \frac{\omega}{s^2 + \omega^2}$$

This verifies the result previously established in Eq. (5.34).

Multiplication by t

The Laplace transform of $tf(t)$ is

$$\mathscr{L}[tf(t)] = -\frac{d}{ds} F(s) \tag{5.43}$$

To verify this theorem, note that

$$-\frac{d}{ds} F(s) = -\frac{d}{ds} \int_0^\infty f(t)e^{-st}\, dt = \int_0^\infty [tf(t)]e^{-st}\, dt$$

The right side is the Laplace transform of the term in brackets, $[tf(t)]$. To illustrate this theorem, consider the unit step function $f(t) = u(t)$, for which $F(s) = 1/s = s^{-1}$. The Laplace transform of $tf(t) = tu(t) = t$ is

$$\mathscr{L}[t] = -\frac{d}{ds} s^{-1} = \frac{1}{s^2}$$

For $f(t) = t$ and $F(s) = 1/s^2$, the Laplace transform of $tf(t) = t^2$ is

$$\mathscr{L}[t^2] = -\frac{d}{ds} (s^{-2}) = \frac{2}{s^3}$$

In general, it follows that

$$\mathscr{L}[t^n] = \frac{n!}{s^{n+1}}$$

Multiplication by e^{at}

The Laplace transform of $e^{at}f(t)$ is

$$\mathscr{L}[e^{at}f(t)] = F(s - a) \tag{5.44}$$

This theorem is verified directly by replacing s by $(s - a)$ in the general expression for $F(s)$. That is,

$$F(s - a) = \int_0^\infty f(t)e^{-(s-a)t}\, dt = \int_0^\infty [e^{at}f(t)]e^{-st}\, dt$$

Thus, $F(s - a)$ is the Laplace transform of the term in brackets, $[e^{at}f(t)]$. To illustrate the application of this theorem, let $f(t) = t^n$. For $F(s) = n!/s^{n+1}$, it follows that

$$\mathscr{L}[e^{at}t^n] = \frac{n!}{(s - a)^{n+1}}$$

As another application, consider the function $f(t) = \sin \omega t$. For $F(s) = \omega/(s^2 + \omega^2)$, it follows that

$$\mathscr{L}[e^{at} \sin \omega t] = \frac{\omega}{(s - a)^2 + \omega^2}$$

Change of Scale

The Laplace transform of $f(at)$ is

$$\mathscr{L}[f(at)] = \frac{1}{a} F\left(\frac{s}{a}\right) \tag{5.45}$$

To verify this theorem, first note that

$$\frac{1}{a} F\left(\frac{s}{a}\right) = \frac{1}{a} \int_0^\infty f(t) e^{-(s/a)t} \, dt$$

Next, replacing t by at, in which case dt is replaced by $a \, dt$, gives

$$\frac{1}{a} F\left(\frac{s}{a}\right) = \int_0^\infty [f(at)] e^{-st} \, dt$$

The right side is the transform of $f(at)$. To illustrate the application of this theorem, consider the function $f(t) = \sin t$, for which $F(s) = 1/(s^2 + 1)$. Thus,

$$\mathscr{L}[\sin \omega t] = \frac{1}{\omega} F\left(\frac{s}{\omega}\right) = \frac{1}{\omega} \frac{1}{(s/\omega)^2 + 1} = \frac{\omega}{s^2 + \omega^2}$$

Transform of an Integral

In using the Laplace transform method to solve integrodifferential equations, it is necessary to obtain the transform of an integral. The procedure for obtaining the equation for the transform of an integral is similar to that for a differential. In the general expression for integration by parts let $u = \int f(t) \, dt$ and $v = -e^{-st}/s$; then $du = f(t)$ and $dv = e^{-st} \, dt$. This yields

$$\int_0^\infty \left[\int f(t) \, dt \right] e^{-st} \, dt = -\frac{e^{-st}}{s} \int f(t) \, dt \bigg|_0^\infty + \frac{1}{s} \int_0^\infty f(t) e^{-st} \, dt$$

The left side is the Laplace transform of the integral of $f(t)$. The value of the first term on the right side is zero at the upper limit. Thus,

$$\mathscr{L}\left[\int f(t) \, dt \right] = \frac{\int f(t) \, dt}{s} \bigg|_{t=0} + \frac{1}{s} F(s)$$

or

$$\mathscr{L}[f^{(-1)}(t)] = \frac{F(s)}{s} + \frac{f^{(-1)}(0)}{s} \tag{5.46}$$

where $f^{(-1)}(t) = \int f(t) \, dt$ and $f^{(-1)}(0) = \int f(t) \, dt \big|_{t=0}$ is the initial value of the integral.

By the repeated application of this procedure, it is found that

$$\mathscr{L}[f^{(-2)}(t)] = \frac{F(s)}{s^2} + \frac{f^{(-1)}(0)}{s^2} + \frac{f^{(-2)}(0)}{s}$$

$$\mathscr{L}[f^{(-n)}(t)] = \frac{F(s)}{s^n} + \frac{f^{(-1)}(0)}{s^n} + \frac{f^{(-2)}(0)}{s^{n-1}} + \cdots + \frac{f^{(-n)}(0)}{s} \tag{5.47}$$

where $f^{(-n)}(t) = \int \cdots \int f(t) \, dt^n$.

Linearity Theorem

The linearity characteristic of Laplace transformations is a very useful property. If k is a constant or a variable which is independent of both t and s, then it follows from Eq. (5.25) that

$$\mathscr{L}[kf(t)] = k\mathscr{L}[f(t)] = kF(s) \qquad (5.48)$$

Another important linearity property is

$$\mathscr{L}[f_1(t) \pm f_2(t)] = F_1(s) \pm F_2(s) \qquad (5.49)$$

Final-Value Theorem

This theorem enables one to obtain the value $f(t)$ of a time function at $t = \infty$ directly from the Laplace transform $F(s)$. This is in effect the same type of information which is obtained from a steady-state analysis.

To develop the final-value theorem, first write the equation for the transform of a derivative in the form

$$\int_0^\infty f'(t)e^{-st}\, dt = sF(s) - f(0)$$

As s approaches zero, then $e^{-st} \approx 1$. Thus,

$$\int_0^\infty f'(t)\, dt = \lim_{s \to 0} sF(s) - f(0)$$

The left side may be written in the form

$$\int_0^\infty f'(t)\, dt = f(t)\Big|_0^\infty = \lim_{t \to \infty} [f(t)] - f(0)$$

The desired final-value theorem is obtained by equating the right sides of the preceding expressions. Thus,

$$\lim_{t \to \infty} [f(t)] = \lim_{s \to 0} sF(s) \qquad (5.50)$$

For functions such as $\sin t$, t^n, or e^{at} where $a > 0$, $\int_0^\infty f'(t)\, dt$ does not exist. When the poles of $sF(s)$ are located on or to the right of the imaginary axis, this integral does not exist. The final-value theorem is not applicable in such cases.

Initial-Value Theorem

With the aid of the initial-value theorem, the value $f(t)$ of a time function at $t = (0+)$ may be computed directly from the transform $F(s)$ for the function. It is to be noted that $f(0+)$ is not the initial value $f(0)$, but rather the value of the function at a time slightly greater than zero.

The derivation of the initial-value theorem follows. For $t \approx 0$, then $e^{-st} \approx 1$; thus the equation for the transform of a derivative can be written as

$$\mathcal{L}[f'(t)] = \int_0^{0+} f'(t)(1)\, dt + \int_{0+}^{\infty} f'(t)e^{-st}\, dt = sF(s) - f(0)$$

As s approaches infinity, $e^{-st} \approx 0$. Thus, as $s \to \infty$, the second integral vanishes. Hence,

$$f(0+) - f(0) = \lim_{s \to \infty} sF(s) - f(0)$$

or

$$f(0+) = \lim_{s \to \infty} sF(s) \tag{5.51}$$

This is the mathematical formulation of the initial-value theorem. Application of the initial-value theorem to the step function shown in Fig. 5.8a gives

$$f(0+) = [sF(s)]_{s=\infty} = \left.\frac{s[h_c + f(0)]}{s}\right|_{s=\infty} = h_c + f(0) \tag{5.52}$$

Thus, a step change of height h_c is seen to occur at $t = 0$.

Illustrative example 5.3 Use the Laplace transformation method to verify the solution of Illustrative example 5.1. That is,

$$(D^2 + 3D + 2)y(t) = f(t) \tag{5.13}$$

where $f(t) = 4t$ and all initial conditions are zero.

SOLUTION Transforming each term yields, for the transformed equation,

$$[s^2 Y(s) - sy(0) - y'(0)] + 3[sY(s) - y(0)] + 2Y(s) = F(s)$$

For $y(0) = y'(0) = 0$, the transformed equation becomes

$$(s^2 + 3s + 2)Y(s) = F(s)$$

Comparison of this transformed equation with the differential equation shows that when all the initial conditions are zero, the transformed equation is obtained by replacing D, $y(t)$, and $f(t)$ in the differential equation by s, $Y(s)$, and $F(s)$ respectively. The transformed equation is now written in the form

$$Y(s) = \frac{F(s)}{(s+1)(s+2)} = \frac{4}{s^2(s+1)(s+2)}$$

$$= \frac{C_2}{s^2} + \frac{C_1}{s} + \frac{K_1}{s+1} + \frac{K_2}{s+2}$$

where $F(s) = \mathscr{L}[f(t)] = \mathscr{L}[4t] = 4/s^2$. The partial-fraction expansion constants are evaluated as follows:

$$C_2 = \lim_{s \to 0} \left[s^2 \frac{4}{s^2(s + 1)(s + 2)} \right] = \left[\frac{4}{(s + 1)(s + 2)} \right]_{s=0} = 2$$

$$C_1 = \lim_{s \to 0} \left[\frac{d}{ds} \frac{4}{s^2 + 3s + 2} \right] = \left[\frac{-4(2s + 3)}{(s^2 + 3s + 2)^2} \right]_{s=0} = -3$$

$$K_1 = \lim_{s \to -1} \left[(s + 1) \frac{4}{s^2(s + 1)(s + 2)} \right] = \left[\frac{4}{s^2(s + 2)} \right]_{s=-1} = 4$$

$$K_2 = \lim_{s \to -2} \left[(s + 2) \frac{4}{s^2(s + 1)(s + 2)} \right] = \left[\frac{4}{s^2(s + 1)} \right]_{s=-2} = -1$$

The resulting transformed equation is

$$Y(s) = \frac{2}{s^2} - \frac{3}{s} + \frac{4}{s + 1} - \frac{1}{s + 2}$$

Inverting yields

$$y(t) = 4e^{-t} - e^{-2t} + 2t - 3 \tag{5.14}$$

This is the same result as was obtained by classical methods [Eq. (5.14)]. Differentiation was involved in the evaluation of C_1. An alternative technique which eliminates the need for differentiation is to evaluate all constants which do not require differentiation (for example, C_2, K_1, and K_2) and then write $Y(s)$ in terms of the unknown constant. Thus,

$$Y(s) = \frac{4}{s^2(s + 1)(s + 2)} = \frac{2}{s^2} + \frac{C_1}{s} + \frac{4}{s + 1} - \frac{1}{s + 2}$$

This result must hold for any value of s. For $s = 1$, it becomes

$$Y(1) = \frac{4}{(1)(2)(3)} = \frac{2}{1} + \frac{C_1}{1} + \frac{4}{2} - \frac{1}{3}$$

Hence $C_1 = \frac{2}{3} - 2 - 2 + \frac{1}{3} = -3$

Illustrative example 5.4 Use the method of Laplace transforms to verify the solution of Illustrative example 5.2. That is,

$$(D + 1)^2(D + 2)y(t) = (D^3 + 4D^2 + 5D + 2)y(t) = f(t) \tag{5.23}$$

where $f(t) = e^{-t}$ and all the initial conditions are zero.

SOLUTION Transforming each term in the differential equation yields, for the transformed equation,

$$[s^3 Y(s) - s^2 y(0) - sy'(0) - y''(0)] + 4[s^2 Y(s) - sy(0) - y'(0)]$$
$$+ 5[sY(s) - y(0)] + 2Y(s) = F(s)$$

For $y(0) = y'(0) = y''(0) = 0$, it follows that

$$(s^3 + 4s^2 + 5s + 2)Y(s) = F(s)$$

The transformed equation may now be written in the form

$$Y(s) = \frac{F(s)}{(s + 1)^2(s + 2)} = \frac{1}{(s + 1)^3(s + 2)}$$

$$= \frac{C_3}{(s + 1)^3} + \frac{C_2}{(s + 1)^2} + \frac{C_1}{s + 1} + \frac{K_1}{s + 2}$$

where $F(s) = \mathscr{L}[f(t)] = \mathscr{L}[e^{-t}] = 1/(s + 1)$. The partial-fraction constants are

$$C_3 = \lim_{s \to -1} \left[(s + 1)^3 \frac{1}{(s + 1)^3(s + 2)} \right] = \left[\frac{1}{(s + 2)} \right]_{s = -1} = 1$$

$$C_2 = \lim_{s \to -1} \left[\frac{d}{ds} \frac{1}{s + 2} \right] = \left[\frac{-1}{(s + 2)^2} \right]_{s = -1} = -1$$

$$C_1 = \lim_{s \to -1} \left[\frac{1}{2!} \frac{d^2}{ds^2} \frac{1}{s + 2} \right] = \left[\frac{1}{(s + 2)^3} \right]_{s = -1} = 1$$

$$K_1 = \lim_{s \to -2} \left[(s + 2) \frac{1}{(s + 1)^3(s + 2)} \right] = \left[\frac{1}{(s + 1)^3} \right]_{s = -2} = -1$$

The resulting partial-fraction expansion is

$$Y(s) = \frac{1}{(s + 1)^3} - \frac{1}{(s + 1)^2} + \frac{1}{s + 1} - \frac{1}{s + 2}$$

Inverting yields

$$y(t) = 0.5t^2e^{-t} - te^{-t} + e^{-t} - e^{-2t} = (1 - t + 0.5t^2)e^{-t} - e^{-2t} \quad (5.24)$$

This checks the result obtained by classical methods [Eq. (5.24)].

The alternative technique for evaluating C_1 and C_2 without the need for differentiation is to substitute the constants $C_3 = 1$ and $K_1 = -1$, which do not require differentiation, into the expression for $Y(s)$. Thus,

$$Y(s) = \frac{1}{(s + 1)^3(s + 2)} = \frac{1}{(s + 1)^3} + \frac{C_2}{(s + 1)^2} + \frac{C_1}{s + 1} - \frac{1}{s + 2}$$

This expression is valid for any value of s. For $s = 0$ and for $s = 1$, we obtain

$$Y(0) = \tfrac{1}{2} = 1 + C_2 + C_1 - \tfrac{1}{2} \qquad \text{or} \qquad C_1 + C_2 = 0$$

$$Y(1) = \frac{1}{2^3(3)} = \frac{1}{24} = \frac{1}{8} + \frac{C_2}{4} + \frac{C_1}{2} - \frac{1}{3} \qquad \text{or} \qquad 2C_1 + C_2 = 1$$

Solving these equations simultaneously verifies that $C_1 = 1$ and $C_2 = -1$.

Illustrative example 5.5 The block-diagram representation for a proportional control system is shown in Fig. 5.9. For $A = 1$, $K_1 = 1$, $K_2 = 2$, $K_H = 0.5$, $B = -5$, $\tau_1 = \frac{1}{6}$, and $\tau_2 = 1.0$, this system is the same as that discussed in Illustrative example 4.1. Determine the response of this system when all initial conditions are zero for each of the following cases:

(a) v is a step function of constant value v, and $u = 0$.
(b) u is a step function of constant value u, and $v = 0$.

SOLUTION The differential equation of operation is

$$c(t) = \frac{AK_1K_2v + BK_2(1 + \tau_1 D)u}{(1 + \tau_1 D)(1 + \tau_2 D) + K_1K_2K_H} \tag{5.53}$$

Because the initial conditions are zero, the transformed equation is obtained by replacing D by s, $c(t)$ by $C(s)$, u by $U(s)$, and v by $V(s)$. Thus,

$$
\begin{aligned}
C(s) &= \frac{AK_1K_2V(s) + BK_2(1 + \tau_1 s)U(s)}{(1 + \tau_1 s)(1 + \tau_2 s) + K_1K_2K_H} \\[2mm]
&= \frac{(AK_1K_2/\tau_1\tau_2)V(s) + (BK_2/\tau_1\tau_2)(1 + \tau_1 s)U(s)}{s^2 + (1/\tau_1 + 1/\tau_2)s + (K_1K_2K_H + 1)/\tau_1\tau_2}
\end{aligned}
$$

Substitution of numerical values into this transformed equation gives

$$C(s) = \frac{12V(s) - 10(6 + s)U(s)}{s^2 + 7s + 12} \tag{5.54}$$

(a) For $V(s) = v/s$ and $U(s) = 0$, the preceding becomes

$$C(s) = \frac{12v}{s(s + 3)(s + 4)} = \left[\frac{1}{s} - \frac{4}{s + 3} + \frac{3}{s + 4}\right]v$$

Inverting yields, for the response,

$$c(t) = (1 - 4e^{-3t} + 3e^{-4t})v$$

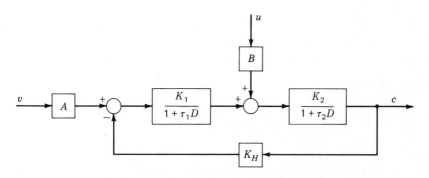

Figure 5.9 Proportional control system.

Note that as time approaches infinity, $c = v$. Because $c = v$ for steady-state operation the coefficient of the v term in Eq. (4.1) must be unity. This fact is substantiated by Eq. (4.12), which describes the steady-state behavior of this system.

(b) For $V(s) = 0$ and $U(s) = u/s$, the transformed equation is

$$C(s) = \frac{-10(6 + s)u}{s(s + 3)(s + 4)} = \left[-\frac{5}{s} + \frac{10}{s + 3} - \frac{5}{s + 4} \right]u$$

Inverting yields

$$c(t) = (-5 + 10e^{-3t} - 5e^{-4t})u$$

For steady-state operation, $c = -5u$. As is verified by Eq. (4.12), the coefficient of the u term is -5.

The steady-state behavior does not depend on the values of τ_1 and τ_2 but the transient behavior does. For example, if $\tau_1 = \frac{1}{15}$ and $\tau_2 = \frac{1}{2}$, the transformed equation becomes

$$C(s) = \frac{60V(s) - 20(15 + s)U(s)}{(s + 5)(s + 12)} \tag{5.55}$$

For $V(s) = v/s$ and $U(s) = 0$, the response is

$$c(t) = [1 - (12/7)e^{-5t} + (5/7)e^{-12t}]v$$

For $V(s) = 0$ and $U(s) = u/s$, the response is

$$c(t) = [-5 + (40/7)e^{-5t} - (5/7)e^{-12t}]u$$

Illustrative example 5.6 The block-diagram representation for an integral-type speed control system such as that discussed in Sec. 4.4 is shown in Fig. 5.10. The output speed is n_o, the desired (input) speed is n_{in}, and the load torque is u. For $\tau = 0.25$, $K_1 = 1.0$, $K_2 = 0.75$, and all initial conditions zero, determine the response of the system for each of the following cases:

(a) n_{in} is a step function of constant value n_{in}, and $u = 0$.
(b) u is a step function of constant value u, and $n_{in} = 0$.

SOLUTION The differential equation of operation is

$$n_o = \frac{K_1 K_2 n_{in} - K_2 Du}{D(1 + \tau D) + K_1 K_2} = \frac{(K_1 K_2/\tau)n_{in} - (K_2 D/\tau)u}{D^2 + (1/\tau)D + (K_1 K_2/\tau)} \tag{5.56}$$

Figure 5.10 Integral control system.

Because the initial conditions are zero, the transformed equation is obtained by replacing D by s, n_o by $N_o(s)$, n_{in} by $N_{in}(s)$, and u by $U(s)$. Thus,

$$N_o(s) = \frac{(K_1 K_2/\tau)N_{in}(s) + (K_2 s/\tau)U(s)}{s^2 + (1/\tau)s + K_1 K_2/\tau}$$

Substitution of $\tau = 0.25$, $K_1 = 1.0$, and $K_2 = 0.75$ gives

$$N_o(s) = \frac{3N_{in}(s) - 3sU(s)}{s^2 + 4s + 3} \tag{5.57}$$

(a) For $N_{in}(s) = n_{in}/s$ and $U(s) = 0$, the preceding becomes

$$N_o(s) = \frac{3n_{in}}{s(s + 1)(s + 3)} = \left(\frac{1.0}{s} - \frac{1.5}{s + 1} + \frac{0.5}{s + 3} \right) n_{in}$$

Inverting yields for the response

$$n_o(t) = (1.0 - 1.5e^{-t} + 0.5e^{-3t})n_{in}$$

Note that as time approaches infinity, $n_o = n_{in}$. Because $n_o = n_{in}$ for steady-state operation, the coefficient of the v term in Eq. (4.1) must be unity. Checking shows that

$$\frac{AK_{G_1}K_{G_2}}{1 + K_{G_1}K_{G_2}K_H} = \frac{AK_{G_2}}{1/K_{G_1} + K_{G_2}K_H} = \frac{A}{K_H} = 1$$

where $K_{G_1} = K_1/D|_{D=0} = \infty$, $A = 1$, and $K_H = 1$.

(b) For $N_{in}(s) = 0$ and $U(s) = u/s$, the transformed equation is

$$N_o(s) = \frac{-3u}{(s + 1)(s + 3)} = \left(\frac{1.5}{s + 3} - \frac{1.5}{s + 1} \right) u$$

Inverting yields

$$n_o(t) = 1.5(e^{-3t} - e^{-t})u$$

Note that as time approaches infinity, $n_o = 0$. Because this is an integral-type control system, there is no steady-state error due to the disturbance u. For this case the coefficient of the u term in Eq. (4.1) must be zero. Checking shows that

$$\frac{BK_{G_2}}{1 + K_{G_1}K_{G_2}K_H} = \frac{(-1)(0.75)}{1 + (\infty)(0.75)(1)} = 0$$

where $B = -1$, $K_H = 1$, $K_{G_2} = [K_2/(1 + \tau D)]_{D=0} = K_2 = 0.75$, and $K_{G_1} = (K_1/D)_{D=0} = \infty$.

Principle of Superposition

The superposition principle is of fundamental importance. This principle states that if there is more than one excitation, the solution is the sum of the responses due to each excitation taken separately. Thus, if the step change in part (a) and the step change in part (*b*) of Illustrative examples 5.5 and 5.6 were to occur at the same time, the total response would be the sum of the solutions for part (*a*) and part (*b*). It also follows from the principle of superposition that if the excitation is the sum of functions such as a step function plus an impulse, then the response is the sum of the responses due to each taken separately.

5.4 INITIAL CONDITIONS

The initial state of a system is specified by the initial conditions. Because initial conditions are the state of the system as *t* approaches zero from a negative direction, to be more precise these initial condition terms should be designated $f(0-)$, $f'(0-)$, $f''(0-)$, etc. However, for the sake of simplicity, initial conditions are designated as $f(0), f'(0), f''(0)$, etc., throughout this text.

The initial value at time $t = 0-$ for the function shown in Fig. 5.11*a* is indicated by $f(0)$. For $t > 0$, the excitation is he^{-t}. For $t = 0+$, the value of the function is h. Thus, a discontinuity or step change $h - f(0)$ occurs at the origin. The Laplace transform for $f(t)$ is

$$F(s) = \mathscr{L}[f(t)] = \int_0^\infty he^{-t}e^{-st}\, dt = h\int_0^\infty e^{-(s+1)t}\, dt = \frac{h}{s+1}$$

The inverse is

$$f(t) = \mathscr{L}^{-1}[F(s)] = he^{-t} \qquad t > 0$$

The symbol $\mathscr{L}^{-1}$ is used to indicate the inverse Laplace transform. Note that the inverse Laplace transform contains no information about the initial condition. Because the inverse yields the response for $t > 0$, the Laplace transform method is capable of solving problems in which there is a discontinuity at the origin, i.e., where $f(0-) \neq f(0+)$.

The derivative $f'(t)$ shown in Fig. 5.11*b* consists of the impulse at the origin whose area $[h - f(0)]$ is equal to the step change in $f(t)$ and the function $(d/dt)(he^{-t}) = -he^{-t}$. The Laplace transform for $f'(t)$ is

$$\mathscr{L}[f'(t)] = [h - f(0)] - \frac{h}{s+1}$$

This result may be verified by application of the formula for the transform of a derivative. That is,

$$\mathscr{L}[f'(t)] = sF(s) - f(0) = s\,\frac{h}{s+1} - f(0) = \left[\frac{sh}{s+1} - h\right] + [h - f(0)]$$

$$= [h - f(0)] - \frac{h}{s+1}$$

Note that the initial condition at time $t = 0-$ must be substituted into the equation for the transform of a derivative.

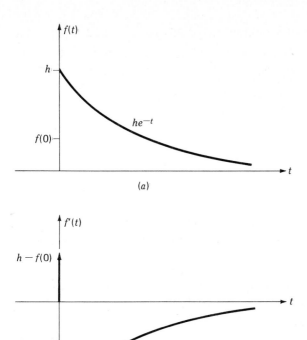

Figure 5.11 (*a*) Function with a discontinuity and (*b*) its derivative.

In taking the Laplace transform of a differential equation, initial-condition terms arise in transforming derivative terms. The initial conditions at time $t = 0-$ are substituted into the resulting transformed equation. Inverting this transformed equation yields the response for $t > 0$. Values at time $t = 0+$ are obtained by letting $t = 0+$ in the response equation. Because it is not necessary that the $0-$ values equal the $0+$ values, the transform method can be used for solving differential equations in which there are discontinuities at the origin. Classical methods are not applicable for such discontinuities. The following examples will help clarify this situation. For a more rigorous presentation of this distinction, see Ref. 1.

Illustrative example 5.7 Use the Laplace transform method to determine the solution of the following differential equation:

$$\frac{dy}{dt} + y = u_1(t) \tag{5.58}$$

The initial condition is $y(0) = 0$.

SOLUTION The transform of the differential equation is

$$[sY(s) - y(0)] + Y(s) = 1$$

For $y(0) = 0$, the transformed equation becomes

$$Y(s) = \frac{1}{s + 1}$$

Inverting yields the response

$$y(t) = e^{-t} \qquad t > 0 \tag{5.59}$$

The value at time $t = 0+$ is

$$y(0+) = e^{-(0+)} = 1$$

This result may also be obtained by application of the initial-value theorem,

$$y(0+) = \lim_{s \to \infty} sY(s) = \left. \frac{s}{s + 1} \right|_{s = \infty} = \left. \frac{1}{1 + 1/s} \right|_{s = \infty} = 1$$

As illustrated in Fig. 5.12, the response function $y(t)$ has a discontinuity at the origin. The initial value $y(0) = y(0-) = 0$ is indicated by the $\times$ to the left of the origin. The value of $y(0+)$ is 1.0. To solve such problems by classical methods, it is necessary to use independent methods to determine the state of the system at time $t = 0+$. Because the system is continuous for $t > 0$, the solution may then be found by classical procedures. To determine the state of the system at $t = 0+$, each term of the differential equation is integrated from $t = 0-$ to $t = 0+$. Thus,

$$\int_{0-}^{0+} \frac{dy}{dt} \, dt + \int_{0-}^{0+} y \, dt = \int_{0-}^{0+} u_1(t) \, dt$$

The value of the first term is $y(0+) - y(0-)$. Because y is a bounded function, the integral from $0-$ to $0+$ vanishes. The right side is the area of the impulse at the origin, which is unity. Thus,

$$y(0+) - y(0-) = 1 \qquad \text{or} \qquad y(0+) = y(0-) + 1 = 1$$

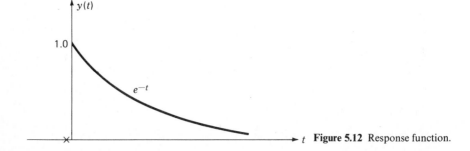

Figure 5.12 Response function.

For $t > 0$, the excitation $u_1(t)$ is zero. Thus, for $t > 0$ the differential equation is

$$\frac{dy}{dt} + y = 0 \qquad y(0+) = 1$$

Because $f(t) = 0$, the particular solution is zero. The general solution consists of the complementary solution only:

$$y(t) = k_1 e^{-t}$$

The constant k_1 is evaluated from the boundary condition that $y(0+) = 1$ when $t = 0+$. Thus,

$$1 = k_1 e^{-(0+)} = k_1$$

The resulting response is

$$y(t) = e^{-t} \qquad t > 0$$

This verifies the solution obtained by transform methods [Eq. (5.59)].

Illustrative example 5.8 Determine the solution of the differential equation:

$$y(t) = \frac{D + 4}{(D + 2)(D + 3)} f(t) \tag{5.60}$$

The forcing function is $f(t) = 2e^{-t}$ and all the initial conditions are zero.

SOLUTION Because all initial conditions are zero, the substitution of s for D, $Y(s)$ for $y(t)$, and $F(s)$ for $f(t)$ yields the transformed equation,

$$Y(s) = \frac{s + 4}{(s + 2)(s + 3)} F(s) = \frac{2(s + 4)}{(s + 1)(s + 2)(s + 3)}$$

$$= \frac{K_1}{s + 1} + \frac{K_2}{s + 2} + \frac{K_3}{s + 3}$$

where $F(s) = \mathscr{L}(2e^{-t}) = 2/(s + 1)$. The partial-fraction–expansion constants are $K_1 = 3$, $K_2 = -4$, and $K_3 = 1$. Thus,

$$Y(s) = \frac{3}{s + 1} - \frac{4}{s + 2} + \frac{1}{s + 3}$$

Inverting yields, for the response,

$$y(t) = 3e^{-t} - 4e^{-2t} + e^{-3t} \tag{5.61}$$

The derivative is

$$y'(t) = -3e^{-t} + 8e^{-2t} - 3e^{-3t}$$

The substitution of $t = 0+$ into the preceding expressions shows that $y(0+) = 0$ and $y'(0+) = 2$. These values may also be verified by application of the initial-value theorem:

$$y(0+) = \lim_{s \to \infty} s\mathscr{L}[y(t)] = sY(s)\Big|_{s=\infty} = \frac{2s(s+4)}{(s+1)(s+2)(s+3)}\Big|_{s=\infty}$$

$$= \frac{2(1/s + 4/s^2)}{(1 + 1/s)(1 + 2/s)(1 + 3/s)}\Big|_{s=\infty} = 0$$

$$y'(0+) = \lim_{s \to \infty} s\mathscr{L}[y'(t)] = s[sY(s) - y(0)]_{s=\infty} = s^2 Y(s)\Big|_{s=\infty} = 2$$

To effect a classical solution, it is first necessary to use independent means to determine the state of the system at $t = 0+$. The differential equation is written in the form

$$y''(t) + 5y'(t) + 6y(t) = f'(t) + 4f(t)$$

Integrating each term twice and then evaluating between the limits $(0-)$ to $(0+)$ shows that $y(0+) = y(0-) = 0$. Next, integrating each term once and then evaluating between the limits $(0-)$ to $(0+)$ shows that $y'(0+) = y'(0-) + f(0+) - f(0-) = f(0+) = 2$. This second-order differential equation is now expressed as the sum of two first-order differential equations. That is,

$$y(t) = \frac{D+4}{(D+2)(D+3)} f(t) = \left(\frac{2}{D+2} - \frac{1}{D+3}\right) f(t)$$

For $r_1 = -2$ and $r_2 = -3$, the complementary solution is

$$y_c = k_1 e^{-2t} + k_2 e^{-3t}$$

For $K_1 = 2$ and $K_2 = -1$, the particular solution is

$$y_p = 2e^{-2t} \int e^{2t}(2e^{-t}) \, dt - e^{-3t} \int e^{3t}(2e^{-t}) \, dt$$

$$= 2e^{-2t}(2e^t) - e^{-3t}(e^{2t}) = 4e^{-t} - e^{-t} = 3e^{-t}$$

Adding the complementary and the particular solutions yields, for the general solution,

$$y(t) = 3e^{-t} + k_1 e^{-2t} + k_2 e^{-3t}$$

Differentiation gives

$$y'(t) = -3e^{-t} - 2k_1 e^{-2t} - 3k_2 e^{-3t}$$

Evaluation of $y(t)$ and $y'(t)$ at $t = 0+$, in which case $y(0+) = 0$ and $y'(0+) = 2$, gives $k_1 = -4$ and $k_2 = 1$. This checks the result given by Eq. (5.61).

Illustrative example 5.9 Determine the solution of Eq. (5.60) for the case in which $f(t) = 0$, $y(0) = 2$, and $y'(0) = -5$.

SOLUTION For $f(t) = 0$, the differential equation becomes

$$(D^2 + 5D + 6)y(t) = 0 \qquad (5.62)$$

Transforming each term gives

$$[s^2 Y(s) - sy(0) - y'(0)] + 5[sY(s) - y(0)] + 6Y(s) = 0$$

$$(s^2 + 5s + 6)Y(s) = (s + 5)y(0) + y'(0) = 2s + 5$$

$$Y(s) = \frac{2s + 5}{(s + 2)(s + 3)} = \frac{1}{s + 2} + \frac{1}{s + 3}$$

Inverting yields, for the response,

$$y(t) = e^{-2t} + e^{-3t} \qquad t > 0 \qquad (5.63)$$

Integrodifferential Equations

A resistor R and a capacitor C in series are shown in Fig. 5.13. The total voltage drop is the sum of the voltage drop $e_r = Ri$ across the resistor and the voltage drop $e_c = \int i\, dt/C$ across the capacitor:

$$e = Ri + \frac{1}{C} \int i\, dt \qquad (5.64)$$

The Laplace transform for this integrodifferential equation is

$$E(s) = RI(s) + \frac{1}{sC} \left[I(s) + \left(\int i\, dt \right)_{t=0} \right] = \frac{1}{sC}(1 + RCs)I(s) + \frac{e_c(0)}{s}$$

where $e_c(0) = \int i\, dt/C|_{t=0}$ is the initial voltage across the capacitor. Solving for $I(s)$ gives

$$I(s) = \frac{C[sE(s) - e_c(0)]}{1 + RCs} \qquad (5.65)$$

The process of obtaining the transform of the integral term may be avoided by first differentiating Eq. (5.64) and then transforming this result. Thus,

$$\frac{de}{dt} = R\frac{di}{dt} + \frac{1}{C}i$$

Figure 5.13 Series RC circuit.

The corresponding transform is

$$sE(s) - e(0) = R[sI(s) - i(0)] + \frac{I(s)}{C}$$

or $$\frac{1}{C}(1 + RCs)I(s) = sE(s) - [e(0) - Ri(0)] = sE(s) - e_c(0)$$

The voltage drop e_c across the capacitor is the applied voltage e minus the voltage drop Ri across the resistor. Solving for $I(s)$ yields the same result as that given by Eq. (5.65).

5.5 GENERAL PROCEDURES

Let it be desired to determine the solution of the following differential equation:

$$(1 + \tau D)y(t) = Kf(t) \qquad (5.66)$$

For convenience in performing the partial-fraction expansion, it is desirable to have the coefficient of the highest power of D in $L_n(D)$ unity. Thus,

$$\left(D + \frac{1}{\tau}\right)y(t) = \frac{K}{\tau}f(t) \qquad (5.67)$$

Transforming each term of this differential equation and solving for $Y(s)$ gives

$$Y(s) = \frac{(K/\tau)F(s) + y(0)}{s + 1/\tau} \qquad (5.68)$$

The symbol $F(s)$ is the transform of the input. The nature of the expression $F(s)$ depends upon the particular input to the system, such as a step function, exponential, sinusoidal, etc.

All Initial Conditions Zero

Let it be desired to determine the response of this system to a step-function input $f(t)$ when all the initial conditions are zero. A plot of the input $f(t)$ is shown in Fig. 5.14a. The initial value is $f(0) = 0$, and then a step change h_c occurs so that the height of this function is $h = h_c$. The substitution of $F(s) = h_c/s$ and $y(0) = 0$ into Eq. (5.68) gives

$$Y(s) = \frac{Kh_c/\tau}{s(s + 1/\tau)}$$

The partial-fraction expansion is

$$Y(s) = Kh_c\left(\frac{1}{s} - \frac{1}{s + 1/\tau}\right)$$

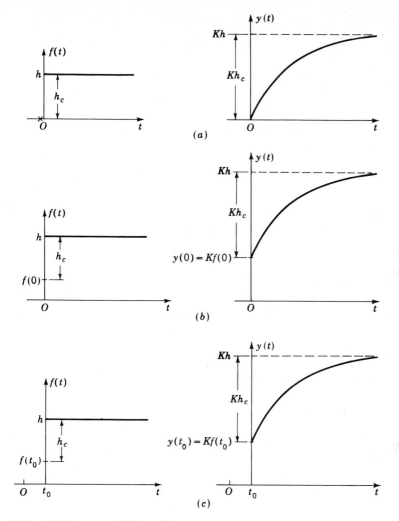

Figure 5.14 Response of a first-order system.

Inverting yields, for the response,

$$y(t) = Kh_c(1 - e^{-t/\tau}) \qquad t > 0 \tag{5.69}$$

A graph of this response is also shown in Fig. 5.14a.

Initially at a Steady-State Operating Condition

A system is initially at a steady-state operating condition if the initial value of all the time derivatives is zero (that is, $D^n y = D^m f = 0$, $n = 1, 2, 3, \ldots$, and $m = 1, 2, 3, \ldots$). When this system is initially at a steady-state operating condition, then from Eq. (5.67) it follows that $y(0) = Kf(0)$.

Let us now determine the response for the case in which the initial value of the input is $f(0)$, as shown in Fig. 5.14b. The height h of the step input is the initial value $f(0)$ plus the change h_c. Thus,

$$F(s) = \frac{h}{s} = \frac{h_c + f(0)}{s} = \frac{h_c + y(0)/K}{s}$$

The substitution of this value of $F(s)$ into Eq. (5.68) gives

$$Y(s) = \frac{Kh_c/\tau}{s(s + 1/\tau)} + \frac{y(0)}{s}$$

The partial-fraction expansion is

$$Y(s) = Kh_c \left(\frac{1}{s} - \frac{1}{s + 1/\tau} \right) + \frac{y(0)}{s} \tag{5.70}$$

Inverting yields

$$y(t) = Kh_c(1 - e^{-t/\tau}) + y(0) \qquad t > 0 \tag{5.71}$$

The first term on the right side is seen to be the response due to the step change h_c alone when all the initial conditions are zero [that is, Eq. (5.69)]. The response shown in Fig. 5.14b is the same as that shown in Fig. 5.14a except that it is raised by the initial value $y(0)$.

In general, it may be shown that for a system which is initially at a steady-state operating condition, the response $y(t)$ is equal to the initial value $y(0)$ plus the response for the case in which all the initial conditions are zero.

Time Delay

To obtain the transform of the derivative of a delayed function, replace $f(\tau)$ by $df(\tau)/d\tau$ in Eq. (5.36). Thus

$$\mathscr{L}\left[\frac{df(\tau)}{d\tau}\right] = e^{-t_o s} \int_0^\infty \frac{df(\tau)}{d\tau} e^{-s\tau} \, d\tau = e^{-t_o s}[sF(s) - f(\tau)]_{\tau=0}$$

$$= [sF(s) - f(t_o)]e^{-t_o s} \tag{5.72}$$

where $f(\tau)|_{\tau=0} = f(t)|_{t=t_o} = f(t_o)$. In general, it follows that

$$\mathscr{L}[f^n(\tau)] = [s^n F(s) - s^{n-1}f(t_o) - \cdots - f^{n-1}(t_o)]e^{-t_o s} \tag{5.73}$$

Thus, to convert a transform to a delayed transform, multiply each term by $e^{-t_o s}$ and evaluate the initial conditions at time t_o, which is the starting point for the delayed function.

Let it be desired to determine the response for the case in which the input of Fig. 5.14b is delayed by time t_o, as shown in Fig. 5.14c. Multiplying each term in Eq. (5.70) by $e^{-t_o s}$ and evaluating the initial conditions at time t_o gives, for the transformed equation,

$$Y(s)e^{-t_o s} = Kh_c \left(\frac{1}{s} - \frac{1}{s + 1/\tau} \right) e^{-t_o s} + \frac{y(t_o)}{s} e^{-t_o s} \qquad (5.74)$$

The operator $e^{-t_o s}$ merely means that t should be replaced by $t - t_o$ in the regular transform. Thus, the inverse of Eq. (5.74) is

$$y(t - t_o) = Kh_c(1 - e^{-(t - t_o)/\tau}) + y(t_o) \qquad t > t_o \qquad (5.75)$$

This is the same as Eq. (5.71) except that t has been replaced by $t - t_o$ and the starting point is at time t_o rather than at $t = 0$. The response shown in Fig. 5.14c is the same as that shown in Fig. 5.14b except that it is delayed by a time t_o. It is not necessary that a system be initially at a steady-state operating condition to effect a time shift. For most problems in which a time shift occurs, it is more convenient to work the problem initially as though there were no time shift, and then replace t by $(t - t_o)$ to obtain the desired result.

The general procedure used to solve differential equations by Laplace transforms may be summarized as follows:

1. Transform each term of the differential equation from the time domain to the s domain, and then solve for $Y(s)$.
2. Substitute the value of the initial conditions and the transform of the input into the expression obtained in step 1.
3. Perform a partial-fraction expansion.
4. Invert each term back to the time domain to obtain the desired time response.

Much simplification in carrying out the algebraic manipulations of a Laplace transform solution is afforded for the following special cases:

1. All initial conditions zero. When all the initial conditions are zero, then the transform of a derivative is $\mathcal{L}[D^n f(t)] = s^n F(s)$. Thus, it follows that the transformed equation is obtained by substituting s for D, $Y(s)$ for $y(t)$, and $F(s)$ for $f(t)$ in the original differential equation.
2. System initially at a steady-state operating condition. When all the initial derivatives are zero, then evaluating Eq. (5.1) at $t = 0$ shows that $by(0) = af(0)$ or $y(0) = (a/b)f(0)$. For this case, the response $y(t)$ is obtained by adding the initial value $y(0)$ to the response for the case in which all the initial conditions are zero (i.e., case 1).
3. Time shift. A time shift is effected by substituting $t - t_o$ for t. It is not necessary that all the initial conditions be zero nor that the system be initially at a steady-state operating condition in order to effect a time shift.

Illustrative example 5.10 Determine the response of the system described by the following differential equation:

$$(D^2 + 7D + 12)y(t) = 6(D + 2)f(t) \qquad (5.76)$$

The forcing function $f(t)$ is a unit step function, and all initial conditions are zero.

SOLUTION Because the initial conditions are zero, substitution of s for D, $Y(s)$ for $y(t)$, and $F(s)$ for $f(t)$ yields

$$Y(s) = \frac{6(s + 2)F(s)}{s^2 + 7s + 12} = \frac{6(s + 2)}{s(s + 3)(s + 4)}$$

where $F(s) = 1/s$ for a unit step function. Performing a partial-fraction expansion gives

$$Y(s) = \frac{1}{s} + \frac{2}{s + 3} - \frac{3}{s + 4}$$

Inverting yields for the time response

$$y(t) = 1 + 2e^{-3t} - 3e^{-4t} \tag{5.77}$$

Let it now be desired to determine the response of this system for the case in which the system is initially at a steady-state operating condition (i.e., initially all derivative terms are zero). The initial value of $f(t)$ is $f(0) = 5$. The unit step function is superimposed upon the constant value $f(0) = 5$, so that for $t > 0$, the input is $f(0) + 1 = 5 + 1 = 6$. The initial steady-state value of y is obtained by letting all derivative terms be zero in the differential equation. Thus, $12y(0) = 6(2)f(0)$ or $y(0) = f(0) = 5$. Adding the initial value of y to the response for the case in which all initial conditions are zero yields, for the response,

$$y(t) = (1 + 2e^{-3t} - 3e^{-4t}) + y(0) = (1 + 2e^{-3t} - 3e^{-4t}) + 5 \tag{5.78}$$

This response may also be verified by transforming the differential equation termwise:

$$[s^2 Y(s) - sy(0) - y'(0)] + 7[s Y(s) - y(0)] + 12Y(s)$$

$$= 6[sF(s) - f(0)] + 12F(s)$$

Substituting $f(0) = y(0)$, $y'(0) = 0$, and $F(s) = [f(0) + 1]/s$ gives

$$(s^2 + 7s + 12)Y(s) = (s + 7)y(0) + 6 + \frac{12[y(0) + 1]}{s}$$

Thus,

$$Y(s) = \frac{6(s + 2) + (s^2 + 7s + 12)y(0)}{s(s^2 + 7s + 12)} = \frac{6(s + 2)}{s(s^2 + 7s + 12)} + \frac{y(0)}{s}$$

Inverting verifies the response obtained directly by application of the vertical shifting theorem [Eq. (5.78)].

Illustrative example 5.11 The system shown in Fig. 5.15 is in a state of equilibrium, with $r = 1$ and $d = 0$. A step-function disturbance $d(t) = u(t)$ is then initiated at time $t = 0$. Determine the response $c(t)$ for $t > 0$.

Figure 5.15 Feedback control system.

SOLUTION The differential equation of operation for this system is

$$(D^2 + 5D + 6)c(t) = 6r(t) + 3Dd(t) \tag{5.79}$$

The expression for steady-state operation at $t = 0$ is obtained by letting $D = 0$ in the differential equation of operation. Thus, $6c(0) = 6r(0)$ or $c(0) = r(0) = 1$. Transforming each term in the differential equation gives

$$[s^2C(s) - sc(0) - c'(0)] + 5[sC(s) - c(0)] + 6C(s) = 6R(s) + 3[sD(s) - d(0)]$$

For $c(0) = r(0) = 1$, $c'(0) = 0$, $d(0) = 0$, $R(s) = 1/s$, and $D(s) = 1/s$, the resulting transformed equation becomes

$$(s^2 + 5s + 6)C(s) = \frac{6}{s} + \frac{3}{s}s + s + 5 = \frac{s^2 + 8s + 6}{s}$$

Hence,

$$C(s) = \frac{s^2 + 8s + 6}{s(s + 2)(s + 3)} = \frac{1}{s} + \frac{3}{s + 2} - \frac{3}{s + 3}$$

Inverting yields, for the response,

$$c(t) = 1 + 3e^{-2t} - 3e^{-3t} \tag{5.80}$$

Because the system is initially at a steady-state operating condition, this result may be obtained directly by adding the initial value $c(0) = 1$ to the response of the system to the step change $d(t) = u(t)$ for the case in which all initial conditions are zero. The transformed equation when all initial conditions are zero is

$$C(s) = \frac{3sD(s)}{s^2 + 5s + 6} = \frac{3}{s^2 + 5s + 6} = \frac{3}{s + 2} - \frac{3}{s + 3}$$

where $D(s) = 1/s$.

Inverting yields, for the response due to the step change only,

$$c(t) = 3e^{-2t} - 3e^{-3t}$$

Thus, adding the initial value $c(0) = 1$ verifies the response previously obtained.

5.6 PIECEWISE CONTINUOUS FUNCTIONS

As is illustrated in Fig. 5.16, a piecewise continuous function is characterized by the fact that the equation for the function changes from interval to interval. For example, in Fig. 5.16a, for $0 < t \le t_o$, $f(t) = at$; and for $t > t_o$, $f(t) = at_o$. For the first interval $0 < t \le t_o$, the input function is inclined at the slope a. Such an inclined straight line is called a ramp function. For $t > t_o$, the input is seen to be a step function. The solution of such problems is effected by starting with the first interval and successively solving for the response in each interval.

Illustrative example 5.12 Determine the solution of the following differential equation:

$$\frac{dy}{dt} + y = f(t) \tag{5.81}$$

where $f(t)$ is the function shown in Fig. 5.17a and all the initial conditions are zero. For $a = 1$ and $t_o = 1$, the function shown in Fig. 5.16a becomes identical to that shown in Fig. 5.17a.

SOLUTION The transform of the differential equation is

$$sY(s) - y(0) + Y(s) = F(s)$$

Solving for $Y(s)$ gives

$$Y(s) = \frac{F(s) + y(0)}{s + 1} \tag{5.82}$$

(a)

(b)

(c)

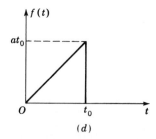

(d)

Figure 5.16 Piecewise continuous functions.

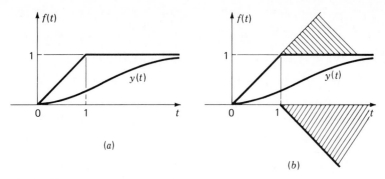

Figure 5.17 A piecewise continuous function.

For the interval $0 < t \leq 1$, the input is the ramp function $f(t) = t$, for which $F(s) = 1/s^2$. Thus, the transformed equation becomes

$$Y(s) = \frac{1/s^2 + y(0)}{s + 1} \tag{5.83}$$

For $y(0) = 0$, it follows that

$$Y(s) = \frac{1}{s^2(s + 1)} = \frac{C_2}{s^2} + \frac{C_1}{s} + \frac{K_1}{s + 1} = \frac{1}{s^2} - \frac{1}{s} + \frac{1}{s + 1}$$

Inverting yields, for the response for the first interval,

$$y(t) = t - 1 + e^{-t} \qquad 0 < t \leq 1 \tag{5.84}$$

For the second interval $(t > 1)$, the input is a unit step function. The initial condition for this second interval is obtained by evaluating the response for the preceding interval at the transition point, $t = 1$. Thus

$$y(1) = [(t - 1) + e^{-t}]_{t=1} = 0.368$$

Because all terms in the differential equation are delayed a time $t_o = 1$, it is more convenient to determine the response for the second interval as though there were no delay and then replace t by $t - t_o = t - 1$ in the final answer. The substitution of $y(0) = 0.368$ and $F(s) = 1/s$ into Eq. (5.82), which is the transform of the differential equation, gives

$$Y(s) = \frac{F(s) + y(0)}{s + 1} = \frac{1/s + 0.368}{s + 1} = \frac{1 + 0.368s}{s(s + 1)}$$

$$= \frac{K_1}{s} + \frac{K_2}{s + 1} = \frac{1}{s} - \frac{0.632}{s + 1}$$

where

$$K_1 = \frac{1 + 0.368s}{s + 1}\bigg|_{s=0} = 1$$

$$K_2 = \frac{1 + 0.368s}{s}\bigg|_{s=-1} = -0.632$$

Inverting gives

$$y(t) = 1 - 0.632e^{-t}$$

Replacing t by $t - t_o = t - 1$ yields, for the response for $t > 1$,

$$y(t - 1) = 1 - 0.632e^{-(t-1)} \qquad t > 1 \qquad (5.85)$$

The resulting response is indicated by the $y(t)$ curve in Fig. 5.17a.

In this method, the initial conditions for a new interval are obtained by evaluating the equation for the preceding interval at the time when the preceding interval ceases and the new one begins.

An alternate method for solving piecewise continuous problems is to regard the input as being the sum of separate functions, as is illustrated in Fig. 5.17b. The sum of the ramp function which begins at $t = 0$ and the equal but opposite ramp function which begins at $t_o = 1$ is seen to yield the function of Fig. 5.17a. It is possible to represent any piecewise continuous function as the sum of other functions which are continuous.

The transform for the first ramp function in Fig. 5.17b is $1/s^2$ and that for the delayed ramp function is $-(1/s^2)e^{-t_o s}$. Thus, the transform $F(s)$ of the input is

$$F(s) = \frac{1}{s^2} - \frac{1}{s^2} e^{-t_o s}$$

Substitution of this result into Eq. (5.82) yields, for the transformed equation,

$$Y(s) = \frac{1/s^2 + y(0)}{s + 1} - \frac{1/s^2}{s + 1} e^{-t_o s} \qquad (5.86)$$

Because of the delaying factor $e^{-t_o s} = e^{-s}$, the second term on the right side should be ignored for $t < t_o = 1$. Thus for $0 < t \leq 1$, the transform is the same as that given by Eq. (5.83). The corresponding response for $0 < t \leq 1$ is given by Eq. (5.84). For $t > 1$, the response is the sum of that due to the second term of Eq. (5.86) and that already obtained for the first term [Eq. (5.84)]. The inverse of $(1/s^2)/(s + 1)$ is given by Eq. (5.84). To take into account the time shift, replace t by $t - t_o = t - 1$. Thus, for $t > 1$ the total response is

$$y(t) = [t - 1 + e^{-t}] - [(t - 1) - 1 + e^{-(t-1)}]$$
$$= 1 + e^{-t} - e^{-(t-1)} = 1 - 0.632e^{-(t-1)} \qquad t > 1 \qquad (5.85)$$

where $e^{-t} = e^{-1}e^{-(t-1)} = 0.368e^{-(t-1)}$.

This verifies the result obtained by the first method [Eq. (5.85)].

An advantage of this second method of solution is that initial conditions appear only in the transformed expression for the first interval. However, in this latter method, the amount of computational effort increases with the number of separate functions required to make up the overall piecewise continuous function. Thus, the choice of the first or second method depends on the particular problem to be solved.

5.7 CONVOLUTION INTEGRAL

This integral may be used to determine the response of a system to any arbitrary input $f(t)$. Consider the transformed equation

$$Y(s) = G(s)F(s)$$

The inverse Laplace transform is

$$y(t) = \mathscr{L}^{-1}[G(s)F(s)] = \mathscr{L}^{-1}\left[G(s)\int_0^\infty f(\lambda)e^{-\lambda s}\,d\lambda\right] = \int_0^\infty f(\lambda)\mathscr{L}^{-1}[G(s)e^{-\lambda s}]\,d\lambda$$

where $F(s) = \int_0^\infty f(\lambda)e^{-\lambda s}\,d\lambda$. Recalling from the real translation theorem that $\mathscr{L}[f(t-t_0)] = F(s)e^{-t_0 s}$ where $f(t-t_0) = 0$ for $t_0 > t$, it follows that

$$\mathscr{L}^{-1}[F(s)e^{-t_0 s}] = \begin{cases} f(t-t_0) & t_0 < t \\ 0 & t_0 > t \end{cases}$$

Similarly,

$$\mathscr{L}^{-1}[G(s)e^{-\lambda s}] = \begin{cases} g(t-\lambda) & \lambda < t \\ 0 & \lambda > t \end{cases}$$

Subdividing the integral from 0 to ∞ into the sum of the intervals from 0 to t and from t to ∞ shows that

$$\int_0^\infty f(\lambda)\mathscr{L}^{-1}[G(s)e^{-\lambda s}]\,d\lambda = \int_0^t f(\lambda)g(t-\lambda)\,d\lambda + \int_t^\infty f(\lambda)(0)\,d\lambda$$

In the first integral on the right side λ is less than t so the inverse is $g(t-\lambda)$. In the second integral, λ is greater than t so the inverse is zero. Because this last integral vanishes

$$y(t) = \mathscr{L}^{-1}[G(s)F(s)] = \int_0^t f(\lambda)g(t-\lambda)\,d\lambda \tag{5.87}$$

This is the convolution integral.

The function $g(t-\lambda)$ is obtained by designating the inverse of $G(s)$ as $g(\lambda)$ and then replacing λ by $t-\lambda$. Similarly, $f(\lambda)$ is obtained by replacing t by λ in the equation for the forcing function $f(t)$. Note that the input $f(\lambda)$ is used directly (without transforming) in this integral procedure for determining the response $y(t)$. The physical significance of $g(t)$ is determined by noting that when $f(t)$ is a unit impulse, then $Y(s) = G(s)$. Thus, $g(t) = \mathscr{L}^{-1}[G(s)]$ is the unit impulse response of the system.

Interchanging $F(s)$ and $G(s)$ in the preceding development yields the alternate form:

$$y(t) = \mathscr{L}^{-1}[F(s)G(s)] = \int_0^t g(\lambda)f(t-\lambda)\,d\lambda \tag{5.88}$$

The convolution integral may be used to obtain the inverse of the product of two transforms. For example, for $F(s) = 1/s^2$ and $G(s) = 1/(s+1)$, then $f(\lambda) = \lambda$, $g(\lambda) = e^{-\lambda}$, and $g(t-\lambda) = e^{-(t-\lambda)}$.

Thus,

$$\mathcal{L}^{-1}[G(s)F(s)] = \mathcal{L}^{-1}\left(\frac{1}{s^2}\frac{1}{s+1}\right) = \int_0^t \lambda e^{-(t-\lambda)} \, d\lambda = e^{-t}\int_0^t \lambda e^{\lambda} \, d\lambda$$

$$= e^{-t}[e^{\lambda}(\lambda - 1)]_0^t = e^{-t} + t - 1$$

Using the alternate form $f(t - \lambda) = (t - \lambda)$ and $g(\lambda) = e^{-\lambda}$ yields the same result. It is important to note that $\mathcal{L}^{-1}[F(s)G(s)]$ is not equal to the product $\mathcal{L}^{-1}[F(s)]\mathcal{L}^{-1}[G(s)]$. For the preceding, $\mathcal{L}^{-1}[F(s)] = t$ and $\mathcal{L}^{-1}[G(s)] = e^{-t}$. This product te^{-t} is not the correct answer.

As another example, let it be desired to determine the inverse of $[\omega/(s^2 + \omega^2)]^2$. By letting

$$F(s) = G(s) = \frac{\omega}{s^2 + \omega^2}$$

then $f(\lambda) = \sin \omega\lambda$ and $g(t - \lambda) = \sin \omega(t - \lambda)$. Application of the convolution integral gives

$$\mathcal{L}^{-1}\left(\frac{\omega}{s^2 + \omega^2}\right)^2 = \int_0^t [\sin \omega\lambda \sin \omega(t - \lambda)] \, d\lambda$$

Using the identity $\sin \alpha \sin \beta = [\cos (\alpha - \beta) - \cos (\alpha + \beta)]/2$ in which $\alpha = \omega\lambda$ and $\beta = \omega(t - \lambda)$ gives

$$\mathcal{L}^{-1}\left(\frac{\omega}{s^2 + \omega^2}\right)^2 = \frac{1}{2}\int_0^t [\cos \omega(2\lambda - t) - \cos \omega t] \, d\lambda$$

$$= \frac{1}{2}\left[\frac{\sin \omega t}{2\omega} - t \cos \omega t\right]$$

Illustrative example 5.13 Use the convolution integral to verify the solution of Illustrative example 5.3. That is,

$$(D^2 + 3D + 2)y(t) = f(t) \tag{5.13}$$

where $f(t) = 4t$ and all initial conditions are zero.

SOLUTION From Illustrative example 5.3, the Laplace transform for $Y(s)$ is

$$Y(s) = \frac{1}{s^2 + 3s + 2}F(s) + \frac{(s + 3)y(0) + y'(0)}{s^2 + 3s + 2}$$

$$= G(s)F(s) + \frac{(s + 3)y(0) + y'(0)}{s^2 + 3s + 2} \tag{5.89}$$

where $G(s) = 1/(s^2 + 3s + 2)$. This is the general form that is obtained in solving any differential equation by the convolution integral. The first term on the right side of Eq. (5.89) yields the response due to the forcing function and the last term yields the response due to the initial conditions. Note that the initial condition term is displayed separately and may be inverted directly. When the initial conditions are zero and $f(t)$ is a unit impulse, then $Y(s) = G(s)$. Thus, $g(t) = \mathscr{L}^{-1}[G(s)]$ is the response of the system to a unit impulse when all the initial conditions are zero. More simply, $g(t)$ is referred to as the impulse response. For $y(0) = y'(0) = 0$, the inverse Laplace transform of Eq. (5.89) is

$$y(t) = \mathscr{L}^{-1}[G(s)F(s)] = \int_0^t f(\lambda)g(t - \lambda)\, d\lambda$$

where

$$g(t) = \mathscr{L}^{-1}[G(s)] = \mathscr{L}^{-1}\left(\frac{1}{s^2 + 3s + 2}\right) = \mathscr{L}^{-1}\left(\frac{1}{s + 1} - \frac{1}{s + 2}\right)$$

$$= e^{-t} - e^{-2t}$$

For $f(t) = 4t$, then $f(\lambda) = 4\lambda$. Thus,

$$y(t) = \int_0^t (4\lambda)[e^{-(t - \lambda)} - e^{-2(t - \lambda)}]\, d\lambda$$

$$= 4e^{-t} \int_0^t \lambda e^{\lambda}\, d\lambda - 4e^{-2t} \int_0^t \lambda e^{2\lambda}\, d\lambda$$

$$= 4e^{-t} - e^{-2t} + 2t - 3 \tag{5.14}$$

This checks the result given by Eq. (5.14). With the aid of the convolution integral, the response of the system to any forcing function $f(t)$ may be obtained by knowing the response of the system to a unit impulse $g(t)$.

Illustrative example 5.14 Use the convolution integral to determine the solution of Illustrative example 5.12. The differential equation is

$$\frac{dy}{dt} + y = f(t) \qquad f(t) = \begin{cases} t & 0 < t \le 1 \\ 1 & t > 1 \end{cases}$$

A plot of the forcing function $f(t)$ is shown in Fig. 5.17a.

SOLUTION The Laplace transform of the differential equation is

$$sY(s) - y(0) + Y(s) = F(s)$$

or

$$Y(s) = \frac{1}{s + 1} F(s) + \frac{y(0)}{s + 1}$$

$$= G(s)F(s) + \frac{y(0)}{s + 1} \tag{5.90}$$

where $G(s) = 1/(s + 1)$. Note that $G(s)$ is the Laplace transform that would result if all the initial conditions were zero and $f(t)$ were a unit impulse. Thus,

$g(t) = \mathcal{L}^{-1}[G(s)]$ is the impulse response of the system. For $y(0) = 0$, the inverse Laplace transform of Eq. (5.90) is

$$y(t) = \mathcal{L}^{-1}[G(s)F(s)] = \int_0^t f(\lambda)g(t - \lambda)\, d\lambda$$

where

$$g(t) = \mathcal{L}^{-1}[G(s)] = \mathcal{L}^{-1}\left(\frac{1}{s+1}\right) = e^{-t}$$

For $0 < t \leq 1, f(t) = t$, in which case $f(\lambda) = \lambda$. Thus,

$$y(t) = \int_0^t \lambda e^{-(t-\lambda)}\, d\lambda = e^{-t}\int_0^t \lambda e^{\lambda}\, d\lambda$$

$$= t - 1 + e^{-t} \qquad 0 < t \leq 1 \tag{5.84}$$

This checks the result given by Eq. (5.84).

For $t > 1$, then $f(\lambda) = \lambda$ for $0 < \lambda \leq 1$ and $f(\lambda) = 1$ for $\lambda > 1$. Thus

$$y(t) = \int_0^1 \lambda e^{-(t-\lambda)}\, d\lambda + \int_1^t (1)e^{-(t-\lambda)}\, d\lambda$$

$$= e^{-t}[e^{\lambda}(\lambda - 1)]_0^1 + e^{-t}[e^{\lambda}]_1^t = 1 - 0.632e^{-(t-1)} \qquad t > 1 \tag{5.85}$$

This checks the previously attained result given by Eq. (5.85).

5.8 ERROR COEFFICIENTS

For the system shown in Fig. 5.18, the error signal is $e(t) = r(t) - H(D)c(t) = r(t) - G(D)H(D)e(t)$. When all initial conditions are zero, the transformed equation is

$$E(s) = \frac{R(s)}{1 + G(s)H(s)}$$

Application of the final-value theorem yields for the steady-state error

$$e_{ss} = \lim_{t \to \infty} e(t) = \lim_{s \to 0} sE(s)$$

For a unit step input $[R(s) = 1/s]$, the steady-state error is

$$e_{ss} = \lim_{s \to 0} s \frac{1/s}{1 + G(s)H(s)} = \lim_{s \to 0} \frac{1}{1 + G(s)H(s)} = \frac{1}{1 + K_p} \tag{5.91}$$

where $K_p = \lim_{s \to 0} G(s)H(s)$ is called the positional error constant.

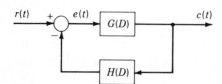

Figure 5.18 Feedback control system.

For a unit ramp input $[r(t) = t, R(s) = 1/s^2]$, the steady-state error is

$$e_{ss} = \lim_{s \to 0} s \frac{1/s^2}{1 + G(s)H(s)} = \lim_{s \to 0} \frac{1}{s + sG(s)H(s)} = \lim_{s \to 0} \frac{1}{sG(s)H(s)} = \frac{1}{K_v} \quad (5.92)$$

where $K_v = \lim_{s \to 0} sG(s)H(s)$ is the velocity error constant.

For a unit parabolic input $[r(t) = t^2/2, R(s) = 1/s^3]$, the steady-state error is

$$e_{ss} = \lim_{s \to 0} \frac{1}{s^2 + s^2 G(s)H(s)} = \lim_{s \to 0} \frac{1}{s^2 G(s)H(s)} = \frac{1}{K_a} \quad (5.93)$$

where $K_a = \lim_{s \to 0} s^2 G(s)H(s)$ is the acceleration error constant. The error constant provides a convenient method for determining the steady-state error to a unit step, unit ramp, or unit parabolic input. Because the response due to initial conditions dies out as time approaches infinity, the preceding results are valid even if the initial conditions are not zero.

Illustrative example 5.15 The system shown in Fig. 5.9 is discussed in Illustrative example 5.5. For this system, determine the steady-state error to a unit step function input.

SOLUTION From Fig. 5.9, it follows that

$$G(s) = \frac{K_1}{1 + \tau_1 s} \frac{K_2}{1 + \tau_2 s} \qquad H(s) = K_H$$

The positional error constant is

$$K_p = \lim_{s \to 0} G(s)H(s) = K_1 K_2 K_H = (1.0)(2.0)(0.5) = 1.0$$

Thus, the steady-state error due to a unit step input is

$$e_{ss} = \frac{1}{1 + K_p} = 0.5$$

The corresponding steady-state values for c and v may be determined by noting from Fig. 5.9 that $e = r - K_H c = r - 0.5c$ and $r = Av = v$. For a unit step input, $r = 1$, and thus $c = 2(r - e_{ss}) = 2(1 - 0.5) = 1.0$ and $v = r = 1.0$. This agrees with the result of Illustrative example 5.5, in which it was found that for a constant input, $c = v$. As demonstrated by this example, the presence of a steady-state error signal does not necessarily imply that the controlled variable is not equal to the command signal.

Illustrative example 5.16 For the system shown in Fig. 5.19a, determine the response to a unit step function, a unit ramp function, and a unit parabolic function when all initial conditions are zero. What is the steady-state error to each of these inputs?

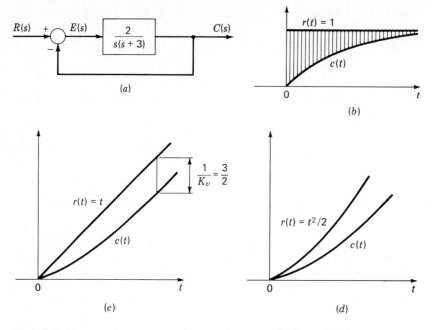

Figure 5.19 Response of a system to unit step, unit ramp, and unit parabolic inputs.

SOLUTION The differential equation of operation is

$$(D^2 + 3D + 2)c(t) = 2r(t)$$

The transformed equation is

$$C(s) = \frac{2R(s)}{(s + 1)(s + 2)} \tag{5.94}$$

For a unit step function input, $R(s) = 1/s$. Carrying out the partial-fraction expansion and inverting yields for the time response

$$c(t) = 1 - 2e^{-t} + e^{-2t}$$

The corresponding input $r(t) = u(t)$ and response $c(t)$ are shown in Fig. 5.19b. The error $e(t) = r(t) - c(t)$ is $e(t) = 1 - (1 - 2e^{-t} + e^{-2t}) = 2e^{-t} - e^{-2t}$. Taking the limit as t approaches infinity yields, for the steady-state error,

$$e_{ss} = \lim_{t \to \infty} e(t) = \lim_{t \to \infty} (2e^{-t} - e^{-2t}) = 0$$

The positional error constant is

$$K_p = \lim_{s \to 0} G(s)H(s) = \frac{2}{s(s + 3)}\bigg|_{s = 0} = \infty$$

Thus, the steady-state error to a unit step function is

$$e_{ss} = \frac{1}{1 + K_p} = \frac{1}{1 + \infty} = 0$$

For a unit ramp input, $r(t) = t$. Substitution of $R(s) = 1/s^2$ into Eq. (5.94) gives

$$C(s) = \frac{2}{s^2(s + 1)(s + 2)} = \frac{1}{s^2} - \frac{3/2}{s} + \frac{2}{s + 1} - \frac{1/2}{s + 2}$$

Inverting yields, for the response,

$$c(t) = t - \tfrac{3}{2} + 2e^{-t} - \tfrac{1}{2}e^{-2t}$$

A plot of the input $r(t) = t$ and the response $c(t)$ is shown in Fig. 5.19c. The error $e(t) = r(t) - c(t)$ is

$$e(t) = t - (t - \tfrac{3}{2} + 2e^{-t} - \tfrac{1}{2}e^{-2t}) = \tfrac{3}{2} + 2e^{-t} + \tfrac{1}{2}e^{-2t}$$

Taking the limit as t approaches infinity yields, for the steady-state error,

$$e_{ss} = \lim_{t \to \infty} e(t) = \tfrac{3}{2}$$

The velocity error constant is

$$K_v = \lim_{s \to 0} sG(s)H(s) = \frac{2s}{s(s + 3)}\bigg|_{s=0} = \frac{2}{3}$$

Thus, the steady-state error to a unit ramp function is

$$e_{ss} = \frac{1}{K_v} = \frac{3}{2}$$

For a unit parabolic input, $r(t) = t^2/2$. Substitution of $R(s) = 1/s^3$ into Eq. (5.94) gives

$$C(s) = \frac{2}{s^3(s + 1)(s + 2)} = \frac{1}{s^3} - \frac{3/2}{s^2} + \frac{7/4}{s} - \frac{2}{s + 1} + \frac{1/4}{s + 2}$$

Inverting yields, for the response,

$$c(t) = \frac{t^2}{2} - \frac{3}{2}t + \frac{7}{4} - 2e^{-t} + \frac{1}{4}e^{-2t}$$

The error, $e(t) = r(t) - c(t)$, is

$$e(t) = \tfrac{3}{2}t - \tfrac{7}{4} + 2e^{-t} - \tfrac{1}{4}e^{-2t}$$

In the limit as t approaches infinity, the error is

$$e_{ss} = \lim_{t \to \infty} e(t) = \left[\frac{3}{2}t - \frac{7}{4} + 2e^{-t} - \frac{1}{4}e^{-2t}\right]_{t=\infty} = \infty$$

A plot of the input $r(t) = t^2/2$ and the response $c(t)$ is shown in Fig. 5.19*d*. Note that the error keeps growing as t increases. The acceleration error constant is

$$K_a = \lim_{s \to 0} s^2 G(s)H(s) = \frac{2s^2}{s(s+3)}\bigg|_{s=0} = \frac{2}{(1+3/s)}\bigg|_{s=0} = \frac{2}{1+\infty} = 0$$

Thus, the steady-state error to a unit parabolic input is

$$e_{ss} = \frac{1}{K_a} = \frac{1}{0} = \infty$$

Consider now the case in which the input is a unit step function $r(t) = u(t)$. The transform of the error signal is

$$E(s) = \frac{R(s)}{1 + G(s)H(s)} = \frac{1/s}{1 + G(s)H(s)}$$

Taking the limit as s approaches zero shows that

$$\lim_{s \to 0} E(s) = \lim_{s \to 0} \frac{1}{s + sG(s)H(s)} = \lim_{s \to 0} \frac{1}{sG(s)H(s)} = \frac{1}{K_v}$$

From the definition of the Laplace transform, it follows that

$$\lim_{s \to 0} E(s) = \lim_{s \to 0} \int_0^\infty e(t)e^{-st}\, dt = \int_0^\infty e(t)\, dt \qquad (5.95)$$

The right side is the time integral of the error to a unit step input. This integral, shown shaded in Fig. 5.19*b*, is

$$\int_0^\infty e(t)\, dt = \frac{1}{K_v} = \frac{3}{2}$$

This result may also be checked by integration. Thus

$$\int_0^\infty e(t)\, dt = \int_0^\infty (2e^{-t} - e^{-2t})\, dt = [-2e^{-t} + \tfrac{1}{2}e^{-2t}]_0^\infty$$

$$= 0 - (-2 + \tfrac{1}{2}) = \tfrac{3}{2}$$

As demonstrated in the preceding, many interesting features and characteristics of systems may be obtained directly from the transformed equation without the need to invert and obtain the time response.

PROBLEMS

5.1 Determine the partial-fraction expansion for the following operators:

(a) $\dfrac{L_m(D)}{L_n(D)} = \dfrac{1}{(D+2)(D+3)}$

(b) $\dfrac{L_m(D)}{L_n(D)} = \dfrac{D+4}{(D+2)(D+3)}$

(c) $\dfrac{L_m(D)}{L_n(D)} = \dfrac{3D+7}{(D+1)(D+2)(D+3)}$

(d) $\dfrac{L_m(D)}{L_n(D)} = \dfrac{D^2 + 4D + 5}{(D+1)(D+2)(D+3)}$

5.2 Determine the partial-fraction expansion for the following operators:

(a) $\dfrac{L_m(D)}{L_n(D)} = \dfrac{3}{(D + 2)(D + 5)}$

(b) $\dfrac{L_m(D)}{L_n(D)} = \dfrac{D + 8}{(D + 2)(D + 5)}$

(c) $\dfrac{L_m(D)}{L_n(D)} = \dfrac{18D + 30}{D(D + 2)(D + 5)}$

(d) $\dfrac{L_m(D)}{L_n(D)} = \dfrac{D^2 + 4D + 10}{D(D + 2)(D + 5)}$

5.3 Determine the partial-fraction expansion for the following operators:

(a) $\dfrac{L_m(D)}{L_n(D)} = \dfrac{1}{(D + 1)^2(D + 2)}$

(b) $\dfrac{L_m(D)}{L_n(D)} = \dfrac{2D + 1}{(D + 1)(D + 2)^2}$

(c) $\dfrac{L_m(D)}{L_n(D)} = \dfrac{3D + 4}{(D + 1)^2(D + 2)^2}$

(d) $\dfrac{L_m(D)}{L_n(D)} = \dfrac{2D + 3}{(D + 1)^3(D + 2)}$

5.4 Consider the differential equation

$$\frac{dy}{dt} + y = f(t)$$

Use classical techniques to determine the solution for each of the following cases:

(a) $f(t) = 0$ and $y(0) = 1$

(b) $f(t) = 1$ and $y(0) = 0$

(c) $f(t) = t$ and $y(0) = 0$

(d) $f(t) = e^{-t}$ and $y(0) = 0$

5.5 Consider the differential equation

$$\frac{dy}{dt} + 2y + 10f(t)$$

Using classical methods, determine the solution for each of the following cases:

(a) $f(t) = 1$ and $y(0) = 0$

(b) $f(t) = 0$ and $y(0) = 1$

(c) $f(t) = e^{-t}$ and $y(0) = 0$

5.6 Consider the differential equation

$$y(t) = \frac{12}{(D + 1)(D + 3)} f(t)$$

Using classical methods, determine the solution when $f(t) = 1$ for each of the following cases:

(a) $y(0) = y'(0) = 0$

(b) $y(0) = 2$ and $y'(0) = 0$

5.7 Same as Prob. 5.6 except $f(t) = e^{-t}$.

5.8 Consider the differential equation

$$y = \frac{1}{(D + 1)^2} f(t)$$

Using classical methods, determine the solution for each of the following cases:

(a) $f(t) = 1$ and $y(0) = y'(0) = 0$

(b) $f(t) = 1$, $y(0) = 2$, and $y'(0) = 0$

(c) $f(t) = 0$, $y(0) = 2$, and $y'(0) = 0$

5.9 Consider the differential equation

$$y(t) = \frac{12}{(D + 1)^2(D + 3)} f(t)$$

Using classical methods, determine the solution for each of the following cases:

(a) $f(t) = 1$ and $y(0) = y'(0) = y''(0) = 0$

(b) $f(t) = 1$, $y(0) = 8$, and $y'(0) = y''(0) = 0$

(c) $f(t) = 0$, $y(0) = 8$, and $y'(0) = y''(0) = 0$

5.10 Use Eq. (5.25) to determine the Laplace transform of each of the following functions:
(a) $f(t) = t$ (b) $f(t) = t^2$ (c) $f(t) = \cos t$

5.11 Given that the Laplace transform of e^{-t} is $1/(s + 1)$ use appropriate theorems to determine the Laplace transform of each of the following functions:
(a) $f(t) = te^{-t}$ (b) $f(t) = t^2 e^{-t}$ (c) $f(t) = e^{-3t}$

5.12 Given that the Laplace transform of $\sin t$ is $1/(s^2 + 1)$, use appropriate theorems to determine the Laplace transform of each of the following functions:
(a) $f(t) = t \sin t$ (b) $f(t) = e^{-t} \sin t$ (c) $f(t) = \sin 3t$

5.13 Determine the Laplace transform for each of the following functions:
(a) $f(t) = 0$ for $t < 2$ and $f(t) = e^{-(t-2)}$ for $t > 2$
(b) $f(t) = 0$ for $t < \pi$ and $f(t) = \sin(t - \pi)$ for $t > \pi$

5.14 Use the equation for obtaining the Laplace transform of the second derivative of $f(t)$ to determine the transform of
(a) $f(t) = \cos \omega t$ (b) $f(t) = \sinh \omega t$

5.15 Use the equation for obtaining the Laplace transform of the integral of $f(t)$ to determine the transform of
(a) $t^2 = \int (2t) \, dt$ (b) $\sin \omega t = \int (\omega \cos \omega t) \, dt$
(c) $\cos \omega t = \int (-\omega \sin \omega t) \, dt$

5.16 Use the Laplace transform method to determine the solution of Prob. 5.4.

5.17 Use the Laplace transform method to determine the solution of Prob. 5.5.

5.18 Use the Laplace transform method to determine the solution of Prob. 5.6.

5.19 Use the Laplace transform method to determine the solution of Prob. 5.7.

5.20 Use the Laplace transform method to determine the solution of Prob. 5.8.

5.21 Use the Laplace transform method to determine the solution of Prob. 5.9.

5.22 The operation of an amplifier is described by the differential equation

$$\frac{dy}{dt} + 0.2y = f(t)$$

Determine the response $y(t)$ to a unit step function $f(t) = u(t)$. All the initial conditions are zero. Identify the time constant and steady-state gain.

5.23 Consider the differential equation

$$\frac{dy}{dt} + y = u(t) + e^{-t} \qquad y(0) = 0$$

where $u(t)$ is a unit step function. Determine the transform of the derivative $\mathcal{L}[y'(t)] = [sY(s) - y(0)]$, and then invert to obtain $y'(t)$.

5.24 For the system shown in Fig. P5.24, determine K_1 and K_2 such that the system will have a steady-state gain of 2 and a time constant of 0.4 seconds.

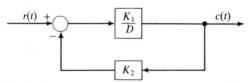

Figure P5.24

5.25 For the system shown in Fig. P5.25, determine K and a such that the system will have a steady-state gain of 1 and a time constant of 0.25 seconds.

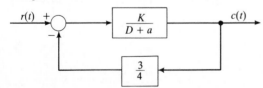

Figure P5.25

5.26 The differential equation of operation for a system is

$$y(t) = \frac{12(D + 1)}{(D + 3)(D + 4)} f(t)$$

Determine the response $y(t)$ when $f(t) = u(t)$ is a unit step function and all initial conditions are zero.

5.27 Determine the response $y(t)$ for the system of Prob. 5.26 when $f(t) = 0$, $y(0) = 1$, and $y'(0) = 0$.

5.28 Determine the response $y(t)$ for the system of Prob. 5.26 when $f(t) = 0$, $y(0) = 0$, and $y'(0) = 1$.

5.29 The differential equation of operation for a system is

$$c(t) = \frac{D + 6}{D^2 + 5D + 6} r(t)$$

Determine the response $c(t)$ for each of the following cases:

    (a) $r(t) = u(t)$, $c(0) = \dot{c}(0) = 0$

    (b) $r(t) = e^{-2t}$, $c(0) = \dot{c}(0) = 0$

    (c) $r(t) = 0$, $c(0) = 1$, and $\dot{c}(0) = 0$

5.30 For the system shown in Fig. P5.30, determine the response $c(t)$ for each of the following cases:

    (a) $r(t) = u(t)$ and $c(0) = \dot{c}(0) = 0$

    (b) $r(t) = 0$, $c(0) = 1$, and $\dot{c}(0) = 1$

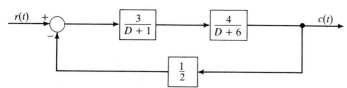

Figure P5.30

5.31 For the system shown in Fig. P5.31, determine the response $c(t)$ for each of the following cases:

    (a) $r(t) = u(t)$ and $c(0) = \dot{c}(0) = 0$

    (b) $r(t) = 0$, $c(0) = 1$, and $\dot{c}(0) = 1$

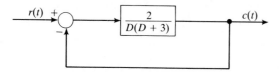

Figure P5.31

5.32 For the system shown in Fig. P5.32, determine the response $c(t)$ for each of the following cases:

    (a) $r(t) = u(t)$ and $c(0) = \dot{c}(0) = 0$

    (b) $r(t) = 0$, $c(0) = 1$, and $\dot{c}(0) = -2$

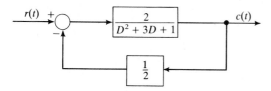

Figure P5.32

5.33 For the system shown in Fig. P5.33, determine the response $c(t)$ when

    (a) $r(t) = u(t)$, $d(t) = 0$, and $c(0) = \dot{c}(0) = 0$

    (b) $r(t) = 0$, $d(t) = u(t)$, $c(0) = 1$, and $\dot{c}(0) = -1$

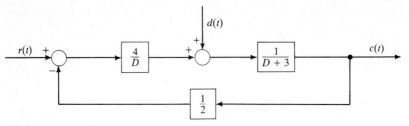

Figure P5.33

5.34 For the feedback control system shown in Fig. P5.34 all the initial conditions are zero and $K = 1.5$. Determine the response for each of the following cases:

    (a) $r(t)$ is a unit step function and $d(t) = 0$.
    (b) $d(t)$ is a unit step function and $r(t) = 0$.
    (c) Both $r(t)$ and $d(t)$ are unit step functions.

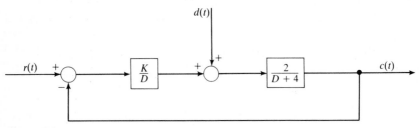

Figure P5.34

5.35 Same as Prob. 5.34 except $K = 2$.

5.36 For the control system shown in Fig. P5.36, determine the response $c(t)$ for each of the following cases:

    (a) $r(t) = u(t)$, $d(t) = 0$, and $c(0) = \dot{c}(0) = 0$
    (b) $r(t) = 0$, $d(t) = u_1(t)$, and $c(0) = \dot{c}(0) = 0$
    (c) $r(t) = d(t) = 0$, $c(0) = 1$, and $\dot{c}(0) = 0$

For each case use the initial-value theorem to check the result for $c(0+)$.

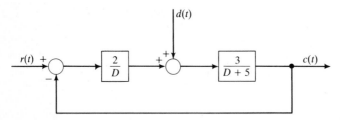

Figure P5.36

5.37 Using the Laplace transform method, determine the solution of the following differential equations for the case in which $f(t)$ is a step function of height $h = 6$; all initial conditions are zero:

    (a) $y(t) = \dfrac{D + 5}{(D + 1)(D + 3)} f(t)$ (b) $y(t) = \dfrac{D + 5}{(D + 1)^2(D + 3)} f(t)$

5.38 Verify the solution of Prob. 5.37 by using the classical method.

5.39 A switch is inserted in series in the circuit of Fig. 5.13. For $t < 0$, the switch is open and all initial conditions are zero. For $t > 0$, the switch is closed and the voltage source $e = 100$ V is applied to the circuit. For $R = 10^6 \ \Omega = 1 \ \text{M}\Omega$ and $C = 10^{-6} \ \text{F} = 1 \ \mu\text{F}$, determine the equation for the resulting current i that flows through the circuit and the voltage e_c across the capacitor.

5.40 Using the Laplace transform method, a student found the solution of the differential equation

$$\ddot{y} + 4\dot{y} + 3y = 4u_1(t) \qquad y(0) = \dot{y}(0) = 0$$

to be

$$y = 2e^{-t} - 2e^{-3t}$$

Being rather thorough, the student decided to check the answer by checking the initial conditions and checking whether the solution satisfied the differential equation. The results were $y(0) = 2 - 2 = 0$, $\dot{y}(0) = -2 + 6 = 4$, and $\ddot{y} + 4\dot{y} + 3y = 0$. Since this solution satisfied neither the differential equation nor the one initial condition, the student concluded that there had been an error in determining y. Is the error in the solution or the conclusion? Explain.

5.41 A mass-spring-damper system is shown in Fig. P5.41.
(a) Write the differential equation.
(b) Use Laplace transforms to solve this equation when $x(0) = 0$, $x'(0) = 1$, $f = 0$, $M = 1$, $B = 3$, and $K = 2$.

Figure P5.41

5.42 Using Laplace transforms, determine the response $y(t)$ for the following differential equations for the case in which all the initial conditions are zero and the input $f(t)$ is an impulse of area $k = 2$ which occurs at time $t = 0$:

(a) $y(t) = \dfrac{2D + 3}{(D + 2)(D + 3)} f(t)$

(b) $y(t) = \dfrac{2D + 3}{D(D + 2)(D + 3)} f(t)$

Use the initial-value theorem to determine $y(0+)$ and the final-value theorem to determine $y(\infty)$.

5.43 Same as Prob. 5.42 except that the system is initially at a steady-state operating condition, with $f(0) = 2$.

5.44 The dynamics of a system are described by the differential equation

$$y(t) = \frac{10(2D + 1)}{(D + 2)(D + 5)} f(t)$$

Use the Laplace transform method to determine the response $y(t)$ when all the initial conditions are zero and the forcing function $f(t)$ is

(a) A unit impulse $u_1(t)$

(b) A unit step function $u(t)$

(c) An exponential e^{-t}

5.45 Use the classical method to verify the solution of Prob. 5.44.

5.46 Same as Prob. 5.44 except that the system is initially at a steady-state operating condition with $f(0) = 3$.

5.47 The system shown in Fig. P5.47 is in a state of equilibrium, with $r = 3$ and $d = 0$. A step-function disturbance $d(t) = 6u(t)$ is then initiated at time $t = 0$. Determine the response $c(t)$ for $t > 0$.

Figure P5.47

5.48 The system shown in Fig. P5.47 is in a state of equilibrium, with $r = 0$ and $d = 6$. A step-function change $r(t) = 3u(t)$ is then initiated at time $t = 0$. Determine the response $c(t)$ for $t > 0$.

5.49 The system shown in Fig. P5.49 is initially at equilibrium, with $r = 1$ and $d = 0$. A step-function disturbance $d(t) = u(t)$ is then initiated at time $t = 0$. Determine the response $c(t)$ for $t > 0$.

Figure P5.49

5.50 A system is described by the differential equation

$$\frac{dy}{dt} + y = f(t) \qquad f(t) = \begin{cases} 1 & 0 < t < 1 \\ 0 & t > 1 \end{cases}$$

All initial conditions are zero. Use the Laplace transform method to determine the response for $0 < t < 1$ and for $t > 1$.

5.51 Same as Prob. 5.50 except $f(t)$ is the function shown in Fig. P5.51.

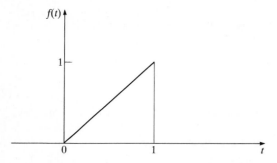

Figure P5.51

5.52 The following functions may be written as products of $1/s$, $1/s^2$, or $1/(s + 1)$. Use the convolution integral to determine the inverse transform of

(a) $Y(s) = 1/s^3$

(b) $Y(s) = 1/s^4$

(c) $Y(s) = 1/s^2(s + 1)$

(d) $Y(s) = 1/(s + 1)^2$

5.53 The following functions may be written as products of $1/s$, $1/(s^2 + 1)$, or $s/(s^2 + 1)$. Use the convolution integral to determine the inverse transform of

(a) $Y(s) = \dfrac{1}{s(s^2 + 1)}$

(b) $Y(s) = \dfrac{1}{(s^2 + 1)^2}$

(c) $Y(s) = \dfrac{s}{(s^2 + 1)^2}$

(d) $Y(s) = \dfrac{s^2}{(s^2 + 1)^2}$

5.54 Same as Prob. 5.50 except use the convolution integral.

5.55 Same as Prob. 5.51 except use the convolution integral.

5.56 For the system shown in Fig. P5.56 use the Laplace transform method to determine the response $y(t)$ when $f(t) = u_1(t)$ is a unit impulse and all initial conditions are zero. Next use the convolution integral to determine the response for each of the following cases:

    (a) $f(t) = u(t)$ and $y(0) = 0$ (b) $f(t) = e^{-t}$ and $y(0) = 1$

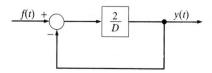

Figure P5.56

5.57 The operation of a control system is described by the differential equation

$$y(t) = \frac{20}{D^2 + 9D + 20} f(t)$$

Use the convolution integral to determine the response for each of the following cases:

    (a) $f(t) = u(t)$ and $y(0) = \dot{y}(0) = 0$ (b) $f(t) = u(t)$ and $y(0) = \dot{y}(0) = 1$

5.58 Same as Prob. 5.57 except $f(t) = e^{-4t}$.

5.59 The operation of a control system is described by the differential equation

$$y(t) = \frac{10(D + 2)}{D^2 + 9D + 20} f(t)$$

Use the convolution integral to determine the response for each of the following cases:

    (a) $f(t) = u(t)$ and $y(0) = \dot{y}(0) = 0$

    (b) $f(t) = u(t)$, $y(0) = 0$, and $\dot{y}(0) = 1$

5.60 Same as Prob. 5.59 except $f(t) = e^{-4t}$.

5.61 Determine the position, velocity, and acceleration error constants and then determine the steady-state error to a unit step, a unit ramp, and a unit parabolic input for the system shown (a) in Fig. P5.47 and (b) in Fig. P5.49.

REFERENCE

1. Raven, F. H.: "Mathematics of Engineering Systems," McGraw-Hill Book Company, New York, 1966.

SIX

TRANSIENT RESPONSE

Because an actual system may be subjected to all types and varieties of input excitations $f(t)$, it becomes impractical to calculate the system response for every possible excitation. In this chapter, it is shown that a very good measure of the transient behavior may be obtained directly from the zeros of the characteristic function (i.e., roots of the characteristic equation). This criterion for evaluating transient performance is obtained by considering the essential characteristics of a general system of order n.

The general operational representation for a differential equation of order n is

$$y(t) = \frac{a_m D^m + a_{m-1} D^{m-1} + \cdots + a_1 D + a_0}{D^n + b_{n-1} D^{n-1} + \cdots + b_1 D + b_0} f(t) \qquad (6.1)$$

The transform of each term is

$$\mathscr{L}[D^n y(t)] = s^n Y(s) - I(s)_n$$

$$b_{n-1} \mathscr{L}[D^{n-1} y(t)] = b_{n-1} s^{n-1} Y(s) - I(s)_{n-1}$$

$$\cdots \cdots \cdots \cdots \cdots \cdots \cdots \cdots \cdots \cdots$$

$$a_m \mathscr{L}[D^m f(t)] = a_m s^m F(s) - I(s)_m$$

$$a_{m-1} \mathscr{L}[D^{m-1} f(t)] = a_{m-1} s^{m-1} F(s) - I(s)_{m-1}$$

$$\cdots \cdots \cdots \cdots \cdots \cdots \cdots \cdots \cdots \cdots$$

where $I(s)_n$, $I(s)_{n-1}$, ... are the initial conditions associated with each transform.

Transforming each term of Eq. (6.1) accordingly and collecting terms yields

$$Y(s) = \frac{(a_m s^m + a_{m-1} s^{m-1} + \cdots + a_1 s + a_0)F(s) + I(s)}{s^n + b_{n-1} s^{n-1} + \cdots + b_1 s + b_0}$$

$$= \frac{L_m(s)F(s) + I(s)}{L_n(s)} \tag{6.2}$$

where $I(s) = I(s)_n + I(s)_{n-1} + \cdots - I(s)_m - I(s)_{m-1} - \cdots$ is the sum of all the initial conditions. By comparison of Eqs. (6.1) and (6.2), it is to be noted that the characteristic function in the s domain, $L_n(s)$, is the same as that in the D domain, $L_n(D)$. The numerator also has the same form, with the exception that the initial conditions $I(s)$ are added in the s domain. Comparison of Eqs. (6.1) and (6.2) shows that when all the initial conditions are zero, the transform is obtained by merely substituting s for D, $Y(s)$ for $y(t)$, and $F(s)$ for $f(t)$ in the operational form of the differential equation.

For this case,

$$Y(s) = \frac{L_m(s)}{L_n(s)} F(s) \tag{6.3}$$

where $L_m(s)/L_n(s)$ is called the transfer function. It is to be noted that $L_m(s)$ and $L_n(s)$ are obtained directly from the differential equation of operation for the system. Thus, the transfer function contains basic information concerning the essential characteristics of a system without regard to initial conditions or excitation.

Control engineers oftentimes write differential equations using the s notation rather than the operator D. This yields directly the form given by Eq. (6.3). This form suffices for most control work. When it is necessary to determine the effect of initial conditions, then s is replaced by D to obtain the differential equation. Transforming the differential equation yields the resulting transformed equation which contains the initial condition terms.

The term $F(s)$ in Eq. (6.2) is the general representation for the transform of the input signal or forcing function. This term may be written as

$$F(s) = \frac{N_{F(s)}}{D_{F(s)}}$$

where $N_{F(s)}$ is the numerator of $F(s)$ and $D_{F(s)}$ is the denominator of $F(s)$. For example, for a unit step function, $F(s) = 1/s$, and thus $N_{F(s)} = 1$ and $D_{F(s)} = s$. Substitution of the preceding representation $F(s) = N_{F(s)}/D_{F(s)}$ into Eq. (6.2) yields the following general transformed form for $Y(s)$:

$$Y(s) = \frac{L_m(s)N_{F(s)} + I(s)D_{F(s)}}{L_n(s)D_{F(s)}} = \frac{A(s)}{B(s)} \tag{6.4}$$

where $A(s)$ and $B(s)$ are polynomials in s.

6.1 INVERSE TRANSFORMATIONS

By an inverse transformation is meant the process of inverting a function from the s domain back to the time domain. The inverse transform $\mathscr{L}^{-1}$ of a function $F(s)$ is defined by the equation

$$\mathscr{L}^{-1}[F(s)] = \frac{1}{2\pi j} \int_C F(s)e^{ts}\, ds = f(t) \tag{6.5}$$

where C is a suitably chosen contour in the s domain.[1] This integral method of evaluating the inverse transform is not employed when the much simpler process of entering a transform table with the given $F(s)$ and reading directly the desired $f(t)$ can be utilized, as is the case for ordinary control analysis. A partial listing of commonly used transforms was given in Table 5.1.

The transform table may be used to obtain the Laplace transform $F(s)$ of a given function of time or to obtain the inverse transform $f(t)$ for a given function of s. This process is analogous to using a logarithmic table for obtaining the logarithm of a number or to using the same table for the opposite process of obtaining antilogarithms.

At first it would appear that the listing of transforms given in Table 5.1 would have to be extended considerably so that it would be applicable to the wide range of problems encountered in the design of systems. However, this is not the case. The listing given in Table 5.1 is adequate for the solution of most ordinary problems that arise in control engineering. The reason for this is that relatively few different types of terms appear in the differential equation after it has been expanded by a partial-fraction expansion. In particular, the zeros of $B(s)$ are either distinct or repeated.

Distinct Zeros

The transformed function $B(s)$ is the denominator of Eq. (6.4). When the zeros of $B(s)$ are distinct, the denominator $B(s)$ can be factored in the form

$$B(s) = (s - r_1)(s - r_2) \cdots (s - r_n) \tag{6.6}$$

where $r_1, r_2, \ldots, r_n$ are n distinct zeros of $B(s)$.

The partial-fraction expansion of Eq. (6.4) is of the form

$$Y(s) = \frac{K_1}{s - r_1} + \frac{K_2}{s - r_2} + \cdots + \frac{K_i}{s - r_i} + \cdots + \frac{K_n}{s - r_n} \tag{6.7}$$

where $K_1, K_2, \ldots, K_n$ are n constants. Each constant K_i may be evaluated by the method used to obtain Eq. (5.6). That is, first multiply both sides of Eq. (6.7) by $s - r_i$; then take the limit as s approaches r_i. After performing these operations, the only term remaining on the right side of Eq. (6.7) is K_i. Thus,

$$K_i = \lim_{s \to r_i} [(s - r_i)Y(s)] \tag{6.8}$$

The inverse transform of Eq. (6.7) is obtained directly from the transform table and is

$$y(t) = K_1 e^{r_1 t} + K_2 e^{r_2 t} + \cdots + K_n e^{r_n t} \tag{6.9}$$

Equation (6.9) shows that each distinct zero of $B(s) = L_n(s)D_{F(s)}$ yields an exponential-type term $K_i e^{r_i t}$ in the response function. The exponent r_i is the corresponding zero of $B(s)$. Each zero $r_1, r_2, \ldots, r_n$ must be negative in order that each term $K_i e^{r_i t}$ in $y(t)$ be a decaying function. If any zero of $B(s)$ is positive, $y(t)$ will increase without bound as t increases to infinity. A constant term results if $r_i = 0$, because $K_i e^{(0)t} = K_i$.

Repeated Zeros

For the case in which $B(s)$ has a multiple or repeated zero r which occurs q times, $B(s)$ may be factored in the form

$$B(s) = (s - r)^q (s - r_1)(s - r_2) \cdots (s - r_{n-q}) \tag{6.10}$$

The corresponding partial-fraction expansion for $Y(s)$ is

$$Y(s) = \frac{C_q}{(s - r)^q} + \frac{C_{q-1}}{(s - r)^{q-1}} + \cdots + \frac{C_1}{s - r} + \frac{K_1}{s - r_1} + \frac{K_2}{s - r_2} + \cdots + \frac{K_{n-q}}{s - r_{n-q}} \tag{6.11}$$

The constant coefficients for the multiple terms are evaluated as follows:

$$C_q = \lim_{s \to r} [(s - r)^q Y(s)]$$

$$C_{q-1} = \lim_{s \to r} \left\{ \frac{d}{ds} [(s - r)^q Y(s)] \right\} \tag{6.12}$$

$$C_{q-k} = \lim_{s \to r} \left\{ \frac{1}{k!} \frac{d^k}{ds^k} [(s - r)^q Y(s)] \right\}$$

From the transform table, the inverse transform of Eq. (6.11) is found to be

$$y(t) = \left[\frac{C_q t^{q-1}}{(q-1)!} + \frac{C_{q-1} t^{q-2}}{(q-2)!} + \cdots + \frac{C_2 t}{1!} + C_1 \right] e^{rt}$$

$$+ K_1 e^{r_1 t} + \cdots + K_{n-q} e^{r_{n-q} t} \tag{6.13}$$

Each response term associated with the repeated zero $(s - r)^q$ is seen to be multiplied by the exponential factor e^{rt}. If the value of r is positive, $y(t)$ will become infinite as time increases. For negative values of r, a decreasing exponential results, and thus the response term due to the repeated zero eventually vanishes.

Illustrative example 6.1 Let it be desired to determine the time response $y(t)$ for the transformed equation

$$Y(s) = \frac{11s + 28}{(s + 2)^2(s + 5)} = \frac{C_2}{(s + 2)^2} + \frac{C_1}{s + 2} + \frac{K_1}{s + 5} \qquad (6.14)$$

The constants are evaluated as follows:

$$C_2 = \lim_{s \to -2} \frac{11s + 28}{s + 5} = 2$$

$$C_1 = \lim_{s \to -2} \left(\frac{d}{ds} \frac{11s + 28}{s + 5} \right) = \lim_{s \to -2} \frac{(s + 5)11 - (11s + 28)}{(s + 5)^2} = 3$$

$$K_1 = \lim_{s \to -5} \frac{11s + 28}{(s + 2)^2} = -3$$

Thus,

$$Y(s) = \frac{2}{(s + 2)^2} + \frac{3}{s + 2} - \frac{3}{s + 5}$$

By use of Table 5.1, the inverse transform of the preceding equation is found to be

$$y(t) = (2t + 3)e^{-2t} - 3e^{-5t} \qquad (6.15)$$

6.2 COMPLEX CONJUGATE ZEROS

Complex zeros of $B(s)$ always occur in pairs, and furthermore these zeros are always conjugates of one another. That is, they have the same real part but equal and opposite imaginary parts. Thus, if the polynomial $B(s)$ has a complex zero $a + jb$, the complex conjugate $a - jb$ will also be a zero of $B(s)$. Although the preceding discussion of distinct zeros is also applicable to complex conjugate zeros, the following analysis brings out more clearly the fact that a pair of complex conjugate zeros in $B(s)$ combine to introduce an exponentially damped sinusoidal term in $y(t)$.

A pair of complex conjugate zeros when multiplied together yield the following quadratic:

$$[s - (a + jb)][s - (a - jb)] = s^2 - 2as + (a^2 + b^2) \qquad (6.16)$$

For any given quadratic term, the values of a and b may be computed by equating coefficients of like terms as follows. Consider the expression

$$s^2 + 4s + 9$$

The coefficient 4 of the s term is equal to $-2a$, so that $-2a = 4$ or $a = -2$. Similarly, equating the constant terms gives $a^2 + b^2 = 9$ or $b = \sqrt{9 - 4} = \sqrt{5}$. Thus, the complex conjugate zeros are $a \pm jb = -2 \pm j\sqrt{5}$. If in the determination of b it is found that b is an imaginary number, the two zeros are real and unequal rather than complex conjugates. For example, consider the quadratic

$$s^2 + 8s + 12$$

The value of a is equal to -4, so that $b = \sqrt{12 - 16} = j\sqrt{4} = j2$. For this case the zeros are $a \pm jb = -4 \pm (j^2 2) = -4 \mp 2 = -6, -2$. Because the case of real zeros has been previously discussed, it is assumed in the following analysis that b is real, so that the zeros are complex conjugates.

For complex conjugate zeros $B(s)$ may be factored in the form

$$B(s) = [s - (a + jb)][s - (a - jb)](s - r_1)\cdots(s - r_{n-2}) \tag{6.17}$$

The partial-fraction expansion for $Y(s) = A(s)/B(s)$ is of the form

$$Y(s) = \frac{K_c}{s - (a + jb)} + \frac{K_{-c}}{s - (a - jb)} + \frac{K_1}{s - r_1} + \cdots + \frac{K_{n-2}}{s - r_{n-2}} \tag{6.18}$$

The inverse transform of Eq. (6.18) is

$$y(t) = K_c e^{(a + jb)t} + K_{-c} e^{(a - jb)t} + K_1 e^{r_1 t} + \cdots + K_{n-2} e^{r_{n-2} t} \tag{6.19}$$

The constants K_c and K_{-c} associated with the complex conjugate zeros are evaluated as usual for distinct zeros by the application of Eq. (6.8). That is,

$$
\begin{aligned}
K_c &= \lim_{s \to a + jb} \left\{ [s - (a + jb)] \frac{A(s)}{[s - (a + jb)][s - (a - jb)](s - r_1)\cdots(s - r_{n-2})} \right\} \\
&= \lim_{s \to a + jb} \left[\frac{1}{2jb} \frac{A(s)}{(s - r_1)\cdots(s - r_{n-2})} \right] = \frac{1}{2jb} K(a + jb) \tag{6.20}
\end{aligned}
$$

where

$$
\begin{aligned}
K(a + jb) &= \lim_{s \to a + jb} \frac{A(s)}{(s - r_1)\cdots(s - r_{n-2})} \\
&= \left[(s^2 - 2as + a^2 + b^2) \frac{A(s)}{B(s)} \right]_{s = a + jb}
\end{aligned}
$$

Similarly, the constant K_{-c} is obtained as follows:

$$
\begin{aligned}
K_{-c} &= \lim_{s \to a - jb} \left\{ [s - (a - jb)] \frac{A(s)}{[s - (a + jb)][s - (a - jb)](s - r_1)\cdots(s - r_{n-2})} \right\} \\
&= \lim_{s \to a - jb} \frac{A(s)}{(-2jb)(s - r_1)\cdots(s - r_{n-2})} = -\frac{1}{2jb} K(a - jb) \tag{6.21}
\end{aligned}
$$

where

$$
\begin{aligned}
K(a - jb) &= \lim_{s \to a - jb} \frac{A(s)}{(s - r_1)\cdots(s - r_{n-2})} \\
&= \left[(s^2 - 2as + a^2 + b^2) \frac{A(s)}{B(s)} \right]_{s = a - jb}
\end{aligned}
$$

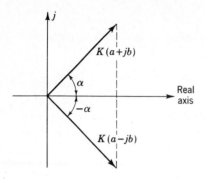

Figure 6.1 Vector representation for $K(a + jb)$ and $K(a - jb)$.

The constants $K(a + jb)$ and $K(a - jb)$ are complex conjugate numbers. These complex numbers may be represented as shown graphically in Fig. 6.1, whence

$$K(a + jb) = |K(a + jb)|e^{j\alpha}$$
$$K(a - jb) = |K(a + jb)|e^{-j\alpha}$$

(6.22)

where $|K(a + jb)| = |K(a - jb)|$ is the length of either vector, α is the angle of the vector $K(a + jb)$, and $-\alpha$ is the angle of the vector $K(a - jb)$.

The constants K_c and K_{-c}, which are also complex conjugate numbers, may be written in the form

$$K_c = \frac{1}{2jb}|K(a + jb)|e^{j\alpha}$$

(6.23)

$$K_{-c} = -\frac{1}{2jb}|K(a + jb)|e^{-j\alpha}$$

Substitution of K_c and K_{-c} from Eq. (6.23) into Eq. (6.19) gives

$$y(t) = \frac{1}{b}|K(a + jb)|e^{at}\frac{e^{j(bt + \alpha)} - e^{-j(bt + \alpha)}}{2j} + K_1 e^{r_1 t} + \cdots + K_{n-2}e^{r_{n-2}t}$$

or $\quad y(t) = \frac{1}{b}|K(a + jb)|e^{at}\sin(bt + \alpha) + K_1 e^{r_1 t} + \cdots + K_{n-2}e^{r_{n-2}t}$ (6.24)

Illustrative example 6.2 Determine the inverse transformation of the following transformed equation:

$$Y(s) = \frac{75}{(s^2 + 4s + 13)(s + 6)}$$

SOLUTION Equating coefficients to obtain the value of a and b for the quadratic yields $-2a = 4$, or $a = -2$, and $a^2 + b^2 = 13$, or $b = \sqrt{13 - 4} = 3$.

Evaluation of $K(a + jb)$ gives

$$K(a + jb) = \left[(s^2 - 2as + a^2 + b^2) \frac{A(s)}{B(s)} \right]_{s = a + jb}$$

$$= \left(\frac{75}{s + 6} \right)_{s = -2 + j3} = \frac{75}{4 + j3} \qquad (6.25)$$

As shown in Fig. 6.2, the vector whose real part is 4 and whose imaginary part is 3 may be expressed in the polar form

$$4 + j3 = 5 \underline{/36.9^\circ}$$

Hence, Eq. (6.25) becomes

$$K(a + jb) = \frac{75}{5 \underline{/36.9^\circ}} = 15 \underline{/-36.9^\circ}$$

Thus,

$$|K(a + jb)| = 15$$

and

$$\alpha = \measuredangle\, K(a + jb) = -36.9^\circ$$

The general form of the inverse transformation is

$$y(t) = \frac{1}{b} |K(a + jb)| e^{at} \sin(bt + \alpha) + K_1 e^{r_1 t}$$

Evaluation of K_1 gives

$$K_1 = \lim_{s \to -6} \frac{75}{s^2 + 4s + 13} = \frac{75}{25} = 3$$

Thus the desired result is

$$y(t) = 5e^{-2t} \sin(3t - 36.9^\circ) + 3e^{-6t} \qquad (6.26)$$

Application of the relationship $\sin(\alpha + \beta) = \sin \alpha \cos \beta + \cos \alpha \sin \beta$ in which $\alpha = 3t$ and $\beta = -36.9^\circ$ yields the alternate form

$$y(t) = e^{-2t}(4 \sin 3t - 3 \cos 3t) + 3e^{-6t} \qquad (6.27)$$

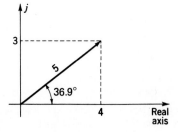

Figure 6.2 Vector representation for $(4 + j3)$.

This form of the result may be obtained directly. The response term due to a pair of complex conjugate roots $a \pm jb$ may be written in the form

$$y(t) = \frac{|K(a+jb)|}{b} e^{at} \sin(bt + \alpha)$$

$$= \frac{e^{at}}{b} |K(a+jb)|(\cos \alpha \sin bt + \sin \alpha \cos bt)$$

$$= \frac{e^{at}}{b} (A \sin bt + B \cos bt) \qquad (6.28)$$

where $\quad A = |K(a+jb)| \cos \alpha$
$\quad\quad\quad B = |K(a+jb)| \sin \alpha$

In Fig. 6.3 is shown the vector $K(a+jb)$. Note that the horizontal component is $A = |K(a+jb)| \cos \alpha$ and the vertical component is $B = |K(a+jb)| \sin \alpha$. For the preceding example $|K(a+jb)| = 15$ and $\alpha = -36.9°$; hence $A = 15 \cos(-36.9°) = 12$ and $B = 15 \sin(-36.9°) = -9$. For $a = -2$ and $b = 3$, application of Eq. (6.28) yields for the response due to the complex conjugate roots

$$e^{-2t}(4 \sin 3t - 3 \cos 3t)$$

Another method for obtaining the response due to complex conjugate roots results by writing $Y(s)$ in the form

$$Y(s) = \frac{75}{(s^2 + 4s + 13)(s + 6)} = \frac{As + B}{(s + 2)^2 + 3^2} + \frac{K_1}{s + 6}$$

Evaluation of the constants yields $A = -3$, $B = 6$, and $K_1 = 3$, whence

$$Y(s) = 4 \frac{3}{(s + 2)^2 + 3^2} - 3 \frac{(s + 2)}{(s + 2)^2 + 3^2} + \frac{3}{s + 6}$$

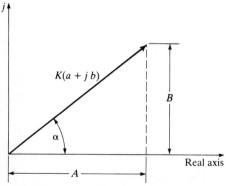

Figure 6.3 Horizontal component A and vertical component B of vector $K(a+jb)$.

Inverting yields the result for $y(t)$ given by Eq. (6.27). It should be pointed out that for higher-order systems the algebraic manipulations get very cumbersome with this latter method, whereas the algebraic manipulations required to evaluate $K(a + jb)$ do not appreciably increase in complexity for higher-order systems.

The general form of the response due to a pair of complex conjugate roots is

$$y(t) = \frac{|K(a + jb)|}{b} e^{at} \sin (bt + \alpha) \qquad (6.29)$$

The exponential factor a is the real part of the complex conjugate roots. The imaginary part b is the frequency of oscillation of the exponentially damped sinusoid. Thus, b is referred to as the damped frequency of oscillation or damped natural frequency. The period of each oscillation is $2\pi/b$. The envelope of this sinusoid is $(1/b) |K(a + jb)| e^{at}$. To have the exponential term decreasing with time, it is necessary that a be negative. For the case in which $a = 0$, a sinusoid of constant amplitude $(1/b) |K(a + jb)|$ results. For $a = 0$, Eq. (6.29) becomes

$$y(t) = \frac{1}{b} |K(a + jb)| \sin (bt + \alpha) \qquad (6.30)$$

In Fig. 6.4 is graphically illustrated the types of time-response terms that result from complex conjugate roots. When the roots lie to the left of the imaginary axis ($a < 0$), a decreasing sinusoid results; when the roots are on the imaginary axis ($a = 0$), a sinusoid of constant amplitude results; and when the roots are to the right of the imaginary axis ($a > 0$), an increasing sinusoid results.

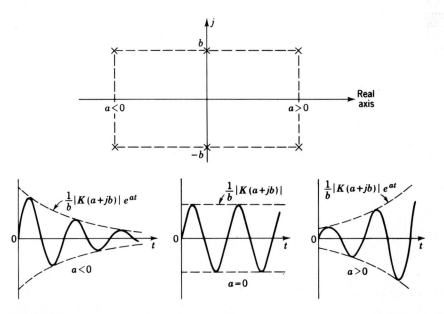

Figure 6.4 Response terms that result from complex conjugate zeros.

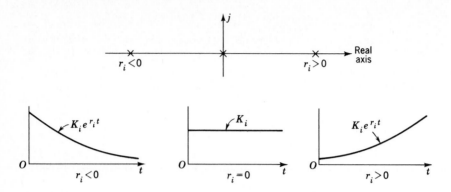

Figure 6.5 Response terms that result from real zeros.

In Fig. 6.5 is shown a plot of the types of response terms that result from real roots. These results follow directly from Eq. (6.9). Because the exponential factor is the value of the root, a negative root ($r_i < 0$) yields an exponentially decreasing term, while a positive root ($r_i > 0$) yields an exponentially increasing term. A root at the origin ($r_i = 0$) results in a constant term. Even though a root may be repeated, the exponential term dominates the response.

6.3 DAMPING RATIO AND NATURAL FREQUENCY

In Fig. 6.6 is shown a pair of complex conjugate roots. The root located at $a + jb$ is described by rectangular coordinates in which a is the horizontal component and b is the vertical component. This root is also described by polar coordinates

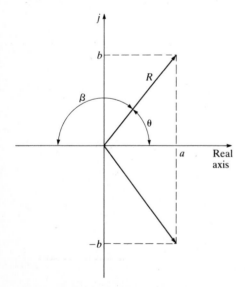

Figure 6.6 Polar representation $Re^{j\theta}$ of the vector $(a + jb)$.

in which R is the distance from the origin to the root and θ is the angle from the real axis to R. The rectangular and polar coordinates are related by the equations

$$a = R \cos \theta$$

$$b = R \sin \theta \tag{6.31}$$

$$a^2 + b^2 = R^2$$

In terms of polar coordinates, the quadratic form is

$$s^2 - 2as + a^2 + b^2 = s^2 - (2R \cos \theta)s + R^2$$

The angle β which is measured from the negative real axis to R is such that $\theta = \pi - \beta$. Noting that $\cos \theta = \cos (\pi - \beta) = -\cos \beta$ yields, for the quadratic form,

$$s^2 + (2R \cos \beta)s + R^2 \tag{6.32}$$

When $\beta = 0$, the quadratic form becomes

$$s^2 + 2Rs + R^2 = (s + R)^2 \tag{6.33}$$

For this case the roots are located on the negative real axis at $-R$. The roots are no longer complex conjugate, but rather repeated and real. The response ceases to be sinusoidal. When $\beta = 0$, the system is said to be critically damped.

The coefficient of the s term $(2R \cos \beta)$ in the quadratic form [Eq. (6.32)] is a measure of the amount of damping in a system. The damping ratio ζ is defined as the ratio of the actual amount of damping in a system $(2R \cos \beta)$ to the amount of damping when the system is critically damped $(2R)$.

Thus,

$$\zeta = \frac{2R \cos \beta}{2R} = \cos \beta \tag{6.34}$$

When $\beta = 90°$, the quadratic form becomes

$$s^2 + R^2$$

For this case, the roots are located on the imaginary axis at $\pm jR$. Because $a = R \cos 90° = 0$ and $b = R \sin 90° = R$, the resulting response is

$$\frac{1}{b} |K(a + jb)| e^{at} \sin (bt + \alpha) = \frac{|K(a + jb)|}{R} \sin (Rt + \alpha)$$

This is a pure sinusoidal oscillation of constant amplitude. The frequency of the oscillation is $R = \omega_n$, where ω_n is the natural frequency. Replacing R by ω_n and $\cos \beta$ by ζ in Eq. (6.32) yields the standard quadratic form

$$s^2 + 2\zeta\omega_n s + \omega_n^2 \tag{6.35}$$

The natural frequency ω_n is the frequency at which the system would oscillate if there were no damping. The natural frequency ω_n is also called the undamped natural frequency. The actual frequency at which the system does oscillate is b.

The frequency b is called the damped natural frequency. Because $a^2 + b^2 = R^2 = \omega_n^2$ and $a = R \cos \theta = -R \cos \beta = -\omega_n \zeta$, it follows that

$$b = \sqrt{\omega_n^2 - a^2} = \omega_n \sqrt{1 - \zeta^2} \qquad (6.36)$$

By knowing the natural frequency ω_n and the damping ratio ζ, then the damped natural frequency may be determined in accordance with Eq. (6.36).

For any given quadratic expression, numerical values of ζ and ω_n are computed by equating coefficients. For the quadratic $s^2 + 25$, it follows that $\omega_n^2 = 25$ or $\omega_n = 5$ and $2\zeta\omega_n = 0$ so that $\zeta = 0$. Similarly, it is found that $a = 0$ and $b = 5$. For the quadratic $s^2 + 6s + 25$, it is found that $\omega_n = 5$, $\zeta = 0.6$, $a = -3$, and $b = 4$. For the quadratic $s^2 + 10s + 25$, it follows that $\omega_n = 5$, $\zeta = 1.0$, $a = -5$, and $b = 0$. For the quadratic $s^2 + 15s + 25$, the values are $\omega_n = 5$, $\zeta = 1.5$, $a = -7.5$, and $b = j5.59$. Because b is imaginary the root location is at $a \pm jb = -7.5 \pm j(j5.59) = -7.5 \mp 5.59 = -13.09, -1.91$. The root location for the four preceding quadratic terms is shown in Fig. 6.7. Note that when the damping ratio ζ is greater than one, the roots are no longer complex conjugate but are real. The response is no longer sinusoidal, but rather exponential.

For the critically damped case ($s^2 + 10s + 25$), the coefficient of the s term is 10. The damping ratio ζ is the ratio of the actual coefficient of the s term to the value when the system is critically damped. Thus, for the quadratic ($s^2 + 6s + 25$), the damping ratio is $\zeta = 6/10 = 0.6$. Similarly, for the quadratic ($s^2 + 15s + 25$), the damping ratio is $\zeta = 15/10 = 1.5$.

In Fig. 6.8 is shown a general plot of a pair of complex conjugate roots. When a is negative, so that a decreasing exponential results, the roots are to the left of the imaginary axis, so that $0 < \beta < 90°$ in which case $1 > \zeta > 0$. For positive values of a, the roots are to the right of the imaginary axis, so that $90° < \beta < 180°$ and $0 > \zeta > -1$. A positive value of ζ yields a decreasing sinu-

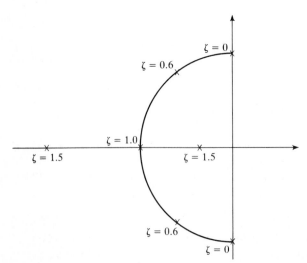

Figure 6.7 Root location for various damping ratios.

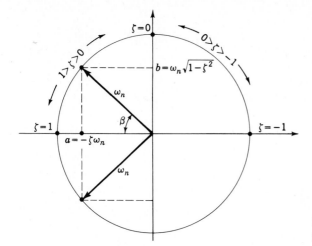

Figure 6.8 General plot of complex conjugate zeros.

soid, while a negative value results in an increasing sinusoid. For ζ equal to zero ($\beta = 90°$), a sinusoid of constant amplitude is obtained. When the damping ratio ζ is greater than 1, then two negative real roots result. When ζ is less than -1, then two positive real roots result.

Illustrative example 6.3 Let it be desired to determine the general equation for the transient response of a second-order system to a unit step function change which occurs at $t = 0$. The operational form of the differential equation is

$$y(t) = \frac{\omega_n^2}{D^2 + 2\zeta\omega_n D + \omega_n^2} f(t) \tag{6.37}$$

Assume that all the initial conditions are zero.

SOLUTION The transform for this differential equation is

$$Y(s) = \frac{\omega_n^2 F(s)}{s^2 + 2\zeta\omega_n s + \omega_n^2} = \frac{\omega_n^2}{(s^2 + 2\zeta\omega_n s + \omega_n^2)s} \tag{6.38}$$

where $F(s) = 1/s$ is the transform for the step input.

For complex conjugate zeros (that is, $-1 < \zeta < 1$), the response has the general form

$$y(t) = \frac{1}{b} |K(a + jb)| e^{at} \sin(bt + \alpha) + K_1 \tag{6.39}$$

Evaluation of $K(a + jb)$ gives

$$K(a + jb) = \frac{\omega_n^2}{s}\bigg|_{s = a + jb} = \frac{\omega_n^2}{a + jb} = \frac{\omega_n^2}{\sqrt{a^2 + b^2}\,\underline{/\tan^{-1} b/a}}$$

Thus,

$$|K(a + jb)| = \frac{\omega_n^2}{\sqrt{a^2 + b^2}} = \omega_n \tag{6.40}$$

and

$$\alpha = \measuredangle K(a + jb) = -\tan^{-1}\frac{b}{a} = \tan^{-1}\frac{-b}{a}$$

$$= \tan^{-1}\frac{-\sqrt{1 - \zeta^2}}{-\zeta} \tag{6.41}$$

where

$$\sqrt{a^2 + b^2} = \omega_n$$

$$a = -\zeta\omega_n$$

$$b = \omega_n\sqrt{1 - \zeta^2}$$

The constant K_1 is evaluated as follows:

$$K_1 = \lim_{s \to 0} \frac{\omega_n^2}{s^2 + 2\zeta\omega_n s + \omega_n^2} = 1$$

Thus, the desired transient response is

$$y(t) = \frac{1}{\sqrt{1 - \zeta^2}} e^{-\zeta\omega_n t} \sin\left[(\omega_n\sqrt{1 - \zeta^2})t + \alpha\right] + 1 \qquad -1 < \zeta < 1 \tag{6.42}$$

For the case in which $\zeta = 1$, the quadratic term in Eq. (6.38) is

$$s^2 + 2\omega_n s + \omega_n^2 = (s + \omega_n)^2$$

Thus the partial-fraction expansion for $Y(s)$ is

$$Y(s) = \frac{C_2}{(s + \omega_n)^2} + \frac{C_1}{s + \omega_n} + \frac{K_1}{s} \tag{6.43}$$

The constants C_2, C_1, and K_1 are evaluated as follows:

$$C_2 = \lim_{s \to -\omega_n} \frac{\omega_n^2}{s} = -\omega_n$$

$$C_1 = \lim_{s \to -\omega_n} \left(\frac{d}{ds}\frac{\omega_n^2}{s}\right) = \frac{-\omega_n^2}{s^2}\bigg|_{s = -\omega_n} = -1$$

$$K_1 = \lim_{s \to 0} \frac{\omega_n^2}{s^2 + 2\omega_n s + \omega_n^2} = 1$$

Thus,

$$y(t) = (C_2 t + C_1)e^{-\omega_n t} + K_1$$

or

$$y(t) = 1 - (\omega_n t + 1)e^{-\omega_n t} \tag{6.44}$$

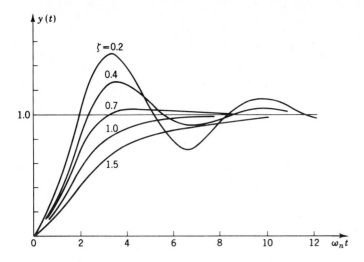

Figure 6.9 Response of a second-order system to a unit step input.

In a similar manner, the equation of the time response for $\zeta > 1$ or $\zeta < -1$ may also be derived. For these cases the zeros are real, and thus exponential terms are contributed to the response.

In Fig. 6.9 is shown the response $y(t)$ to a step change in the input for various values of the damping ratio ζ. It is to be noted that for $\zeta < 0.4$, there is an excessive amount of overshooting and oscillations. For $\zeta > 1$, an excessive amount of time is required to reach the new operating condition. Thus, for most control work, it is desired to have $0.4 < \zeta < 1$.

As shown in Fig. 6.6, the form $a \pm jb$ is the rectangular representation for complex conjugate roots. That is, in rectangular coordinates, the real part a is the horizontal component and the imaginary part b is the vertical component. Similarly, the polar form for complex conjugate roots is specified by the radius $\omega_n = \sqrt{a^2 + b^2}$ and the angle $\zeta = \cos \beta$. For computational purposes the response is usually obtained most readily by using the rectangular form. It is then an easy matter to obtain the polar quantities ω_n and ζ. The polar form has the advantage of providing a more general insight into the actual behavior of the system. For example, Fig. 6.9 shows that the form of the response is determined by ζ. The speed of response is governed by ω_n. That is, Fig. 6.9 shows that for $\zeta = 0.2$, the response $y(t)$ first crosses the value $y(t) = 1$ when $\omega_n t = 2$. Thus, if $\omega_n = 0.1$, the time is $t = 2/\omega_n = 2/0.1 = 20$ s. For $\omega_n = 10$, the time is $t = 2/\omega_n = 2/10 = 0.2$ s. Hence, the larger the value of ω_n, the faster is the speed of response.

Logarithmic Decrement

For an exponentially damped sinusoid, as shown in Fig. 6.10, the amplitude of the sinusoid after each oscillation changes in a geometric series. At time t_1 the

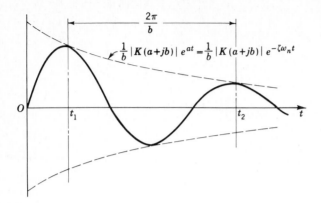

Figure 6.10 Logarithmic decrement.

amplitude is $(1/b)|K(a + jb)|e^{at_1}$. The time required to complete one period is $T = 2\pi/b$, and thus the time after one oscillation is completed is $t_2 = t_1 + T = t_1 + 2\pi/b$. The new amplitude is $(1/b)|K(a + jb)|e^{a(t_1 + 2\pi/b)}$. The ratio of amplitudes is

$$\frac{(1/b)|K(a + jb)|e^{at_1}}{(1/b)|K(a + jb)|e^{at_1}e^{2\pi a/b}} = e^{-2\pi a/b} = \exp\left[\frac{2\pi\zeta}{\sqrt{(1 - \zeta^2)}}\right] \tag{6.45}$$

The natural logarithm of this amplitude ratio is $-2\pi a/b = 2\pi\zeta/\sqrt{1 - \zeta^2}$, which is called the logarithmic decrement. The amplitude ratio after one oscillation is thus seen to be a function of the damping ratio only.

6.4 COMPUTER SOLUTION

The digital computer provides a very powerful tool for investigating the behavior of control systems. The solution of complex systems may be obtained on a small personal digital computer with rather simple programs. As is shown in the following, to solve a differential equation it is only necessary to integrate numerically. In Fig. 6.11 is shown a plot of the derivative $\dot{x}(t)$. The value of $x(t)$ at time t is

$$x(t) = \int \dot{x}(t)\, dt = \int_{t_0}^{t} \dot{x}(t)\, dt + C$$

Evaluation at time $t = t_0$ shows that $C = x(t_0)$. Thus

$$x(t) = \int_{t_0}^{t} \dot{x}(t)\, dt + x(t_0)$$

The preceding integral is the area under the curve of $\dot{x}(t)$ from t_0 to t. When the time increment $t - t_0$ is small, the area is closely approximated by the area of the

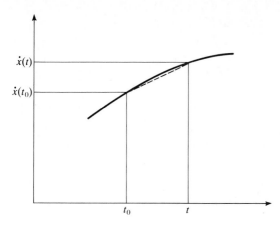

Figure 6.11 Trapezoidal approximation of area under a curve.

trapezoid. Thus,

$$x(t) = 0.5(t - t_0)[\dot{x}(t_0) + \dot{x}(t)] + x(t_0)$$

The computer statement for this integration is

$$X = 0.5*H*(DXO + DX) + XO$$

where

$$X = x(t) = \text{value of } x \text{ at time } t$$
$$XO = x(t_0) = \text{value of } x \text{ at time } t_0$$
$$DX = \dot{x}(t) = \text{value of } \dot{x} \text{ at time } t$$
$$DXO = \dot{x}(t_0) = \text{value of } \dot{x} \text{ at time } t_0$$
$$H = t - t_0 = \text{time increment}$$

The computer programs in this text are written in FORTRAN. They can also be written in other languages such as BASIC, PASCAL, C, etc.

Illustrative example 6.4 Write the program for solving the differential equation

$$y(t) = \frac{D + 5}{D^2 + 2D + 5} f(t)$$

The forcing function is $f(t) = u(t)$ and all initial conditions are zero.

SOLUTION When the numerator has a differential operator, the first step is to rewrite the differential equation such that the numerator operator is unity. That is,

$$x(t) = \frac{1}{D^2 + 2D + 5} f(t)$$

Hence

$$y(t) = (D + 5) \frac{1}{D^2 + 2D + 5} f(t) = (D + 5)x(t)$$

The differential equation for x is rewritten in the form

$$\ddot{x} + 2\dot{x} + 5x = f(t)$$

This differential equation is now solved for the highest-order derivative

$$\ddot{x} = f(t) - 2\dot{x} - 5x$$

Solving for the highest-order derivative is a fundamental step in the solution of a differential equation on a computer. The lower-order derivatives are then obtained by successive integration of the highest-order derivative. For this example problem,

$$\dot{x} = \int \ddot{x}\, dt \qquad \text{and} \qquad x = \int \dot{x}\, dt$$

The computer statements for the differential equation are

$$\text{D2X} = \text{F} - 2.0^*\text{DX} - 5.0^*\text{X}$$

and
$$\text{Y} = 5.0^*\text{X} + \text{DX}$$

where
$$\begin{aligned}
\text{X} &= x(t)\\
\text{DX} &= \dot{x}(t)\\
\text{D2X} &= \ddot{x}(t)\\
\text{F} &= f(t)\\
\text{Y} &= y(t)
\end{aligned}$$

The computer statements for performing the integrations are

$$\text{DX} = 0.5^*\text{H}^*(\text{D2XO} + \text{D2X}) + \text{DXO}$$
$$\text{X} = 0.5^*\text{H}^*(\text{DXO} + \text{DX}) + \text{XO}$$

The following program may now be written:

```
      T = 0.0
      X = 0.0
      DX = 0.0
      D2X = 0.0
      H = 0.01
      DO 25 K = 0,400
      F = 1.0
      Y = 5.0*X + DX
      IF (MOD(K,10).NE.0) GO TO 10
      WRITE (6,5) T,Y
    5 FORMAT (2F10.2)
   10 T = T + H
      XO = X
      DXO = DX
      D2XO = D2X
```

```
       D2X = F − 2.0*DX − 5.0*X
       DX = 0.5*H*(D2XO + D2X) + DXO
        X = 0.5*H*(DXO + DX) + XO
 25 CONTINUE
```

The first four statements of this program initialize the system. That is, these statements set the initial values $t_0 = 0$, $x(t_0) = 0$, $\dot{x}(t_0) = 0$, and $\ddot{x}(t_0) = 0$.

The next statement sets the integration interval $H = t - t_0 = 0.01$ s. The following statement starts the DO loop. The process is repeated 400 times so the final value will be the response when $T = 400H = 400(0.01) = 4$ s. The next statement is the forcing function. For a unit step function $F = 1.0$. The statement $Y = 5.0*X + DX$ computes the present value of the output $y(t)$. The first time through, $y(t) = y(t_0)$ is the value of $y(t)$ at time t_0. The IF MOD statement causes corresponding values of T and Y to be printed every time the remainder of dividing K by 10 is zero (i.e., when $K = 0, 10, 20, 30$, etc.). The statement $T = T + H$ sets the value of time at the end of the next integration integral. The statements $XO = X$, $DXO = DX$, and $D2XO = D2X$ set the values of $x(t_0)$, $\dot{x}(t_0)$, and $\ddot{x}(t_0)$ at the beginning of the new interval as the corresponding values $x(t)$, $\dot{x}(t)$, and $\ddot{x}(t)$ at the end of the preceding interval. The following statement computes the new value of the second derivative $\ddot{x}(t) = D2X$ at the end of this interval. The next statement integrates $\ddot{x}(t)$ over the interval to obtain the value of the first derivative $\dot{x}(t) = DX$ at the end of the interval. Similarly, the statement following integrates $\dot{x}(t)$ over the interval to obtain the value $x(t) = X$ at the end of the interval. The last statement sends the computer back to the beginning of the DO loop.

It is an easy matter to determine the response of this system to any forcing function by using the appropriate forcing function statement. For example, for $f(t) = 3e^t$, then

```
       F = 3.0*EXP(T)
```

When the forcing function is an impulse, then the impulse is approximated by a pulse. A unit impulse would be approximated by a pulse of width $H = 0.01$ and height $F = 100$. Thus,

```
       DO 25 K = 0,400
       F = 0.0
       IF(K.EQ.0)F = 100
```

Note that for the first interval $(K = 0)$, $F = 100$. For all other intervals, $F = 0$. Trapezoidal integration is very accurate and yields good results in solving differential equations on digital computers.

One of the most accurate methods for numerical integration is the fourth-order Runge-Kutta. The program for solving the preceding example problem using the fourth-order Runge-Kutta method is

```
      EXTERNAL DERIV
      COMMON F
      DIMENSION X(2),DX(2),WK1(2),WK2(2)
      NDR = 2
      T = 0.0
      H = 0.01
      DT = H
      X(1) = 0.0
      X(2) = 0.0
      DO 20 K = 0,400
      F = 1.0
      Y = 5.0*X(1)+X(2)
      IF (MOD(K,10).NE.0) GO TO 10
      WRITE (6,5) T,Y
    5 FORMAT (2F10.2)
   10 CALL RGKT(X,DX,T,DT,WK1,WK2,NDR,DERIV)
   20 CONTINUE
      STOP
      END

      SUBROUTINE DERIV (X,DX,T)
      COMMON F
      DIMENSION X(2),DX(2)
      DX(1) = X(2)
      DX(2) = F − 2.0*X(2)−5.0*X(1)
      RETURN
      END

      SUBROUTINE RGKT(Y,YD,T,DT,WK1,WK2,NDR,DERIV)
      DIMENSION Y(NDR),YD(NDR),WK1(NDR),WK2(NDR)
      DT2 = 0.5*DT
      M = 1
   10 TP = T+FLOAT(M/2)*DT2
      CALL DERIV(Y,YD,TP)
      DO 20 I = 1, NDR
      GO TO (12,14,16,18)M
   12 WK1(I) = Y(I)
      E = YD(I)*DT2
      WK2(I) = E
      GO TO 20
   14 E = YD(I)*DT2
      WK2(I) = WK2(I)+2.0*E
      GO TO 20
```

```
16 E = YD(I)*DT
   WK2(I) = WK2(I)+E
   GO TO 20
18 E = (WK2(I)+YD(I)*DT2)/3.0
20 Y(I) = WK1(I)+E
   M = M+1
   IF (M.LE.4) GO TO 10
   T = TP
   RETURN
   END
```

In this program NDR is the order of the differential equation which is 2 in this case. The statements before the DO loop initialize the system. Note that $x(t)$ is designated as X(1) and $\dot{x}(t)$ is designated as X(2). The forcing function statement is F = 1.0. The next statement computes the value of $y(t)$ from $x(t)$ and $\dot{x}(t)$. The IF MOD statement causes corresponding values of T and Y to be printed every time the remainder of dividing K by 10 is zero (i.e., when K = 0, 10, 20, ...). The CALL RGKT statement transfers the program to the RUNGE-KUTTA sub-routine (i.e., SUBROUTINE RGKT). The differential equation to be solved is

$$\ddot{x} = f(t) - 2\dot{x} - 5x$$

The CALL DERIV statement causes SUBROUTINE DERIV to be entered. The following statements are the formulation of the differential equation to be solved. That is,

```
DX(1) = X(2)
DX(2) = F-2.0*X(2)-5.0*X(1)
```

where $x(t) = $ X(1), $\dot{x}(t) = $ DX(1) = X(2), and $\ddot{x}(t) = $ DX(2). Note that the left side is always a term of the form DX(1), DX(2), ..., DX(N) and the right side never has a term that appears on the left side. In summary, by knowing the differential equation and the values at the beginning of the integration interval, the Runge-Kutta subroutine determines the values at the end of the interval.

In writing the computer program for a control system, first write the equation for the input and then proceed through the block diagram in the same manner in which the signals flow through the actual system. Thus, the computer program is written for the input-output relationship for each block in the block diagram representation for the system, and then these programs are hooked up in the same order in which the blocks are connected in the actual system. This technique is illustrated in the following example.

Illustrative example 6.5 Determine the computer diagram for the control system shown in Fig. 6.12. The input is $r(t) = u(t)$, a unit step function, and all initial conditions are zero.

SOLUTION The equation for the first summer is

$$a = r - d$$

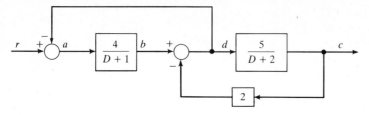

Figure 6.12 Multiloop control system.

The differential equation for the block following this summer is

$$b = \frac{4}{D + 1} a \qquad \text{or} \qquad \dot{b} = 4a - b$$

Integration of $\dot{b}$ yields b. The corresponding computer statements are

```
A=R−D
DB=4.0*A−B
B=0.5*H*(DBO+DB)+BO
```

The equation for the next summer is

$$d = b - 2c$$

The differential equation for the block following this summer is

$$c = \frac{5}{D + 2} d \qquad \text{or} \qquad \dot{c} = 5d - 2c$$

Integration of $\dot{c}$ yields c. The corresponding computer statements are

```
D=B−2.0*C
DC=5.0*D−2.0*C
C = 0.5*H*(DCO + DC) + CO
```

Because all initial conditions are zero, then $c = \dot{c} = r = 0$ at time $t = 0$. It follows that because $\dot{c} = 5d - 2c$, then $d = 0$ at time $t = 0$. Similarly, from the relationships $b = d + 2c$, $a = r - d$, and $\dot{b} = 4a - b$, it follows that $b = 0$, $a = 0$, and $\dot{b} = 0$ at time $t = 0$.

The resulting computer program using trapezoidal integration is

```
T=0.0
A=0.0
B=0.0
DB=0.0
C = 0.0
DC=0.0
```

```
        D=0.0
        H=0.01
        DO 25 K=0,400
        R=1.0
        IF (MOD(K,10).NE.0) GO TO 10
        WRITE(6,5) T,C
      5 FORMAT(2F10.2)
     10 T=T+H
        A=R-D
        BO=B
        DBO=DB
        DB=4.0*A-B
        B=0.5*H*(DBO+DB) + BO
        D=B-2.0*C
        CO=C
        DCO=DC
        DC=5.0*D-2.0*C
        C=0.5*H*(DCO+DC)+CO
     25 CONTINUE
```

When using the fourth-order Runge-Kutta method to integrate, the differential equation for the first block is written in the form

$$DB(1) = 4.0*A - B(1)$$

where $b = B(1)$ and $\dot{b} = DB(1)$. Similarly, the differential equation for the second block is expressed in the form

$$DC(1) = 5.0*D - 2.0*C(1)$$

where $c = C(1)$ and $\dot{c} = DC(1)$. The resulting computer program is

```
        EXTERNAL DERVB, DERVC
        DIMENSION B(1),C(1)
        COMMON A,D
        NDR=1
        T1=0.0
        T2=0.0
        H=0.01
        DT=H
        B(1)=0.0
        C(1)=0.0
        D=0.0
        DO 25 K=0,400
        IF (MOD(K,10).NE.0) GO TO 10
        WRITE (6,5) T1,C(1)
      5 FORMAT (2F10.2)
```

```
   10  R = 1.0
       A = R − D
       CALL RGKT(B,DB,T1,DT,WK1,WK2,1,DERVB)
       D = B(1) − 2.0*C(1)
       CALL RGKT(C,DC,T2,DT,WK1,WK2,1,DERVC)
   25  CONTINUE
       END

       SUBROUTINE DERVB(B,DB,T1)
       COMMON A,D
       DIMENSION B(1),DB(1)
       DB(1) = 4.0*A − B(1)
       RETURN
       END

       SUBROUTINE DERVC(C,DC,T2)
       COMMON A,D
       DIMENSION C(1),DC(1)
       DC(1) = 5.0*D − 2.0*C(1)
       RETURN
       END

       SUBROUTINE RGKT(Y,YD,T,DT,WK1,WK2,NDR,DERIV)
       DIMENSION Y(NDR),YD(NDR),WK1(NDR),WK2(NDR)
       DT2 = 0.5*DT
       M = 1
   10  TP = T + FLOAT(M/2)*DT2
       CALL DERIV(Y,YD,TP)
       DO 20 I = 1,NDR
        ⋮
       RETURN
       END
```

The statements in the main program before the DO loop initialize the system. Following the WRITE statement is the $R = 1$ statement which sets the input as a unit step function. The $A = R − D$ statement is the equation for the first summer. The CALL RGKT statement transfers the program to the SUBROUTINE RGKT. Only the first few lines of this program are written because this subroutine is always the same. It was given in full in the Runge-Kutta program for Illustrative example 6.4. When SUBROUTINE RGKT gets to the CALL DERIV statement it will go back to the CALL RGKT statement and then to the subroutine indicated by the last name in the parenthesis, which is DERVB. Note that SUBROUTINE DERVB is the differential equation for the first block ($\dot{b} = 4a − b$). SUBROUTINE RGKT starts with the value of b at the beginning of the integration interval and integrates $\dot{b}$ to determine the value of b at the end of the interval. Next the statement for the next summing point $D = B(1) − 4.0*C(1)$ is executed. The next statement CALL RGKT again transfers the computer to

SUBROUTINE RGKT. When the CALL DERIV statement is reached it will go back to the CALL RGKT statement and then go to the subroutine indicated by the last name in parenthesis, which is DERVC. SUBROUTINE DERVC is the differential equation for the second block ($\dot{c} = 5d - 2c$). SUBROUTINE RGKT starts with the value of c at the beginning of the integration interval and integrates $\dot{c}$ to determine the value of c at the end of the interval.

The digital computer is easily programmed to handle nonlinear systems. For example, suppose that the equation relating the output y of a block to the input $f(t)$ is

$$\dddot{y} + \dot{y}\ddot{y} + y^2 = f(t)$$

Solving for the highest-order derivative yields

$$\dddot{y} = f(t) - y^2 - \dot{y}\ddot{y}$$

Letting $y = X$, $\dot{y} = DX$, $\ddot{y} = D2X$, $\dddot{y} = D3X$, and $f(t) = F$, this becomes

$$D3X = F - X**2 - DX*D2X$$

Trapezoidal integration may now be used to integrate $\dddot{y} = D3X$ to obtain $\ddot{y} = D2X$. Integration of $\ddot{y} = D2X$ yields $\dot{y} = DX$, and finally integration of $\dot{y} = DX$ yields y.

To use the Runge-Kutta method, let $y = X(1)$, $\dot{y} = DX(1) = X(2)$, $\ddot{y} = DX(2) = X(3)$, and $\dddot{y} = DX(3)$. The resulting form of the differential equation to be used in SUBROUTINE DERIV is

```
DX(1) = X(2)
DX(2) = X(3)
DX(3) = F - X(1)**2 - X(2)*X(3)
```

6.5 TRANSIENT RESPONSE SPECIFICATIONS

A typical response of a second-order system to a step-function input is shown in Fig. 6.13. Depending on the performance criterion for the particular system under investigation, various specifications may be employed. The rise time t_r is the time at which the response first attains its final steady-state value. The percent maximum overshoot is 100 times the maximum amount by which the response overshoots its final steady-state value divided by the final steady-state value. The settling time t_s is the time required before the response does not oscillate more than some small percentage such as 2 or 5 percent from the final steady-state value. For the case of a step input of height h, the right side of Eq. (6.42) is multiplied by h. That is,

$$y(t) = h + \frac{h}{\sqrt{1 - \zeta^2}} e^{-\zeta\omega_n t} \sin\left[(\omega_n\sqrt{1 - \zeta^2})t + \alpha\right] \qquad (6.46)$$

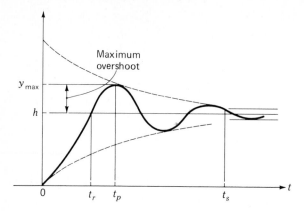

Figure 6.13 System response to a step input.

where α is given by Eq. (6.41). The time t_r such that $y(t) = h$ occurs when the sine term vanishes, i.e., when

$$(\omega_n\sqrt{1 - \zeta^2})t_r + \alpha = 0$$

Thus,

$$t_r = \frac{-\alpha}{\omega_n\sqrt{1 - \zeta^2}} = \frac{-\tan^{-1}(-\sqrt{1 - \zeta^2}/-\zeta)}{\omega_n\sqrt{1 - \zeta^2}} = \frac{\tan^{-1}(\sqrt{1 - \zeta^2}/-\zeta)}{\omega_n\sqrt{1 - \zeta^2}}$$

For a given value of ζ, the rise time is inversely proportional to the natural frequency ω_n. A plot of the normalized rise time $\omega_n t_r$ versus damping ratio ζ is shown in Fig. 6.14a.

Because $\tan \alpha = -\sqrt{1 - \zeta^2}/-\zeta$, it follows that $\cos \alpha = -\zeta$ and $\sin \alpha = -\sqrt{1 - \zeta^2}$. Thus, Eq. (6.46) may be expressed in the form

$$y(t) = h - \frac{he^{-\zeta\omega_n t}}{\sqrt{1 - \zeta^2}}[\zeta \sin(\omega_n\sqrt{1 - \zeta^2})t + \sqrt{1 - \zeta^2} \cos(\omega_n\sqrt{1 - \zeta^2})t] \quad (6.47)$$

The derivative dy/dt is

$$\frac{dy}{dt} = \frac{\omega_n he^{-\zeta\omega_n t}}{\sqrt{1 - \zeta^2}} \sin(\omega_n\sqrt{1 - \zeta^2})t$$

Peak values of $y(t)$ occur when $dy/dt = 0$, that is, when

$$(\omega_n\sqrt{1 - \zeta^2})t_p = \pi, 2\pi, 3\pi, \ldots$$

The maximum value of $y(t)$ occurs when

$$t_p = \frac{\pi}{\omega_n\sqrt{1 - \zeta^2}} \quad (6.48)$$

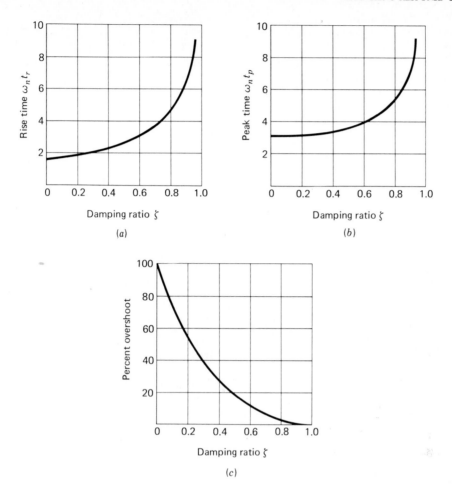

Figure 6.14 Performance criteria.

A plot of the normalized peak time $\omega_n t_p$ versus damping ratio ζ is shown in Fig. 6.14b. Substitution of $(\omega_n \sqrt{1 - \zeta^2})t = \pi$ into Eq. (6.47) yields, for the maximum value,

$$y_{max} = h + h \exp\left(-\zeta\pi/\sqrt{1 - \zeta^2}\right) \tag{6.49}$$

Subtracting the final steady-state value h, dividing by h, and multiplying by 100 yields, for the maximum percent overshoot,

$$PO = 100 \exp\left(-\zeta\pi/\sqrt{1 - \zeta^2}\right) \tag{6.50}$$

Note that the maximum value and percent overshoot depend on the damping ratio ζ only. A plot of percent overshoot versus damping ratio ζ is shown in Fig. 6.14c.

The envelopes to the response curve are shown dotted in Fig. 6.13. The sine term in Eq. (6.46) causes the response to oscillate between these envelopes. The maximum value for the sine term is $+1$ and the minimum value is -1. The difference between each envelope and the final value is $he^{-\zeta\omega_n t}/\sqrt{1-\zeta^2}$. Each envelope is within 5 percent of the final value when

$$\frac{he^{-\zeta\omega_n t}}{h\sqrt{1-\zeta^2}} = \frac{e^{-\zeta\omega_n t}}{\sqrt{1-\zeta^2}} \leq 0.05$$

Because $e^{-3} = 0.05$, it is customary to approximate the settling time such that the response is within 5 percent of its final value when $\zeta\omega_n t \geq 3$. Thus,

$$t_s \geq \frac{3}{\zeta\omega_n} \tag{6.51}$$

Similarly, because $e^{-4} = 0.02$, the settling time such that the response is within 2 percent of the final value is approximated by

$$t_s \geq \frac{4}{\zeta\omega_n} \tag{6.52}$$

Illustrative example 6.6 For the system shown in Fig. 6.15, determine the natural frequency, damping ratio, damped natural frequency, rise time, percent overshoot, and approximate 5 percent settling time.

SOLUTION The characteristic equation for this system is $s^2 + 5s + 25 = 0$. Comparison with the general form given by Eq. (6.35) shows that $\omega_n^2 = 25$ or $\omega_n = 5$ and $2\zeta\omega_n = 5$ or $\zeta = 5/10 = 0.5$. The damped natural frequency is $b = \omega_n\sqrt{1-\zeta^2} = 5\sqrt{3/4} = 4.33$ rad/s. The rise time is

$$t_r = \frac{\tan^{-1}(\sqrt{1-\zeta^2}/-\zeta)}{\omega_n\sqrt{1-\zeta^2}}$$

$$= \frac{\tan^{-1}[(\sqrt{3}/2)/(-1/2)]}{4.33} = \frac{\tan^{-1}(\sqrt{3}/-1)}{4.33} = \frac{2\pi/3}{4.33} = 0.483 \text{ s} \tag{6.53}$$

The angle whose tangent is $\sqrt{3}/-1$ is located in the second quadrant and is $120°$, which is $2\pi/3$ rad. Application of Eq. (6.50) yields, for the percent overshoot,

$$PO = 100 \exp(-\zeta\pi/\sqrt{1-\zeta^2}) = 100 \exp(-\pi/\sqrt{3}) = 16.3\% \tag{6.54}$$

The approximate settling time is

$$t_s \approx \frac{3}{\zeta\omega_n} = \frac{3}{(0.5)(5)} = 1.2 \text{ s} \tag{6.55}$$

Figure 6.15 Feedback control system.

6.6 GENERAL FORM OF TRANSIENT RESPONSE

The general form of the transient response can be ascertained directly from the zeros of the transformed function $B(s)$. For example, suppose that the zeros of $B(s)$ are those plotted in Fig. 6.16. It follows that $B(s)$ may be factored in the form

$$B(s) = [s - (a_1 + jb_1)][s - (a_1 - jb_1)](s - jb_2)(s + jb_2)(s - 0)(s - r_2)(s - r)^2$$

Performing a partial-fraction expansion on $Y(s) = A(s)/B(s)$ yields

$$Y(s) = \frac{K_{c_1}}{s - (a_1 + jb_1)} + \frac{K_{-c_1}}{s - (a_1 - jb_1)} + \frac{K_{c_2}}{s - jb_2} + \frac{K_{-c_2}}{s + jb_2}$$

$$+ \frac{K_1}{s} + \frac{K_2}{s - r_2} + \frac{C_2}{(s - r)^2} + \frac{C_1}{s - r} \qquad (6.56)$$

Taking the inverse transform of the preceding expression yields

$$y(t) = \frac{1}{b_1} |K(a_1 + jb_1)| e^{a_1 t} \sin (b_1 t + \alpha_1)$$

$$+ \frac{1}{b_2} |K(jb_2)| \sin (b_2 t + \alpha_2) + K_1 + K_2 e^{r_2 t} + (C_2 t + C_1)e^{rt} \qquad (6.57)$$

It is to be noted that a pair of complex conjugate zeros yields an exponentially varying sinusoidal term. A pair of complex conjugate zeros on the imaginary axis yields a sinusoid with a constant amplitude. The zero at the origin contributes a constant term. Distinct or multiple zeros on the real axis yield exponential terms.

The term $B(s) = L_n(s)D_{F(s)}$ consists of the zeros of $L_n(s)$ of the characteristic function for the system plus the zeros $D_{F(s)}$ corresponding to the denominator of the transform of the input excitation. If any zero of the characteristic function for

Figure 6.16 Graphical representation of zeros of $B(s)$.

the system $L_n(s)$ lies to the right of the imaginary axis, the response contains an increasing time function and will increase without bound. Thus, if any zero of $L_n(s)$ lies in the right half-plane (i.e., to the right of the imaginary axis), then the system is basically unstable. Whether a system is stable or unstable is a basic property of the system $L_n(s)$ itself and not the particular input, or excitation, to the system.

The zeros of $D_{F(s)}$ yield response terms associated with the particular excitation to the system. Take, for example, a ramp function $D_{F(s)} = s^2$, which gives a response term of the form $C_1 t$. It should be noted that the input which is a ramp function eventually becomes infinite, and thus the output of the system has been forced to infinity because of the particular input. As is illustrated by this example, the zeros of $D_{F(s)}$ do not affect the basic stability of a system but merely yield response terms appropriate to the particular excitation.

Illustrative example 6.7 The differential equation of operation for a control system is

$$y(t) = \frac{360(D^2 + D + 1)}{(D^2 + 2D + 5)(D^2 + 6D + 9)(D^2 + 6D + 8)} f(t) \qquad (6.58)$$

Determine the general form of the response equation when the input excitation $f(t)$ is a unit step function.

SOLUTION Because of the unit step function input, $N_{F(s)} = 1$, $D_{F(s)} = s$, so that $B(s) = L_n(s)D_{F(s)} = L_n(s)s$. Thus,

$$Y(s) = \frac{360(s^2 + s + 1)(1) + sI(s)}{(s^2 + 2s + 5)(s^2 + 6s + 9)(s^2 + 6s + 8)s} = \frac{A(s)}{B(s)} \qquad (6.59)$$

The partial-fraction expansion for $Y(s)$ gives

$$Y(s) = \frac{K_c}{s - (-1 + j2)} + \frac{K_{-c}}{s - (-1 - j2)} + \frac{C_2}{(s + 3)^2} + \frac{C_1}{s + 3} + \frac{K_1}{s + 2} + \frac{K_2}{s + 4} + \frac{K_3}{s}$$

Thus, the general form of the time solution is

$$y(t) = \tfrac{1}{2}|K(a + jb)|e^{-t} \sin(2t + \alpha) + (C_2 t + C_1)e^{-3t}$$
$$+ K_1 e^{-2t} + K_2 e^{-4t} + K_3 \qquad (6.60)$$

It should be noted in all cases that the exponent of each exponential term is equal to the horizontal distance from the imaginary axis to the zero of interest. That is, the exponential factor is equal to the numerical value of the real part of the zero. The terms due to zeros which are located far to the left of the imaginary axis have large exponential decaying factors and tend to decrease very rapidly to negligible quantities. Thus, zeros closer to the imaginary axis usually have more effect on the transient behavior. Accordingly, the analysis of complicated control systems is oftentimes approximated by omitting from the characteristic function zeros which do not affect substantially the performance of the system.

6.7 RESPONSE TO AN EXTERNAL DISTURBANCE

In this section, it is shown that the characteristic function for the differential equation which relates the output of a system to a change in the external disturbance is the same as that for the differential equation which relates the output to a change in the desired input. In Fig. 6.17 is shown the general representation for a feedback control system, in which $d(t)$ represents the external disturbance.

As previously discussed, the effect of the input $r(t)$ and external disturbance $d(t)$ on the output or controlled variable $c(t)$ may be considered individually and then each result added by superposition to obtain the total variation in $c(t)$. The block diagram which relates the input $r(t)$ to the output $c(t)$ without regard to the external disturbance is shown in Fig. 6.18. The equation relating $r(t)$ and $c(t)$ is

$$c(t) = \frac{G_1(D)G_2(D)}{1 + G_1(D)G_2(D)H(D)} \, r(t)$$

By using N_{G_1} to designate the numerator of $G_1(D)$ and D_{G_1} to designate the denominator of $G_1(D)$, etc., the preceding expression may be written in the form

$$c(t) = \frac{N_{G_1} N_{G_2} D_H}{D_{G_1} D_{G_2} D_H + N_{G_1} N_{G_2} N_H} \, r(t) \tag{6.61}$$

The characteristic function $L_n(D)$ for the system is $D_{G_1} D_{G_2} D_H + N_{G_1} N_{G_2} N_H$.

The block diagram which relates the external disturbance $d(t)$ to the output when $r(t)$ is considered zero is shown in Fig. 6.19. The equation relating $d(t)$ and $c(t)$ is

$$c(t) = \frac{G_2(D)}{1 + G_1(D)G_2(D)H(D)} \, d(t)$$

$$= \frac{N_{G_2} D_{G_1} D_H}{D_{G_1} D_{G_2} D_H + N_{G_1} N_{G_2} N_H} \, d(t) \tag{6.62}$$

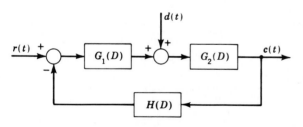

Figure 6.17 General representation for a feedback control system.

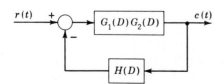

Figure 6.18 Block diagram for consideration of the input $r(t)$.

Figure 6.19 Block diagram for consideration of the external disturbance $d(t)$.

It can be shown that, if an excitation enters any place in the loop, the differential equation relating the disturbance and the output will always have the same characteristic function $L_n(D)$, which is the product of all the denominator terms plus the product of all the numerator terms. The constants which appear in the partial-fraction expansion depend upon where the disturbance enters the system. Thus, there is but one characteristic function for a system, and this function gives basic information as to the transient behavior of the system.

Impulse Response

The response of a system to a unit impulse excitation provides a good indication or measure of the general transient behavior of the system. The unit impulse is in effect a momentary disturbance which upsets the initial state of equilibrium of the system. In time, a stable system will return again to its equilibrium state.

Substitution of $F(s) = 1$ into Eq. (6.2) yields the transformed equation for a unit impulse excitation:

$$Y(s) = \frac{L_m(s) + I(s)}{L_n(s)} \tag{6.63}$$

It is to be noted that the basic form of the response of a system to a unit impulse is determined entirely by the zeros of the characteristic function $L_n(s)$. If any zero lies to the right of the imaginary axis, the output increases without bound. Thus the system is basically unstable. For an unstable system, it is impossible to achieve any equilibrium state, because as soon as the power is turned on, the output continually increases with time.

In Fig. 6.20 is shown the response to a unit impulse of the second-order system whose operational equation is

$$(D^2 + 2\zeta\omega_n D + \omega_n^2)y(t) = \omega_n^2 f(t) \tag{6.64}$$

It is to be seen from Fig. 6.20 that for $\zeta < 0.4$ there is a considerable amount of oscillation before the system again reaches equilibrium operation. Also, for $\zeta > 1.0$ a considerable amount of time is required for the system to return to its

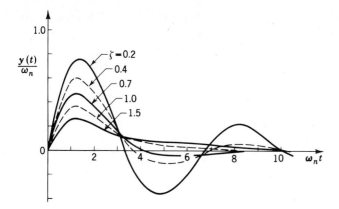

Figure 6.20 Response of a second-order system to a unit impulse.

initial state. Thus, it seems desirable to have $0.4 < \zeta < 1.0$. This is the result that was obtained from a consideration of the step-function response. Because the scale of the horizontal axis is $\omega_n t$, the speed of response is governed by the natural frequency ω_n, as was described in Sec. 6.3.

6.8 ROUTH'S STABILITY CRITERION

A major difficulty in using the Laplace transform method for determining the transient response of a feedback control system is that it necessitates determining the roots of the characteristic equation. The general form for the characteristic equation has previously been shown to be

$$D_{G_1} D_{G_2} D_H + N_{G_1} N_{G_2} N_H = 0 \qquad (6.65)$$

In determining the block diagram for a system, the terms N_{G_1}, N_{G_2}, N_H, D_{G_1}, D_{G_2}, and D_H are usually obtained in factored form. Because of the plus sign in Eq. (6.65), the roots of N_{G_1}, N_{G_2}, etc., are not the roots of the characteristic equation. Thus, it becomes necessary to determine the roots of the general polynomial represented by Eq. (6.65). This presents no difficulty for first- and second-order systems. The root of a first-order system is immediately obvious, and the two roots of a quadratic equation are readily solved. However, a third-order system requires determining the three roots of a cubic, a fourth-order system necessitates solving a quartic, etc.

Routh's criterion[2] is a method for determining whether or not any of the roots of the characteristic equation are in the right half-plane. The application of this criterion is as follows. First write the characteristic equation in the general form

$$b_n s^n + b_{n-1} s^{n-1} + b_{n-2} s^{n-2} + \cdots + b_2 s^2 + b_1 s + b_0 \qquad (6.66)$$

Next arrange the coefficients of the characteristic equation according to the following schedule:

$$
\begin{array}{cccccc}
b_n & b_{n-2} & b_{n-4} & b_{n-6} & \cdots \\
b_{n-1} & b_{n-3} & b_{n-5} & b_{n-7} & \cdots \\
c_1 & c_2 & c_3 & c_4 & \cdots \\
d_1 & d_2 & d_3 & \cdots \\
\cdots & \cdots & \cdots \\
e_1 & e_2 & 0 \\
f_1 & f_2 & 0 \\
g_1 & 0 \\
h_1 & 0
\end{array}
\tag{6.67}
$$

After first arranging the b coefficients as shown, the row of c terms is evaluated as follows:

$$
c_1 = \frac{b_{n-1}b_{n-2} - b_n b_{n-3}}{b_{n-1}}
$$

$$
c_2 = \frac{b_{n-1}b_{n-4} - b_n b_{n-5}}{b_{n-1}}
\tag{6.68}
$$

$$
c_3 = \frac{b_{n-1}b_{n-6} - b_n b_{n-7}}{b_{n-1}}
$$

Examination of Eq. (6.68) reveals the pattern for evaluating the c coefficients. By dropping down a row, the same pattern is used to obtain the d coefficients. That is,

$$
d_1 = \frac{c_1 b_{n-3} - b_{n-1}c_2}{c_1}
$$

$$
d_2 = \frac{c_1 b_{n-5} - b_{n-1}c_3}{c_1}
\tag{6.69}
$$

This process is continued until one more row is obtained than the order of the differential equation. Thus, a third-order equation has four rows, a fourth-order equation has five rows, etc. To illustrate, consider the characteristic equation

$$
s^4 + 3s^3 + s^2 + 6s + 2 = 0
\tag{6.70}
$$

The first two rows of the following array are obtained directly from the coefficients of Eq. (6.70), and the remaining rows are computed as just described:

$$
\begin{array}{ccccc}
s^4 & 1 & 1 & 2 & 0 \\
s^3 & 3 & 6 & 0 \\
s^2 & -1 & 2 & 0 \\
s^1 & 12 & 0 \\
s^0 & 2 & 0
\end{array}
\tag{6.71}
$$

The left column identifies the row. The first row is indicated by s^4, the second by s^3, etc. The exponent in the first term s^4 is the order of the polynomial. Because there is one more row than the order of the polynomial, the last row is always indicated by s^0.

Routh's criterion states that the number of changes of sign of the coefficients in the first column of the array is equal to the number of roots of the characteristic equation that are located to the right of the imaginary axis. For the preceding example, the signs of the numbers in the first column are seen to go from plus to minus and then back to plus again, so that there are two changes of sign. Thus, there are two roots located to the right of the imaginary axis.

The number of coefficients in the first column is $n + 1$, where n is the order of the characteristic equation. If the Routh array is continued past the last row, it is found that only zeros result. The reason that it is necessary to keep track of the number of pertinent rows is that if the coefficient of the last row is zero, then the characteristic equation has a root at the origin. If the coefficients of the last two pertinent rows are zero, then the characteristic equation has a double root at the origin, etc.

A Zero in the First Column

When one of the coefficients in the first column is zero, it may be replaced by a very small number ε for the purpose of computing the remaining coefficients in the array. Thus, consider the characteristic equation

$$s^5 + 2s^4 + 4s^3 + 8s^2 + 10s + 6 = 0 \qquad (6.72)$$

The Routh array is

s^5	1	4	10	0
s^4	2	8	6	0
s^3	$\varepsilon \approx 0$	7	0	
s^2	$8 - 14/\varepsilon$	6	0	
s^1	7	0		
s^0	6	0		

where $d_1 = (8\varepsilon - 14)/\varepsilon \approx 8 - 14/\varepsilon$ is a very large negative number when ε is positive and is a very large positive number when ε is negative. Regardless of whether ε is taken as a small positive number or a small negative number, there are two sign changes in the first column. Thus, the characteristic equation has two roots located to the right of the imaginary axis.

A Row of Zeros

When two or more of the roots of the characteristic equation are symmetrically located about the origin, then a row of Routh's array will contain all zeros. This situation occurs when the characteristic equation has a pair of real roots with opposite signs ($\pm r$), complex conjugate roots on the imaginary axis ($\pm j\omega$), or a pair of complex conjugate roots with opposite real parts $(a \pm jb, -a \pm jb)$. The row preceding the row of zeros is the polynomial of such factors. This polynomial is called the auxiliary equation.

Thus consider the characteristic equation

$$s^6 + 6s^5 + 10s^4 + 12s^3 + 13s^2 - 18s - 24 = 0 \qquad (6.73)$$

The Routh array is

s^6	1	10	13	-24	0
s^5	6	12	-18	0	
s^4	8	16	-24	0	
s^3	0	0	0		

The s^3 row contains all zeros. The auxiliary equation $A(s)$ is obtained from the coefficients of the preceding row as follows:

$$A(s) = 8s^4 + 16s^2 - 24 = 8(s^4 + 2s^2 - 3) = 8(s^2 - 1)(s^2 + 3)$$

The auxiliary equation is always an even order. The power of s for the first coefficient (8) is s^4, the power of s for the next coefficient (16) is s^2, and the power of s for the last coefficient (-24) is $s^0 = 1$. In general, the power of s for the first coefficient is the order of the row, and the power decreases by 2 for each succeeding coefficient. Dividing the polynomial by the auxiliary equation yields

$$
\begin{array}{r}
s^2 + 6s + 8 \\
s^4 + 2s^2 - 3 \overline{)\, s^6 + 6s^5 + 10s^4 + 12s^3 + 13s^2 - 18s - 24} \\
\underline{s^6 \qquad\quad + 2s^4 \qquad\quad - 3s^2} \\
6s^5 + 8s^4 + 12s^3 + 16s^2 - 18s \\
\underline{6s^5 \qquad\quad + 12s^3 \qquad\quad - 18s} \\
8s^4 \qquad\quad + 16s^2 \qquad - 24 \\
\underline{8s^4 \qquad\quad + 16s^2 \qquad - 24}
\end{array}
$$

Thus, the original characteristic equation may be written in the form

$$s^6 + 6s^5 + 10s^4 + 12s^3 + 13s^2 - 18s - 24 = (s^4 + 2s^2 - 3)(s^2 + 6s + 8) = 0$$

The roots of $(s^4 + 2s^2 - 3) = (s^2 - 1)(s^2 + 3)$ consist of the pair of real and opposite roots $(s^2 - 1) = (s - 1)(s + 1)$ and the pair of complex conjugate roots on the imaginary axis $(s^2 + 3) = (s + j\sqrt{3})(s - j\sqrt{3})$.

As another example, consider the characteristic equation

$$s^6 + 3s^5 + 2s^4 + 4s^2 + 12s + 8 = 0 \qquad (6.74)$$

The Routh array is

s^6	1	2	4	8
s^5	3	0	12	0
s^4	2	0	8	0
s^3	0	0	0	

The auxiliary equation is

$$A(s) = 2s^4 + 8 = 2(s^4 + 4)$$

Dividing the characteristic equation by $s^4 + 4$ yields

$$
\require{enclose}
\begin{array}{r}
s^2 + 3s + 2 \\
s^4 + 4\overline{\smash{)}s^6 + 3s^5 + 2s^4 + 0s^3 + 4s^2 + 12s + 8} \\
\underline{s^6 \qquad\qquad\qquad\; + 4s^2} \\
3s^5 + 2s^4 \qquad\qquad + 12s \\
\underline{3s^5 \qquad\qquad\qquad + 12s} \\
2s^4 \qquad\qquad\qquad + 8 \\
\underline{2s^4 \qquad\qquad\qquad + 8}
\end{array}
$$

The factored form for the original characteristic equation is

$$s^6 + 3s^5 + 2s^4 + 4s^2 + 12s + 8 = (s^4 + 4)(s^2 + 3s + 2) = 0$$

The roots of $(s^4 + 4) = (s^2 + 2s + 2)(s^2 - 2s + 2)$ consist of the complex pair located at $-1 \pm j$ and the pair located at $+1 \pm j$. Such a set of complex conjugate roots is called a quadripole.

It is not necessary to use Routh's criterion if any of the coefficients of the characteristic equation are zero or negative. When this is so, it can be shown that at least one root is located on, or to the right of, the imaginary axis.

To determine how many roots of the characteristic equation lie to the right of some vertical line a distance σ from the imaginary axis (i.e., the number of roots that have a real part greater than σ), transform the characteristic equation by substituting $s + \sigma$ for s, and then apply Routh's criterion as just described. The number of changes of sign in the first column for this new function is equal to the number of roots which are located to the right of the vertical line through σ.

6.9 SUMMARY

In this chapter it is shown that the transient response of a system is governed primarily by the location of the zeros of $B(s) = L_n(s)D_{F(s)}$. The zeros of $D_{F(s)}$ yield response terms appropriate to the particular excitation to the system. The characteristic function $L_n(s)$ is a basic property of the system itself. When all the zeros of $L_n(s)$ are located in the left half-plane, the system is stable (i.e., for any bounded input the response is also bounded). If any zero of $L_n(s)$ is located in the right half-plane the system is unstable (i.e., the response is always unbounded). The imaginary axis is the borderline between stable and unstable systems.

Complex imaginary zeros are undesirable because they yield constant sinusoids. A zero of $L_n(s)$ at the origin is also undesirable because it indicates an integration of the input. Note that an integrator in the feedforward elements (which integrates the error signal) does not yield a zero of $L_n(s)$ at the origin

Table 6.1 Location of zeros and corresponding response functions

Zeros of $B(s) = L_n(s)D_{F(s)}$	Type of response
Left half-plane (distinct or repeated)	Decaying exponential and/or decaying sinusoid
Right half-plane (distinct or repeated)	Increasing exponential and/or increasing sinusoid
Imaginary axis:	
Distinct	Constant and/or constant sinusoid
Repeated	Increasing time function and/or increasing sinusoid

because the zeros of $D_{G(s)}$ are not the zeros of $L_n(s)$. When the characteristic function $L_n(s)$ has a zero at the origin, a constant input (i.e., a step function) yields an unbounded time term $C_1 t$ in the output which is the integral of the input.

The basic form of the response due to repeated zeros is the same as that for distinct zeros, with the exception that repeated zeros on the imaginary axis yield increasing time terms rather than time terms with constant amplitudes. Because $B(s) = L_n(s)D_{F(s)}$, the zeros of $L_n(s)$ and $D_{F(s)}$ act independently to yield the time response unless one or more of the zeros of $L_n(s)$ and $D_{F(s)}$ are the same. This introduces repeated zeros in $B(s)$, which affects the basic form of the response equation only if the zeros are on the imaginary axis. For example, a repeated complex imaginary zero results in an increasing sinusoid rather than a constant sinusoid. Similarly, a repeated zero at the origin yields an increasing time function rather than a constant. To ensure stability, zeros of $L_n(s)$ should be excluded not only from the right half-plane but also from the imaginary axis.

Table 6.1 summarizes the types of response terms associated with the zeros of $B(s)$.

PROBLEMS

6.1 Determine the time response $y(t)$ for each of the following transformed equations:

(a) $Y(s) = \dfrac{4}{s(s + 1)(s + 2)}$

(b) $Y(s) = \dfrac{s + 4}{s(s + 1)(s + 2)}$

(c) $Y(s) = \dfrac{4}{s(s - 1)(s - 2)}$

(d) $Y(s) = \dfrac{s + 4}{s(s - 1)(s - 2)}$

6.2 Determine the time response $y(t)$ for each of the following transformed equations:

(a) $Y(s) = \dfrac{s + 3}{(s + 2)^2}$

(b) $Y(s) = \dfrac{s + 1}{(s + 2)^3}$

(c) $Y(s) = \dfrac{s^2 + 2}{(s + 2)^3}$

(d) $Y(s) = \dfrac{s^3 + 2s^2 + 4s + 2}{(s + 2)^4}$

6.3 Invert each of the following to determine the response $y(t)$:

(a) $Y(s) = \dfrac{2s + 1}{(s + 2)^2(s + 5)}$

(b) $Y(s) = \dfrac{5s + 1}{(s + 2)(s + 5)^2}$

(c) $Y(s) = \dfrac{3(s - 1)}{(s + 2)^2(s + 5)^2}$

(d) $Y(s) = \dfrac{3(2s + 1)}{(s + 2)^3(s + 5)}$

6.4 Use the Laplace transform method to determine the solution of each of the following differential equations when all of the initial conditions are zero:

(a) $(D + 1)(D + 2)y = e^{-t}$

(b) $(D + 1)(D + 2)y = e^{-2t}$

(c) $(D + 1)(D + 2)y = e^{-t} - e^{-2t}$

6.5 Use the Laplace transform method to determine the solution of the following differential equation when $f(t) = u_1(t)$ is a unit impulse, $y(0) = 1$, and $\dot{y}(0) = 0$:

$$(D + 1)^2 y(t) = f(t)$$

Use the initial-value theorem to check $y(0+)$ and $\dot{y}(0+)$.

6.6 Use the Laplace transform method to determine the solution of the following differential equation for the case in which $f(t) = u(t)$ is a unit step function and all initial conditions are zero:

$$y(t) = \dfrac{4(D + 1)}{(D + 2)^2} f(t)$$

6.7 Use the Laplace transform method to determine the solution of the following differential equation when $f(t) = e^{-t}$ and all initial conditions are zero:

$$y(t) = \dfrac{D + 4}{(D + 1)(D + 2)} f(t)$$

6.8 Same as Prob. 6.7 except $f(t) = e^{-2t}$.

6.9 Same as Prob. 6.7 except $f(t) = e^{-t} + e^{-2t}$.

6.10 For the system shown in Fig. P6.10, determine the response $c(t)$ for each of the following cases:

(a) $r(t) = u(t)$ and $c(0) = \dot{c}(0) = 0$

(b) $r(t) = e^{-t}$ and $c(0) = \dot{c}(0) = 1$

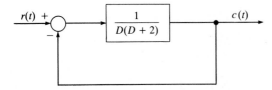

r(t) +

$\dfrac{1}{D(D + 2)}$

c(t)

Figure P6.10

6.11 For the system shown in Fig. P6.11, determine K and a such that the response to a unit impulse $[r(t) = u_1(t)]$ has the form

$$C(t) = C_1 e^{-t} + C_2 e^{-4t}$$

Evaluate C_1 and C_2 when all the initial conditions are zero.

R(s) +

$\dfrac{K}{s(s + a)}$

C(s)

Figure P6.11

6.12 Same as Prob. 6.11, except that it is desired that the impulse response have the form

$$c(t) = (C_0 + C_1 t)e^{-2t}$$

Evaluate C_0 and C_1 when all the initial conditions are zero.

6.13 Same as Prob. 6.11 except that it is desired that the impulse response have the form

$$c(t) = Ce^{-t} \sin (2t + \alpha)$$

Evaluate C and α when all the initial conditions are zero.

6.14 Use the Laplace transform method to solve the following differential equation:

$$y(t) = \frac{25(D + 1)}{D^2 + 10D + 25} f(t)$$

All the initial conditions are zero, and $f(t)$ is a unit step function. Use the initial-value theorem to check $y(0+)$ and $y'(0+)$.

6.15 Use the Laplace transform method to solve the following differential equations:

(a) $\dfrac{dy}{dt} + y = 2 \sin t$ (b) $\dfrac{dy}{dt} + y = 2 \cos t$

All the initial conditions are zero.

6.16 A system is described by the differential equation

$$(D^2 + 1)y = f(t)$$

All the initial conditions are zero. A unit step input $f(t) = u(t)$ is applied for π seconds [$f(t) = 1$ for $0 < t < \pi$ and $f(t) = 0$ for $t > \pi$]. Determine the response for $0 < t < \pi$ and for $t > \pi$.

6.17 Use the Laplace transform method to solve the following differential equation:

$$y(t) = \frac{25(D + 1)}{D^2 + 8D + 25} f(t)$$

All the initial conditions are zero, and $f(t)$ is a unit step function. Use the initial-value theorem to check $y(0+)$ and $y'(0+)$.

6.18 Same as Prob. 5.34 except $K = 4$.

6.19 Consider the differential equation

$$y(t) = \frac{5(D + 1)}{D^2 + 2D + 5} f(t)$$

Determine the response $y(t)$ for each of the following cases:

(a) $f(t) = u(t)$ and $y(0) = \dot{y}(0) = 0$ (b) $f(t) = 0$ and $y(0) = \dot{y}(0) = 1$

Specify the damping ratio, the natural frequency, and the damped natural frequency.

6.20 For the system shown in Fig. P6.20, determine K_1, K_2, and a such that the system will have a steady-state gain of 1.0, a damping ratio $\zeta = 0.5$, and a natural frequency $\omega_n = 4.0$.

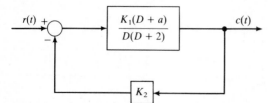

Figure P6.20

6.21 For the system shown in Fig. P6.21, determine K_1, K_2, and a such that the system will have a steady-state gain of 1.0, a damping ratio $\zeta = 0.6$, and a natural frequency $\omega_n = 5.0$.

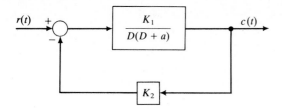

Figure P6.21

6.22 For the system shown in Fig. P6.22, the forcing function is $r(t) = e^{-2t}$, $c(0) = 1$, and $\dot{c}(0) = 0$. Determine the response $c(t)$ for each of the following cases:

  (a) $K = 1$ (b) $K = 10$

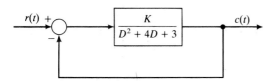

Figure P6.22

6.23 For the system shown in Fig. P6.23, the forcing function is $r(t) = e^{-t}$ and all of the initial conditions are zero. Determine the response $c(t)$ for each of the following cases:

  (a) $K = 1$ (b) $K = 2$

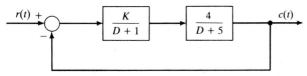

Figure P6.23

6.24 The block-diagram representation of a hydraulic system which provides the power for a numerically controlled machine tool is shown in Fig. P6.24. The forcing function is $r(t) = u(t)$ and all of the initial conditions are zero. Determine the response $c(t)$ when

  (a) $K = 5$ (b) $K = 9$ (c) $K = 25$

Figure P6.24

6.25 For the system shown in Fig. P6.25, determine the response $c(t)$ for each of the following cases:

  (a) $r(t) = 4u(t)$, $d(t) = 0$, and $c(0) = \dot{c}(0) = 0$
  (b) $r(t) = 0$, $d(t) = 4u(t)$, and $c(0) = \dot{c}(0) = 0$
  (c) $r(t) = d(t) = 0$, $c(0) = 4$, and $\dot{c}(0) = 0$

What is the damping ratio, natural frequency, and damped natural frequency for this system?

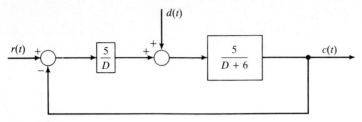

Figure P6.25

6.26 The block diagram for a feedback control system is shown in Fig. P6.26. Determine the response $c(t)$ for each of the following cases:

(a) $r(t) = 5u(t)$, $d(t) = 0$, and $c(0) = \dot{c}(0) = 0$

(b) $r(t) = 0$, $d(t) = 5u(t)$, and $c(0) = \dot{c}(0) = 0$

(c) $r(t) = d(t) = 0$, $c(0) = 1$, and $\dot{c}(0) = 0$

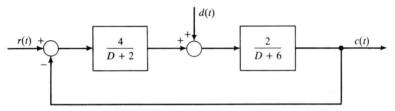

Figure P6.26

6.27 For the system shown in Fig. 5.10, $K_1 = 1.0$, $K_2 = 0.75$, $\tau = 0.25$, and all initial conditions are zero. Determine the response for each of the following cases:

(a) n_{in} is a unit impulse and $u = 0$ (b) u is a unit impulse and $n_{in} = 0$

6.28 Same as Prob. 6.27 except that $K_2 = 1.25$ rather than 0.75.

6.29 Determine the damping ratio, undamped natural frequency, and damped natural frequency for the system shown in Fig. P6.29. What is the response $c(t)$ of this system to a unit step function excitation $r(t) = u(t)$ when all initial conditions are zero?

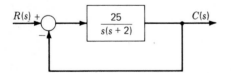

Figure P6.29

6.30 To improve the transient behavior of the system of Fig. P6.29 a controller with proportional and derivative action is added, as shown in Fig. P6.30. Determine the value of K such that the resulting system will have a damping ratio of 0.5. What is the response $c(t)$ of this resulting system to a unit step function excitation $r(t) = u(t)$ when all initial conditions are zero?

Figure P6.30

6.31 To improve the dynamic performance of the system of Fig. P6.29 derivative action (Ks) is added to the feedback path, as shown in Fig. P6.31. Determine the value of K such that the resulting system will have a damping ratio of 0.5. What is the response $c(t)$ of this resulting system to a unit step function excitation $r(t) = u(t)$ when all initial conditions are zero?

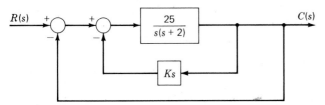

Figure P6.31

6.32 For the system shown in Fig. P6.29, determine the rise time t_r, the peak time t_p, the 2 percent settling time t_s, and the percent overshoot.

6.33 Change the term $s + 2$ in Fig. P6.29 to $s + 10$ and then determine the rise time t_r, the peak time t_p, the 2 percent settling time t_s, and the percent overshoot.

6.34 Write a computer program to determine the response $c(t)$ for each of the systems shown in Fig. P6.34. The forcing function is $r(t) = u(t)$ and all initial conditions are zero. Use

   (a) Trapezoidal integration
   (b) Fourth-order Runge-Kutta integration

Figure P6.34

6.35 Write a computer program to determine the response $c(t)$ for each of the systems shown in Fig. P6.35. The forcing function is $r(t) = u(t)$ and all initial conditions are zero. Use

   (a) Trapezoidal integration
   (b) Fourth-order Runge-Kutta integration

Figure P6.35

6.36 Write a computer program to determine the response $c(t)$ for each of the systems shown in Fig. P6.36. The forcing function is $r(t) = u(t)$ and all initial conditions are zero. Use

   (a) Trapezoidal integration
   (b) Fourth-order Runge-Kutta integration

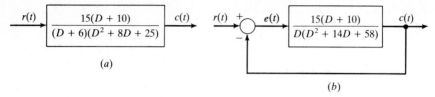

(a)

(b)

Figure P6.36

6.37 Write a computer program to determine the response $c(t)$ for the system shown in Fig. P6.37. The forcing function is $r(t) = u(t)$ and all initial conditions are zero. Use
   (a) Trapezoidal integration
   (b) Fourth-order Runge-Kutta integration

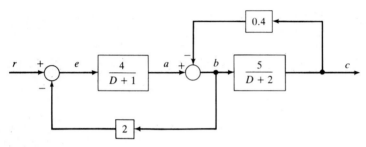

Figure P6.37

6.38 Write a computer program to determine the response of the system shown in Fig. P6.38. Determine the response when $r(t) = d(t) = 2u(t)$ and all initial conditions are zero. To check that superposition holds, determine the response to $r(t)$ and $d(t)$ separately, and then add the results. Use
   (a) Trapezoidal integration
   (b) Fourth-order Runge-Kutta integration

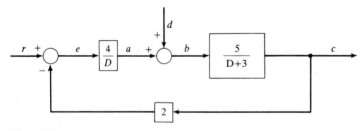

Figure P6.38

6.39 Write a computer program to evaluate the integral of e^t. Use
   (a) Trapezoidal integration
   (b) Fourth-order Runge-Kutta integration
Let t vary from 0 to 10 in increments of $\Delta t = 0.1$. Print out corresponding values of t, e^t, and the integral of e^t when $t = 0, 1, 2, \ldots, 10$.

6.40 The characteristic function for a control system is known to be

$$L_n(D) = (D + 5)(D^2 + 4D + 13)$$

Determine the general form of the equation for the response $y(t)$ of this system when the input excitation $f(t)$ is

    (a) An impulse   (b) A step function

Is the general form of these response expressions affected by the initial conditions?

6.41 Determine the characteristic function for the control system shown in Fig. P6.41. Use block-diagram algebra to move the constants A and B into the main loop. Does this affect the characteristic function?

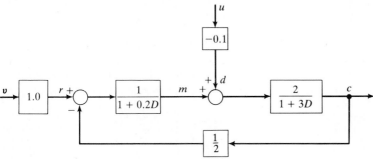

Figure P6.41

6.42 The transfer function $G(s)$ for a unity-feedback control system is

$$G(s) = \frac{K}{s(s^2 + 12s + 45)}$$

When $K = 50$, the roots of the characteristic equation are -5, -5, and -2. Thus,

$$s(s^2 + 12s + 45) + 50 = (s + 5)^2(s + 2)$$

Determine the response $c(t)$ of this system for the case in which $r(t) = 0$, $c(0) = 0$, $\dot{c}(0) = 0$, and $\ddot{c}(0) = 45$.

6.43 Same as Prob. 6.42 except $K = 104$. For this case the roots of the characteristic equation are -8, $-2 + j3$, and $-2 - j3$. Thus,

$$s(s^2 + 12s + 45) + 104 = (s + 8)(s^2 + 4s + 13)$$

6.44 (a) Apply Routh's criterion to determine an equation for K in terms of a, b, and c such that the cubic equation given below will have no roots with positive real parts:

$$as^3 + bs^2 + cs + K = 0$$

    (b) Same as part (a) except for the equation

$$as^3 + bs^2 + Ks + c = 0$$

6.45 The characteristic equation for a feedback control system is

$$(s + 2)(s^2 + 4s + 8) + K = 0$$

Use Routh's criterion to determine the range of values of K for which the system is stable.

6.46 The characteristic equation for a control system is

$$s(s^2 + 4s + 10) + K(s + 5) = 0$$

Use Routh's criterion to determine the range of values of K such that the system is stable.

6.47 The characteristic equation for a control system is

$$s(s^2 + 8s + a) + 4(s + 8) = 0$$

Use Routh's criterion to obtain the range of values of a for which the system is stable.

6.48 For each of the characteristic functions given below:

$$s^3 + 2s^2 + 5s + 24$$
$$s^4 + 2s^3 + 6s^2 + 2s + 5$$
$$s^4 + 3s^3 + 4s^2 + 6s$$

(a) Determine the number of zeros that lie on or to the right of the imaginary axis.
(b) Determine the number of zeros that have a real part greater than or equal to -4.

6.49 The characteristic equation for a certain system is

$$s(s^3 + 4s^2 + 2s + 3) + K(s + 1) = 0$$

Using Routh's criterion, determine the range of values of K for which the system is stable.

6.50 Apply Routh's criterion to determine the number of roots that lie in the right half-plane for each of the following characteristic equations:

(a) $s^3 + 5s^2 + 6s = 0$ (b) $s^3 + s^2 - s - 1 = 0$
(c) $s^3 + 2s^2 + 4s + 8 = 0$ (d) $s^4 + 5s^3 + 6s^2 = 0$
(e) $s^4 + 5s^3 + 5s^2 - 5s - 6 = 0$ (f) $s^4 + 5s^3 + 7s^2 + 5s + 6 = 0$
(g) $s^5 + s^4 + 5s^3 + 5s^2 + 4s + 4 = 0$ (h) $s^5 + 4s^4 + 6s^3 + 24s^2 + 25s + 100 = 0$

Identify the roots that are symmetrically located about the origin.

6.51 Determine the range of values of K such that the system whose characteristic equation is given below is stable:

$$s(s^2 + 2s + 5) + K(s + 4) = 0$$

Determine the factored form of the characteristic equation when K is such that the characteristic equation has a pair of roots on the imaginary axis.

6.52 For the system shown in Fig. P6.52, determine the range of values of K such that the system is stable. Determine the factored form of the characteristic equation when K has its maximum value in this range.

$R(s)$ $+$ $\dfrac{K(s + 9)}{s(s + 2)(s + 5)}$ $C(s)$

Figure P6.52

6.53 For the system shown in Fig. P6.53, determine the range of values of a such that the system is stable. Determine the factored form of the characteristic equation when a is such that the characteristic equation has a pair of roots on the imaginary axis.

$R(s)$ $+$ $\dfrac{10(s + a)}{s(s + 2)(s + 5)}$ $C(s)$

Figure P6.53

6.54 The characteristic equation for a system is

$$s(s^2 + 6s + 13) + K = 0$$

(a) Determine the value of K such that the characteristic equation has a pair of complex conjugate roots on the imaginary axis. Write the factored form of the characteristic equation for this case.

(b) Determine the value of K such that the characteristic equation has a pair of complex conjugate roots whose real part is -1. Write the factored form of the characteristic equation for this case.

6.55 The characteristic equation for a feedback control system is

$$s(s^2 + 8s + 20) + K = 0$$

(a) Determine the value of the gain K such that the characteristic equation has a pair of roots on the imaginary axis. Write the factored form of the characteristic equation for this case.

(b) Determine the value of the gain K such that the characteristic equation has a pair of roots on the vertical axis which passes through (-1). Write the factored form of the characteristic equation for this case.

6.56 For the system shown in Fig. P6.56, determine the range of values of K such that the system is stable.

(a) Write the factored form of the characteristic equation when K is such that the characteristic equation has a pair of roots on the imaginary axis.

(b) Determine the value of K such that the characteristic equation has a pair of complex conjugate roots whose real part is equal to -1. What is the factored form of the characteristic equation for this case?

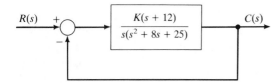

Figure P6.56

REFERENCES

1. Raven, F. H.: "Mathematics of Engineering Systems," Chap. 6, McGraw-Hill Book Company, New York, 1966.
2. Routh, E. J.: "Dynamics of a System of Rigid Bodies," 3d ed., The Macmillan Company, New York, 1877.

SEVEN

THE ROOT-LOCUS METHOD

The root-locus method was developed by W. R. Evans.[1,2] This method enables one to determine the roots of the characteristic equation (i.e., the zeros of the characteristic function) by knowing the factored form of the feedforward and feedback elements of a control system. As is discussed in Chap. 6, the transient behavior of a system is governed primarily by the roots of the characteristic equation for the system. Neither the initial conditions nor the particular excitation affects the basic operation of a system.

7.1 SIGNIFICANCE OF ROOT LOCI

The characteristic equation for the control system shown in Fig. 7.1a is

$$s(s + 4) + K = s^2 + 4s + K = 0 \tag{7.1}$$

The roots of the characteristic equation depend upon the value of K, which is the static loop sensitivity. As is shown in Fig. 7.1a, the static loop sensitivity K is the product of all the constant terms in the control loop when the coefficient of each s term is unity.

Because Eq. (7.1) is a quadratic, the roots are $r_{1,2} = a \pm jb$, in which $-2a = 4$ or $a = -2$ and $b = \sqrt{K - a^2}$. Thus,

For $K > 4$: $r_{1,2} = -2 \pm j\sqrt{K - 4}$

For $K = 4$: $r_1 = r_2 = -2$ $\tag{7.2}$

For $K < 4$: $r_{1,2} = -2 \pm j\sqrt{K - 4} = -2 \mp \sqrt{4 - K}$

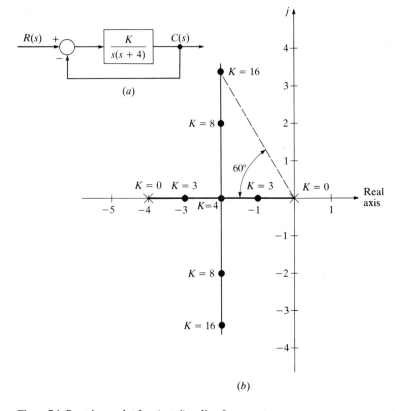

Figure 7.1 Root-locus plot for $s(s + 4) + K = 0$.

In this last case, the roots are real and unequal. The heavy lines in Fig. 7.1b are a plot of the roots of this characteristic equation for various values of K. When $K = 0$, the roots are $r_{1, 2} = 0, -4$; when $K = 3$, the roots are $-1, -3$; when $K = 4$, the roots are $r_1 = r_2 = -2$; when $K = 8$, the roots are $r_{1, 2} = -2 \pm j2$; etc.

Such a plot of the roots of the characteristic equation for each value of K as K varies from 0 to ∞ is a root-locus plot. From such a root-locus plot, it is an easy matter to select the value of K to yield the desired roots of the characteristic equation. For example, let it be desired to have a damping ratio $\zeta = 0.5$. As discussed in the preceding chapter, $\beta = \cos^{-1} \zeta = \cos^{-1} 0.5 = 60°$. As is shown in Fig. 7.1b, the line inclined at the angle $\beta = 60°$ intersects the root-locus plot at a value of $K = 16$. From this plot, the corresponding roots are

$$r_{1, 2} = -2 \pm j\sqrt{12} = -2 \pm j2\sqrt{3}$$

Thus, the factored form of the characteristic equation is

$$(s - r_1)(s - r_2) = [s - (-2 + j2\sqrt{3})][s - (-2 - j2\sqrt{3})] = s^2 + 4s + 16 \quad (7.3)$$

From the preceding expression, it follows that $\omega_n^2 = 16$ or $\omega_n = 4$, and $2\zeta\omega_n = 4$ or $\zeta = 4/2\omega_n = \frac{1}{2}$.

Illustrative example 7.1 For the system shown in Fig. 7.1a, determine the response $c(t)$ when $r(t) = 0$ and $c(0) = \dot{c}(0) = 1$ for each of the following cases:
   (a) $K = 3$ (b) $K = 4$ (c) $K = 8$

SOLUTION From Fig. 7.1a, it follows that the Laplace transform for $C(s)$ is

$$C(s) = \frac{K}{s(s+4)+K} R(s) \qquad (7.4)$$

This is the transformed equation when all initial conditions are zero. Replacing s by D, $C(s)$ by $c(t)$, and $R(s)$ by $r(t)$ yields the differential equation for the system,

$$c(t) = \frac{K}{D(D+4)+K} r(t)$$

or $\qquad (D^2 + 4D + K)c(t) = Kr(t)$

For $r(t) = 0$, the Laplace transform is

$$s^2C(s) - sc(0) - \dot{c}(0) + 4[sC(s) - c(0)] + KC(s) = 0$$

Hence,

$$(s^2 + 4s + K)C(s) = (s+4)c(0) + \dot{c}(0)$$

For $c(0) = \dot{c}(0) = 1$, this becomes

$$(s^2 + 4s + K)C(s) = s + 5$$

   (a) For $K = 3$, the transformed equation for $C(s)$ is

$$C(s) = \frac{s+5}{s^2+4s+3} = \frac{s+5}{(s+1)(s+3)} = \frac{2}{s+1} - \frac{1}{s+3}$$

Inverting yields, for the response,

$$c(t) = 2e^{-t} - e^{-3t}$$

It is to be noted that for $K = 3$, the factored form of the characteristic equation is

$$s^2 + 4s + 3 = (s+1)(s+3) = 0$$

Note from the root-locus plot shown in Fig. 7.1b that when $K = 3$ one root is located at -1 and the other root is located at -3.
   (b) For $K = 4$, the transform for $C(s)$ is

$$C(s) = \frac{s+5}{s^2+4s+4} = \frac{s+5}{(s+2)^2} = \frac{3}{(s+2)^2} + \frac{1}{s+2}$$

Inverting yields, for the response,

$$c(t) = (1 + 3t)e^{-2t}$$

For $K = 4$, the characteristic equation has repeated roots at -2. That is,

$$s^2 + 4s + 4 = (s + 2)^2 = 0$$

Note from the root-locus plot that when $K = 4$, the characteristic equation has a repeated root at -2.

(c) For $K = 8$, the transform for $C(s)$ is

$$C(s) = \frac{s + 5}{s^2 + 4s + 8}$$

This characteristic equation has complex conjugate roots located at $-2 \pm j2$. Evaluating $K(a + jb)$ yields

$$K(a + jb) = (s + 5)\Big|_{s = -2 + j2} = 3 + j2 = \sqrt{13}\,\underline{/33.7^\circ}$$

Hence

$$c(t) = \frac{|K(a + jb)|}{b} e^{at} \sin (bt + \alpha) = \frac{\sqrt{13}}{2} e^{-2t} \sin (2t + 33.7^\circ)$$

When $K(a + jb)$ is written in the form $A + jB$, then the result may be expressed in the equivalent form

$$c(t) = \frac{e^{at}}{b} (A \sin bt + B \cos bt) = \frac{e^{-2t}}{2} (3 \sin 2t + 2 \cos 2t)$$

where A is the real part of $K(a + jb)$ and B is the imaginary part. Note from the root-locus plot that when $K = 8$, the characteristic equation has complex conjugate roots at $-2 + j2$ and $-2 - j2$.

It is to be noted from Fig. 7.1b that for $0 < K < 4$ the roots of the characteristic equation are real. For example, when $K = 3$ the roots are located at -1 and -3. For real roots the system response is the sum of the exponential terms associated with each root. For $K > 4$, the roots of the characteristic equation are complex conjugates. For the characteristic equation $s^2 + 4s + K$, it follows that $\omega_n = \sqrt{K}$ and $2\zeta\omega_n = 4$ or $\zeta = 2/\sqrt{K}$. The larger the value of K the smaller is the damping ratio and the more oscillatory is the response. Many systems are such that they become unstable for large values of K. The larger the value of K the better is the steady-state accuracy. Thus, K must be sufficiently large to insure good steady-state accuracy but not too large such that the system will be too oscillatory or, worse yet, unstable.

In Fig. 7.2 is shown a general feedback control system. The transfer function $G(s)$ for the feedforward elements may be written in the form $G(s) = N_G/D_G$, where N_G is the numerator of $G(s)$ and D_G is the denominator. Similarly, the

Figure 7.2 General feedback control system.

transfer function for the feedback elements $H(s)$ may be written in the form $H(s) = N_H/D_H$. The characteristic equation is

$$D_G D_H + N_G N_H = 0$$

Dividing through by $D_G D_H$ yields the form

$$1 + G(s)H(s) = 1 + \frac{N_G N_H}{D_G D_H} = 1 + \frac{K(s - z_1)(s - z_2) \cdots (s - z_m)}{(s - p_1)(s - p_2) \cdots (s - p_n)} = 0 \quad (7.5)$$

where $G(s)H(s) = N_G N_H/D_G D_H$ is the open-loop transfer function. The zeros of the open-loop transfer function $(z_1, z_2, \ldots, z_m)$ are the roots of $N_G N_H = 0$. The poles of the open-loop transfer function $(p_1, p_2, \ldots, p_n)$ are the roots of $D_G D_H = 0$.

The preceding equation may be written in the form

$$\frac{(s - p_1)(s - p_2) \cdots (s - p_n)}{(s - z_1)(s - z_2) \cdots (s - z_m)} = -K \quad (7.6)$$

In constructing root-locus plots, the roots $p_1, p_2, \ldots, p_n$ of $D_G D_H = 0$ are plotted as ×'s (read ex's) and the roots $z_1, z_2, \ldots, z_m$ of $N_G N_H = 0$ are plotted as ○'s (read circles). In Fig. 7.3a is shown a typical × located at p_i. The vector from the origin to the × is p_i. The vector from the origin to any point s is s. The vector from the × at p_i to the s is $(s - p_i)$. Similarly, in Fig. 7.3b is shown a typical circle located at z_i. The vector from the origin to the circle is z_i. The vector from z_i to s is $(s - z_i)$. The numerator terms in Eq. (7.6) are the vectors from all of the ×'s to a trial point s. Similarly, the denominator terms in Eq. (7.6) are the vectors from

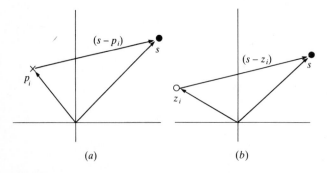

Figure 7.3 Representation for (a) vector $(s - p_i)$, (b) vector $(s - z_i)$.

all of the $\bigcirc$'s to the trial point s. The magnitude of Eq. (7.6) is

$$\frac{|s-p_1||s-p_2|\cdots|s-p_n|}{|s-z_1||s-z_2|\cdots|s-z_m|}=\frac{\displaystyle\prod_{i=1}^{n}|s-p_i|}{\displaystyle\prod_{i=1}^{m}|s-z_m|}=|-K| \tag{7.7}$$

This is referred to as the magnitude condition. The magnitude condition is used to obtain the value of the gain K at any point on a root-locus plot. From Eq. (7.7), it follows that the magnitude of the gain K at any point on a root-locus plot is equal to the product of the distances from all the $\times$'s to the point divided by the product of distances from all the $\bigcirc$'s to the point. When there are no $\bigcirc$'s, the magnitude of the gain is simply the product of the distances from all the $\times$'s to the point. For example, at the point $s=-1$ in Fig. 7.1b, the distance from the $\times$ at the origin to the point at -1 is 1. The distance from the $\times$ at -4 to the point at -1 is 3. Because there are no $\bigcirc$'s, the gain is the product of the distances from all the $\times$'s to the point, which in this case is $K=(1)(3)=3$. At the point $s=-2+j2$, the distance from the $\times$ at the origin to the point is $\sqrt{(-2)^2+(2)^2}=\sqrt{8}$. The distance from the $\times$ at -4 to the point $-2+j2$ is $\sqrt{2^2+2^2}=\sqrt{8}$. The gain is the product of these distances, which is $K=\sqrt{8}\sqrt{8}=8$.

Taking the angle of each vector represented in Eq. (7.6) shows that

$$[\measuredangle(s-p_1)+\measuredangle(s-p_2)+\cdots+\measuredangle(s-p_n)]$$
$$-[\measuredangle(s-z_1)+\measuredangle(s-z_2)+\cdots+\measuredangle(s-z_m)]=\measuredangle-K$$

For positive values of K, the point $-K$ lies on the negative axis, as is shown in Fig. 7.4. The vector $-K$ is the vector from the origin to the point $(-K)$ on the negative axis. The angle of this vector is 180°, (180° ± 360°), (180° ± 720°), etc. Thus, the preceding equation becomes

$$\sum_{i=1}^{n}\measuredangle(s-p_i)-\sum_{i=1}^{m}\measuredangle(s-z_i)=180°\pm k360° \qquad k=0,1,2,\ldots \tag{7.8}$$

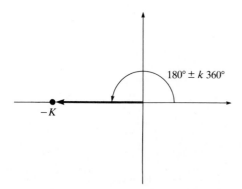

Figure 7.4 Vector representation for $-K$.

This is called the angle condition. The left side is the sum of the angles from all the $\times$'s to a point s minus the sum of the angles from all the $\bigcirc$'s to point s. In order that a point s be a point on a root-locus plot, it is necessary that the sum of the angles from all the $\times$'s to the point minus the sum of the angles from all the $\bigcirc$'s to the point equal $180° \pm k360°$.

7.2 CONSTRUCTION OF LOCI

By determining certain points and asymptotes, the loci may be sketched in quite readily. From Eq. (7.6), it follows that when $s = p_i$ then $K = 0$. Thus, one locus begins or originates at each $\times$. The number of $\times$'s is equal to the order n of the characteristic equation. A second-order equation has two loci, a third-order equation has three loci, etc. It also follows from Eq. (7.6) that when $s = z_i$ then $K = \infty$. Thus, one loci ends or terminates at each of the m $\bigcirc$'s. Usually there are more $\times$'s than $\bigcirc$'s. The remaining $n - m$ loci terminate along asymptotes at infinity.

The first place to start investigating the location of loci is along the real axis. In Fig. 7.5a is shown an $\times$ which is located on the real axis and a trial point s which lies to the right of the $\times$. The angle of the vector from the $\times$ to the point s is $0°$. Figure 7.5b is the same as Fig. 7.5a, except that the point s is located to the left of the $\times$. The angle of the vector from the $\times$ to point s is now $180°$. As the trial point s moves from the right side of an $\times$ or $\bigcirc$ to the left side, the angle of the vector changes from $0°$ to $180°$. In Fig. 7.5c is shown a pair of complex conjugate $\times$'s and a trial point s on the real axis. The angle from the lower $\times$ to the trial point is θ. The angle from the upper $\times$ to the trial point is $-\theta$. The sum

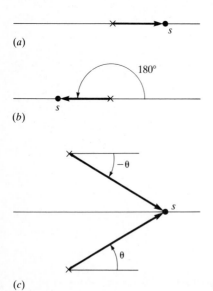

(a)

(b)

(c)

Figure 7.5 Angle to a trial point s on the real axis.

Figure 7.6 Determination of loci on the real axis.

$\theta + (-\theta)$ is zero. Thus, complex conjugate ×'s or ○'s do not affect the location of the loci on the real axis.

In Fig. 7.6 is shown the ×'s and ○'s which are located on the real axis for a characteristic equation. To the far right of all the ×'s and ○'s, the angle from each × and from each ○ to the trial point is 0°. There is no loci in this region because the sum of the angles from all the ×'s to the trial point minus the sum of the angles from all the ○'s to the trial point is zero. In the region between the origin and −1, the angle of the vector from the × at the origin to the trial point has changed by 180°. All other angles are still zero. There is a loci in the region from the origin to −1 because the sum of the angles from all the ×'s to the trial point minus the sum of the angles from all the ○'s to the trial point is 180°. In the region between −1 and −2, the angle of the vector from the ○ at −1 to the trial point has changed by 180°. The angle condition is no longer satisfied, so there is no loci in this region. It is to be noted that every time a trial point moves from the right of an × or ○ to the left, the angle changes by 180°. As shown in Fig. 7.6, the location of the loci on the real axis alternates between ×'s and ○'s. As is illustrated in Fig. 7.7, there is never a locus to the right of the first × or ○ on the real axis, but there is always a locus to the left of the first × or ○, there is never a locus to the left of the second × or ○, there is always a locus to the left of the third × or ○, never left of the fourth, always left of the fifth, and so on, alternating. Note in Fig. 7.1*b* that there is an × at the origin and an × at −4. The loci on the real axis lies between these ×'s.

The next step in the construction of the loci is to determine the asymptotes as *s* approaches infinity, i.e., the location of the loci for large values of *s*. In Fig. 7.8

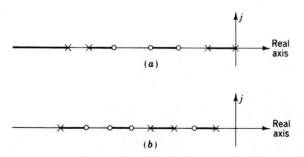

Figure 7.7 Loci on the real axis.

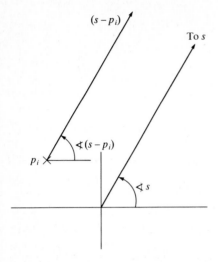

Figure 7.8 Trial point s located at infinity.

is shown an $\times$ located at point p_i. The point s is near infinity and cannot be seen on this plot. However, following the vector which is marked (to s) leads to the point s. In order that the vector $(s - p_i)$ meets the point s at infinity it is necessary that these two vectors be parallel. Thus, the angle of the vector $(s - p_i)$ must be equal to the angle of the vector s. That is,

$$\angle (s - p_i) \approx \angle s$$

Similarly, it follows that for very large values of s,

$$\angle (s - z_i) \approx \angle s$$

For very large values of s, the angle condition [Eq. (7.8)] becomes

$$\lim_{s \to \infty} \sum_{i=1}^{n} \angle (s - p_i) - \lim_{s \to \infty} \sum_{i=1}^{m} \angle (s - z_i) = \sum_{i=1}^{n} \angle s - \sum_{i=1}^{m} \angle s$$

$$= (n - m) \angle s = 180° \pm k360°$$

Thus, the angles of the asymptotes are

$$\angle s = \frac{180° \pm k360°}{n - m} \qquad k = 0, 1, 2, 3, \ldots \tag{7.9}$$

where n = highest power of s in $D_G D_H$ = number of $\times$'s
   m = highest power of s in $N_G N_H$ = number of $\bigcirc$'s

The number of distinct asymptotes is equal to $n - m$. Although it would appear from Eq. (7.9) that there are more asymptotes, the angles repeat for values of k after $n - m$ distinct angles have been determined. For $n - m = 1, 2, 3$, and 4,

Eq. (7.9) becomes

$$\measuredangle s = 180° \pm k360° \qquad\qquad n - m = 1$$

$$\measuredangle s = \frac{180° \pm k360°}{2} = 90° \pm k180° \qquad n - m = 2$$

$$\measuredangle s = \frac{180° \pm k360°}{3} = 60° \pm k120° \qquad n - m = 3$$

$$\measuredangle s = \frac{180° \pm k360°}{4} = 45° \pm k90° \qquad n - m = 4$$

In Fig. 7.9 is shown a plot of the asymptotes for the cases in which $n - m = 1, 2, 3,$ and 4. For $n - m = 1$ the angle is $180°$; for $n - m = 2$ the angles are $\pm 90°$; for $n - m = 3$ the angles are $\pm 60°$ and $180°$; and for $n - m = 4$ the angles are $\pm 45°$ and $\pm 135°$.

In order to locate the asymptotes, it is necessary to know where they intersect the real axis. The point σ_c where the asymptotes cross the real axis is determined from the general equation

$$\sigma_c = \frac{\sum \text{roots of } D_G D_H - \sum \text{roots of } N_G N_H}{(\text{number of roots of } D_G D_H) - (\text{number of roots of } N_G N_H)}$$

$$= \frac{(p_1 + p_2 + \cdots + p_n) - (z_1 + z_2 + \cdots + z_m)}{n - m} \qquad (7.10)$$

To verify Eq. (7.10), first write the characteristic equation [that is, Eq. (7.6)] in the form

$$(s - p_1)(s - p_2) \cdots (s - p_n) + K(s - z_1)(s - z_2) \cdots (s - z_m) = 0$$

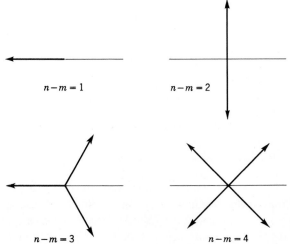

$n-m = 1$

$n-m = 2$

$n-m = 3$

$n-m = 4$

Figure 7.9 Asymptotes for $n - m = 1, 2, 3,$ and 4.

Solving for $-K$ gives

$$-K = \frac{(s-p_1)(s-p_2)\cdots(s-p_n)}{(s-z_1)(s-z_2)\cdots(s-z_m)} = \frac{s^n - (p_1 + p_2 + \cdots + p_n)s^{n-1} + \cdots}{s^m - (z_1 + z_2 + \cdots + z_m)s^{m-1} + \cdots}$$

(7.11)

It is to be noted that the coefficient of the second term in a polynomial is the negative of the sum of the roots.

Each numerator factor and each denominator factor may be written in the following form:

$$s - p_i = (s - \sigma_c) - (p_i - \sigma_c)$$

$$s - z_i = (s - \sigma_c) - (z_i - \sigma_c)$$

In Fig. 7.10, the point s is indicated by a dot, the point p_i by an $\times$, the point z_i by a $\bigcirc$, and σ_c by a $+$. The various vector quantities are indicated on this diagram. For a given value of p_i the vector $p_i - \sigma_c$ remains constant. Hence, as s becomes infinite, $s - \sigma_c \gg p_i - \sigma_c$. Thus, for very large s,

$$s - p_i \approx s - \sigma_c$$

Similarly, as s becomes infinite it follows that

$$s - z_i \approx s - \sigma_c$$

Thus,

$$-K \approx \frac{(s - \sigma_c)^n}{(s - \sigma_c)^m} = (s - \sigma_c)^{n-m} = s^{n-m} - (n - m)\sigma_c\, s^{n-m-1} + \cdots$$

Performing the division indicated in Eq. (7.11) gives

$$-K = s^{n-m} - [(p_1 + p_2 + \cdots + p_n) - (z_1 + z_2 + \cdots + z_m)]s^{n-m-1} + \cdots$$

Equating the coefficients of the second terms in the two preceding equations for $-K$ and solving for σ_c verifies the result given by Eq. (7.10). For the case of Fig. 7.1b, there is an $\times$ at the origin, an $\times$ at -4, and no $\bigcirc$'s. For $n = 2$ and $m = 0$, then $n - m = 2$. From Fig. 7.9, for $n - m = 2$ the angle of the asymptotes is $\pm 90°$. Application of Eq. (7.10) in which $p_1 = 0$ and $p_2 = -4$ yields

$$\sigma_c = \frac{[(0) + (-4)] - [0]}{2} = -2$$

For the case of Fig. 7.1b, the root-locus plot lies on the asymptotes. In Fig. 7.1b, it is to be noted that there is a loci on the real axis between the origin and -4. Because there are no $\bigcirc$'s, the two loci must terminate at infinity along asymptotes. Thus, it is necessary that the loci break away from the real axis. The location of the point σ_b at which the loci breaks away from the real axis is distinguished by the fact that the gain K has its maximum value on the real axis at the breakaway point. Note that at the origin and at -4 the value of K is zero, at -1 and at -3 the value of K is 3, and at -2 the value of K is 4. Differentiating the equation for K with respect to s and then setting dK/ds equal to zero

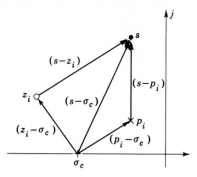

Figure 7.10 Asymptote crossing σ_c.

yields the value of s at which K is a maximum. The characteristic equation for Fig. 7.1b is

$$s(s + 4) + K = 0$$

Solving for $-K$ shows that

$$-K = s^2 + 4s$$

Differentiating and setting $-dK/ds$ equal to zero yields

$$-\frac{dK}{ds} = 2(s + 2) = 0$$

Thus, the value of s at which the loci breaks away from the real axis is $s = \sigma_b = -2$. As is illustrated in Fig. 7.1b, two loci always break away from the real axis in a direction perpendicular to the axis.

In Fig. 7.11a is shown the block diagram for a feedback control system and in Fig. 7.11b is shown the corresponding root-locus plot. The characteristic equation for this system is

$$(s^2 + 8s + 25) + K(s + 1)(s + 5) = 0$$

This system has a pair of complex conjugate $\times$'s at $-4 \pm j3$, a $\bigcirc$ at -1 and a $\bigcirc$ at -5. Because there is a loci on the real axis between -1 and -5 and because one loci terminates at each of the $\bigcirc$'s, it is necessary that the loci break into the real axis. A break-in point is similar to a breakaway point except the loci breaks into the real axis at a break-in point whereas it breaks away from the real axis at a breakaway point. At a break-in point, K has its minimum value on the real axis. Setting dK/ds equal to zero yields the value $s = \sigma_b$ at which the break-in point occurs. Thus,

$$-K = \frac{s^2 + 8s + 25}{s^2 + 6s + 5}$$

$$-\frac{dK}{ds} = \frac{(s^2 + 6s + 5)(2s + 8) - (s^2 + 8s + 25)(2s + 6)}{(s^2 + 6s + 5)^2}$$

$$= -\frac{2(s^2 + 20s + 55)}{(s^2 + 6s + 5)^2} = 0$$

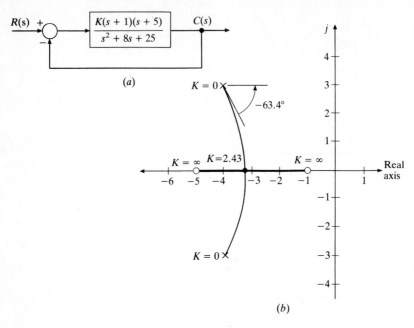

(a)

(b)

Figure 7.11 Root-locus plot for $(s^2 + 8s + 25) + K(s + 1)(s + 5) = 0$.

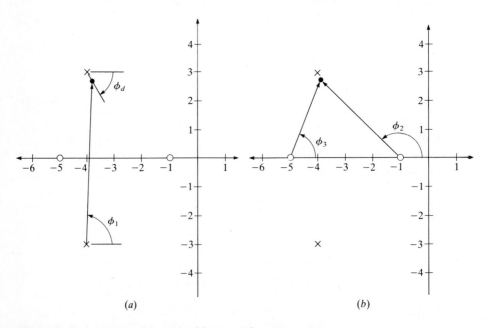

(a)   (b)

Figure 7.12 Determination of the angle of departure ϕ_d.

The value of s which makes the numerator go to zero is the location of the break-in point, $s = \sigma_b = -3.29$. Substitution of this value of s into the equation for K shows that the value of K at the break-in point is $K = 2.43$.

As is shown in Fig. 7.11b, the angle at which the loci departs or leaves the $\times$ at $-4 + j3$ is $-63.4°$. This angle is called the angle of departure. The value of the angle of departure is obtained by taking a trial point s which is located close to the $\times$ at $-4 + j3$ and then applying the angle condition. In Fig. 7.12a, the angle ϕ_1 is the angle from the $\times$ at $-4 - j3$ to the point indicated by the dot which is close to the $\times$ at $-4 + j3$. The unknown angle of departure ϕ_d is the angle from the $\times$ at $-4 + j3$ to the nearby point. In Fig. 7.12b, the angle ϕ_2 is the angle from the $\bigcirc$ at -1 to the point near the $\times$ at $-4 + j3$, and the angle ϕ_3 is the angle from the $\bigcirc$ at -5 to this point. Application of the angle condition shows that

$$(\phi_1 + \phi_d) - (\phi_2 + \phi_3) = 180° \pm k360°$$

where $\phi_1 + \phi_d$ is the sum of the angles from the $\times$'s to the trial point and $\phi_2 + \phi_3$ is the sum of the angles from the $\bigcirc$'s to the trial point.

As the trial point approaches the $\times$ at $-4 + j3$, then ϕ_1, ϕ_2, and ϕ_3 become

$$\phi_1 = \tan^{-1} \frac{6}{0} = 90°$$

$$\phi_2 = \tan^{-1} \frac{3}{-3} = 135°$$

$$\phi_3 = \tan^{-1} \frac{3}{1} = 71.6°$$

Different values of k merely add or subtract $360°$ from the angle. For $k = 0$, the preceding angle condition becomes

$$(90° + \phi_d) - (135° + 71.6°) = 180°$$

Hence

$$\phi_d = 296.6° = -63.4°$$

Because complex roots always occur as conjugate pairs, root-locus plots are always symmetrical about the real axis. Thus, there is no need to obtain the angle of departure from the $\times$ at $-3 - j4$. The angle of departure must be $63.4°$.

In Fig. 7.13a is shown the block diagram for a system and in Fig. 7.13b is shown the corresponding root-locus plot. This is the same system as shown in Fig. 7.11 except that the $\times$'s and $\bigcirc$'s have been interchanged. Because there are two $\times$'s and two $\bigcirc$'s, one loci begins at each $\times$ and one loci terminates at each

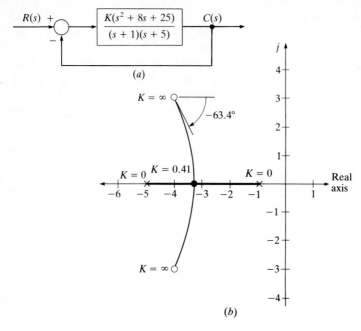

Figure 7.13 Root-locus plot for $(s + 1)(s + 5) + K(s^2 + 8s + 25) = 0$.

○. The breakaway point is obtained as follows:

$$-K = \frac{s^2 + 6s + 5}{s^2 + 8s + 25}$$

$$-\frac{dK}{ds} = \frac{(s^2 + 8s + 25)(2s + 6) - (s^2 + 6s + 5)(2s + 8)}{(s^2 + 8s + 25)^2}$$

$$= \frac{2(s^2 + 20s + 55)}{(s^2 + 8s + 25)^2} = 0$$

The value of s for which $dK/ds = 0$ is the breakaway point $s = \sigma_b = -3.29$. As is illustrated in Fig. 7.13b, the angle at which the loci approaches the ○ at $-4 + j3$ is $-63.4°$. This angle is called the angle of arrival. In Fig. 7.14a, the angle from the × at -1 to a point near the circle at $-4 + j3$ is ϕ_1. The angle from the × at -5 to this point is ϕ_2. In Fig. 7.14b, the angle from the ○ at $-4 - j3$ to the trial point is ϕ_3. The unknown angle of arrival ϕ_a is the angle from the circle at $-4 + j3$ to the trial point which is nearby. Application of the angle condition shows that

$$(\phi_1 + \phi_2) - (\phi_3 + \phi_a) = 180° \pm k360°$$

where $(\phi_1 + \phi_2)$ is the sum of the angles from the ×'s to the trial point and $(\phi_3 + \phi_a)$ is the sum of the angles from the ○'s to the trial point.

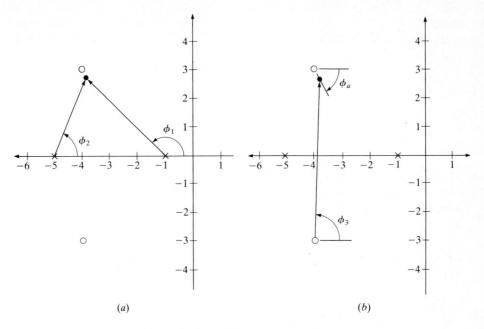

Figure 7.14 Determination of the angle of arrival ϕ_a.

As the trial point s approaches the circle at $-4 + j3$, the values of ϕ_1, ϕ_2, and ϕ_3 are

$$\phi_1 = \tan^{-1} \frac{3}{-3} = 135°$$

$$\phi_2 = \tan^{-1} \frac{3}{1} = 71.6°$$

$$\phi_3 = \tan^{-1} \frac{6}{0} = 90°$$

For $k = 0$, the preceding angle condition becomes

$$(135° + 71.6°) - (90° + \phi_a) = 180°$$

Hence

$$\phi_a = -63.4°$$

For constructing the path of the loci in the complex plane a trial point is selected in the region and then the angle condition is applied. It is a good idea to start by choosing all trial points on the same horizontal line. For a trial point on one side of the loci the angle condition will yield an angle greater than 180°, and for a trial point on the other side an angle less than 180° is obtained. This information indicates in what direction a new trial point should be taken. The use of a spirule, which is a commercially available device, saves much effort in finding the angle of a trial point. As is explained in Sec. 7.7, the digital computer can be used to obtain the root-locus plot for a given characteristic equation. Points on the loci corresponding to various values of the gain K are automatically obtained by the digital computer.

7.3 GENERAL PROCEDURE

In the preceding section, it was shown that the loci could be sketched in quite accurately by knowing a few critical parameters such as breakaway or break-in points, location on the real axis, asymptotes, angle of departure from complex conjugate roots, etc. The general procedure for constructing root loci is summarized as follows:

1. Origin. When K is zero, the roots of the characteristic equation are the roots of $D_G D_H$. Thus, each locus originates at a root of $D_G D_H$ (designated by ×'s) and the number of individual loci is equal to n, the number of roots of $D_G D_H$.
2. Terminus. As K becomes very large, m loci (m is the number of roots of $N_G N_H$) will approach the m roots of $N_G N_H$. That is, one locus will terminate at each of the m roots of $N_G N_H$ (designated by ○'s). The remaining $n - m$ loci will approach infinity along asymptotes.
3. Asymptotes. The angle at which each of the $n - m$ loci approaches infinity is determined from Eq. (7.9). That is,

$$\angle s = \frac{180° \pm k360°}{n - m} \qquad (7.9)$$

The point σ_c at which the asymptotes intersect or cross the real axis is computed by Eq. (7.10). That is,

$$\sigma_c = \frac{(p_1 + p_2 + \cdots + p_n) - (z_1 + z_2 + \cdots + z_m)}{n - m} \qquad (7.10)$$

4. Loci on real axis. Complex conjugate roots of $D_G D_H$ or $N_G N_H$ have no effect on the location of loci on the real axis. The place at which the loci are located along the real axis is determined by considering only roots of $D_G D_H$ and $N_G N_H$ which lie on the real axis. As is illustrated in Fig. 7.7, there is never a locus to the right of the first × or ○ on the real axis but there is always a locus to the left of the first × or ○, there is never a locus to the left of the second × or ○, there is always a locus to the left of the third × or ○, never left of fourth, always left of fifth, and so on, alternating.
5. Angle of departure. The angle of departure of a locus from a complex conjugate root of $D_G D_H$ is obtained by selecting a trial point very close to this root and applying the angle condition.
6. Angle of arrival. The angle at which a locus will terminate at a complex conjugate root of $N_G N_H$ is determined by taking a trial point which is close to this

root and applying the angle condition. This process is similar to that used to obtain the angle of departure from a root of $D_G D_H$.

7. Breakaway or break-in points. The point σ_b at which the locus breaks away from or breaks into the real axis is determined by finding the real value $s = \sigma_b$ at which $dK/ds = 0$.

The occurrence of a breakaway or break-in point can be recognized from a consideration of the $\times$'s and $\bigcirc$'s which lie on the real axis. Every locus begins at

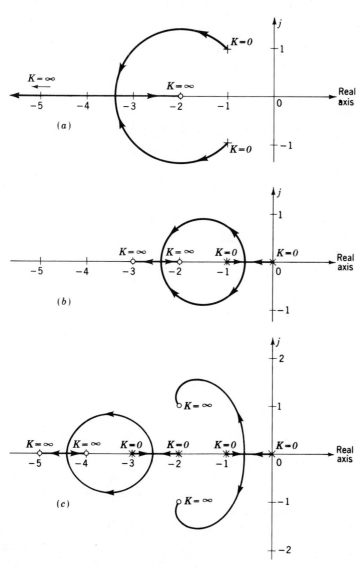

Figure 7.15 Root-locus plots.

an × and terminates at a ○ or along an asymptote at ∞. Thus, there must be a breakaway point between any two adjacent ×'s on the real axis which are connected by a locus. Similarly, a break-in point is required if a ○ on the real axis is not connected to an adjacent × on the real axis by a locus. Thus, if the locus is not located entirely on the real axis between an adjacent × or ○, it is necessary that it come into the real axis from elsewhere. The preceding rules may be verified for the root-locus plots shown in Fig. 7.15.

Illustrative example 7.2 Construct the root-locus plot for the control system shown in Fig. 7.16a.

SOLUTION The characteristic equation is

$$s(s + 4)(s + 6) + K = 0 \tag{7.12}$$

As shown in Fig. 7.16b, there is an × at the origin, an × at -4, and an × at -6. There are no ○'s. On the real axis, there is a loci between the origin and -4, and from -6 to $-\infty$. Because there are no ○'s, all three loci must terminate at infinity along asymptotes. Application of Eq. (7.10) yields, for

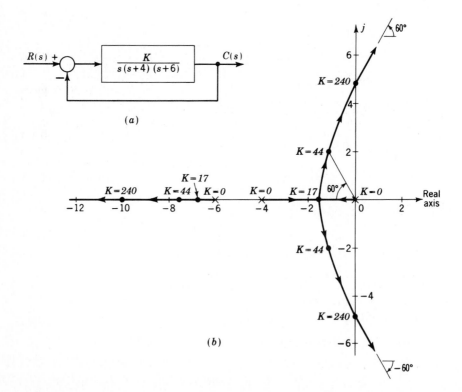

Figure 7.16 Root-locus plot for $s(s + 4)(s + 6) + K = 0$.

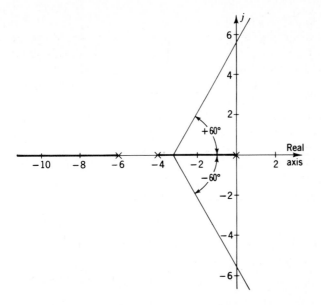

Figure 7.17 Location of loci on the real axis and the asymptotes.

the point σ_c at which the asymptotes intersect the real axis,

$$\sigma_c = \frac{0 + (-4) + (-6)}{3} = -\frac{10}{3} = -3\tfrac{1}{3}$$

For $n - m = 3$, the angles of the asymptotes are $\pm 60°$ and $180°$. The information obtained thus far is shown in Fig. 7.17. To determine the breakaway point, note that

$$-K = s(s + 4)(s + 6) = s^3 + 10s^2 + 24s \tag{7.13}$$

Differentiating and setting $-dK/ds$ equal to zero,

$$-\frac{dK}{ds} = 3s^2 + 20s + 24 = 0$$

whence $\sigma_b = -1.57$.

The resulting root-locus plot may now be drawn as shown in Fig. 7.16b.

A plot of the roots of the characteristic equation as K varies from 0 to ∞ yields very valuable information. In particular, it permits the characteristic equation to be written in factored form. That is,

$$s(s + 4)(s + 6) + K = (s - r_1)(s - r_2)(s - r_3) \tag{7.14}$$

The right side is the factored form of the characteristic equation. Because the number of roots $r_1, r_2, \ldots, r_n$ is equal to the order of the equation, the number of loci is also equal to the order of the equation, which in this case is 3. For each value of K, there corresponds a particular value of r_1, r_2, and r_3. Thus, for each value of K from 0 to ∞, the root-locus plot is the plot of values of r_1, r_2, and r_3.

The three loci for Eq. (7.14) are drawn in Fig. 7.16b. One locus starts at $r_1 = -6$ for $K = 0$ and proceeds out along the negative real axis as K increases. Another locus starts at $r_2 = -4$ and goes to the right along the real axis to the breakaway point $\sigma_b = -1.57$, and then leaves the real axis and proceeds out along the 60° asymptote toward infinity. The third locus starts at $r_3 = 0$ and moves along the negative real axis to the breakaway point $\sigma_b = -1.57$. This locus then leaves the real axis and proceeds toward infinity along the $-60°$ asymptote. In the construction of the loci, the three loci are determined without regard to which is considered the r_1, the r_2, or the r_3 locus. That is, from the similarity of terms in Eq. (7.14), it is seen that the particular subscripts 1, 2, and 3 may be used interchangeably.

The value K at which a root-locus plot crosses the imaginary axis may be obtained by application of Routh's criterion. The application of Routh's criterion to the characteristic equation

$$s(s + 4)(s + 6) + K = s^3 + 10s^2 + 24s + K = 0$$

gives the following array:

s^3	1	24	0
s^2	10	K	0
s^1	$\dfrac{240 - K}{10} \approx \varepsilon$	0	
s^0	K		

The value of K which makes the s^1 row vanish is $K = 240$. The auxiliary equation is $10s^2 + K = 10(s^2 + 24)$. Thus, the characteristic equation has complex conjugate roots located at $r_{1,2} = \pm j\sqrt{24} = \pm j4.9$. Dividing the characteristic equation by the auxiliary equation $(s^2 + 24)$ yields, for the factored form,

$$s^3 + 10s^2 + 24s + 240 = (s^2 + 24)(s + 10)$$

For each value of K the corresponding roots of the characteristic equation may be determined directly from the root-locus plot. These roots in turn govern the transient behavior. From the root-locus plot, the designer may select the value of K such that the system will have a desired transient response. For example, let it be desired to have a damping ratio of 0.5. As shown in Fig. 7.16b, the intersection of the line drawn at the angle $\beta = \cos^{-1} 0.5 = 60°$ with the loci is at the point $s = -1.2 + j2.1$. As is illustrated in Fig. 7.18, application of the magnitude condition yields, for the value of K at this point,

$$K = |s||s + 4||s + 6| = \sqrt{-1.2^2 + 2.1^2}\sqrt{2.8^2 + 2.1^2}\sqrt{4.8^2 + 2.1^2} = 44$$

For $a = -1.2$ and $b = 2.1$, the corresponding quadratic term is $s^2 + 2.4s + 5.85$. Dividing the characteristic equation by this quadratic shows that when $K = 44$,

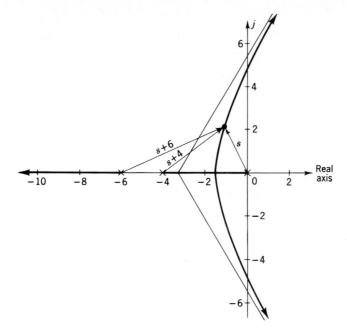

Figure 7.18 Application of the magnitude condition.

the factored form of the characteristic equation is

$$s^3 + 10s^2 + 24s + 44 = (s^2 + 2.4s + 5.85)(s + 7.6)$$

To illustrate the application of the preceding concepts, let it be desired to determine the response $c(t)$ of the system shown in Fig. 7.16a for the case in which $K = 44$, the input $r(t)$ is a unit impulse, and all the initial conditions are zero.

Replacing s by D in the block diagram yields, for the differential equation,

$$c(t) = \frac{Kr(t)}{D(D + 4)(D + 6) + K} = \frac{Kr(t)}{D^3 + 10D^2 + 24D + K}$$

The transform is

$$C(s) = \frac{KR(s) + [s^2c(0) + s\dot{c}(0) + \ddot{c}(0)] + 10[sc(0) + \dot{c}(0)] + 24c(0)}{s^3 + 10s^2 + 24s + K}$$

For $K = 44$, $R(s) = 1$, and all initial conditions zero,

$$C(s) = \frac{44}{(s + 7.6)(s^2 + 2.4s + 5.85)}$$

The general form of the response is

$$c(t) = K_1 e^{-7.6t} + \frac{|K(a + jb)|}{b} e^{at} \sin(bt + \alpha)$$

The value of K_1 is

$$K_1 = \lim_{s \to -7.6} \frac{44}{s^2 + 2.4s + 5.85} = \frac{44}{81.85} = 0.54$$

The value of $K(a + jb)$ is

$$K(a + jb) = \lim_{s \to -1.2 + j2.1} \frac{44}{s + 7.6} = \frac{44}{6.4 + j2.1}$$

$$= \frac{44}{45.37}(6.4 - j2.1) = 6.2 - j2.0 = 6.532 \,\underline{/-18.2°}$$

Thus the desired solution is

$$c(t) = 0.54e^{-7.6t} + 3.11e^{-1.2t}\sin(2.1t - 18.2°)$$

or

$$c(t) = 0.54e^{-7.6t} + \frac{e^{-1.2t}}{2.1}(6.2\sin 2.1t - 2.0\cos 2.1t)$$

It is to be noted that regardless of the particular problem being solved, the general form of the transient response is determined by the roots of the characteristic equation. For a given value of K these roots are ascertained directly from the root-locus plot.

Illustrative example 7.3 Construct the root-locus plot for the system shown in Fig. 7.19a.

SOLUTION The characteristic equation is

$$s(s^2 + 12s + 45) + K = 0$$

As shown in Fig. 7.19b, there is an × at the origin and a complex conjugate pair of ×'s at $-6 \pm j3$. On the real axis there is a loci from the origin to minus infinity. For three ×'s and no ○'s, all three loci terminate along asymptotes at infinity. The angles of the asymptotes are $+60°$, $-60°$, and $180°$. Application of Eq. (7.10) yields, for the point σ_c where the asymptotes intersect the real axis,

$$\sigma_c = \frac{[(-6 + j3) + (-6 - j3) + (0)] - 0}{3 - 0} = \frac{-12}{3} = -4$$

As is shown in Fig. 7.20, to determine the angle of departure from the × at the point $-6 + j3$, a trial point is selected close to this ×. Application of the angle condition gives

$$\phi_1 + \phi_2 + \phi_d = 180°$$

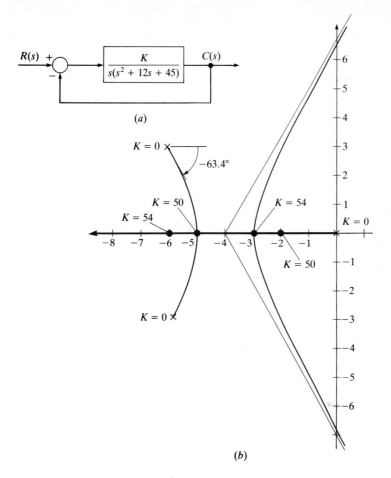

Figure 7.19 Root-locus plot for $s(s^2 + 12s + 45) + K = 0$.

In the limit as the trial point approaches the $\times$ at $-6 + j3$, the values of ϕ_1 and ϕ_2 are

$$\phi_1 = \tan^{-1} \frac{3}{-6} = 153.4°$$

$$\phi_2 = \tan^{-1} \frac{6}{0} = 90°$$

Substitution of these values into the angle condition gives

$$153.4 + 90° + \phi_d = 180°$$

Thus,

$$\phi_d = -63.4°$$

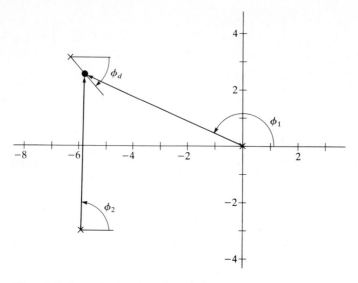

Figure 7.20 Determination of the angle of departure.

This angle of departure is such as to lead the loci toward the real axis where there is a loci. There is a tendency for loci in the complex plane to be drawn to loci on the real axis. To check for break-in and breakaway points, first write

$$-K = s^3 + 12s^2 + 45s$$

Differentiation shows that

$$-\frac{dK}{ds} = 3(s^2 + 8s + 15) = 0$$

The values of s for which $dK/ds = 0$ are

$$\sigma_b = \frac{-8 \pm \sqrt{64 - 60}}{2} = -4 \pm 1 = -5, -3$$

As shown in Fig. 7.19b, the point $\sigma_b = -5$ is a break-in point and the point $\sigma_b = -3$ is a breakaway point. The value of K at which the loci breaks into the real axis is $K = 50$. The value of K at which the loci breaks away from the real axis is $K = 54$. For a given value of K there are always three roots of the characteristic equation. The factored form is

$$s(s^2 + 12s + 45) + K = (s - r_1)(s - r_2)(s - r_3)$$

For $0 < K < 50$, one root lies on the loci between the $\times$ at $-6 + j3$ and the break-in point. The conjugate root lies on the loci between the $\times$ at $-6 - j3$ and the break-in point. The third root lies on the real axis between the $\times$ at the origin and the -2 point. For $50 < K < 54$, one root lies on the real axis between the break-in point at -5 and the point -6. The second root lies between the break-in point and the breakaway point. The third root lies between the -2 point and the breakaway point. For $K > 54$, one root lies on the real axis between -6 and minus infinity. Another root lies on the loci which leaves the breakaway point and approaches the 60° asymptote. The third root lies on the conjugate loci.

Illustrative example 7.4 The characteristic equation for a system is

$$(s + 2)(s + 4)(s + a) + K = 0$$

Determine a such that the characteristic equation will have a pair of complex conjugate roots located at $(-2 \pm j2\sqrt{3})$. Write the factored form of the characteristic equation for this case.

SOLUTION In Fig. 7.21, the ×'s are located at -2, -4, and the unknown point $-a$. In order that the point indicated by the dot $(-2 + j2\sqrt{3})$ be on the root-locus plot, the angle condition must be satisfied. That is,

$$\phi_1 + \phi_2 + \phi_3 = 180°$$

where $\quad \phi_1 = \tan^{-1} \dfrac{2\sqrt{3}}{0} = 90°$

$$\phi_2 = \tan^{-1} \dfrac{2\sqrt{3}}{2} = 60°$$

$$\phi_3 = \tan^{-1} \dfrac{2\sqrt{3}}{a - 2}$$

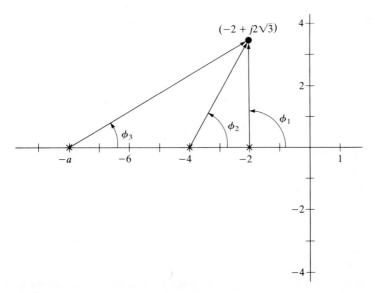

Figure 7.21 Determination of a such that the point $(-2 + j2\sqrt{3})$ lies on the root-locus plot.

Thus,

$$\tan^{-1} \frac{2\sqrt{3}}{a-2} = 180° - 90° - 60° = 30°$$

Taking the tangent of both sides shows that

$$\frac{2\sqrt{3}}{a-2} = \tan 30° = \frac{1}{\sqrt{3}}$$

Solving for a gives

$$a - 2 = 2(3) = 6$$

Thus,

$$a = 8$$

Application of the magnitude condition yields, for the value of K at the point $(-2 + j2\sqrt{3})$,

$$K = |s + 2||s + 4||s + 8|$$
$$= 2\sqrt{3} \sqrt{(2\sqrt{3})^2 + 2^2} \sqrt{(2\sqrt{3})^2 + 6^2} = 96$$

The quadratic form of the complex conjugate roots at $-2 \pm j2\sqrt{3}$ is $(s^2 + 4s + 16)$. Dividing the characteristic equation $(s + 2)(s + 4)(s + 8) + 96$ by this factor yields the third factor $(s + 10)$. Thus, the factored form of the characteristic equation for this case is

$$(s + 2)(s + 4)(s + 8) + 96 = (s + 10)(s^2 + 4s + 16)$$

7.4 NEWTON'S METHOD AND THE REMAINDER THEOREM

When the equation for dK/ds is a cubic, quartic, or higher-degree polynomial, solving for the break-in or breakaway point is a very laborious process. Newton's method is a technique for obtaining a close approximation to the break-in or breakaway point for such cases.

Let $P(s)$ be the numerator of dK/ds. Thus when $P(s) = 0$, then $dK/ds = 0$. In Fig. 7.22 is shown a plot of $P(s)$ versus s. The tangent to this curve at $s = s_1$ is

$$P'(s_1) = \frac{P(s_1)}{s_1 - s_2}$$

Solving for s_2 gives

$$s_2 = s_1 - \frac{P(s_1)}{P'(s_1)} \tag{7.15}$$

The point s_2, which is the intersection of the tangent and the horizontal axis, is a closer approximation to σ_b than is s_1. After evaluating s_2, then applying the same

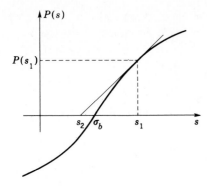

Figure 7.22 Typical plot of $P(s)$ versus s.

process yields a closer approximation s_3, etc. Usually one trial yields sufficient accuracy for constructing the root-locus plot. It is now shown how $P(s_1)/P'(s_1)$ may be obtained by use of the remainder theorem.

Dividing $P(s)$ by $(s - s_1)$ gives

$$\frac{P(s)}{s - s_1} = Q(s) + \frac{R_1}{s - s_1}$$

where R_1 is the remainder that results when $P(s)$ is divided by $s - s_1$. Multiplying through by $s - s_1$ gives

$$P(s) = (s - s_1)Q(s) + R_1 \tag{7.16}$$

Taking the limit as s approaches s_1 shows that

$$P(s_1) = R_1 \tag{7.17}$$

By similarly dividing $Q(s)$ by $s - s_1$, it follows that

$$\frac{Q(s)}{s - s_1} = T(s) + \frac{R_2}{s - s_1} \tag{7.18}$$

where R_2 is the remainder that results when $Q(s)$ is divided by $s - s_1$. Multiplying through Eq. (7.18) by $s - s_1$ and then taking the limit as s approaches s_1 shows that

$$Q(s_1) = R_2$$

To show that $Q(s_1) = P'(s_1)$, first differentiate Eq. (7.16) with respect to s:

$$P'(s) = (s - s_1)Q'(s) + Q(s)$$

Taking the limit as s approaches s_1 verifies the fact that

$$P'(s_1) = Q(s_1) = R_2 \tag{7.19}$$

Equation (7.15) may now be expressed in the form

$$s_2 = s_1 - \frac{R_1}{R_2} \tag{7.20}$$

To illustrate the application of this method, let us assume as a first trial that the breakaway point for Fig. 7.16 occurs at $s_1 = -2.0$. From Eq. (7.13), the polynomial $P(s)$ is

$$P(s) = 3s^2 + 20s + 24$$

Division by $s - s_1 = s + 2$ yields

$$
\begin{array}{r}
3s + 14 \\
s + 2 \overline{\smash{\big)}\, 3s^2 + 20s + 24} \\
\underline{3s^2 + 6s} \\
14s + 24 \\
\underline{14s + 28} \\
-4
\end{array}
$$

Hence, $Q(s) = 3s + 14$ and $R_1 = -4$. Dividing $Q(s)$ by $s - s_1$ gives

$$
\begin{array}{r}
3 \\
s + 2 \overline{\smash{\big)}\, 3s + 14} \\
\underline{3s + 6} \\
8
\end{array}
$$

Hence, $T(s) = 3$ and $R_2 = 8$. Application of Eq. (7.20) yields the closer approximation:

$$s_2 = -2 - \frac{-4}{8} = -1.5$$

The preceding divisions are simplified by the use of synthetic division, as is illustrated below:

$$
\begin{array}{rrrl}
3 & 20 & 24 &)-2 \\
& -6 & -28 & \\
\hline
3 & 14 & \boxed{-4} &
\end{array}
$$

The first row consists of the coefficients of the polynomial $P(s)$ and the root -2. The first coefficient 3 is brought below the line; then multiplying 3 by -2 yields -6 as shown. Adding 20 and -6 gives 14 as shown. Multiplying 14 by -2 yields the -28 term. Finally, adding 24 and -28 yields the remainder $R_1 = -4$, which is circled. It is to be noted that the coefficients of $Q(s)$ appear to the left of the circled remainder. The remainder R_2 is now obtained by similarly applying synthetic division to the $Q(s)$ polynomial. That is,

$$
\begin{array}{rrl}
3 & 14 &)-2 \\
& -6 & \\
\hline
3 & \boxed{8} &
\end{array}
$$

Thus, the remainder R_2 is 8 and $T(s)$ is simply 3.

Let us now take -1.5 as the new trial breakaway point and then employ synthetic division to obtain the next closer approximation. Thus,

$$
\begin{array}{r}
3.00 \quad\ \ 20.00 \quad\ \ 24.00 \quad\ \ )-1.50 \\
-4.50 \ \ -23.25 \quad\quad\quad\\
\hline
3.00 \quad\ \ 15.50 \quad\ \ \boxed{0.75} \quad\quad\ \\
-4.50 \quad\quad\quad\quad\quad\quad\\
\hline
3.00 \quad\ \ \boxed{11.00} \quad\quad\quad\quad\ \
\end{array}
$$

Thus, $R_1 = 0.75$, $R_2 = 11.00$, $Q(s) = 3s + 15.50$, and $T(s) = 3$. The closer approximation is

$$
s_3 = s_2 - \frac{R_1}{R_2} = -1.50 - \frac{0.75}{11.00} = -1.57
$$

7.5 LOCI EQUATIONS

In studying control systems, one frequently encounters the case in which there are two $\times$'s and one $\bigcirc$, as shown in Fig. 7.23. It is now shown that the locus follows a circular path in the complex plane. The characteristic equation for this case is

$$
(s - p_1)(s - p_2) + K(s - z) = 0 \tag{7.21}
$$

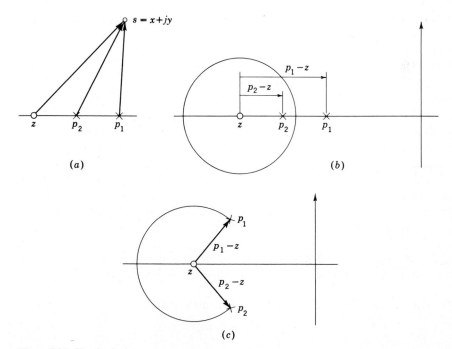

Figure 7.23 Circular loci.

As shown in Fig. 7.23a, a trial point s may be expressed in the form $s = x + jy$. The substitution of $s = x + jy$ in the preceding equation gives

$$(x^2 + 2jxy - y^2) - (p_1 + p_2 - K)(x + jy) + p_1 p_2 - Kz = 0$$

In order that this equation equal zero, it is necessary that both the real parts and the imaginary parts (that is, j components) be zero. Thus,

$$x^2 - y^2 - (p_1 + p_2 - K)x + p_1 p_2 - Kz = 0$$

$$2xy - (p_1 + p_2 - K)y = 0$$

To eliminate K between the two preceding equations, first solve each for K. That is,

$$K = \frac{(x^2 - y^2) - (p_1 + p_2)x + p_1 p_2}{z - x} \tag{7.22}$$

$$K = (p_1 + p_2) - 2x \tag{7.23}$$

The term K is now eliminated by equating the right sides of the preceding expressions. Thus,

$$x^2 + y^2 = p_1 p_2 + [2x - (p_1 + p_2)]z$$

Completing the square gives

$$(x - z)^2 + y^2 = (p_1 - z)(p_2 - z) \tag{7.24}$$

This is recognized as the equation of a circle with center at z. The radius is the square root of the product of the distance from z to p_1 and the distance from z to p_2. That is,

$$R = \sqrt{(p_1 - z)(p_2 - z)} \tag{7.25}$$

A typical root-locus plot for the case in which p_1 and p_2 are real is shown in Fig. 7.23b. For the case in which p_1 and p_2 are complex conjugate roots, then the distance from z to p_1 equals the distance from z to p_2. Thus, the radius is simply the distance from z to either of the complex conjugate roots, as is shown in Fig. 7.23c.

Equation (7.23) may be used to determine the gain K for points on the loci in the complex plane (i.e., off the real axis). For loci on the real axis, $y = 0$. Thus, letting $y = 0$ in Eq. (7.22) yields an equation for evaluating the gain K for loci points on the real axis.

Illustrative example 7.5 Construct the root-locus plot for the system shown in Fig. 7.24a and then determine the gain K at $x = -4$ and at $x = -8$.

SOLUTION The characteristic equation for this system is

$$s(s + 2) + K(s + 4) = 0$$

Thus, $z = -4$, $p_1 = 0$, and $p_2 = -2$.

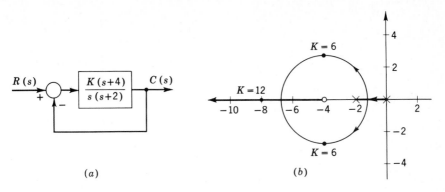

Figure 7.24 Root-locus plot for $s(s + 2) + K(s + 4) = 0$.

The location of the loci on the real axis may be drawn as shown in Fig. 7.24b. Because there are two $\times$'s and a $\bigcirc$, then in the complex plane (i.e., off the real axis) the loci is a circle with center at $z = -4$ and radius

$$R = \sqrt{(p_1 - z)(p_2 - z)} = \sqrt{(4)(2)} = \sqrt{8}$$

The application of Eq. (7.23) to determine the gain K at $x = -4$ gives

$$K = (p_1 + p_2) - 2x = (0 - 2) - 2(-4) = 6$$

For the loci on the real axis at $x = -8$, application of Eq. (7.22) in which $y = 0$ gives

$$K = \frac{x^2 - (p_1 + p_2)x + p_1 p_2}{z - x} = \frac{64 - (-2)(-8)}{-4 - (-8)} = \frac{48}{4} = 12$$

The preceding method for obtaining the equation for a circular locus may be applied to any locus. The method is to replace s by $x + jy$ in the characteristic equation. Because both the real and imaginary parts must be zero, two equations are obtained. One equation gives values of gain K for loci on the real axis and the other equation gives values of gain K for loci in the complex plane. Eliminating K between these two equations yields a general equation in x and y for the loci. Corresponding values of x and y that satisfy this equation are points on the loci. In the preceding example the equation was recognized as that of a circle.

7.6 VARIATION OF PARAMETERS

Thus far the discussion of root locus has been concerned with the case in which the gain K is the variable parameter. By algebraically rearranging the characteristic equation, the effect of the change of any parameter can be investigated.

To illustrate this procedure, consider the system shown in Fig. 7.25a. The characteristic equation for this system is

$$s(s + 4) + K = 0$$

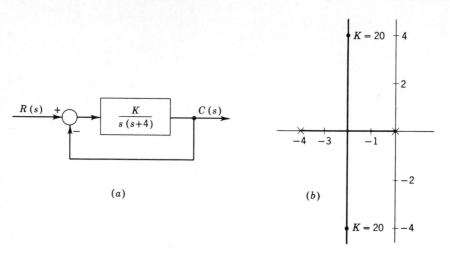

Figure 7.25 Root-locus plot for $s(s + 4) + K = 0$.

The basic root-locus plot is shown in Fig. 7.25b. For $K = 20$, it is to be noted that the roots of the characteristic equation are $-2 \pm j4$. Thus

$$s(s + 4) + 20 = [s - (-2 - j4)][s - (-2 + j4)] = 0$$

Let it now be desired to investigate the effect of varying the parameter 4 when $K = 20$. Replacing the parameter 4 by $4 + \Delta$, where Δ is the change in the parameter, yields, for the new characteristic equation,

$$s[s + (4 + \Delta)] + 20 = [s(s + 4) + 20] + \Delta s$$

$$= [s - (-2 - j4)][s - (-2 + j4)] + \Delta s = 0 \quad (7.26)$$

This equation has the familiar form except that Δ is now the variable rather than K. The new root-locus plot begins at the value of the roots on the basic plot for $K = 20$. Equation (7.26) shows that the ×'s for the new plot are at $s = -2 \pm j4$ and there is a ○ at the origin. The corresponding root-locus plot for positive values of Δ is shown in Fig. 7.26a.

In investigating the effect of a change in parameter, it is desired to know the effect of decreasing as well as increasing the parameter. Thus far, root-locus plots have been constructed for positive values of K only. From Eqs. (7.6) and (7.8), it follows that when K (or Δ) is positive, the summation of the angles is $180° \pm k360°$. When K (or Δ) is a negative number, then $(-K)$ is a positive number which may be represented by a point on the positive real axis. For negative K the angle is

$$\sphericalangle (-K) = 0° \pm k360°$$

Thus, for negative values of K the angle condition becomes

$$\sum_{i=1}^{n} (s - p_i) - \sum_{i=1}^{m} (s - z_i) = 0° \pm k360° \qquad k = 0, 1, 2, 3, \ldots \quad (7.27)$$

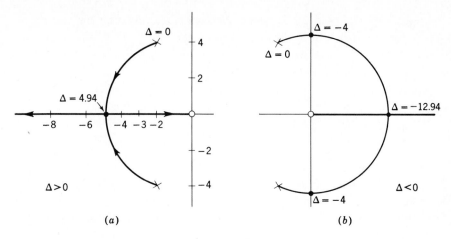

Figure 7.26 Root-locus plot for $[s(s + 4) + 20] + \Delta s = 0$.

The angles for the asymptotes are

$$\measuredangle s = \frac{0° \pm k360°}{n - m} \qquad k = 0, 1, 2, 3, \ldots \qquad (7.28)$$

For $n - m = 1, 2, 3$, and 4, Eq. (7.28) becomes

$$\measuredangle s = 0° \pm k360° \qquad n - m = 1$$
$$\measuredangle s = 0° \pm k180° \qquad n - m = 2$$
$$\measuredangle s = 0° \pm k120° \qquad n - m = 3$$
$$\measuredangle s = 0° \pm k90° \qquad n - m = 4$$

The corresponding asymptotes for $n - m = 1, 2, 3$, and 4 are shown in Fig. 7.27. For $n - m = 1$ the angle is 0°; for $n - m = 2$ the angles are 0° and 180°; for $n - m = 3$ the angles are 0° and $\pm 120°$; and for $n - m = 4$ the angles are 0°,

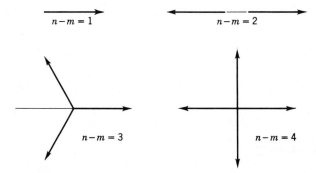

Figure 7.27 Asymptotes for negative K.

$\pm 90°$, and $180°$. The point σ_c at which the asymptotes intersect the real axis is obtained by Eq. (7.10) in the same manner as for positive K. That is,

$$\sigma_c = \frac{\sum\limits_{i=1}^{n} p_i - \sum\limits_{i=1}^{m} z_i}{n - m} \qquad (7.10)$$

The following rules for constructing a root-locus plot for negative values of K (or Δ) are very similar to those given in Sec. 7.3 for positive values of K:

1. Origin. One locus begins at each $\times$, as was the case for positive K.
2. Terminus. One locus ends at each $\bigcirc$, as was the case for positive K. The remaining $n - m$ loci terminate at infinity along asymptotes.
3. Asymptotes. The angle at which each of the $n - m$ loci approaches infinity is

$$\angle s = \frac{0° \pm k360°}{n - m}$$

The corresponding asymptotes for $n - m = 1$, 2, 3, and 4 are shown in Fig. 7.27. The point σ_c at which the asymptotes intersect the real axis is obtained by Eq. (7.10) in the same manner as for positive K.
4. Loci on the real axis. The location of loci on the real axis is determined by the $\times$'s and $\bigcirc$'s that lie on the real axis. The summation of angles at any point on the real axis is either $0°$ or $180°$. For positive K the loci are regions on the real axis where the angle is $180°$. The remaining regions on the real axis are regions where the angle is $0°$, and thus these are the regions of the loci for negative values of K. In Fig. 7.28 is shown the location of loci on the real axis for negative K for the same $\times$ and $\bigcirc$ pattern as Fig. 7.7. The loci now appear at all regions on the real axis that did not have loci in Fig. 7.7. For the case of negative K, a locus always appears to the right of the first $\times$ or $\bigcirc$ on the real axis, there is never a locus to the left of the first $\times$ or $\bigcirc$, there is always a locus to the left of the second $\times$ or $\bigcirc$, there is never a locus to the left of the third $\times$ or $\bigcirc$, and so on, alternating.
5. Angle of departure. Because the summation of the angles is now $0°$ rather than $180°$, the angle of departure for negative K is always $180°$ different from that obtained for positive K.

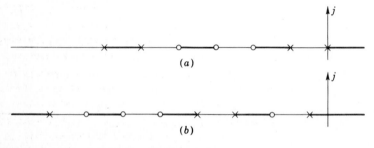

Figure 7.28 Location of loci on the real axis for negative K.

6. Angle of arrival. The angle of arrival for negative K is always $180°$ different from that obtained for positive K.
7. Breakaway or break-in points. These are determined in the same way as for positive K. In determining the breakaway or break-in points for positive K, oftentimes extra or extraneous values were obtained. Actually, these values are the ones for the case of negative K.

In Fig. 7.26b is shown the root-locus plot for negative values of Δ. Because there are two $\times$'s and one $\bigcirc$, the loci form a circle in the complex plane. The system becomes unstable when the locus crosses the imaginary axis, which occurs at $\Delta = -4$. The preceding concepts are further illustrated in the following example.

Illustrative example 7.6 Suppose that for the system shown in Fig. 7.16 the gain K must be maintained at 240 to achieve satisfactory steady-state operation. For $K = 240$ the roots of the characteristic equation are $s = \pm j\sqrt{24}$ and $s = -10$. Hence, the factored form of the characteristic equation is

$$s(s + 4)(s + 6) + 240 = (s + 10)(s^2 + 24)$$

Determine whether or not this system can be stabilized and good dynamic behavior achieved by changing the parameter 4 in the term $s + 4$.

SOLUTION Replacing 4 by $4 + \Delta$ yields, for the new characteristic equation,

$$s[(s + 4) + \Delta](s + 6) + 240 = [s(s + 4)(s + 6) + 240] + \Delta s(s + 6)$$
$$= (s + 10)(s^2 + 24) + \Delta s(s + 6)$$
$$= 0$$

To construct the root-locus plot for positive values of Δ, first draw the loci on the real axis as shown in Fig. 7.29a. Next the angle of departure from the $\times$ at $s = j\sqrt{24}$ is

$$\left(\phi_d + 90° + \tan^{-1}\frac{\sqrt{24}}{10}\right) - \left(90° + \tan^{-1}\frac{\sqrt{24}}{6}\right) = 180°$$

Solving for ϕ_d gives $\phi_d = 193.1°$.

A break-in point is seen to occur between 0 and -6 on the real axis. As a first trial, let s_1 be the midpoint -3. To apply Newton's method, first solve the characteristic equation for Δ:

$$\Delta = -\frac{(s + 10)(s^2 + 24)}{s(s + 6)}$$

Differentiation to obtain $d\Delta/ds$ gives

$$\frac{d\Delta}{ds} = \frac{s^4 + 12s^3 + 36s^2 - 480s - 1440}{(s^2 + 6s)^2} = 0$$

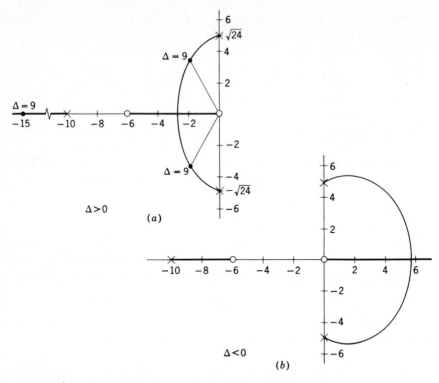

Figure 7.29 Root-locus plot for $[s(s + 4)(s + 6) + 240] + s(s + 6)\Delta = 0$.

The polynomial $P(s)$ is taken as the numerator of $d\Delta/ds$. Thus, application of double synthetic division to obtain the remainders yields

$$
\begin{array}{rrrrrr}
1 & 12 & 36 & -480 & -1440 & \underline{\;|-3} \\
 & -3 & -27 & -27 & 1521 & \\
\hline
1 & 9 & 9 & -507 & \enclose{circle}{81} & \\
 & -3 & -18 & 27 & & \\
\hline
1 & 6 & -9 & \enclose{circle}{-480} & &
\end{array}
$$

Thus, $R_1 = 81$, $R_2 = -480$, $Q(s) = s^3 + 9s^2 + 9s - 507$, and $T(s) = s^2 + 6s - 9$. The value of s_2 is

$$
s_2 = -3 - \frac{81}{-480} = -2.83
$$

The remaining portion of the root-locus plot for positive values of Δ may now be sketched in as shown in Fig. 7.29a.

To construct the root-locus plot for negative Δ, first draw the loci on the real axis as shown in Fig. 7.29b. The angle of departure for negative values of Δ is $180°$ different from that for positive Δ; hence

$$
\phi_d = 193.1° - 180.0° = 13.1°
$$

It is seen that a break-in point must occur on the positive real axis. Taking $s_1 = 5$ as the first trial and then applying double synthetic division gives

$$
\begin{array}{rrrrrr}
1 & 12 & 36 & -480 & -1440 & \underline{)5} \\
 & 5 & 85 & 605 & 625 & \\ \hline
1 & 17 & 121 & 125 & \boxed{-815} & \\
 & 5 & 110 & 1155 & & \\ \hline
1 & 22 & 231 & \boxed{1280} & &
\end{array}
$$

Thus, $R_1 = -815$, $R_2 = 1280$, $Q(s) = s^3 + 17s^2 + 121s + 125$, and $T(s) = s^2 + 22s + 231$. Application of Eq. (7.24) to obtain the closer approximation to the break-in point gives

$$
s_2 = 5.0 - \frac{-815}{1280} = 5.64
$$

From the root-locus plot it is seen that good dynamic behavior is obtained for $\Delta = 9$, in which case the three roots are $-2 \pm j\sqrt{12}$ and -15. For the complex conjugate roots, $\zeta = 0.5$ and $\omega_n = 4$.

7.7 COMPUTER SOLUTION

Many digital computers have subprograms for obtaining the roots of a polynomial. On such computers it is an easy matter to obtain all the roots of a characteristic equation for various values of K. To illustrate this procedure, consider the characteristic equation

$$
s(s^2 + 8s + 25) + K(s + 2)(s + 4) = s^3 + (K + 8)s^2 + (6K + 25)s + 8K = 0
$$

This is a third-order system. Thus, for each value of K there are three roots. Although the format for calling up the subprogram for obtaining the roots will vary from computer to computer, a typical program is

```
        DO 10 K = 0,25
        FK = FLOAT(K)
        COEF(1) = 1.0
        COEF(2) = FK + 8.0
        COEF(3) = 6.0*FK + 25.0
        COEF(4) = 8.0*FK
        CALL ROOTS (COEF,3,ROOTR,ROOTI)
        WRITE (6,*) K,ROOTR(1),ROOTI(1),ROOTR(2),
        ROOTI(2),ROOTR(3),ROOTI(3)
     10 CONTINUE
```

The DO statement specifies the value of K for which the roots will be obtained. For this example, the roots are obtained for $K = 0, 1, 2, 3, \ldots, 25$. Because K is an integer, the FLOAT(K) statement is needed so that FK will be a real number having the value of K. The next four statements input the values of the coefficients of the polynomial. COEF(1) is the coefficient of s^3, COEF(2) is the coefficient of s^2, COEF(3) is the coefficient of s, and COEF(4) is the constant. The CALL ROOTS statement calls up the subprogram for obtaining the roots of the polynomial. ROOTR is the real part and ROOTI is the imaginary part of the root. The WRITE statement prints out the value of K and the real and imaginary parts of all three roots for that particular value of K. ROOTR(1) is the real part and ROOTI(1) is the imaginary part of the first root, ROOTR(2) is the real part and ROOTI(2) is the imaginary part of the second root, etc.

To obtain a root-locus plot, the real and imaginary parts of each root are stored for each value of K. The real part is the x component of the root location and the imaginary part is the y component. The real and imaginary parts may be stored by putting a $J = 1$ statement before the DO loop and then adding the following statements immediately before the CONTINUE statement:

```
XP(J) = ROOTR(1)
YP(J) = ROOTI(1)
XP(J+1) = ROOTR(2)
YP(J+1) = ROOTI(2)
XP(J+2) = ROOTR(3)
YP(J+2) = ROOTI(3)
J = J+3
```

These points may now be plotted to obtain the root-locus plot.

7.8 SENSITIVITY

The parameters used in the design of control systems vary due to factors such as wear, aging, variations in the operating point, temperature, etc. It is thus desired to know the effect of small variations in these parameters upon the dynamic response of the system. In Sec. 7.6, it was shown how to determine the effect of such variations upon the location of the roots of the characteristic equation. The sensitivity method is described in the following.

The sensitivity S is defined as the percentage change in the system transmittance T compared to the percentage change in the parameter K. That is,

$$S_K^T = \frac{dT/T}{dK/K} = \frac{K}{T}\frac{dT}{dK} = \frac{d \ln T}{d \ln K} \tag{7.29}$$

where $T(s) = C(s)/R(s)$ is the closed-loop transfer function for the system. The term K may represent any parameter, such as gain, time constant, damping ratio, natural frequency, etc. To have the transmittance T which represents the system dynamics insensitive to variations in a parameter, then the sensitivity S should be zero, or as small as possible.

The transfer function for the system of Fig. 7.25 may be expressed in the general form

$$G(s) = \frac{K}{s(s + a)} = \frac{K}{s(s + 1/\tau)}$$

where $a = 1/\tau$ in which τ is the time constant. As indicated in the following, for computational purposes it is easier to work with a. It is an easy matter to convert the resulting answer from a to τ if so desired.

The transmittance T is

$$T(s) = \frac{C(s)}{R(s)} = \frac{G(s)}{1 + G(s)} = \frac{K}{s(s + a) + K}$$

The sensitivity with respect to the gain K is

$$S_K^T = \frac{K}{T}\frac{dT}{dK} = \frac{K[s(s + a) + K]}{K}\frac{[s(s + a) + K] - K}{[s(s + a) + K]^2}$$

$$= \frac{s(s + a)}{s(s + a) + K} \tag{7.30}$$

The sensitivity with respect to the parameter a is

$$S_a^T = \frac{a}{T}\frac{dT}{da} = \frac{a[s(s + a) + K]}{K}\frac{-sK}{[s(s + a) + K]^2}$$

$$= \frac{-as}{s(s + a) + K} \tag{7.31}$$

The steady-state sensitivity is obtained by letting $s = 0$. For both Eqs. (7.30) and (7.31), the steady-state sensitivity obtained by letting $s = 0$ is $S_K^T = S_a^T = 0$. The dynamic sensitivity is obtained by replacing s by $j\omega$ in the sensitivity equations and then making plots of the sensitivity as a function of the frequency ω. The significance of replacing s by $j\omega$ is explained in Chaps. 11 and 12.

By defining sensitivity in a slightly different manner, it is possible to determine the change in the roots of the characteristic equation for small variations of a parameter. Thus, sensitivity is now defined as the rate of change of the root location with respect to the parameter. That is,

$$S_K^s = \frac{ds}{dK} \tag{7.32}$$

where ds/dK is the rate of change of the root location s in the characteristic equation with respect to K. For the system of Fig. 7.25, the characteristic equation is

$$s(s + a) + K = (s - r_1)(s - r_2) = 0$$

The derivative with respect to K is

$$2s \frac{ds}{dK} + a \frac{ds}{dK} + 1 = 0$$

Thus, the sensitivity with respect to the parameter K is

$$S_K^s = \frac{ds}{dK} = \frac{-1}{a + 2s}$$

At the reference operating condition $a = 4$ and $K = 20$, the roots are $r_1 = -2 + j4$ and $r_2 = -2 - j4$. The sensitivity of the location of the root r_1 is obtained by evaluating S_K^s at $s = r_1$. Thus,

$$S_K^{r_1} = \frac{dr_1}{dK} = \frac{-1}{4 + 2(-2 + j4)} = \frac{-1}{j8} = \frac{j}{8}$$

Because $dr_1 = (dr_1/dK)dK = S_K^{r_1} \, dK$, then

$$\Delta r_1 \approx S_K^{r_1} \, \Delta K = \frac{j}{8} \, \Delta K \tag{7.33}$$

If ΔK is 4, then $\Delta r_1 \approx 0.5j$. Hence, for $K = 20 + \Delta K = 24$, then $r_1 \approx (-2 + j4) + 0.5j = -2 + j4.50$. From the characteristic equation, the exact value of r_1 is found to be $-2 + j\sqrt{20} = -2 + j4.47$. In a similar manner, the change in the location of the root r_2 may also be found.

The sensitivity of the location of the roots with respect to the parameter a is

$$S_a^s = \frac{ds}{da}$$

The derivative of the characteristic equation with respect to a is

$$2s \frac{ds}{da} + a \frac{ds}{da} + s \frac{da}{da} = 0$$

Thus, the sensitivity ds/da is

$$S_a^s = \frac{ds}{da} = \frac{-s}{a + 2s}$$

The root sensitivity at $s = r_1$ is

$$S_a^{r_1} = \frac{dr_1}{da} = \frac{-(-2 + j4)}{4 + 2(-2 + j4)} = -\frac{2 + j}{4}$$

Thus,

$$\Delta r_1 \approx -\frac{2 + j}{4} \, \Delta a \tag{7.34}$$

Because $a = 1/\tau$, if the time constant τ is changed from $\frac{1}{4}$ to $\frac{1}{6}$, then $\Delta a = 6 - 4 = 2$. Thus,

$$r_1 \approx (-2 + j4) - \frac{(2 + j)}{4} 2 = -3 + j3.50$$

From the characteristic equation (for $a = 4 + 2 = 6$ and $K = 20$) the exact value of r_1 is found to be $-3 + j\sqrt{11} = -3 + j3.32$.

When the sensitivity is expressed in the polar form, then Eqs. (7.33) and (7.34) become

$$\Delta r_1 \approx 0.125 \ \Delta K \ \underline{/90°}$$

$$\Delta r_1 \approx -\sqrt{5} \ \frac{\Delta a}{4} \ \underline{/206.6°}$$

The angles indicate the direction that the roots leave the reference value for positive variations in ΔK or Δa. The direction is reversed by 180° for negative variations.

If both K and a change simultaneously, then

$$\Delta s \approx \frac{\partial s}{\partial K} \Delta K + \frac{\partial s}{\partial a} \Delta a = S_K^s \ \Delta K + S_a^s \ \Delta a$$

Thus, the effects add when more than one parameter varies. For example, if $\Delta K = 4$ and $\Delta a = 2$, then the change Δr_1 is

$$\Delta r_1 \approx \frac{j}{8} 4 - \frac{2 + j}{4} 2 = -1$$

The new value of r_1 is $(-2 + j4) - 1 = -3 + j4$. From the characteristic equation for $K = 20 + 4 = 24$ and $a = 4 + 2 = 6$, the exact value is found to be $-3 + j\sqrt{15} = -3 + j3.87$.

PROBLEMS

7.1 The root-locus plot for the system of Fig. 7.1a is given in Fig. 7.1b. Determine the response equation $c(t)$ for the case in which $r(t)$ is a unit step function and $K = 8$. All the initial conditions are zero.

7.2 The location of the $\times$'s and $\bigcirc$'s of the characteristic equation for various systems is shown in Fig. P7.2. Determine the characteristic equation for each system and then sketch the root-locus plot. For each case, determine the value of K when the characteristic equation has a root located at -2.

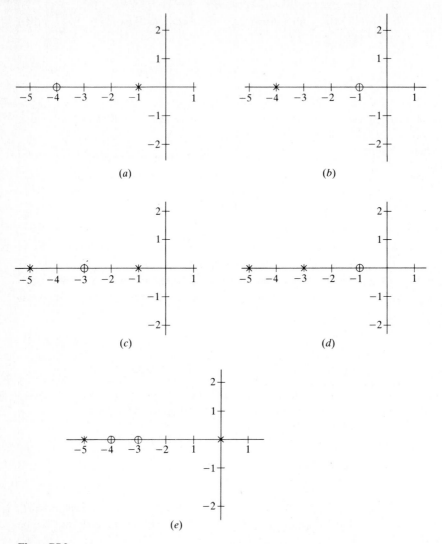

(a)

(b)

(c)

(d)

(e)

Figure P7.2

7.3 For the system shown in Fig. P7.3, write the characteristic equation and then determine the roots for $K = 5$, for $K = 9$, and for $K = 25$. Sketch the root-locus plot. Show on this plot where $K = 5$, 9, and 25. For each case, determine the response $c(t)$ when $r(t) = 0$, $c(0) = 0$, and $\dot{c}(0) = 1$.

$$R(s) \; + \; \xrightarrow{\quad} \; \boxed{\dfrac{K}{s(s+6)}} \; \xrightarrow{\quad} \; C(s)$$

Figure P7.3

7.4 For the system shown in Fig. P.7.4, write the characteristic equation and then determine the roots for $K = 3$, for $K = 4$, and for $K = 20$. Sketch the root-locus plot. Show on this plot where $K = 3$, 4, and 20. For each case, determine the response $c(t)$ when $r(t) = 0$, $c(0) = 1$, and $\dot{c}(0) = 0$.

Figure P7.4

7.5 The block diagram for a control system is shown in Fig. P7.5. From the root-locus plot, it is found that when $K = 3$ the roots of the characteristic equation are located at $-4 \pm j3$. Determine the response $c(t)$ of this system for each of the following cases:

    (a) $r(t) = 0$, $c(0) = 0$, and $\dot{c}(0) = 3$
    (b) $r(t) = e^{-t}$ and $c(0) = \dot{c}(0) = 0$

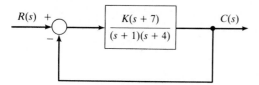

Figure P7.5

7.6 The block diagram for a control system is shown in Fig. P7.6. From the root-locus plot, it is found that when $K = 40$ the roots of the characteristic equation are located at -4 and $-1 \pm j3$. Determine the response $c(t)$ for the case in which $r(t) = 0$, $c(0) = \dot{c}(0) = 0$, and $\ddot{c}(0) = 18$.

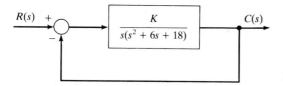

Figure P7.6

7.7 The block diagram for a control system is shown in Fig. P7.7. From the root-locus plot, it is found that when $K = 6$ the roots of the characteristic equation are located at -1, -3, and -4. Determine the response $c(t)$ when $r(t) = u_1(t)$ is a unit impulse and all initial conditions are zero.

Figure P7.7

7.8 Apply the angle condition to determine whether or not the point $s = -2 + j2$ is on the root-locus plot for $s(s + 2) + K(s + 4) = 0$. If it is, determine the value of K at this point and the factored form of the characteristic equation.

7.9 Apply the angle condition to determine whether or not the point $s = -4 + j3$ is on the root-locus plot for $s(s + 4)(s + 7)(s + 8) + K(s + 1) = 0$. If it is, determine the value of K at this point and the factored form of the characteristic equation.

7.10 The location of the ×'s and ○'s of the characteristic equation for various systems is shown in Fig. P7.10. For each system, determine the characteristic equation and then sketch the root-locus plot.

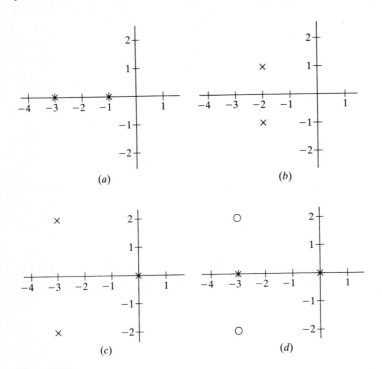

(a)

(b)

(c)

(d)

Figure P7.10

7.11 Sketch the root-locus plot for the system shown in Fig. P7.11. Determine the value of K to yield a damping ratio of 0.5.

Figure P7.11

7.12 The block diagram and corresponding root-locus plot for a control system are shown in Fig. P7.12.

(a) Determine the differential equation relating the output $c(t)$ and the input $r(t)$.

(b) Determine the response $c(t)$ for the case in which all the initial conditions are zero, $K = 20$, and the input excitation $r(t)$ is a unit impulse.

(c) Use Routh's criterion to determine the value of the gain K at which the system becomes unstable.

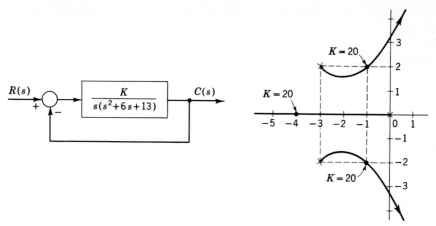

Figure P7.12

7.13 The block diagram for a control system is shown in Fig. P7.13. The controller is represented by the block $G_c(s)$. Determine the characteristic equation and then sketch the corresponding root-locus plot for each of the following cases:

(a) Proportional controller, $G_c(s) = K_1$ and $H(s) = 1$
(b) Integral controller, $G_c(s) = K_1/s$ and $H(s) = 1$
(c) Proportional plus integral controller, $G_c(s) = K_1(1 + 1/s)$ and $H(s) = 1$
(d) Proportional plus rate feedback, $G_c(s) = K_1$ and $H(s) = K_3(1 + s)$

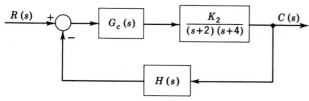

Figure P7.13

7.14 For the control system shown in Fig. P7.14, the box $G_c(s)$ represents the characteristics of the controller. Construct the root-locus plot for each of the following cases:

(a) Proportional controller, $G_c(s) = K_1$
(b) Integral controller, $G_c(s) = K_1/s$
(c) Integral plus proportional controller, $G_c(s) = K_1(1 + 1/s)$

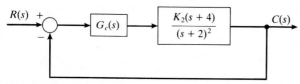

Figure P7.14

7.15 Sketch the root-locus plot for the system shown in Fig. P7.15 for each of the following cases:

(a) $G_1(s) = \dfrac{K_1}{s}$ (b) $G_1(s) = \dfrac{K_1}{s+4}$

For each case, determine the value of $K_1 K_2 K_3$ to yield a damping ratio of 0.5 for the dominant roots (i.e., the ones located nearest the imaginary axis).

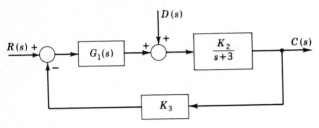

Figure P7.15

7.16 Sketch the root-locus plot for each of the two systems shown in Fig. P7.16. Determine the value of K at which each system becomes unstable. Comment on the effect of adding an integrating element, as is done in case b.

Figure P7.16

7.17 The block diagram and root-locus plot for a control system are shown in Fig. P7.17. Repeated roots occur at $s = -6$.

(a) Determine the value of the gain K such that the characteristic equation has repeated roots.

(b) For the case in which the characteristic equation has repeated roots, determine the response $c(t)$ when $r(t)$ is a unit impulse and all the initial conditions are zero.

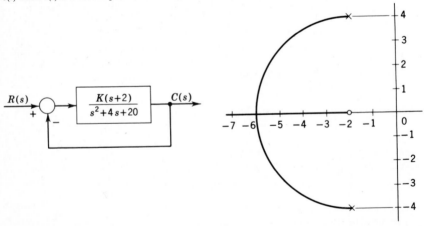

Figure P7.17

7.18 The root-locus plot for the system of Fig. 7.16a is given in Fig. 7.16b. When $K = 17$, the characteristic equation has a repeated root at -1.57 and a distinct root at -6.9. Determine the response $c(t)$ for the case in which $r(t)$ is a unit impulse and $K = 17$. All the initial conditions are zero.

7.19 The characteristic equation for the system shown in Fig. 7.16a is given by Eq. (7.12). If a zero is added to the system, the characteristic equation becomes

$$s(s + 4)(s + 6) + K(s + 2) = 0$$

Sketch the root-locus plot for this new system. Comment on the effect of adding the zero.

7.20 Sketch the root-locus plot for each of the characteristic equations given below:
  (a) $s^2(s + 8) + K = 0$ (b) $s^2(s + 8) + K(s + 2) = 0$

7.21 Construct the root-locus plot for the system shown in Fig. P6.30. Determine the value of K so that the resulting system will have a damping ratio of 0.6.

7.22 Sketch the root-locus plot for each of the following characteristic equations:
  (a) $(s + 1)(s^2 + 2s + 2) + K = 0$
  (b) $(s + 1)(s^2 + 2s + 2) + K(s + 2) = 0$
  (c) $(s + 1)(s^2 + 2s + 2) + K(s + 2)(s + 4) = 0$

7.23 Sketch the root-locus plot for each of the following characteristic equations:
  (a) $s(s + 3) + K(s + 4) = 0$ (b) $(s + 1)^2 + K(s + 3) = 0$

7.24 Sketch the root-locus plot for each of the following characteristic equations:
  (a) $s(s^2 + 8s + 25) + K = 0$
  (b) $s(s^2 + 8s + 25) + K(s + 2) = 0$
  (c) $s(s^2 + 8s + 25) + K(s + 2)(s + 4) = 0$

7.25 Sketch the root-locus plot for each of the following characteristic equations:
  (a) $(s + 2)^2 + K(s + 4) = 0$ (b) $(s + 2)^3 + K(s + 4) = 0$
  (c) $s(s^2 + 10s + 26) + K = 0$ (d) $(s + 3)^2 + K(s^2 + 10s + 26) = 0$

7.26 A certain system has the characteristic equation

$$s(s^2 + 8s + 20) + K(s + 8) = 0$$

Construct the root-locus plot for this system. Determine the value of the gain K when the characteristic equation has a pair of complex conjugate roots whose real part is -2.

7.27 Sketch the root-locus plot for the system shown in Fig. P7.27 for each of the following cases:
  (a) $H(s) = 1$ (b) $H(s) = s + 1$
For each case, determine the value of K at which each system becomes unstable. Comment on the effect of adding derivative action in the feedback path (i.e., case b).

Figure P7.27

7.28 The root-locus method may be used to factor a polynomial. Note that the polynomial $s^2 + 4s + 20$ may be written as $s(s + 4) + K_1 = 0$ or as $s^2 + K_2(s + 5) = 0$. The intersection of these root-locus plots are the roots of $s^2 + 4s + 20 = 0$. Construct the root-locus plots and then determine the roots.

7.29 For each of the characteristic functions given below, sketch the root-locus plot for both positive and negative K:
  (a) $s(s + 3)^2 + K = 0$ (b) $s(s + 2)(s + 4)(s + 6) + K = 0$
  (c) $s(s + 6)^2 + K(s^2 + 6s + 18) = 0$ (d) $s(s + 2)(s + 4) + K(s^2 + 4s + 8) = 0$

7.30 For the control system shown in Fig. P7.30, sketch the root-locus plot for
(a) $K > 0$ (b) $K < 0$

Figure P7.30

7.31 The root-locus plot for the system whose characteristic equation is $s(s + 2) + K(s + 4) = 0$ is shown in Fig. 7.24. When $K = 6$, the roots are $-4 \pm j\sqrt{8}$. Hence, for $K = 6$, the factored form of the characteristic equation is

$$s(s + 2) + 6(s + 4) = s^2 + 8s + 24$$

To investigate the effect of changing the parameter 4, replace 4 by $4 + \Delta$ and then:
(a) Sketch the root-locus plot for positive Δ.
(b) Sketch the root-locus plot for negative Δ.

7.32 Same as Prob. 7.31 except vary the parameter 2 instead of 4.

7.33 A feedback control system and its root-locus plot are shown in Fig. P7.33. This is the same system shown in Fig. 7.19.
(a) When $K = 50$, the roots of the characteristic equation are located at -5, -5, and -2. Replace the parameter 45 in the characteristic equation by $45 + \Delta$ and proceed to construct the root-locus plot for $\Delta > 0$ and the root-locus plot for $\Delta < 0$.
(b) When $K = 104$, the roots of the characteristic equation are located at -8, $-2 + j3$, and $-2 - j3$. Replace the parameter 45 by $45 + \Delta$ and proceed to construct the root-locus plot for $\Delta > 0$ and the root-locus plot for $\Delta < 0$.

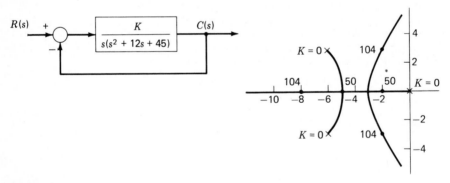

Figure P7.33

7.34 For the system of Prob. 7.12 it is desired to investigate the effect of varying the parameter 13 when $K = 20$. Thus replace the parameter 13 by $13 + \Delta$ in the characteristic equation and then sketch the resulting root-locus plot for $\Delta > 0$ and the root-locus plot for $\Delta < 0$.

7.35 For the system of Prob. 7.12 it is desired to investigate the effect of varying the parameter 6 when $K = 20$. Thus replace the parameter 6 by $6 + \Delta$ and then sketch the resulting root-locus plot for $\Delta > 0$ and the root-locus plot for $\Delta < 0$.

7.36 The feedforward transfer function for a unity-feedback control system is

$$G(s) = \frac{K(s + 8)}{s(s^2 + 8s + 20)}$$

The root-locus plot for this system is shown in Fig. P7.36. When the gain K is 4, the roots of the characteristic equation are -4, $-2 + j2$, and $-2 - j2$. For $K = 4$, investigate the effect of varying the parameter 20 in the characteristic equation. Sketch the resulting root-locus plot for both positive and negative values of Δ.

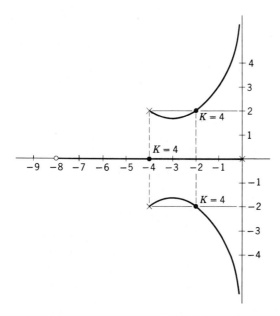

Figure P7.36

7.37 It is desired to design a compensator to stabilize the system shown in Fig. P7.37. The only means available is a constant multiplier K which can be inserted in the feedback loop as shown. Sketch the root-locus plot and then use Routh's criterion to determine the gain K such that the roots of the characteristic equation lie on

    (*a*) The imaginary axis   (*b*) The vertical axis at -3

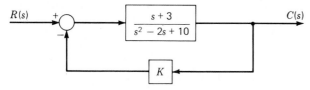

Figure P7.37

7.38 A feedback control system with internal feedback is shown in Fig. P7.38. Determine the overall equation relating the output $C(s)$ to the input $R(s)$ and then

    (*a*) Construct the root-locus plot for the basic system without internal feedback ($\beta = 0$).

    (*b*) Construct the root-locus plot for the actual system in which β is the variable parameter, $K_1 = 2.5$, and $K_2 = 10.0$.

    (*c*) Determine the value of β such that the resulting system will have a damping ratio of $\zeta = 0.6$.

    (*d*) Determine the value of β such that the resulting system will have a damping ratio of $\zeta = 1.0$.

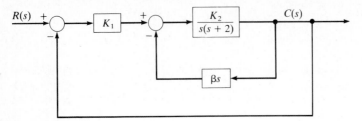

Figure P7.38

7.39 The portion of the system enclosed by the dashed box in Fig. P7.39 represents the plant to be controlled. The root-locus plot for this plant shows that for $K_1 = 2$, one root is located at the origin and the other root is located at -2. For $K_1 = 2$, construct the root-locus plot for the entire system, $0 < K_2 < \infty$. For what values of K_2 does the system have an oscillatory response? What is the smallest possible damping ratio?

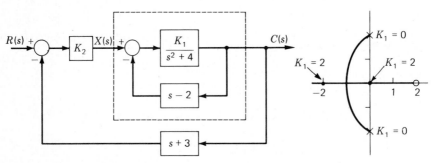

Figure P7.39

7.40 In Fig. P7.40 is shown a feedback control system with internal feedback. Construct the root-locus plot for the basic system without the internal feedback ($\beta = 0$) and then construct the root-locus plot for the modified system in which β is the variable parameter, $K_1 = 2$, and $K_2 = 5$. Determine the value of β such that the resulting system will have a damping ratio of (a) $\zeta = 0.5$ and (b) $\zeta = 1.0$.

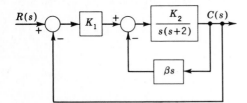

Figure P7.40

7.41 The block diagram for controlling the azimuth angle ϕ of the stable platform of an inertial guidance system is shown in Fig. P7.41a. Construct the root-locus plot and determine the roots of the characteristic equation when $K = 4$.

The use of internal feedback with derivative action (βs) to stabilize this system is shown in Fig. P7.41b. Construct the equivalent block diagram in which the internal feedback loop is eliminated. Construct the root-locus plot for the new characteristic equation ($K = 4$) and then determine the value of β such that the new characteristic equation will have a damping ratio of (a) $\zeta = 0.5$ and (b) $\zeta = 1.0$.

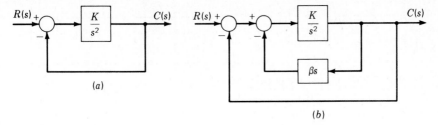

Figure P7.41

7.42 The transfer function for a unity-feedback control system is

$$G(s) = \frac{K}{s^2 + 2\zeta\omega_n s + \omega_n^2} = \frac{15}{s^2 + 4s + 5}$$

The roots of the characteristic equation are $r_{1,2} = -2 \pm j4$. Determine the equation for the change Δr_1 in the root r_1 due to a change in (a) K, (b) ω_n, and (c) ζ.

REFERENCES

1. Evans, W. R.: Graphical Analysis of Control Systems, *Trans. AIEE*, **67**, 547–551, 1948.
2. Evans, W. R.: "Control-System Dynamics," McGraw-Hill Book Company, New York, 1954.

EIGHT

ANALOG COMPUTERS

The use of computers has played a major role in recent advances in the design of automatic control systems. These computers may be divided into two types, analog computers and digital computers.

An analog computer is one in which the equation describing the operation of the computer is analogous to that for the actual system. The most commonly used analog computer is the electronic analog computer, in which voltages at various places within the computer are proportional to the variable terms in the actual system. As is shown in this chapter, the operation of a control system can be simulated by the use of an analog computer.

Basically a digital computer can only add and subtract. Thus it is necessary to reduce all problems to rather elementary mathematical manipulations. This process is called programming. The programming of a problem for solution on a digital computer makes extensive use of the methods of "numerical analysis" to convert the problem to the numerical operations which the computer can perform. It may require weeks or even months to program a problem for a computer, which in turn completes the solution in a few minutes or seconds. A digital computer has been referred to as an "energetic moron" in that it is capable of performing thousands of simple additions and subtractions in a second. A digital computer must store information for use in later computations. This is usually done by means of a magnetic disk, which acts as a memory device.

The input to a digital computer consists of numbers and instructions for the operation of the machine on these numbers. These numbers and instructions may be typed directly into the machine at the computer terminal or they may be read from a magnetic tape.

Because of the ability of both digital and analog computers to solve complicated mathematical equations almost instantaneously, they are often incorporated as part of a control system to compute desired information. This information may then be used immediately to improve the control of the particular system. For example, in an inertial guidance system, the output of three mutually perpendicular accelerometers is fed into a computer which in turn calculates the position of the vehicle. Thus, the output of this computer is the actual position of the vehicle, which is compared with the desired position to yield an error signal for actuating the steering mechanism.

The electronic analog computer is a very powerful tool for investigating the performance of control systems. For more complex systems, the advantages of the analog computer become more apparent. Analog computers are used for many purposes besides that of investigating linear and nonlinear control systems. For example, they are used to solve nonlinear differential equations, partial differential equations, systems of differential or partial differential equations, matrix and eigenvalue problems, operations research problems, etc. New applications and uses for this versatile computing device are continually being discovered.

This chapter is primarily concerned with the use of these computers for simulating control systems. For this purpose, the equation describing the operation of the analog computer is analogous to that which represents the actual physical system. The variable quantities of the actual system, such as the output, input, error, etc., are represented by voltages at various places within the analog computer. Permanent records of these voltages may be obtained by using recording equipment. By using potentiometers or variable capacitors to vary the resistance or capacitance at various places within the computer, the effect upon system performance of changing the corresponding parameters in the actual system (e.g., gain, time constants, damping ratio, etc.) may be determined immediately.

8.1 COMPUTER OPERATIONS

To solve any linear differential equation with constant coefficients, it is necessary only to make use of the processes of integration, summation, and multiplication by a constant. This is illustrated by the block diagram of Fig. 8.1 for the equation

$$M\ddot{y} + B\dot{y} + Ky = f(t) \tag{8.1}$$

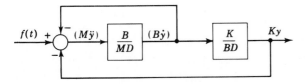

Figure 8.1 Block diagram for $M\ddot{y} + B\dot{y} + Ky = f(t)$.

Figure 8.2 Schematic representation for an operational amplifier.

To set up the block-diagram representation for a differential equation, first solve for the highest-order differential appearing in the original equation. The highest-order term appearing in Eq. (8.1) is $M\ddot{y}$. Solving Eq. (8.1) for $M\ddot{y}$ yields

$$M\ddot{y} = f(t) - B\dot{y} - Ky \tag{8.2}$$

Successive integration of the highest-order differential and multiplication by appropriate constants yields the other lower-order terms. That is,

$$\frac{B}{MD}(M\ddot{y}) = B\dot{y}$$

$$\frac{K}{BD}(B\dot{y}) = Ky \tag{8.3}$$

Each term on the right side of Eq. (8.2) goes into the summer of Fig. 8.1, so that the output of the summer is proportional to the acceleration. Successive integration of this acceleration yields the velocity and displacement.

The heart of the electronic analog computer is the operational amplifier, which is a very-high-gain dc amplifier. This device may be used as an integrator, summer, or multiplier. The particular mathematical operation depends upon the particular network of resistors and capacitors which are placed around it.

In Fig. 8.2 is shown the schematic representation of an operational amplifier. The input voltage is e_i, the output is e_2, and the amplification is $-A$. Thus,

$$e_2 = -Ae_i \tag{8.4}$$

The reason for the minus sign is that the amplifier reverses the phase of the input. For most operational amplifiers, the value of A is very large. Values of A may range from 100×10^3 to 100×10^6.

Multiplication by a Constant

By feeding the input voltage through a resistor R_1 and by putting a resistor R_2 in parallel with the amplifier as shown in Fig. 8.3, a circuit for multiplication by a constant is obtained. Because the input grid of the amplifier draws little or no current (a typical value is 10^{-9} A), then

$$i_1 \approx i_2 \tag{8.5}$$

Figure 8.3 Operational amplifier circuit for multiplying by a constant.

The values of i_1 and i_2 are

$$i_1 = \frac{e_1 - e_i}{R_1} \qquad i_2 = \frac{e_i - e_2}{R_2} \qquad (8.6)$$

Equating i_1 and i_2 yields

$$\frac{e_1 - e_i}{R_1} = \frac{e_i - e_2}{R_2} \qquad (8.7)$$

Usually e_2 is less than 10 V, so for very large values of A, it follows from Eq. (8.4) that $e_i = -e_2/A \approx 0$. The substitution of $e_i \approx 0$ into Eq. (8.7) gives

$$e_2 = -\frac{R_2}{R_1} e_1 \qquad (8.8)$$

The value of the multiplication constant is $-R_2/R_1$. When $R_2 = R_1$, the constant is -1, which means simply that the phase of the input signal has been inverted.

Integration

Replacing the resistor R_2 of Fig. 8.3 with a capacitor, as shown in Fig. 8.4, yields an integration circuit. The current i_2 flowing through this capacitor is

$$i_2 = C_2 D(e_i - e_2) \qquad (8.9)$$

By equating i_1 and i_2 as before, it follows that

$$\frac{e_1 - e_i}{R_1} = C_2 D(e_i - e_2) \qquad (8.10)$$

Because $e_i \approx 0$ the preceding expression reduces to

$$e_2 = \frac{-e_1}{R_1 C_2 D} = \frac{-1}{R_1 C_2} \int_0^t e_1 \, dt + e_2(0) \qquad (8.11)$$

where $e_2(0)$ is the value of e_2 when $t = 0$. In addition to integration, the circuit of Fig. 8.4 also multiplies by the constant $-1/R_1 C_2$.

Figure 8.4 Operational amplifier circuit for integrating.

Summation

The effect of summation is obtained by placing the desired quantities to be added in parallel at the input to the computer circuit. The output will then be the summation of the effect due to each individual input. A general summing circuit is shown in Fig. 8.5a and a circuit to integrate more than one quantity is shown in Fig. 8.5b. A simplified notation for these various computer elements is also shown in Fig. 8.5a and b. This schematic notation saves much effort in drawing the computer diagram for a circuit.

For the summing circuit of Fig. 8.5a, the sum of the currents i_a plus i_b plus i_c is

$$i_a + i_b + i_c = \frac{e_a - e_i}{R_a} + \frac{e_b - e_i}{R_b} + \frac{e_c - e_i}{R_c} \approx \frac{e_a}{R_a} + \frac{e_b}{R_b} + \frac{e_c}{R_c}$$

The current i_2 is

$$i_2 = \frac{e_i - e_2}{R_2} \approx -\frac{e_2}{R_2}$$

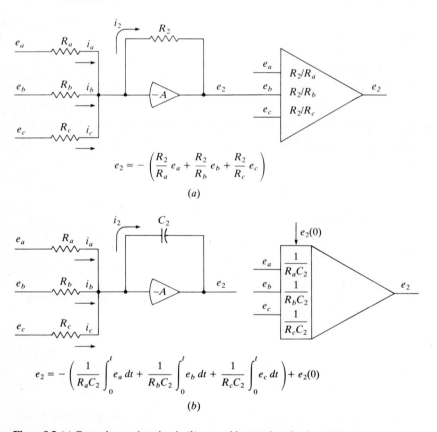

$$e_2 = -\left(\frac{R_2}{R_a} e_a + \frac{R_2}{R_b} e_b + \frac{R_2}{R_c} e_c\right)$$

(a)

$$e_2 = -\left(\frac{1}{R_a C_2}\int_0^t e_a \, dt + \frac{1}{R_b C_2}\int_0^t e_b \, dt + \frac{1}{R_c C_2}\int_0^t e_c \, dt\right) + e_2(0)$$

(b)

Figure 8.5 (a) General summing circuit, (b) general integrating circuit.

Equating the right sides of these equations and solving for e_2 gives

$$e_2 = -\left(\frac{R_2}{R_a}e_a + \frac{R_2}{R_b}e_b + \frac{R_2}{R_c}e_c\right) \tag{8.12}$$

The minus sign results because an amplifier always reverses the sign of the input voltage. Note that the equation for the voltage e_2 may be obtained directly from the schematic diagram by multiplying each input voltage by the value of the ratio of the resistors for that particular input. That is, e_a is multiplied by R_2/R_a, e_b by R_2/R_b, and e_c by R_2/R_c.

For the integrating circuit shown in Fig. 8.5b, the sum of the currents i_a plus i_b plus i_c is

$$i_a + i_b + i_c = \frac{e_a - e_i}{R_a} + \frac{e_b - e_i}{R_b} + \frac{e_c - e_i}{R_c} \approx \frac{e_a}{R_a} + \frac{e_b}{R_b} + \frac{e_c}{R_c}$$

The current i_2 is

$$i_2 = C_2 D(e_i - e_2) \approx -C_2 De_2$$

Equating the right sides of these equations and solving for e_2 shows that

$$e_2 = -\left(\frac{1}{R_a C_2 D}e_a + \frac{1}{R_b C_2 D}e_b + \frac{1}{R_c C_2 D}e_c\right)$$

$$= -\left(\frac{1}{R_a C_2}\int_0^t e_a\,dt + \frac{1}{R_b C_2}\int_0^t e_b\,dt + \frac{1}{R_c C_2}\int_0^t e_c\,dt\right) + e_2(0) \tag{8.13}$$

Note that the schematic representation reveals directly this equation. That is, the equation for the voltage e_2 is obtained by multiplying the integral of each input voltage by the corresponding $(1/RC)$ factor for that input and adding the initial value $e_2(0)$.

8.2 DIRECT PROGRAMMING

Let it be desired to determine the computer diagram for solving the following differential equation:

$$\ddot{y} + b\dot{y} + ay = f(t) \tag{8.14}$$

The initial conditions are designated as $y(0)$ and $\dot{y}(0)$. The expected maximum values for y, $\dot{y}$, $\ddot{y}$, and $f(t)$ are designated as y_m, $\dot{y}_m$, $\ddot{y}_m$, and $f(t)_m$.

Solving Eq. (8.14) for the highest-order derivative $\ddot{y}$ gives

$$\ddot{y} = f(t) - ay - b\dot{y} \tag{8.15}$$

Dividing through by $\ddot{y}_m$ and expressing $f(t)$ as $f(t)_m[f(t)/f(t)_m]$, y as $y_m(y/y_m)$, and $\dot{y}$ as $\dot{y}_m(\dot{y}/\dot{y}_m)$ yields

$$\frac{\ddot{y}}{\ddot{y}_m} = \frac{f(t)_m}{\ddot{y}_m}\left(\frac{f(t)}{f(t)_m}\right) - a\frac{y_m}{\ddot{y}_m}\left(\frac{y}{y_m}\right) - b\frac{\dot{y}_m}{\ddot{y}_m}\left(\frac{\dot{y}}{\dot{y}_m}\right) \tag{8.16}$$

Note that each term in parentheses is the ratio of the variable compared to the maximum value [that is, $f(t)/f(t)_m$, y/y_m, and $\dot{y}/\dot{y}_m$]. Depending on the type of operational amplifiers used in a particular computer, the manufacturer suggests limiting the output voltage to a certain value such as 10 V. If the output voltage exceeds the suggested maximum value, the amplifier begins to saturate. Each amplifier has a light that goes on when it saturates.

The general computer diagram is shown in Fig. 8.6a. Because an amplifier reverses the sign from the input to the output, the sign in front of each scale factor alternates in going from one amplifier to the next. This fact is illustrated in Fig. 8.6a in which the output of the first amplifier is $e_1 = -(\ddot{y}/\ddot{y}_m)e_m$, the output of the second is $e_2 = (\dot{y}/\dot{y}_m)e_m$, and the output of the third is $e_3 = -(y/y_m)e_m$.

A more detailed diagram of the first amplifier is shown in Fig. 8.6b. Multiplying through Eq. (8.16) by e_m where e_m is the suggested maximum voltage gives

$$\frac{\ddot{y}}{\ddot{y}_m}e_m = \frac{f(t)_m}{\ddot{y}_m}\left(\frac{f(t)}{f(t)_m}\right)e_m - a\frac{y_m}{\ddot{y}_m}\left(\frac{y}{y_m}\right)e_m - b\frac{\dot{y}_m}{\ddot{y}_m}\left(\frac{\dot{y}}{\dot{y}_m}\right)e_m \qquad (8.17)$$

As shown in Fig. 8.6b, each term on the right side is inputted to the first amplifier to generate $(\ddot{y}/\ddot{y}_m)e_m$. Because the amplifier reverses sign, the output of the first amplifier is $e_1 = -(\ddot{y}/\ddot{y}_m)e_m$. It should be noted that Eq. (8.17) is merely an alge-

(a)

(b)   (c)

Figure 8.6 Analog computer diagram for $\ddot{y} + b\dot{y} + ay = f(t)$: (a) overall diagram, (b) first amplifier, (c) circuit diagram.

braic rearrangement of Eq. (8.15). That is, the ay and $b\dot{y}$ terms in Eq. (8.15) have been taken to the right side, multiplied and divided by y_m and $\dot{y}_m$ respectively, and then each term has been divided by $\ddot{y}_m$ and multiplied by e_m. The circuit diagram for the first amplifier is shown in Fig. 8.6c. The voltage relationship is

$$e_1 = -\frac{R_2}{R}e - \frac{R_2}{R_a}e_a - \frac{R_2}{R_b}e_b \tag{8.18}$$

Because the output of the second amplifier is $e_2 = (\dot{y}/\dot{y}_m)e_m$, it is necessary to use a sign changer to obtain $e_b = -e_2 = (-\dot{y}/\dot{y}_m)e_m$. The voltage $e_a = e_3 = (-y/y_m)e_m$ may be fed back directly. Multiplying through Eq. (8.17) by -1 gives

$$-\frac{\ddot{y}}{\ddot{y}_m}e_m = -\frac{f(t)_m}{\ddot{y}_m}\left(\frac{f(t)}{f(t)_m}\right)e_m - a\frac{y_m}{\ddot{y}_m}\left(-\frac{y}{y_m}\right)e_m - b\frac{\dot{y}_m}{\ddot{y}_m}\left(-\frac{\dot{y}}{\dot{y}_m}\right)e_m \tag{8.19}$$

A termwise comparison of Eq. (8.18), which is the voltage relationship for the first amplifier, with Eq. (8.19), which is the physical relationship to be solved by the first amplifier, shows that to have $e_1 = -(\ddot{y}/\ddot{y}_m)e_m$, $e = [f(t)/f(t)_m]e_m$, $e_a = (-y/y_m)e_m$, and $e_b = (-\dot{y}/\dot{y}_m)e_m$, then

$$\frac{R_2}{R} = \frac{f(t)_m}{\ddot{y}_m} \qquad \frac{R_2}{R_a} = a\frac{y_m}{\ddot{y}_m} \qquad \frac{R_2}{R_b} = b\frac{\dot{y}_m}{\ddot{y}_m} \tag{8.20}$$

Note that the right sides of these expressions contain all known quantities whence the value for all the resistors for the first amplifier may be obtained. The voltage equation for the second amplifier which is an integrator is

$$e_2 = -\frac{1}{(R_1C_2)_2}\int_0^t e_1\,dt + e_2(0) \tag{8.21}$$

where $(R_1C_2)_2$ is the product of the resistance and capacitance for amplifier 2. Because the integral of acceleration is velocity, it follows that

$$\dot{y} = \int_0^t \ddot{y}\,dt + \dot{y}(0)$$

This may be rewritten in the form

$$\frac{\dot{y}}{\dot{y}_m}e_m = -\frac{\ddot{y}_m}{\dot{y}_m}\int_0^t\left(-\frac{\ddot{y}}{\ddot{y}_m}\right)e_m\,dt + \frac{\dot{y}(0)}{\dot{y}_m}e_m \tag{8.22}$$

Equation (8.21) is the voltage relationship and Eq. (8.22) is the corresponding physical relationship. To have $e_2 = (\dot{y}/\dot{y}_m)e_m$ and $e_1 = -(\ddot{y}/\ddot{y}_m)e_m$, a termwise comparison of Eqs. (8.21) and (8.22) shows that

$$\frac{1}{(R_1C_2)_2} = \frac{\ddot{y}_m}{\dot{y}_m} \qquad \text{and} \qquad e_2(0) = \frac{\dot{y}(0)}{\dot{y}_m}e_m \tag{8.23}$$

Note that the value of $1/(R_1C_2)_2$ is equal to the ratio of the maximum value of the input $\ddot{y}_m$ to the maximum value of the output $\dot{y}_m$. The initial-condition expression could also have been obtained by evaluation of $e_2 = (\dot{y}/\dot{y}_m)e_m$ at time $t = 0$.

The voltage equation for the third amplifier is

$$e_3 = -\frac{1}{(R_1 C_2)_3} \int_0^t e_2 \, dt + e_3(0) \tag{8.24}$$

Because the integral of velocity is displacement,

$$y = \int_0^t \dot{y} \, dt + y(0)$$

This may be rewritten in the form

$$-\frac{y}{y_m} e_m = -\frac{\dot{y}_m}{y_m} \int_0^t \left(\frac{\dot{y}}{\dot{y}_m}\right) e_m \, dt - \frac{y(0)}{y_m} e_m \tag{8.25}$$

A termwise comparison of the voltage relationship [Eq. (8.24)] with the physical relationship [Eq. (8.25)] shows that to have $e_3 = -(y/y_m)e_m$ and $e_2 = (\dot{y}/\dot{y}_m)e_m$, then

$$\frac{1}{(R_1 C_2)_3} = \frac{\dot{y}_m}{y_m} \quad \text{and} \quad e_3(0) = -\frac{y(0)}{y_m} e_m \tag{8.26}$$

The value of $1/(R_1 C_2)_3$ is equal to the ratio of the maximum value of the input $\dot{y}_m$ to the maximum value of the output y_m. In general, the value of $1/R_1 C_2$ for an integrator is equal to the ratio of the maximum value of the input (the physical quantity to be integrated) to the maximum value of the output. Evaluating the voltage relationship $e_3 = -(y/y_m)e_m$ at time $t = 0$ verifies the result shown for the preceding initial-condition relationship.

Each integrator may be initially biased by a dc voltage to give the effect of initial conditions, as is represented diagrammatically at the top of each integrator in Fig. 8.6a. In practice, the initial condition is obtained by placing a source of constant potential such as a battery of potential $e_2(0)$ in parallel with the capacitor, as shown in Fig. 8.7. For $t < 0$, the switch is up so that the battery can charge the capacitor to the desired initial value. At $t = 0$, the switch moves down to disconnect the battery so that the integrator circuit functions as previously discussed. There is never any need to apply an initial voltage $e(0)$ to account for initial conditions when an amplifier is used to multiply by a constant. Only integration requires initial conditions.

Figure 8.7 Circuit for obtaining initial conditions.

The relationships $e_1 = -(\ddot{y}/\ddot{y}_m)e_m$, $e_2 = (\dot{y}/\dot{y}_m)e_m$, and $e_3 = -(y/y_m)e_m$ may be rewritten in the form

$$\frac{e_1}{e_m} = -\frac{\ddot{y}}{\ddot{y}_m} \qquad \frac{e_2}{e_m} = \frac{\dot{y}}{\dot{y}_m} \qquad \frac{e_3}{e_m} = -\frac{y}{y_m} \qquad (8.27)$$

For numerical purposes, suppose $\ddot{y}_m = 20$, $\dot{y}_m = 10$, and $y_m = 5$. If $e_m = 10$ V for the analog computer, then when $e_1 = 7.5$ V the corresponding value of $\ddot{y}$ is -15. For $e_2 = 7.5$ V, then $\dot{y} = 7.5$. For $e_3 = 7.5$ V, then $y = -3.75$. Consider now the case of an analog computer for which $e_m = 100$ V. When $e_1 = 75$ V, the corresponding value of $\ddot{y}$ is -15. For $e_2 = 75$ V, then $\dot{y} = 7.5$. For $e_3 = 75$ V, then $y = -3.75$. From Eq. (8.27) it is seen that the ratio of the actual voltage to e_m is equal to the ratio of the actual value of the parameter divided by the maximum value of the parameter. With this method, the computer diagram is independent of the value of e_m for the particular computer being used.

It is not necessary to rearrange algebraically the differential equation and then compare it to the voltage relationship in order to obtain the value of the resistors for the first amplifier. A much simpler technique is now described. Consider the differential equation

$$\ddot{y} + b\dot{y} + ay = f(t) \qquad (8.14)$$

Solving for $\ddot{y}$, dividing through by $\ddot{y}_m$, multiplying through by e_m, and then expressing each variable as a ratio compared to its maximum value gives

$$\frac{\ddot{y}}{\ddot{y}_m}e_m = \frac{f(t)_m}{\ddot{y}_m}\left(\frac{f(t)}{f(t)_m}\right)e_m + a\frac{y_m}{\ddot{y}_m}\left(-\frac{y}{y_m}\right)e_m + b\frac{\dot{y}_m}{\ddot{y}_m}\left(-\frac{\dot{y}}{\dot{y}_m}\right)e_m \qquad (8.17)$$

Because an amplifier reverses sign, when each term on the right side is fed into the first amplifier the output is the negative of the left side; that is, $e_1 = -(\ddot{y}/\ddot{y}_m)e_m$. By referring to Fig. 8.6b, it is to be noted that the ratio of resistors R_2/R for the input voltage $e = [f(t)/f(t)_m]e_m$ is the coefficient $f(t)_m/\ddot{y}_m$ of the term $[f(t)/f(t)_m]e_m$ in Eq. (8.17). Similarly, the ratio of resistors R_2/R_a for the input voltage $e_a = -(y/y_m)e_m$ is the coefficient $ay_m/\ddot{y}_m$ of the $(-y/y_m)e_m$ term. Finally, the ratio of resistors R_2/R_b for the input voltage $e_b = -e_2 = (-\dot{y}/\dot{y}_m)e_m$ is the coefficient $b\dot{y}_m/\ddot{y}_m$ of the $(-\dot{y}/\dot{y}_m)e_m$ term. The need for the sign changer is easily detected because in the computer diagram of Fig. 8.6a the voltage $e_2 = (\dot{y}/\dot{y}_m)e_m$ is generated whereas in Eq. (8.17) the term $(-\dot{y}/\dot{y}_m)e_m$ is obtained. The value of $1/(R_1C_2)_2$ for the second amplifier is $\ddot{y}_m/\dot{y}_m$ and the value of $1/(R_1C_2)_3$ for the third amplifier is $\dot{y}_m/y_m$.

Illustrative example 8.1 Let it be desired to determine the computer diagram for solving the following differential equation:

$$\ddot{y} + 2\dot{y} + 10y = f(t) \qquad (8.28)$$

The initial conditions are $y(0) = 3$ m and $\dot{y}(0) = 5$ m/s. The maximum values are $y_m = 5$ m, $\dot{y}_m = 10$ m/s, $\ddot{y}_m = 20$ m/s^2, and $f(t)_m = 8$ N.

SOLUTION Solving Eq. (8.28) for the highest-order derivative yields

$$\ddot{y} = f(t) - 10y - 2\dot{y} \tag{8.29}$$

Dividing through by $\ddot{y}_m$ and expressing $f(t)$, y, and $\dot{y}$ in terms of ratios compared to their respective maximum values gives

$$\frac{\ddot{y}}{\ddot{y}_m} = \frac{f(t)_m}{\ddot{y}_m}\left(\frac{f(t)}{f(t)_m}\right) + 10\,\frac{y_m}{\ddot{y}_m}\left(-\frac{y}{y_m}\right) + 2\,\frac{\dot{y}_m}{\ddot{y}_m}\left(-\frac{\dot{y}}{\dot{y}_m}\right) \tag{8.30}$$

In Fig. 8.8a is shown the analog computer diagram for this differential equation. It suffices to write the ratio of the physical quantity to its maximum value throughout the computer diagram. That is, although the output from the first amplifier is $e_1 = -(\ddot{y}/\ddot{y}_m)e_m$, only $-\ddot{y}/\ddot{y}_m$ is written. Similarly, the output from the second amplifier is $e_2 = (\dot{y}/\dot{y}_m)e_m$, but only $\dot{y}/\dot{y}_m$ is written, etc.

Each term in parentheses on the right side of Eq. (8.30) represents a voltage which is inputted to the first amplifier. The coefficient for each term

(a)

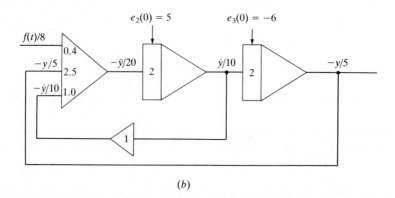

(b)

Figure 8.8 Analog computer diagram for $\ddot{y} + 2\dot{y} + 10y = f(t)$.

is the ratio of resistors for that signal. That is, the ratio of resistors R_2/R for the $[f(t)/f(t)_m]$ signal is $f(t)_m/\ddot{y}_m$, the ratio R_2/R_a for the $-y/y_m$ signal is $10y_m/\ddot{y}_m$, and the ratio R_2/R_b for the $-\dot{y}/\dot{y}_m$ signal is $2\dot{y}_m/\ddot{y}_m$. The need for the sign changer is apparent because the signal $\dot{y}/\dot{y}_m$ occurs as the output of the second amplifier in the computer diagram whereas $-\dot{y}/\dot{y}_m$ occurs in Eq. (8.30). With the right side of Eq. (8.30) as the input to the first amplifier, the output is the negative of the left side $(-\ddot{y}/\ddot{y}_m)$. The value of $1/R_1C_2$ for the second amplifier is $\ddot{y}_m/\dot{y}_m$ and that for the third is $\dot{y}_m/y_m$. Because $e_2 = (\dot{y}/\dot{y}_m)e_m$ and $e_3 = -(y/y_m)e_m$, the initial biasing voltages are $e_2(0) = [\dot{y}(0)/\dot{y}_m]e_m$ and $e_3(0) = -[y(0)/y_m]e_m$.

Substituting numerical values for y_m, $\dot{y}_m$, $\ddot{y}_m$, and $f(t)_m$ into Eq. (8.30) gives

$$\frac{\ddot{y}}{20} = \frac{8}{20}\left(\frac{f(t)}{8}\right) + 10\frac{5}{20}\left(-\frac{y}{5}\right) + 2\frac{10}{20}\left(-\frac{\dot{y}}{10}\right)$$

$$= 0.4\left(\frac{f(t)}{8}\right) + 2.5\left(-\frac{y}{5}\right) + 1.0\left(-\frac{\dot{y}}{10}\right) \tag{8.31}$$

The resulting computer diagram is shown in Fig. 8.8b. Note that $\ddot{y}_m/\dot{y}_m = 20/10 = 2$ and $\dot{y}_m/y_m = 10/5 = 2$. Similarly $e_2(0) = [\dot{y}(0)/\dot{y}_m]e_m = (5/10)10 = 5$ and $e_3(0) = -[y(0)/y_m]e_m = -(3/5)10 = -6$.

It is a relatively easy matter to determine the differential equation that is being solved by a particular computer diagram. For example, multiplying each input for the first amplifier of Fig. 8.8b by the ratio of resistors for each input gives

$$0.4\left(\frac{f(t)}{8}\right) + 1.0\left(\frac{-\dot{y}}{10}\right) + 2.5\left(\frac{-y}{5}\right) \tag{8.32}$$

Because the amplifier reverses sign, the output $-(\ddot{y}/20)$ is the negative of the preceding input. That is,

$$\frac{-\ddot{y}}{20} = -\frac{f(t)}{20} + \frac{2\dot{y}}{20} + \frac{10y}{20} \tag{8.33}$$

Rearranging yields the original differential equation

$$\ddot{y} + 2\dot{y} + 10y = f(t) \tag{8.28}$$

Because $e_2(0)/e_m = \dot{y}(0)/\dot{y}_m = \dot{y}(0)/10$ and $e_3(0)/e_m = -y(0)/y_m = -y(0)/5$, then for $e_2(0) = 5$, $\dot{y}(0) = 5$ and for $e_3(0) = -6$, $y(0) = 3$. The input to the second amplifier of Fig. 8.8b is $-\ddot{y}/20$. Because this is an integrator, the output $\dot{y}/10$ is the negative of the integral of the input multiplied by the factor $(1/R_1C_2)_2 = 2$. That is,

$$\frac{\dot{y}}{10} = -2\int\left(-\frac{\ddot{y}}{20}\right)dt$$

Hence

$$\dot{y} = \int \ddot{y}\, dt$$

Similarly, for the third amplifier the output $(-y/5)$ is the negative of the integral of the input $(\dot{y}/10)$ multiplied by the factor $(1/R_1C_2)_3 = 2$. That is,

$$-\frac{y}{5} = -2 \int \frac{\dot{y}}{10}\, dt$$

or

$$y = \int \dot{y}\, dt$$

For each system being studied, there is usually sufficient information available to make a reasonable estimate of the maximum value of each term. If an error is made in predicting the maximum value of a term, then the maximum value of the voltage corresponding to that term will not be e_m. Such a situation is easily detected and corrected as follows. Suppose $e_m = 10$ V and the maximum value of the voltage e_1 is found to be 7.5 V. The maximum value of $\ddot{y}$ is $(7.5/10) = \frac{3}{4}$ of the originally estimated maximum value of 20; hence the actual maximum value is 15. The computer diagram may be revised using $\ddot{y}_m = 15$, in which case the maximum value of the voltage e_1 will be 10 V. It should be noticed that the original solution for which the maximum value of e_1 was 7.5 V and the corresponding $\ddot{y}_m$ is 15 is correct. The reason for revising the computer diagram is to obtain better accuracy by using the full range of 10 V rather than 7.5 V.

The general procedure for setting up an electronic analog computer to solve a differential equation of order n may be summarized as follows:

1. Solve the differential equation for the highest-order derivative. For example, solving Eq. (8.14) for $\ddot{y}$ gives

$$\ddot{y} = f(t) - ay - b\dot{y}$$

2. Divide through the preceding by $\ddot{y}_m$ and then express $f(t)$, y, and $\dot{y}$ in terms of ratios compared to their maximum values. That is,

$$\frac{\ddot{y}}{\ddot{y}_m} = \frac{f(t)_m}{\ddot{y}_m}\left(\frac{f(t)}{f(t)_m}\right) + a\frac{y_m}{\ddot{y}_m}\left(-\frac{y}{y_m}\right) + b\frac{\dot{y}_m}{\ddot{y}_m}\left(-\frac{\dot{y}}{\dot{y}_m}\right) \tag{8.34}$$

The input to the first amplifier is the right side. Because an amplifier reverses sign, the output voltage is $e_1 = -(\ddot{y}/\ddot{y}_m)e_m$ but only $-\ddot{y}/\ddot{y}_m$ is written on the computer diagram. The output voltages of successive integrators are $e_2 = (\dot{y}/\dot{y}_m)e_m$ and $e_3 = (-y/y_m)e_m$, but only $\dot{y}/\dot{y}_m$ and $-y/y_m$ are written on the diagram. The coefficients in front of the parentheses are the ratio of resistors to be used at the input to the first amplifier [that is, $f(t)_m/\ddot{y}_m = R_2/R$, $ay_m/\ddot{y}_m = R_2/R_a$, and $\dot{y}_m/\ddot{y}_m = R_2/R_b$]. When the sign of the term in the computer diagram $(\dot{y}/\dot{y}_m)$ is opposite to that of the corresponding term in the parentheses, $(-\dot{y}/\dot{y}_m)$, a sign changer must be used to multiply the feedback signal by -1. The value of $1/R_1C_2$ for each integrator is equal to the ratio of the maximum value for the input divided by the maximum value for the output [that is, $1/(R_1C_2)_2 = \ddot{y}_m/\dot{y}_m$ and $1/(R_1C_2)_3 = \dot{y}_m/y_m$].

The equation that is being solved by a summing amplifier is obtained by noting that the output is the negative of the sum of each input multiplied by its ratio of resistors. For an integrating amplifier the output is the negative of the integral of the sum of each input multiplied by its $1/RC$ value.

Consider now the differential equation

$$z = \frac{D+4}{D^2 + 2D + 10} f(t) \tag{8.35}$$

This may be expressed in the form

$$z = (D + 4)y = \dot{y} + 4y \tag{8.36}$$

where

$$y = \frac{f(t)}{D^2 + 2D + 10}$$

The differential equation for y is the same as Eq. (8.28) and the corresponding computer diagram is shown in Fig. 8.8. Dividing through Eq. (8.36) by z_m and then expressing y and $\dot{y}$ in terms of ratios compared to their maximum values gives

$$\frac{z}{z_m} = \frac{\dot{y}_m}{z_m}\left(\frac{\dot{y}}{\dot{y}_m}\right) + 4\frac{y_m}{z_m}\left(\frac{y}{y_m}\right) \tag{8.37}$$

For $y_m = 5$, $\dot{y}_m = 10$, and $z_m = 25$ this becomes

$$\frac{z}{25} = \frac{10}{25}\left(\frac{\dot{y}}{10}\right) + 4\frac{5}{25}\left(\frac{y}{5}\right) = 0.4\left(\frac{\dot{y}}{10}\right) + 0.8\left(\frac{y}{5}\right) \tag{8.38}$$

The final computer diagram is shown in Fig. 8.9. The portion for solving Eq. (8.28) is the same as Fig. 8.8b. To obtain $z/25$, the input to the last amplifier is the negative of the right side of Eq. (8.38). That is,

$$0.4\left(-\frac{\dot{y}}{10}\right) + 0.8\left(-\frac{y}{5}\right)$$

The coefficient 0.4 is the ratio of resistors for the input $(-\dot{y}/10)$ and 0.8 is the ratio of resistors for the input $(-y/5)$. Because $\dot{y}/10$ is obtained as an output in the portion of the computer diagram for solving Eq. (8.28), it is necessary to use a sign changer to obtain $-\dot{y}/10$. The term $-y/5$ is obtained directly.

A feedback control system is shown in Fig. 8.10a. The maximum expected values are $c_m = 2$, $\dot{c}_m = 4$, $\ddot{c}_m = 10$, $f(t)_m = 2$, and $r_m = 1$. The initial conditions

Figure 8.9 Analog computer diagram for $(D^2 + 2D + 10)z = (D + 4)f(t)$.

(a)

(b)

(c)

Figure 8.10 Analog computer diagram for a feedback control system.

are $c(0) = 1$ and $\dot{c}(0) = 2$. To construct the analog computer diagram for this control system, first construct the diagram relating $c(t)$ and $f(t)$. Thus,

$$c(t) = \frac{10}{D(D + 2)} f(t)$$

or

$$\ddot{c} + 2\dot{c} = 10 f(t)$$

Solving for $\ddot{c}$, the highest-order derivative,

$$\ddot{c} = 10 f(t) - 2\dot{c}$$

Dividing through by $\ddot{c}_m$ and expressing each term as a ratio compared to its maximum value gives

$$\frac{\ddot{c}}{\ddot{c}_m} = 10 \frac{f(t)_m}{\ddot{c}_m} \left(\frac{f(t)}{f(t)_m} \right) - 2 \frac{\dot{c}_m}{\ddot{c}_m} \left(\frac{\dot{c}}{\dot{c}_m} \right)$$

For $\ddot{c}_m = 10$, $\dot{c}_m = 4$, and $f(t)_m = 2$, this becomes

$$\frac{\ddot{c}}{10} = 2 \left(\frac{f(t)}{2} \right) + 0.8 \left(-\frac{\dot{c}}{4} \right)$$

The resulting computer diagram is shown in Fig. 8.10(b). The input to the first amplifier is the right side. The ratio of resistors for the $f(t)/2$ signal is 2 and that for the $-\dot{c}/4$ signal is 0.8. Because $\dot{c}/4$ appears in the computer diagram, it is necessary to use a sign changer to obtain $-\dot{c}/4$. The output of this amplifier is $-\ddot{c}/10$. The first integration yields the $\dot{c}/\dot{c}_m = \dot{c}/4$ signal and the second integration yields the $-c/c_m = -c/2$ signal. The value of $1/RC$ for the first integration is $\ddot{c}_m/\dot{c}_m = 10/4 = 5/2$ and that for the second integration is $\dot{c}_m/c_m = 4/2 = 2$. The biasing voltages are $e_2(0) = [\dot{c}(0)/4]e_m = (2/4)10 = 5$ V and $e_3(0) = [-c(0)/2]e_m = (-1/2)10 = -5$ V.

The equation for the comparator is

$$f(t) = r(t) - c(t)$$

In terms of maximum values, this becomes

$$\frac{f(t)}{f(t)_m} = \frac{r(t)_m}{f(t)_m} \left(\frac{r(t)}{r(t)_m} \right) - \frac{c(t)_m}{f(t)_m} \left(\frac{c(t)}{c(t)_m} \right)$$

For $f(t)_m = 2$, $r(t)_m = 1$, and $c(t)_m = 2$, this becomes

$$\frac{f(t)}{2} = \frac{1}{2} \left(\frac{r(t)}{1} \right) + \left(-\frac{c(t)}{2} \right) \tag{8.39}$$

The resulting computer diagram for simulating the system shown in Fig. 8.10a may now be completed as shown in Fig. 8.10c. The preceding equation for the comparator is solved by the first amplifier. The input to this amplifier is the right side of Eq. (8.39). The ratio of resistors for the $[r(t)/1] = r(t)$ signal is $\frac{1}{2}$, and that for the $-c(t)/2$ signal is 1. The output is $-f(t)/2$. The input to the diagram shown in Fig. 8.10*b* is $f(t)/2$. To avoid a sign changer, the signs of all the signals are changed in going from Fig. 8.10*b* to 8.10*c*. Thus $f(t)/2$ becomes $-f(t)/2$, $-\ddot{c}/10$ becomes $\ddot{c}/10$, $\dot{c}/4$ becomes $-\dot{c}/4$, $-c/2$ becomes $c/2$, and the signs of the biasing voltages are changed.

8.3 TIME SCALE

For many problems, it is desired that the speed at which the analog computer solves the problem be different from the speed at which the phenomena actually occur. For example, various phenomena of astronomy require years, so obviously it is desirable to increase the speed at which such problems are solved on the computer. For other phenomena which take place very rapidly, it is necessary to slow down the speed at which such problems are simulated by the computer. Letting t represent the time at which a phenomenon actually occurs and the term τ represent the time required for this phenomenon to occur on the computer, $\tau = at$ relates the actual time t to the computer or machine time τ. For $a < 1$ the phenomenon occurs faster in the computer than it does in nature. For example, if $a = 0.1$, something which actually requires 10 s to complete is completed by the computer in $\tau = 0.1t = 0.1 \times 10 = 1$ s. Similarly, if $a > 1$, the phenomenon is slowed down by the computer.

Illustrative example 8.2 Let it be desired to slow down the computer solution of Eq. (8.28) by a factor of 5.

SOLUTION The first step in the solution of this problem is to transform the original equation from a function of actual time t to a function of machine time τ. This is accomplished by noting that

$$\tau = at \tag{8.40}$$

$$\frac{d\tau}{dt} = a \tag{8.41}$$

Thus,

$$\dot{y} = \frac{dy}{dt} = \frac{d\tau}{dt}\frac{dy}{d\tau} = a\frac{dy}{d\tau}$$

$$\ddot{y} = \frac{d^2y}{dt} = \frac{d}{dt}\frac{dy}{dt} = \frac{d\tau}{dt}\frac{d}{d\tau}a\frac{dy}{d\tau} = a^2\frac{d^2y}{d\tau^2} \tag{8.42}$$

Similarly, it may be shown that in general

$$\frac{d^n y}{dt^n} = a^n\frac{d^n y}{d\tau^n} \tag{8.43}$$

Application of the preceding rules to convert the original time expression given by Eq. (8.28) from a function of t to a function of τ gives

$$a^2 \frac{d^2y}{d\tau^2} + 2a \frac{dy}{d\tau} + 10y = f(\tau/a) \tag{8.44}$$

where $f(\tau/a)$ is obtained by substitution of τ/a for t in the original function $f(t)$. Because of the change of variable, the term y in Eq. (8.44) is now a function of τ rather than t. To slow down the solution by a factor of 5, the value of a is 5, so that Eq. (8.44) becomes

$$25 \frac{d^2y}{d\tau^2} + 10 \frac{dy}{d\tau} + 10y = f(\tau/5) \tag{8.45}$$

The transformed initial conditions are

$$\frac{dy}{d\tau}\bigg|_{\tau=0} = \frac{1}{a} \frac{dy}{dt}\bigg|_{t=0} = \frac{1}{5}(5) = 1.0 \text{ m/s}$$

Similarly, the transformed maximum values are

$$\frac{d^2y}{d\tau^2}\bigg|_m = \frac{1}{a^2} \frac{d^2y}{dt^2}\bigg|_m = \frac{20}{25} = 0.8 \text{ m/s}^2$$

$$\frac{dy}{d\tau}\bigg|_m = \frac{1}{a} \frac{dy}{dt}\bigg|_m = \frac{10}{5} = 2.0 \text{ m/s}$$

As is shown in Fig. 8.11, the function $f(\tau/5)$ is obtained by multiplying the original time scale by the factor $a = 5$. Thus, a time scale change does not effect the initial and maximum values of $f(\tau/a)$. Similarly, the initial and maximum values of y are unaffected by a change in time scale. By using the symbolism $\ddot{y}_\tau = d^2y/d\tau^2$, $\dot{y}_\tau = dy/d\tau$, etc., the maximum values are $\ddot{y}_{\tau_m} = 0.8$, $\dot{y}_{\tau_m} = 2.0$, $y_{\tau_m} = 5$, and $f_{\tau_m} = 8$. The initial values are $y_\tau(0) = 3$ and $\dot{y}_\tau(0) = 1$.

The computer diagram is now obtained by application of the general procedure given in the preceding section. Solving Eq. (8.45) for $\ddot{y}_\tau = d^2y/d\tau^2$, dividing through by $\ddot{y}_{\tau_m}$, and then expressing f_τ, y_τ, and $\dot{y}_\tau$ as ratios compared to their

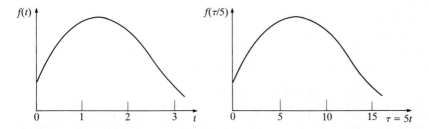

Figure 8.11 Graph of $f(t)$ versus t and $f(\tau/5)$ versus τ.

maximum values gives

$$\frac{\ddot{y}_\tau}{\ddot{y}_{\tau m}} = \frac{f_{\tau m}}{25\ddot{y}_{\tau m}}\left(\frac{f_\tau}{f_{\tau m}}\right) + \frac{10}{25}\frac{y_{\tau m}}{\ddot{y}_{\tau m}}\left(-\frac{y_\tau}{y_{\tau m}}\right) + \frac{10}{25}\frac{\dot{y}_{\tau m}}{\ddot{y}_{\tau m}}\left(-\frac{\dot{y}_\tau}{\dot{y}_{\tau m}}\right) \qquad (8.46)$$

Substitution of numerical values gives

$$\frac{\ddot{y}_\tau}{0.8} = \frac{8}{20}\left(\frac{f_\tau}{8}\right) + 10\frac{5}{20}\left(-\frac{y_\tau}{5}\right) + 10\frac{2}{20}\left(-\frac{\dot{y}_\tau}{2}\right)$$

$$= 0.4\left(\frac{f_\tau}{8}\right) + 2.5\left(-\frac{y_\tau}{5}\right) + 1.0\left(-\frac{\dot{y}_\tau}{2}\right) \qquad (8.47)$$

The resulting computer diagram is shown in Fig. 8.12a. With the right side of Eq. (8.47) as the input to the first amplifier, the output is the negative of the left side, that is, $-\ddot{y}_\tau/0.8$. The coefficient 0.4 is the ratio of resistors for $f_\tau/8$, the coefficient 2.5 is the ratio of resistors for $-y_\tau/5$, and the coefficient 1.0 is the ratio of resistors for $-\dot{y}_\tau/2$. Because $\dot{y}_\tau/2$ appears as the output of the second amplifier in Fig. 8.12a, it is necessary to use a sign changer to obtain the $-\dot{y}_\tau/2$ term which appears in Eq. (8.47). The value of $1/R_1C_2$ for the second amplifier is $\ddot{y}_{\tau m}/\dot{y}_{\tau m} = 0.8/2 = 0.4$, and the initial voltage is $e_2(0) = [\dot{y}_\tau(0)/\dot{y}_{\tau m}]e_m = (1/2)10 = 5$ V. For the third amplifier the value of $1/R_1C_2$ is $\dot{y}_{\tau m}/y_{\tau m} = 2/5 = 0.4$, and the initial voltage is $e_3(0) = -[y_\tau(0)/y_{\tau m}]e_m = -(3/5)10 = -6$ V.

Comparison of Fig. 8.8b which is the computer diagram when there is no time scale change with Fig. 8.12a shows that the gains (0.4, 2.5, 1.0) in the first amplifier are unaffected by the time scale change. Only the gains in the amplifiers which integrate are affected. It is to be noted that the value of the gain in the integrators of Fig. 8.12a is equal to that in Fig. 8.8b divided by the scale constant a (that is, $2/a = 2/5 = 0.4$). In general, the value of the gains of a summing amplifier are unaffected by a time scale change and the new value of the gain in an integrating amplifier is equal to the value for the case of no time scale change divided by the value of a. It is also to be noted that the values of the initial biasing voltages for the integrators are unaffected by the time scale change. Thus, the easiest way to obtain the computer diagram for a time scale change is to first of all obtain the diagram for the case of no time scale change and then divide the gain of each integrator by a. The maximum values are most readily obtained by noting that the maximum value ($y_m = y_{\tau m} = 5$) for the y term is unaffected by the time scale change [that is, $e_3 = (-y/5)e_m = -(y_\tau/5)e_m$]. The ratio $\dot{y}_{\tau m}/y_{\tau m} = 1/(R_1C_2)_3 = 0.4$ is the gain of the third amplifier. Thus, $\dot{y}_{\tau m}$ is equal to the gain of the third amplifier times $y_{\tau m}$ [that is, $\dot{y}_{\tau m} = 0.4(5) = 2$]. Hence, the output of the second amplifier is $e_2 = (\dot{y}_\tau/2)e_m$. Similarly, $\ddot{y}_{\tau m}$ is equal to the gain of the second amplifier times $\dot{y}_{\tau m}$ [that is, $\ddot{y}_{\tau m} = 0.4(2) = 0.8$]. Thus, the output of the first amplifier is $e_1 = (-\ddot{y}_\tau/0.8)e_m$. The values of the initial biasing voltages [$e_2(0) = 5$ and $e_3(0) = -6$] are unaffected by the time scale change.

If it is not necessary to measure $d^2y/d\tau^2$, the first amplifier of Fig. 8.12a which sums and the second amplifier which integrates may be combined into one amplifier which both sums and integrates. The new computer diagram in which

Figure 8.12 Analog computer diagram for $\ddot{y} + 2\dot{y} + 10y = f(t)$ with a time scale change $\tau = 5t$.

the first two amplifiers of Fig. 8.12a have been combined into one amplifier is shown in Fig. 8.12b. The gains (0.16, 1.0, and 0.4) for the new first amplifier are equal to the product of the gains (0.4, 2.5, and 1.0) for the summing amplifier and the gain (0.4) for the integrating amplifier [that is, (0.4)(0.4) = 0.16, (2.5)(0.4) = 1.0, and (1.0)(0.4) = 0.4]. Because one amplifier has been eliminated in going from Fig. 8.12a to Fig. 8.12b, the signals $\dot{y}_\tau/2$ and $-y_\tau/5$ of Fig. 8.12a have opposite signs in Fig. 8.12b. The feedback path from $\dot{y}_\tau/2$ of Fig. 8.12a to the first amplifier has a sign changer but the path from $y_\tau/5$ to the first amplifier does not. Because the signs of these terms have been reversed in Fig. 8.12b, there is no sign changer in the feedback path from $-\dot{y}_\tau/2$ to the new first amplifier but there is a sign changer in the feedback path from $y_\tau/5$ to the new first amplifier. Note also that the signs of all the initial biasing voltages have been switched in going from Fig. 8.12a to Fig. 8.12b. To summarize, the elimination of the one amplifier changes the signs of all the resulting output voltages, and thus the sign of each feedback quantity must be reversed. Similarly, the sign of each initial condition voltage must be changed.

Some operational amplifiers use standard resistors of 1×10^6, 0.25×10^6, and 0.10×10^6 Ω and a standard capacitor of 1×10^{-6} F. Thus, only gains of $1/RC$ equal to 1, 4, or 10 are readily available. It is possible to put two resistors in parallel or series at the input to obtain some other effective value of resistance. For example, two 1×10^6 Ω resistors in series yield a 2×10^6 Ω resistance, while two 1×10^6 Ω resistors in parallel yield a 0.5×10^6 Ω resistance.

In general, it is necessary to use a potentiometer to obtain the desired effective resistance. Figure 8.13 shows a schematic diagram of a potentiometer. The voltage relationship is

$$e_0 = \frac{R_b}{R_a} e_{\text{in}} = k e_{\text{in}} \tag{8.48}$$

where

$$k = \frac{R_b}{R_a} \qquad 0 \le k \le 1$$

The computer diagram of Fig. 8.12b may be modified by the addition of three potentiometers as shown in Fig. 8.12c such that only gains of 1, 4, or 10 are necessary at each amplifier. By comparison of Fig. 8.12b and c, it is to be noted that the effective gain of an amplifier is the product of the gain of the amplifier and the value of k for the potentiometer in front of the amplifier.

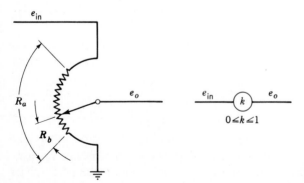

Figure 8.13 Potentiometer.

8.4 SIMULATION

A very important application of the analog computer is the simulation of auto-matic control systems. One method that could be used to simulate a control system would be to determine the overall differential equation and then solve this on the computer. Usually, though, one is interested in determining the effect on the system performance when certain parameters are varied. Using the preceding technique would mean solving a new differential equation for each change.

Because of the similarity between a block diagram and a computer diagram, it is customary to simulate each portion of the system and then interconnect each of these elements. Thus, the effect of changing one of the terms in the original block diagram may be achieved by changing the corresponding quantity in the computer diagram.

In Fig. 8.14 is shown a schematic representation of an operational amplifier in which the input impedance is Z_1 and the parallel impedance is Z_2. The equation of operation for this amplifier is

$$\frac{e_2}{e_1} = -\frac{Z_2}{Z_1} \tag{8.49}$$

When the input impedance is a resistor R_1 and the parallel impedance is a resistor R_2, the preceding expression reduces to the result given by Eq. (8.8). For the case in which $Z_1 = R_1$ and $Z_2 = 1/C_2 D$, the result given by Eq. (8.11) is verified.

In Table 8.1 is shown a number of computer circuits for simulating various transfer functions. For the first circuit, $Z_1 = R_1$ and $Z_2 = 1/(1/R_2 + C_2 D) = R_2/(1 + R_2 C_2 D)$. Substitution of the values of these impedances into Eq. (8.49) yields, for the equation of operation for this computer circuit,

$$e_2 = -\frac{R_2}{R_1(1 + R_2 C_2 D)} e_1 \tag{8.50}$$

In Fig. 8.15*a* is shown a typical block diagram, with the corresponding computer diagram shown in Fig. 8.15*b*.

Illustrative example 8.3 Suppose that the system shown in Fig. 8.15*a* is used to control the angular position of a shaft. For this system, it is known that $K_1 = 10$, $K_2 = 5$, and $\tau = 1.0$. The initial values are $m(0) = 2.0$ and $c(0) = 1.0$. The maximum values have been estimated to be $c(t)_m = 2$ rad, $r(t)_m = 1$ rad, $m(t)_m = 5$ N·m, and $d(t)_m = 10$ N·m. Determine the values of the resistors and capacitors for the computer diagram of Fig. 8.15*b*.

Figure 8.14 General schematic representation for an operational amplifier.

Table 8.1

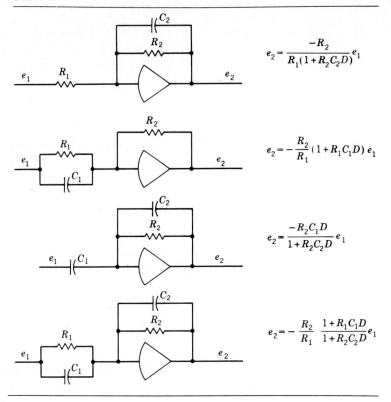

$$e_2 = \frac{-R_2}{R_1(1 + R_2 C_2 D)} e_1$$

$$e_2 = -\frac{R_2}{R_1}(1 + R_1 C_1 D)\, e_1$$

$$e_2 = \frac{-R_2 C_1 D}{1 + R_2 C_2 D} e_1$$

$$e_2 = -\frac{R_2}{R_1}\, \frac{1 + R_1 C_1 D}{1 + R_2 C_2 D} e_1$$

SOLUTION The voltage equation for the first amplifier is

$$e = \frac{1}{R_r C_e} \int_0^t e_r \, dt - \frac{1}{R_c C_e} \int_0^t e_b \, dt + e_m(0) \qquad (8.51)$$

The equation which describes the operation of the corresponding portion of the actual system is

$$m(t) = K_1 \int_0^t [r(t) - c(t)]\, dt + m(0)$$

$$= K_1 \int_0^t r(t)\, dt - K_1 \int_0^t c(t)\, dt + m(0) \qquad (8.52)$$

Multiplying through by the maximum voltage e_m, dividing through by $-m(t)_m$, and then expressing $r(t)$ and $c(t)$ in terms of ratios compared to their maximum values gives

$$-\left(\frac{m}{m_m}\right)e_m = -K_1 \frac{r_m}{m_m} \int_0^t \left(\frac{r}{r_m}\right)e_m \, dt + K_1 \frac{c_m}{m_m} \int_0^t \left(\frac{c}{c_m}\right)e_m \, dt - \left(\frac{m(0)}{m_m}\right)e_m$$

$$(8.53)$$

(a)

(b)

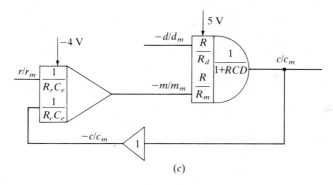

(c)

Figure 8.15 Analog computer diagram for a feedback control system.

In order that $e = -(m/m_m)e_m$, $e_r = (r/r_m)e_m$, and $e_b = -(c/c_m)e_m$, comparison of corresponding terms in Eqs. (8.51) and (8.53) shows that

$$\frac{1}{R_r C_e} = K_1 \frac{r_m}{m_m} = 10 \times \frac{1}{5} = 2 \qquad \frac{1}{R_c C_e} = K_1 \frac{c_m}{m_m} = 10 \times \frac{2}{5} = 4 \quad (8.54)$$

Because $e = -(m/m_m)e_m$, the initial biasing voltage is $e_m(0) = -[m(0)/m_m]e_m = -(2/5)10 = -4$ V. For any system r_m, m_m, c_m, R_r, and R_c are constant. Thus the gain K_1 may be varied independently by varying the capacitance C_e.

If a 1-μF capacitor is used for C_e, then

$$R_r = \frac{1}{2C_e} = \frac{1}{2 \times 10^{-6}} = 500\,000 \; \Omega$$

$$R_c = \frac{1}{4C_e} = \frac{1}{4 \times 10^{-6}} = 250\,000 \; \Omega$$

From Table 8.1, it follows that the voltage equation for the second amplifier is

$$e_c = -\frac{R}{R_m(1 + RCD)} e - \frac{R}{R_d(1 + RCD)} e_d \tag{8.55}$$

The equation for the corresponding portion of the actual system is

$$c(t) = \frac{K_2}{1 + \tau D} [m(t) + d(t)] \tag{8.56}$$

Expressing in terms of ratios compared to maximum values gives

$$\left(\frac{c}{c_m}\right) e_m = \frac{m_m K_2}{c_m(1 + \tau D)} \left(\frac{m}{m_m}\right) e_m + \frac{d_m K_2}{c_m(1 + \tau D)} \left(\frac{d}{d_m}\right) e_m \tag{8.57}$$

Comparison of Eqs. (8.55) and (8.57) shows that to have $e_c = (c/c_m)e_m$, $e = -(m/m_m)e_m$, and $e_d = -(d/d_m)e_m$, then

$$\frac{R}{R_m} = K_2 \frac{m_m}{c_m} = 5 \times \frac{5}{2} = 12.5 \qquad \frac{R}{R_d} = K_2 \frac{d_m}{c_m} = 5 \times \frac{10}{2} = 25 \qquad \tau = RC = 1 \tag{8.58}$$

Because $e_c = (c/c_m)e_m$, the initial biasing voltage is $e_c(0) = [c(0)/c_m]e_m = (1/2)10 = 5$ V.

If a 1-μF capacitor is used for C, then

$$R = \frac{1}{C} = 1\,000\,000 \; \Omega$$

$$R_m = \frac{R}{12.5} = 80\,000 \; \Omega$$

$$R_d = \frac{R}{25} = 40\,000 \; \Omega$$

In Fig. 8.15c is shown the schematic diagram for the analog computer circuit shown in Fig. 8.15b. The first amplifier is an integrator. The input to this integrator is $(1/R_r C_e)(r/r_m)$ plus $(1/R_c C_e)(-c/c_m)$. Because an amplifier changes sign from input to output, the negative of the output is equal to the integral of the input. Thus,

$$\frac{m}{m_m} = \frac{1}{R_r C_e} \int_0^t \frac{r}{r_m} \, dt - \frac{1}{R_e C_e} \int_0^t \frac{c}{c_m} \, dt + \frac{m(0)}{m_m} \tag{8.59}$$

The second amplifier of Fig. 8.15c multiplies the input by $1/(1 + RCD)$. The input is $(R/R_d)(-d/d_m) + (R/R_m)(-m/m_m)$. Because of the sign reversal from input to output, the output (c/c_m) is the negative of the input. Thus,

$$\frac{c}{c_m} = \frac{1}{1 + RCD}\left[\frac{R}{R_m}\left(\frac{m}{m_m}\right) + \frac{R}{R_d}\left(\frac{d}{d_m}\right)\right] \tag{8.60}$$

To obtain the analog computer diagram for simulating a control system, it is not necessary to write both the voltage equation for the circuit and the physical relationship for the system. It suffices to write the physical relationship only. The equation for the first part of the system shown in Fig. 8.15a is

$$m = \frac{K_1}{D}(r - c) = K_1 \int_0^t r \, dt - K_1 \int_0^t c \, dt + m(0) \tag{8.61}$$

Expressing this in terms of ratios compared to maximum values gives

$$\frac{m}{m_m} = K_1 \frac{r_m}{m_m} \int_0^t \left(\frac{r}{r_m}\right) dt + K_1 \frac{c_m}{m_m} \int_0^t \left(-\frac{c}{c_m}\right) dt + \frac{m(0)}{m_m} \tag{8.62}$$

With the right side as the input, the output will be $-(m/m_m)$. By referring to Fig. 8.15c, it is to be noticed that the coefficient of the first integral $(K_1 r_m/m_m)$ is the value $(1/R_r C_e)$ for the signal (r/r_m). Similarly, the coefficient $(K_1 c_m/m_m)$ of the second integral is the value $(1/R_c C_e)$ for the signal $(-c/c_m)$. Thus,

$$\frac{1}{R_r C_e} = K_1 \frac{r_m}{m_m} \qquad \frac{1}{R_c C_e} = K_1 \frac{c_m}{m_m} \tag{8.54}$$

These are the same results given by Eq. (8.54). The equation for the second portion of the system shown in Fig. 8.15a is

$$c = \frac{K_2}{1 + \tau D}(m + d) \tag{8.63}$$

Expressing in terms of ratios compared to maximum values gives

$$\frac{c}{c_m} = \frac{1}{1 + \tau D}\left[K_2 \frac{m_m}{c_m}\left(\frac{m}{m_m}\right) + K_2 \frac{d_m}{c_m}\left(\frac{d}{d_m}\right)\right] \tag{8.64}$$

The $1/(1 + RCD)$ in the schematic diagram for this second amplifier indicates that the input is multiplied by this factor. With the negative of the right side as the input to this amplifier the output is c/c_m. The coefficient $K_2 m_m/c_m$ of the m/m_m term is the value of R/R_m for the $-m/m_m$ signal in Fig. 8.15c. Similarly, the coefficient $K_2 d_m/c_m$ is the value of R/R_d for the $-d/d_m$ signal. Thus,

$$\frac{R}{R_m} = K_2 \frac{m_m}{c_m} \qquad \frac{R}{R_d} = K_2 \frac{d_m}{c_m} \qquad \tau = RC \tag{8.58}$$

These are the same results given by Eq. (8.58). Because C appears only in the equation $\tau = RC$, τ may be varied independently by varying C. A variable capacitor provides a convenient means for varying C. To change the value of K_2, both

R_m and R_d must be changed accordingly. Because the output e_c is to be subtracted from the input e_r, it is necessary to multiply the output by -1, as shown in the feedback path of Fig. 8.15b.

If, in testing, it is found that $e_{c_m} \neq 10$ V, this is evidence that the originally estimated value for c_m is not correct. Because $e_c = (c/c_m)e_m$, the actual value of c_m is now easily determined. When the maximum value of c is near 10 V, there is no need to change the originally estimated value of c_m. However, if the maximum voltage is quite small, then it is desirable to increase the accuracy with which the output voltage can be read by using the new value of c_m. Similarly, if c_m was sufficiently greater than 10 V to cause overloading, then it would be necessary to use the new value of c_m.

A time scale change $\tau = at$ is readily effected by substituting aD for D in the block diagram of Fig. 8.15a.

A major application of analog computers is in the design of systems with nonlinear components. Standard electronic circuits are available for simulating commonly encountered nonlinear effects such as coulomb friction, backlash, dead-zone saturation, continuous nonlinear functions, etc.

PROBLEMS

8.1 (*a*) In Fig. P8.1*a* is shown the analog computer diagram for solving the differential equation

$$\dot{y} + 4y = f(t)$$

Complete this diagram for the case in which $y_m = 1$, $\dot{y}_m = 2$, and $f(t)_m = 1$.

(*a*)

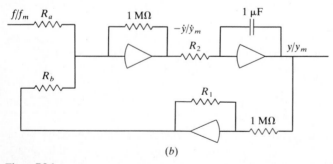

(*b*)

Figure P8.1

(b) In Fig. P8.1b is shown the circuit diagram corresponding to the analog computer diagram shown in Fig. P8.1a. Determine the values of R_a, R_b, R_1, and R_2 for this diagram.

8.2 (a) In Fig. P8.2a is shown the analog computer diagram for solving the differential equation

$$\ddot{y} + 2\dot{y} + 5y = f(t)$$

Complete this diagram for the case in which $y_m = 1$, $\dot{y}_m = 2$, $\ddot{y}_m = 4$, and $f(t)_m = 0.5$.

(b) In Fig. P8.2b is shown the circuit diagram corresponding to the analog computer diagram shown in Fig. P8.2a. Determine the values of R_a, R_b, R_c, R_1, R_2, and R_3 for this diagram.

(a)

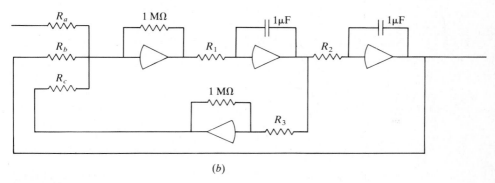

(b)

Figure P8.2

8.3 In running Prob. 8.2 it is found that the maximum voltage obtained for the $-y/y_m$ signal is 5 V rather than 10 V. What is the actual value of y_m? Revise the analog computer diagram so that the maximum voltage corresponding to $-y/y_m$ will be 10 V.

8.4 In running Prob. 8.2 it is found that the maximum voltage obtained for the $\dot{y}/\dot{y}_m$ signal is 5 V rather than 10 V. What is the actual value of $\dot{y}_m$? Revise the analog computer diagram so that the maximum voltage corresponding to $\dot{y}/\dot{y}_m$ will be 10 V.

8.5 In running Prob. 8.2 it is found that the maximum voltage obtained for the $-\ddot{y}/\ddot{y}_m$ signal is 5 V rather than 10 V. What is the actual value of $\ddot{y}_m$? Revise the analog computer diagram so that the maximum voltage corresponding to $-\ddot{y}/\ddot{y}_m$ will be 10 V.

8.6 Determine the computer diagram for the following first-order equation:

$$(0.5D + 1)y = f(t)$$

The initial condition is $y(0) = 0.2$ m, and the maximum expected values are $f(t)_m = 1$ N, $y_m = 0.5$ m, and $\dot{y}_m = 2$ m/s.

8.7 Determine the computer diagram for the following differential equation:

$$\ddot{y} + 6\dot{y} + 5y = f(t)$$

The maximum expected values are $y_m = 0.5$, $\dot{y}_m = 1.0$, $\ddot{y}_m = 2.0$, and $f_m = 5.0$. The initial conditions are $y(0) = 0.2$ and $\dot{y}(0) = -0.2$.

8.8 Determine the computer diagram for the following differential equation:

$$\ddot{y} + 25y = f(t)$$

The maximum expected values are $y_m = 0.1$, $\dot{y}_m = 0.5$, $\ddot{y}_m = 2.5$, and $f_m = 1.0$. The initial conditions are $y(0) = 0$ and $\dot{y}(0) = 0.5$.

8.9 Determine the computer diagram to solve the following differential equation:

$$\ddot{y} + 2y = 5f(t)$$

The initial conditions are $y(0) = 3$ and $\dot{y}(0) = 2$. The maximum expected values are $y_m = 10$, $\dot{y}_m = 20$, $\ddot{y}_m = 50$, and $f(t)_m = 2$.

8.10 Determine the differential equation that is being solved by the computer diagram shown in Fig. P8.10.

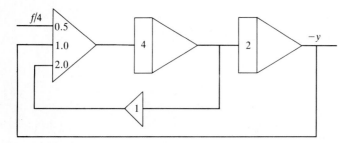

Figure P8.10

8.11 (a) Determine the differential equation that is being solved by the analog computer diagram shown in Fig. P8.11a.

(b) In Fig. P8.11b is shown the circuit diagram for the first amplifier. Determine the equation for R_a, R_b, and R_c.

8.12 Determine the differential equation that is being solved by the computer diagram shown in Fig. P8.12. What are the initial conditions $y(0)$ and $\dot{y}(0)$?

8.13 Determine the differential equation that is being solved by the computer diagram shown in Fig. P8.13. Specify the initial conditions for the differential equation. Revise the computer diagram to speed up the solution by a factor of 2.

8.14 Construct the analog computer diagram for the differential equation

$$\ddot{y} + 10y = 2f(t)$$

The initial conditions are $y(0) = \dot{y}(0) = 2$. The maximum expected values are $y_m = 2$, $\dot{y}_m = 4$, $\ddot{y}_m = 10$, and $f(t)_m = 5$. Also draw the diagram to speed up the solution by a factor of 2.

8.15 Determine the computer diagram for the following third-order differential equation:

$$(D^3 + 2D^2 + 5D + 10)y = f(t)$$

The initial conditions are $y(0) = 0.5$, $\dot{y}(0) = 0$, and $\ddot{y}(0) = 1.0$. The maximum expected values are $f(t)_m = 2.5$, $y_m = 1.0$, $\dot{y}_m = 2.0$, $\ddot{y}_m = 5.0$, and $\dddot{y}_m = 10$.

$$(a)$$

$$(b)$$

Figure P8.11

Figure P8.12

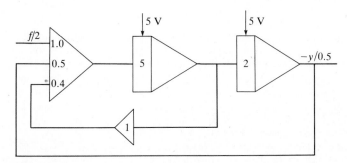

Figure P8.13

8.16 Let it be desired to speed up the computer solution for Prob. 8.6 by a factor of 5. Determine the computer diagram for this case.

8.17 Determine the computer diagram which speeds up the solution of Prob. 8.7 by a factor of 5.

8.18 Determine the computer diagram which slows down the solution of Prob. 8.15 by a factor of 2.

8.19 (a) Determine the computer diagram to solve the following differential equation:

$$\dddot{y} + 8\ddot{y} + 25\dot{y} + 20y = f(t)$$

The initial conditions are $\ddot{y}(0) = 3$, $\dot{y}(0) = 1$, $y(0) = 1$. The maximum expected values are $\ddot{y}_m = 10$, $\ddot{y}_m = 5$, $\dot{y}_m = 2$, $y_m = 1$, and $f(t)_m = 40$.

(b) Determine the computer diagram which will speed up the solution of part (a) by a factor of 2.

8.20 The block diagram representation for a control system is shown in Fig. P8.20a. The analog computer diagram for simulating this system is shown in Fig. P8.20b. The portion enclosed by the dashed box simulates the feedforward portion of the block diagram from e to c. Complete the analog computer diagram for the case in which $c_m = 2$, $\dot{c}_m = 4$, $e_m = 1$, $r_m = 1$, and $c(0) = 1$.

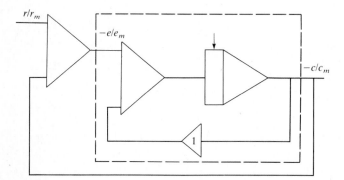

Figure P8.20

8.21 A feedback control system is shown in Fig. P8.21a. The analog computer diagram for simulating this system is shown in Fig. P8.21b. The portion enclosed by the dashed box simulates the feedforward portion of the system from e to c. The initial conditions are $c(0) = 1$ and $\dot{c}(0) = 2$. The maximum values are $c_m = 2$, $\dot{c}_m = 4$, $\ddot{c}_m = 10$, $e_m = 1$, and $r_m = 2$. Complete the analog computer diagram.

8.22 Determine the differential equation that is being solved by the computer diagram shown in Fig. P8.22. What are the initial values $x(0)$ and $\dot{x}(0)$? What are the expected maximum values for x, $\dot{x}$, and $\ddot{x}$? Revise the computer diagram for each of the following cases:

(a) The maximum value of x is found to be twice its expected value.
(b) The maximum value of $\dot{x}$ is found to be twice its expected value.
(c) The maximum value of $\ddot{x}$ is found to be twice its expected value.

Figure P8.21

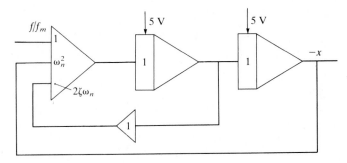

Figure P8.22

8.23 Determine the differential equation that is being solved by the computer diagram shown in Fig. P8.23.

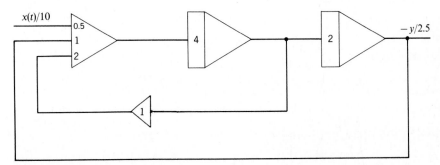

Figure P8.23

8.24 (*a*) Determine the differential equation that is being solved by the analog computer diagram shown in Fig. P8.24.

(*b*) For the first amplifier,

$$\frac{by_m}{\ddot{y}_m} = \frac{R_2}{R_b} \qquad \frac{a\dot{y}_m}{\ddot{y}_m} = \frac{R_2}{R_a}$$

Let $a = 2\zeta\omega_n$ and $b = \omega_n^2$. Solve for ζ and ω_n in terms of these resistors and maximum values. Can either ζ or ω_n be varied independently?

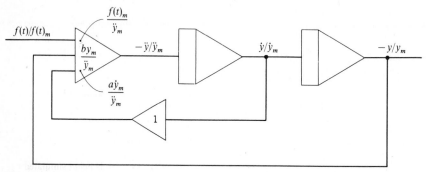

Figure P8.24

8.25 Derive the equation of operation for the amplifier circuit shown in Fig. P8.25.

Figure P8.25

8.26 Construct the computer diagram for simulating the control system shown in Fig. P8.26. The maximum expected values are $r_m = 2.0$ and $c_m = 1.0$.

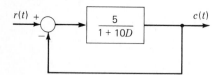

Figure P8.26

8.27 Set up the computer diagram for simulating the system shown in Fig. P8.27 for the case in which $r(t)_m = 2.0$, $c(t)_m = 1.0$, $K = 20$, and $\tau = 0.5$. (Use a 1-μF capacitor.)

Figure P8.27

8.28 A control system and the corresponding computer diagram for simulating the control system are shown in Fig. P8.28. For $r_m = 2$, $c_m = 1$, $C_2 = 1$ μF, $R_2 = 1$ MΩ, $R_r = 0.1$ MΩ, and $R_c = 0.2$ MΩ, determine the corresponding values of K_1, K_2, and τ.

Figure P8.28

8.29 Determine the block-diagram representation for the control system which is being simulated by the analog computer diagram shown in Fig. P8.29.

Figure P8.29

8.30 Construct the computer diagram for simulating the control system shown in Fig. P8.30. For this system $K_1 = 5$, $K_2 = 2$, and $\tau = 1$. The maximum expected values are $r(t)_m = 1.0$ and $c(t)_m = 2.0$. (Let $C_2 = 1$ μF.)

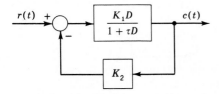

Figure P8.30

8.31 Both the analog computer diagram and the system it is simulating are shown in Fig. P8.31. Determine the values of K_1, K_2, and τ for the system.

Figure P8.31

8.32 (*a*) Determine the block diagram for the system which is being simulated by the analog computer diagram shown in Fig. P8.32.

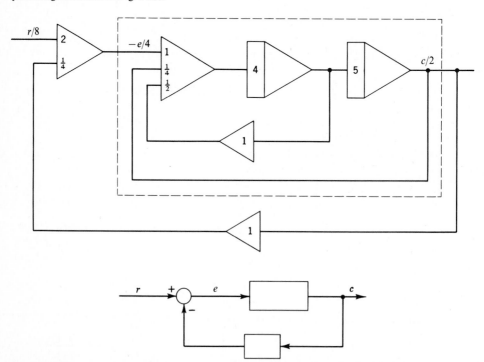

Figure P8.32

Hint: First obtain the block diagram for the portion of the circuit enclosed by dashed lines, i.e., the block diagram relating the input e to the output c.

(b) Draw directly the new computer diagram for speeding up the solution of the above problem by a factor of 2.

8.33 Determine the computer diagram for simulating the control system shown in Fig. P8.33. The maximum expected values are $r_m = 0.5$, $d_m = 1.0$, $m_m = 1.0$, and $c_m = 2.0$.

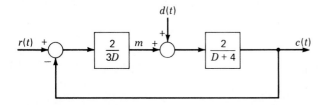

Figure P8.33

8.34 In Fig. P8.34a is shown a control system and in Fig. P8.34b is shown an analog computer diagram for simulating the system. Determine the values of K_1, K_2, K_H, and τ.

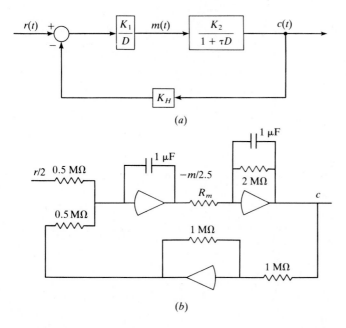

Figure P8.34

8.35 Determine the computer diagram for simulating the control system shown in Fig. P8.35. The maximum expected values are $r(t)_m = 0.5$, $c(t)_m = 1.0$, $d(t)_m = 2.5$, and $m(t)_m = 2.0$. The values of the constants are $K_1 = 10$, $K_2 = 20$, $\tau_1 = 0.2$, and $\tau_2 = 1.0$. (Use 1-μF capacitors.)

Figure P8.35

8.36 A system is described by the two simultaneous equations

$$\dot{x} + 4x + 2y = f(t)$$

$$2x + 4y + \dot{y} = 0$$

Solve the first differential equation for $\dot{x}$ and the second for $\dot{y}$, then proceed to complete the computer diagram shown in Fig. P8.36. That is, indicate the various amplifier gains and hook up the appropriate interconnecting and feedback signals. The maximum expected values are $f_m = 2.0$, $x_m = 0.5$, $\dot{x}_m = 1.0$, $y_m = 1.0$, and $\dot{y}_m = 2.0$. The initial conditions are $x(0) = 0.2$ and $y(0) = 0.1$.

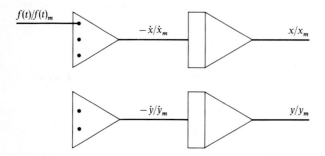

Figure P8.36

8.37 Construct the computer diagram for solving the two simultaneous differential equations

$$\dot{x} + 5x = y + f(t)$$

$$\ddot{y} + 10y = 2x$$

The initial conditions are $x(0) = 1$, $\dot{y}(0) = 0$, and $y(0) = 0.2$. The maximum expected values are $x_m = 4$, $\dot{x}_m = 10$, $y_m = 1$, $\dot{y}_m = 2$, $\ddot{y}_m = 4$, and $f(t)_m = 10$.

NINE

STATE-SPACE METHODS

State-space methods provide a generalized technique for investigating the behavior of control systems. This approach is especially well suited to the analysis of more complex and nonlinear control systems. A feature of state-space methods is that they lend themselves to solution by either digital or analog computers.

9.1 SYSTEM REPRESENTATION

Common techniques employed for obtaining the state-space representation for a control system are direct programming, parallel programming, series programming, and general programming. The choice of the state-space representation depends on the nature of the particular problem to be solved.

Direct Programming

To illustrate this method, consider the differential equation

$$y(t) = \frac{D + 3}{(D + 1)(D + 2)} f(t) \tag{9.1}$$

In direct programming, x_1 is taken as the differential equation in which the numerator operator is 1. That is,

$$x_1 = \frac{1}{(D + 1)(D + 2)} f(t) = \frac{1}{D^2 + 3D + 2} f(t)$$

Thus,

$$y(t) = (D + 3)x_1 = \dot{x}_1 + 3x_1$$

By letting $\dot{x}_1 = x_2$, then

$$y(t) = 3x_1 + x_2 \qquad (9.2)$$

The differential equation for x_1 may be written in the form

$$(D^2 + 3D + 2)x_1 = \ddot{x}_1 + 3\dot{x}_1 + 2x_1 = f(t)$$

Solving for $\ddot{x}_1 = \dot{x}_2$ gives

$$\dot{x}_2 = -2x_1 - 3x_2 + f(t) \qquad (9.3)$$

The block diagram for obtaining the computer solution of Eq. (9.3) is shown in Fig. 9.1a. Note that the output of the summer (indicated by the circle) is $\dot{x}_2$. Successive integration of $\dot{x}_2$ yields in turn $x_2 = \dot{x}_1$ and x_1. The final computer diagram for obtaining $y(t)$ from x_1 and x_2 using Eq. (9.2) is shown in Fig. 9.1b.

The state-space representation for the differential equation [Eq. (9.1)] is given by Eqs. (9.2) and (9.3). These equations may be expressed in the matrix form:

$$\begin{bmatrix} \dot{x}_1 \\ \dot{x}_2 \end{bmatrix} = \begin{bmatrix} 0 & 1 \\ -2 & -3 \end{bmatrix} \begin{bmatrix} x_1 \\ x_2 \end{bmatrix} + \begin{bmatrix} 0 \\ 1 \end{bmatrix} f(t)$$

$$y = \begin{bmatrix} 3 & 1 \end{bmatrix} \begin{bmatrix} x_1 \\ x_2 \end{bmatrix}$$

$$(9.4)$$

(a)

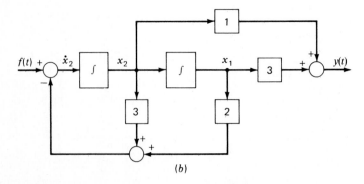

(b)

Figure 9.1 Computer diagram for direct programming.

The x's are referred to as the state variables. Note that the output of each integrator in Fig. 9.1 is a state variable.

The general procedure for obtaining the state-space representation by direct programming is to let x_1 be the original differential equation in which the numerator operator (that is, $D + 3$) is taken as 1. Each derivative of x_1 is then assigned a state variable (for example, $\dot{x}_1 = x_2$, $\ddot{x}_1 = \dot{x}_2 = x_3$, etc.). The number of state variables is equal to the order of the differential equation. For the case in which the original differential equation is such that the numerator operator is 1, then $y = x_1$, $\dot{y} = \dot{x}_1 = x_2$, etc.

Because of the simplicity of the state-space relationships $\dot{x}_1 = x_2$, $\dot{x}_2 = x_3$, ..., etc., direct programming is the most commonly used method for obtaining the state-space representation.

Parallel Programming

In parallel programming the differential equation is expressed in the form of a partial-fraction expansion. Thus,

$$y(t) = \frac{D + 3}{(D + 1)(D + 2)} f(t) = \left(\frac{2}{D + 1} - \frac{1}{D + 2}\right) f(t) \tag{9.5}$$

This result may be written as

$$y(t) = 2x_1 - x_2 \tag{9.6}$$

where

$$x_1 = \frac{f(t)}{D + 1} \qquad x_2 = \frac{f(t)}{D + 2} \tag{9.7}$$

In Fig. 9.2a is shown the computer diagram for generating the basic term

$$x = \frac{f(t)}{D + a}$$

or

$$\dot{x} = f(t) - ax$$

The computer diagram for generating the state variables as defined in Eq. (9.7) is shown to the left of the dotted partition in Fig. 9.2b. To the right of the dotted partition is indicated the manner in which the state variables are combined in Eq. (9.6) to yield the desired solution $y(t)$. Equations (9.6) and (9.7) make up a state-space representation for the system [Eq. (9.1)]. In terms of matrix notation, Eqs. (9.6) and (9.7) become

$$\begin{bmatrix} \dot{x}_1 \\ \dot{x}_2 \end{bmatrix} = \begin{bmatrix} -1 & 0 \\ 0 & -2 \end{bmatrix} \begin{bmatrix} x_1 \\ x_2 \end{bmatrix} + \begin{bmatrix} 1 \\ 1 \end{bmatrix} f(t)$$

$$y(t) = \begin{bmatrix} 2 & -1 \end{bmatrix} \begin{bmatrix} x_1 \\ x_2 \end{bmatrix} \tag{9.8}$$

(a)

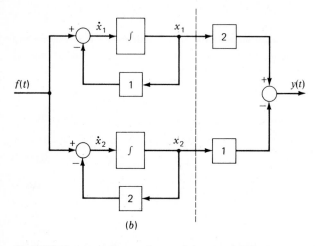

(b)

Figure 9.2 Computer diagram for parallel programming.

The characteristic function for Eq. (9.5) is $(D + 1)(D + 2)$. Note in Eq. (9.8) that -1 and -2, which are the roots of the characteristic equation, are displayed on the principal diagonal of the first matrix to the right of the equal sign. All other elements of this matrix are zero. In general, with parallel programming the roots of the characteristic equation are displayed on the principal diagonal.

Series Programming

Consider the general nth-order differential equation

$$(D^n + a_1 D^{n-1} + \cdots + a_{n-1}D + a_n)y(t)$$
$$= (b_1 D^{n-1} + b_2 D^{n-2} + \cdots + b_{n-1}D + b_n)f(t) \quad (9.9)$$

This differential equation may be written in the matrix form

$$
\begin{bmatrix} \dot{x}_1 \\ \dot{x}_2 \\ \vdots \\ \dot{x}_n \end{bmatrix} =
\begin{bmatrix} -a_1 & 1 & 0 & \cdots & 0 \\ -a_2 & 0 & 1 & \cdots & 0 \\ \multicolumn{5}{c}{\cdots\cdots\cdots\cdots\cdots\cdots} \\ -a_n & 0 & 0 & \cdots & 0 \end{bmatrix}
\begin{bmatrix} x_1 \\ x_2 \\ \vdots \\ x_n \end{bmatrix} +
\begin{bmatrix} b_1 \\ b_2 \\ \vdots \\ b_n \end{bmatrix} f(t) \quad (9.10)
$$

where $y(t) = x_1$.

Figure 9.3 Computer diagram for series programming.

For Eq. (9.1), it follows that $a_1 = 3$, $a_2 = 2$, $b_1 = 1$, and $b_2 = 3$. The resulting matrix representation is

$$\begin{bmatrix} \dot{x}_1 \\ \dot{x}_2 \end{bmatrix} = \begin{bmatrix} -3 & 1 \\ -2 & 0 \end{bmatrix} \begin{bmatrix} x_1 \\ x_2 \end{bmatrix} + \begin{bmatrix} 1 \\ 3 \end{bmatrix} f(t)$$

$$y(t) = \begin{bmatrix} 1 & 0 \end{bmatrix} \begin{bmatrix} x_1 \\ x_2 \end{bmatrix}$$

(9.11)

To prove that this matrix representation satisfies Eq. (9.1) first write the matrix equations in the form

$$\dot{x}_1 = -3x_1 + x_2 + f(t)$$
$$\dot{x}_2 = -2x_1 + 3f(t)$$

Differentiating the first equation and then substituting the resulting expression for $\dot{x}_2$ into the second gives

$$\dot{x}_2 = \ddot{x}_1 + 3\dot{x}_1 - f'(t) = -2x_1 + 3f(t)$$

or $\qquad\qquad \ddot{x}_1 + 3\dot{x}_1 + 2x_1 = f'(t) + 3f(t)$

This is identical to Eq. (9.1), with $x_1 = y(t)$. The block-diagram representation for the matrix equations is shown in Fig. 9.3.

Series programming has the feature that y is always equal to x_1. However, the other state-space relationships tend to be somewhat awkward. In the case of the preceding example $\dot{x}_1 = -3x_1 + x_2 + f(t)$. Note that $\dot{x}_1$ is not equal to x_2 as was the case for direct programming.

General Programming

Consider the differential equation

$$(D^2 + a_1 D + a_2)y(t) = (b_0 D^2 + b_1 D + b_2)f(t) \qquad (9.12)$$

The $b_0 D^2 f(t)$ term is eliminated by letting $y(t) = x_1 + b_0 f(t)$. Thus,

$$\ddot{x}_1 + a_1 \dot{x}_1 + a_2 x_1 = (b_1 - a_1 b_0)f'(t) + (b_2 - a_2 b_0)f(t) \qquad (9.13)$$

Let it now be desired to transform this into the form

$$\begin{bmatrix} \dot{x}_1 \\ \dot{x}_2 \end{bmatrix} = \begin{bmatrix} 0 & 1 \\ -a_2 & -a_1 \end{bmatrix} \begin{bmatrix} x_1 \\ x_2 \end{bmatrix} + \begin{bmatrix} h_1 \\ h_2 \end{bmatrix} f(t) \qquad (9.14)$$

The corresponding matrix relationships are

$$\dot{x}_1 = x_2 + h_1 f(t)$$

$$\dot{x}_2 = -a_2 x_1 - a_1 x_2 + h_2 f(t)$$

Solving the first matrix relationship for x_2 shows that

$$x_2 = \dot{x}_1 - h_1 f(t)$$

Differention gives

$$\dot{x}_2 = \ddot{x}_1 - h_1 f'(t)$$

Substitution of the preceding results for x_2 and $\dot{x}_2$ into the second matrix relationship yields

$$\ddot{x}_1 - h_1 f'(t) = -a_2 x_1 - a_1 [\dot{x}_1 - h_1 f(t)] + h_2 f(t)$$

or $\qquad \ddot{x}_1 + a_1 \dot{x}_1 + a_2 x_1 = h_1 f'(t) + [h_2 + a_1 h_1]f(t)$

This has the same form as Eq. (9.13). Equating the coefficients of the $f'(t)$ and $f(t)$ terms shows that

$$h_1 = b_1 - a_1 b_0$$

$$h_2 = b_2 - a_2 b_0 - a_1 h_1$$

For the case of the differential equation given by Eq. (9.1), $a_1 = 3$, $a_2 = 2$, $b_0 = 0$, $b_1 = 1$, and $b_2 = 3$. Hence, $h_1 = 1$ and $h_2 = 0$. The resulting state-space representation is

$$\begin{bmatrix} \dot{x}_1 \\ \dot{x}_2 \end{bmatrix} = \begin{bmatrix} 0 & 1 \\ -2 & -3 \end{bmatrix} \begin{bmatrix} x_1 \\ x_2 \end{bmatrix} + \begin{bmatrix} 1 \\ 0 \end{bmatrix} f(t) \qquad (9.15)$$

where $y(t) = x_1 + b_0 f(t) = x_1$. The block-diagram representation is shown in Fig. 9.4.

In general, the differential equation

$$(D^n + a_1 D^{n-1} + \cdots + a_{n-1}D + a_n)y(t)$$

$$= (b_0 D^n + b_1 D^{n-1} + \cdots + b_{n-1}D + b_n)f(t) \qquad (9.16)$$

is transformed into the form

$$\begin{bmatrix} \dot{x}_1 \\ \dot{x}_2 \\ \vdots \\ \dot{x}_n \end{bmatrix} = \begin{bmatrix} 0 & 1 & 0 & \cdots & 0 \\ 0 & 0 & 1 & \cdots & 0 \\ \multicolumn{5}{c}{\dotfill} \\ -a_n & -a_{n-1} & -a_{n-2} & \cdots & -a_1 \end{bmatrix} \begin{bmatrix} x_1 \\ x_2 \\ \vdots \\ x_n \end{bmatrix} + \begin{bmatrix} h_1 \\ h_2 \\ \vdots \\ h_n \end{bmatrix} f(t) \qquad (9.17)$$

Figure 9.4 Computer diagram for general programming.

where

$$h_1 = b_1 - a_1 b_0$$

$$h_2 = b_2 - a_2 b_0 - a_1 h_1$$

$$h_3 = b_3 - a_3 b_0 - a_2 h_1 - a_1 h_2 \qquad (9.18)$$

$$\vdots$$

$$h_n = b_n - a_n b_0 - a_{n-1} h_1 - a_{n-2} h_2 - \cdots - a_1 h_{n-1}$$

and the state variables are

$$x_1 = y(t) - b_0 f(t)$$

$$x_2 = \dot{x}_1 - h_1 f(t)$$

$$x_3 = \dot{x}_2 - h_2 f(t) \qquad (9.19)$$

$$\vdots$$

$$x_n = \dot{x}_{n-1} - h_{n-1} f(t)$$

In this method, the state-space representation is obtained by substituting the coefficients of the differential equation into a general form [Eq. (9.17)]. Note that the highest power of D on both sides of Eq. (9.16) is n. Other methods of obtaining the state-space representation are limited to the case in which the highest power of D on the right side is one less $(n - 1)$ than that on the left side (n).

There is no unique state-space representation for a system. Comparison of Eqs. (9.4), (9.8), (9.11), and (9.15) shows that any system whose operation is described by a linear differential equation may be written in the general matrix form

$$\dot{\mathbf{x}} = A\mathbf{x} + \mathbf{b}f(t) \qquad (9.20)$$

$$y = \mathbf{c}^T \mathbf{x} \qquad (9.21)$$

In the following section, it is shown that the preceding matrix form is also applicable to closed-loop systems.

Closed-Loop Systems

The state-space representation may be obtained directly from the block diagram for a system. For the system shown in Fig. 9.5, the input is $u(t)$ and the output is $y(t)$. One state variable $x_1, x_2, \ldots$ is assigned for each order of D in the denominator.

For the $1/(D + 5)$ term, the state variable is x_1. For the $2/(D^2 + 3D + 4)$ term, the state variables are x_2 and x_3, where $x_3 = \dot{x}_2$. The equation for each block is

$$\frac{x_1}{x_2} = \frac{1}{D + 5} \qquad \frac{x_2}{e} = \frac{2}{D^2 + 3D + 4}$$

Thus,

$$\dot{x}_1 = -5x_1 + x_2$$

$$\dot{x}_2 = x_3$$

$$\dot{x}_3 = \ddot{x}_2 = -3\dot{x}_2 - 4x_2 + 2e$$

The equation for e is

$$e = u - y = u - x_1$$

The resulting state-space representation for this system is

$$\begin{bmatrix} \dot{x}_1 \\ \dot{x}_2 \\ \dot{x}_3 \end{bmatrix} = \begin{bmatrix} -5 & 1 & 0 \\ 0 & 0 & 1 \\ -2 & -4 & -3 \end{bmatrix} \begin{bmatrix} x_1 \\ x_2 \\ x_3 \end{bmatrix} + \begin{bmatrix} 0 \\ 0 \\ 2 \end{bmatrix} u$$

The output equation is

$$y = x_1$$

As is illustrated in Fig. 9.5, the output from any block which contains the operator D in the denominator is assigned a state variable. Thus, the output of the $1/(D + 5)$ block is x_1 and the output of the $2/(D^2 + 3D + 4)$ block is assigned x_2. When the order of the denominator is greater than one, additional state variables are assigned ($x_3 = \dot{x}_2$, etc.) such that the number of state variables is equal to the order of D in the denominator. In Fig. 9.6a is shown a system which contains the numerator term $D + 2$. Numerator terms do not require additional state variables. To show this, the numerator term is displayed separately in Fig. 9.6b. The output z serves as a dummy variable. The general procedure when there is a

Figure 9.5 Closed-loop system.

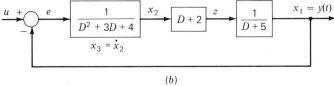

(b)

Figure 9.6 Closed-loop system.

numerator term is to put it in a separate block and place it ahead of the block containing the denominator term. The output from the block containing the numerator term is the dummy variable z. The output from the block containing the denominator term which is the input to the block containing the numerator is a state variable. The equation for each block is

$$\frac{x_1}{z} = \frac{1}{D+5} \qquad \frac{x_2}{e} = \frac{1}{D^2 + 3D + 4}$$

Thus,

$$\dot{x}_1 = -5x_1 + z$$

$$\dot{x}_2 = x_3$$

$$\dot{x}_3 = \ddot{x}_2 = -3\dot{x}_2 - 4x_2 + e$$

Using the relationship $e = u - x_1$ to eliminate e and the relationship $z = \dot{x}_2 + 2x_2 = x_3 + 2x_2$ to eliminate z yields, for the state-space representation for this system,

$$\begin{bmatrix} \dot{x}_1 \\ \dot{x}_2 \\ \dot{x}_3 \end{bmatrix} = \begin{bmatrix} -5 & 2 & 1 \\ 0 & 0 & 1 \\ -1 & -4 & -3 \end{bmatrix} \begin{bmatrix} x_1 \\ x_2 \\ x_3 \end{bmatrix} + \begin{bmatrix} 0 \\ 0 \\ 1 \end{bmatrix} u$$

The output relationship is

$$y = x_1$$

To obtain the state-space representation for the system shown in Fig. 9.7a, the numerator term is placed in a separate block and placed ahead of the denominator term, as shown in Fig. 9.7b. The output of the numerator term is z, which is the system output y. The output of the denominator term $1/(D^2 + 3D + 4)$ is x_1. Because this denominator is of second order in D, the additional state variable $x_2 = \dot{x}_1$ is assigned. The output of the $1/D$ block is x_3.

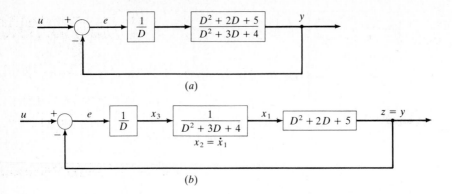

(a)

(b)

Figure 9.7 Closed-loop system.

The equation for the numerator block is

$$z = (D^2 + 2D + 5)x_1 = \ddot{x}_1 + 2\dot{x}_1 + 5x_1 = \dot{x}_2 + 2x_2 + 5x_1$$

The equation for the denominator block is

$$(D^2 + 3D + 4)x_1 = \dot{x}_2 + 3x_2 + 4x_1 = x_3$$

Solving for $\dot{x}_2$ in terms of the state variables gives

$$\dot{x}_2 = -4x_1 - 3x_2 + x_3$$

Substituting this result into the expression for z yields

$$z = x_1 - x_2 + x_3$$

The equation for the $1/D$ block is

$$Dx_3 = \dot{x}_3 = e = \mu - z = -x_1 + x_2 - x_3 + u$$

Thus, the state-space representation for this system is

$$\begin{bmatrix} \dot{x}_1 \\ \dot{x}_2 \\ \dot{x}_3 \end{bmatrix} = \begin{bmatrix} 0 & 1 & 0 \\ -4 & -3 & 1 \\ -1 & 1 & -1 \end{bmatrix} \begin{bmatrix} x_1 \\ x_2 \\ x_3 \end{bmatrix} + \begin{bmatrix} 0 \\ 0 \\ 1 \end{bmatrix} u$$

Because $y = z$, the output relationship is

$$y = \begin{bmatrix} 1 & -1 & 1 \end{bmatrix} \begin{bmatrix} x_1 \\ x_2 \\ x_3 \end{bmatrix}$$

Note that the preceding equations have the general form given by Eqs. (9.20) and (9.21).

A system which has one input u and one output y is referred to as a single-input–single-output system. A system with more than one input is a multiple-input system, and a system with more than one output is a multiple-output

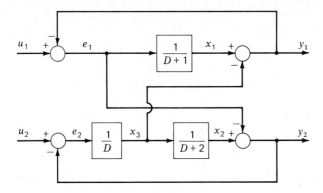

Figure 9.8 Multiple-input–multiple-output system.

system. A system with more than one input and more than one output is a multiple-input–multiple-output system. Matrix methods lend themselves to the analysis of such systems. The system shown in Fig. 9.8 has two inputs and two outputs. The equation for each of the state variables is

$$\frac{x_1}{e_1} = \frac{1}{D+1} \qquad \frac{x_2}{x_3} = \frac{1}{D+2} \qquad \frac{x_3}{e_2} = \frac{1}{D}$$

Thus,

$$\dot{x}_1 = -x_1 + e_1$$
$$\dot{x}_2 = -2x_2 + x_3$$
$$\dot{x}_3 = e_2$$

The equations for e_1 and e_2 are

$$e_1 = u_1 - y_1 = u_1 - (x_1 - x_3) = -x_1 + x_3 + u_1$$
$$e_2 = u_2 - y_2 = u_2 - (x_2 - e_1) = u_2 - x_2 + (-x_1 + x_3 + u_1)$$
$$= -x_1 - x_2 + x_3 + u_1 + u_2$$

Substituting e_1 and e_2 into the system equations and expressing the result in matrix form gives

$$\begin{bmatrix} \dot{x}_1 \\ \dot{x}_2 \\ \dot{x}_3 \end{bmatrix} = \begin{bmatrix} -2 & 0 & 1 \\ 0 & -2 & 1 \\ -1 & -1 & 1 \end{bmatrix} \begin{bmatrix} x_1 \\ x_2 \\ x_3 \end{bmatrix} + \begin{bmatrix} 1 & 0 \\ 0 & 0 \\ 1 & 1 \end{bmatrix} \begin{bmatrix} u_1 \\ u_2 \end{bmatrix}$$

The output relationships are

$$y_1 = x_1 - x_3$$
$$y_2 = x_2 - e_1 = x_1 + x_2 - x_3 - u_1$$

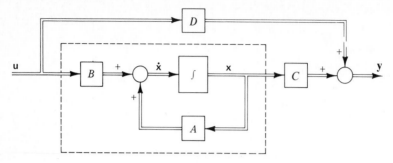

Figure 9.9 General state-space representation for a system.

The corresponding matrix form is

$$\begin{bmatrix} y_1 \\ y_2 \end{bmatrix} = \begin{bmatrix} 1 & 0 & -1 \\ 1 & 1 & -1 \end{bmatrix} \begin{bmatrix} x_1 \\ x_2 \\ x_3 \end{bmatrix} + \begin{bmatrix} 0 & 0 \\ -1 & 0 \end{bmatrix} \begin{bmatrix} u_1 \\ u_2 \end{bmatrix}$$

The general matrix representation for a multiple-input–multiple-output system is

$$\dot{\mathbf{x}} = A\mathbf{x} + B\mathbf{u} \tag{9.22}$$

The general output relationship is

$$\mathbf{y} = C\mathbf{x} + D\mathbf{u} \tag{9.23}$$

The block-diagram representation for these matrix relationships is shown in Fig. 9.9. The portion enclosed by the dashed box is the system equation. The remaining portion is the output relationship. The double-width arrows represent more than one variable. For the case of a single input u, B becomes the column matrix $\mathbf{b}$ and D becomes the column matrix $\mathbf{d}$. For the case of a single output y, C becomes $\mathbf{c}^T$. For the usual case in which there is no feedforward path from the input u to the output, the D term in Eq. (9.23) vanishes. Thus, most systems can be represented in the form given by Eqs. (9.20) and (9.21), in which $\mathbf{b}$ is replaced by B for multiple-input systems and $\mathbf{c}^T$ is replaced by C for the multiple-output systems.

Computer Solution

The state-space representation for a differential equation or for a control system lends itself very well to solution on a digital computer. The solution is obtained by solving the various state-space equations to obtain the $\dot{x}$ terms, and then integrating each $\dot{x}$ term to obtain the corresponding state-variable.

In Sec. 6.4, it was shown how differential equations may be solved numerically. To summarize, in Fig. 9.10 is shown a plot of $\dot{x}$ versus time. The integral of $\dot{x}$ is

$$x = \int \dot{x}\, dt = x_0 + \int_{t_0}^{t_0 + \Delta t} \dot{x}\, dt$$

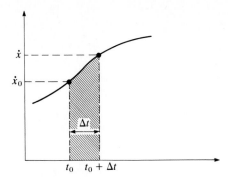

Figure 9.10 Plot of $\dot{x}$ versus time.

where x_0 is the value of x at the beginning of the interval (at time t_0). The area shown shaded is the integral of $\dot{x}$ from time t_0 to time $t_0 + \Delta t$. For a small time increment Δt, this area is closely approximated by the area of the trapezoid:

$$0.5(\dot{x}_0 + \dot{x})\Delta t$$

where $\dot{x}_0$ is the value of $\dot{x}$ at the beginning of the interval and $\dot{x}$ is the value at the end. The equation for the value of x is

$$x = x_0 + 0.5(\dot{x}_0 + \dot{x})\Delta t$$

where x is the value at the end of the interval.

To illustrate this procedure, consider the differential equation given by Eq. (9.1). The state-space representation obtained by direct programming is given by Eq. (9.4). That is,

$$\dot{x}_1 = x_2$$
$$\dot{x}_2 = -2x_1 - 3x_2 + f(t)$$
$$y = 3x_1 + x_2$$

The computer program for obtaining the response y for the case in which $f(t) = u(t)$ is a unit step function and all the initial conditions are zero is

```
X1 = 0.0
X2 = 0.0
DX2 = 0.0
F = 1.0
T = 0.0
DT = 0.01
DO 25 K = 0,400
Y = 3.0*X1 + X2
IF (MOD(K,10).NE.0) GO TO 10
WRITE (6,5) T,Y
5 FORMAT (2F 10.2)
```

```
10 T = T + DT
   X10 = X1
   X20 = X2
   DX20 = DX2
   DX2 = F - 2.0*X1 - 3.0*X2
   X2 = X20 + 0.5*(DX2 + DX20)*DT
   X1 = X10 + 0.5*(X2 + X20)*DT
25 CONTINUE
   END
```

In this program X1, X2, and DX2 represent x_1, x_2, and $\dot{x}_2$ respectively. The first three statements are the initial values. The next statement F = 1.0 is the forcing function $f(t) = u(t)$. The time increment for the integration interval is DT = 0.01 s. The DO loop solves the equations 400 times. Thus the duration of the solution is 400(0.01) = 4 s. The response is Y = 3.0*X1 + X2. Corresponding values of T and Y are now written. The time for the next integration interval is T = T + DT. The value of X1 that is presently stored in the computer becomes the old value (beginning value) for the next interval, X10 = X1. Similarly, the values of X2 and DX2 that are presently stored become the old values (beginning values) for the next interval, X20 = X2 and DX20 = DX2. The new value of DX2 is obtained by the statement DX2 = F - 2.0*X1 - 3.0*X2. The new value of X2 is obtained by the statement X2 = X20 + 0.5*(DX2 + DX20)*DT. Note that the value X20 on the right side is the value of X2 at the beginning of the interval. The value of the derivative at the beginning of the interval is DX20 and the value at the end is DX2; thus the term 0.5*(DX2 + DX20)*DT is the trapezoidal integration of $\dot{x}_2$ during the interval. Similarly, the new value of X1 is obtained by the statement X1 = X10 + 0.5*(X2 + X20)*DT. Because $\dot{x}_1 = x_2$, the last term is the trapezoidal integration of $\dot{x}_1$ during the interval.

Although trapezoidal integration was used in the preceding computer program, any integration method such as the fourth-order, Runge-Kutta method as explained in Section 6.4 can be used.

9.2 SIGNAL-FLOW GRAPHS

A signal-flow graph is a diagram which may be used to represent a control system. The signal-flow-graph representation for a system appears to be quite similar to the block-diagram representation. However, there are important differences. For more complex systems, the signal-flow graph is easier to construct than the block diagram. System equations are more readily obtained from the signal-flow graph than from the block diagram.

The left side of Fig. 9.11 shows the block-diagram representation and the right side shows the equivalent signal-flow graph. Each small circle in a signal-flow graph is called a node, which is a point that represents a variable or signal. The directed line segment joining two nodes is called a branch. The gain between

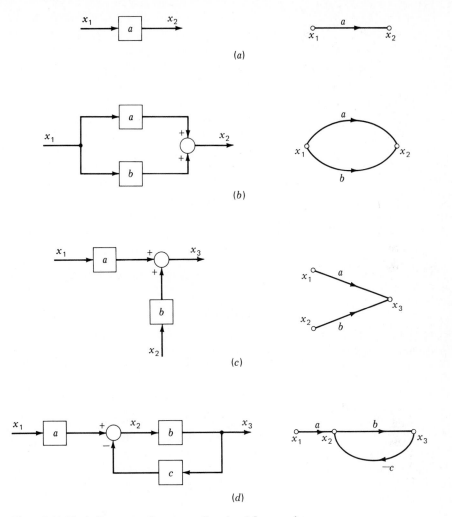

Figure 9.11 Block diagrams and corresponding signal-flow graphs.

two nodes is called the transmittance. In Fig. 9.11a, the variable x_1 is multiplied by the transmittance a to yield the output x_2 (that is, $x_2 = ax_1$). In Fig. 9.11b, the output x_2 is the sum of ax_1 and bx_1 (that is, $x_2 = ax_1 + bx_1$). In Fig. 9.11c, the signal x_3 is the sum of ax_1 and bx_2. In Fig. 9.11d, the signal x_2 is $ax_1 - cx_3$ and the signal x_3 is bx_2.

As illustrated in Fig. 9.12a, the upper branch with a transmittance a and the lower branch with a transmittance b may be combined into one branch with a transmittance $a + b$ [note that $x_2 = ax_1 + bx_1 = (a + b)x_1$]. The equation for the left side of Fig. 9.12b is

$$x_3 = bx_2 = b(ax_1 - cx_3)$$

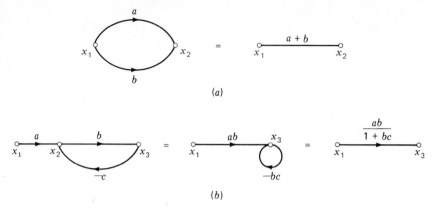

Figure 9.12 Equivalent signal-flow graphs.

The equation for the equivalent signal-flow graph shown in the middle of Fig. 9.12b is

$$x_3 = abx_1 - bcx_3$$

The basis for the signal-flow representation shown on the right side of Fig. 9.12b is obtained by solving the preceding expression for x_3. That is,

$$x_3 = \frac{ab}{1 + bc} x_1$$

The block-diagram representation for two basic control systems is shown on the left side of Fig. 9.13. The corresponding signal-flow graph for each system is shown on the right side of Fig. 9.13.

The relationship between an input node and an output node is obtained by the application of Mason's gain formula.[1] This formula yields directly the overall transmittance from an input node to an output node. That is,

$$T = \frac{x_{\text{out}}}{x_{\text{in}}} = \sum_k \frac{T_k \Delta_k}{\Delta} \tag{9.24}$$

where T_k is the gain (transmittance) of the kth forward path from the input node x_{in} to the output node x_{out} and Δ is the determinant of the graph, which is defined as follows:

$\Delta = 1 - $ (sum of all individual loop gains) $+$ (sum of gain products of all combinations of two nontouching loops) $-$ (sum of gain products of all combinations of three nontouching loops) $+ \cdots$

$\Delta_k = $ determinant of graph in which all loops touching the kth forward path are removed (i.e., set equal to zero)

A forward path is any path which goes from the input node to the output node along which no node is passed through more than once. A loop is any path

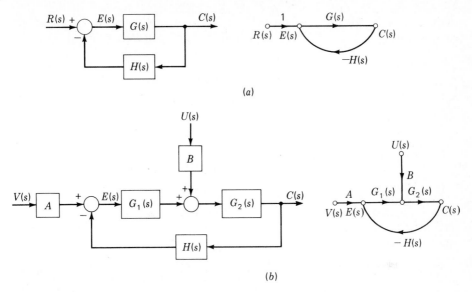

Figure 9.13 Basic systems.

which originates and terminates at the same node along which no node is passed through more than once. Touching loops are loops which have one or more nodes in common. Similarly, a loop which touches the kth forward path is one that has one or more nodes in common with the path.

The first step in applying Mason's gain formula is to identify the forward paths, from the input node to the output node, and all the loops in the system. For the signal-flow graph shown in Fig. 9.13a, there is one forward path from the input node $R(s)$ to the output node $C(s)$, and there is only one loop. Thus,

$$T_1 = (1)G(s) = G(s) \qquad L_1 = G(s)[-H(s)] = -G(s)H(s)$$

The determinant of the graph is

$$\Delta = 1 - L_1 = 1 + G(s)H(s)$$

Because loop L_1 touches the forward path T_1, setting L_1 equal to zero in the equation for Δ gives $\Delta_1 = 1$. Application of Mason's gain formula yields the relationship

$$\frac{C(s)}{R(s)} = \frac{T_1\Delta_1}{\Delta} = \frac{G(s)}{1 + G(s)H(s)} \qquad (9.25)$$

For the signal-flow graph of Fig. 9.13b, there is one forward path from the input node $V(s)$ to the output node $C(s)$, and there is only one loop. Thus,

$$T_1 = AG_1(s)G_2(s) \qquad L_1 = -G_1(s)G_2(s)H(s)$$

The determinant of the graph is

$$\Delta = 1 - L_1 = 1 + G_1(s)G_2(s)H(s)$$

Because loop L_1 touches the forward path T_1, setting L_1 equal to zero in the equation for Δ gives $\Delta_1 = 1$. Application of Mason's gain formula gives

$$\frac{C(s)}{V(s)} = \frac{T_1 \Delta_1}{\Delta} = \frac{AG_1(s)G_2(s)}{1 + G_1(s)G_2(s)H(s)} \tag{9.26}$$

By regarding $U(s)$ as the input node in Fig. 9.13b and $C(s)$ as the output node, it follows that

$$T_1 = BG_2(s)$$

$$L_1 = -G_1(s)G_2(s)H(s)$$

$$\Delta = 1 - L_1 = 1 + G_1(s)G_2(s)H(s)$$

Because the loop L_1 touches the forward path from $U(s)$ to $C(s)$, setting L_1 equal to zero in the expression for Δ yields $\Delta_1 = 1$. The overall relationship between $U(s)$ and $C(s)$ is

$$\frac{C(s)}{U(s)} = \frac{T_1 \Delta_1}{\Delta} = \frac{BG_2(s)}{1 + G_1(s)G_2(s)H(s)} \tag{9.27}$$

An input node represents an independent source or signal entering the system. There are only two input nodes [$V(s)$ and $U(s)$] for the system shown in Fig. 9.13b. Nodes which are not input nodes may be regarded as output nodes. If $E(s)$ is regarded as an output node, it is to be noted that the transmittance of the forward path from $V(s)$ to $E(s)$ is A. Because loop L_1 touches this path, it follows that

$$T_1 = A \qquad\qquad L_1 = -G_1(s)G_2(s)H(s)$$

$$\Delta = 1 - L_1 = 1 + G_1(s)G_2(s)H(s) \qquad \Delta_1 = 1$$

Application of Mason's gain formula yields

$$\frac{E(s)}{V(s)} = \frac{T_1 \Delta_1}{\Delta} = \frac{A}{1 + G_1(s)G_2(s)H(s)} \tag{9.28}$$

The transmittance of the path from $U(s)$ to $E(s)$ is $-BG_2(s)H(s)$. Loop L_1 touches this path. Thus,

$$T_1 = -BG_2(s)H(s) \qquad\qquad L_1 = -G_1(s)G_2(s)H(s)$$

$$\Delta = 1 - L_1 = 1 + G_1(s)G_2(s)H(s) \qquad \Delta_1 = 1$$

Application of Mason's gain formula gives

$$\frac{E(s)}{U(s)} = \frac{T_1 \Delta_1}{\Delta} = -\frac{BG_2(s)H(s)}{1 + G_1(s)G_2(s)H(s)} \tag{9.29}$$

The block diagram and corresponding signal-flow graph for a more complex system are shown in Fig. 9.14. There are two forward paths from the input $R(s)$ to the output $C(s)$:

$$T_1 = G_1 G_2 G_3 G_5 \qquad T_2 = G_4 G_5$$

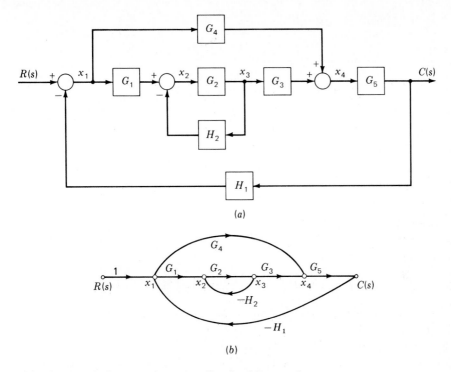

Figure 9.14 Block diagram and corresponding signal-flow graph.

The gains for the three loops in the system are

$$L_1 = -G_1 G_2 G_3 G_5 H_1 \qquad L_2 = -G_2 H_2 \qquad L_3 = -G_4 G_5 H_1$$

The numbering of the forward paths and loops is arbitrary. Because L_2 and L_3 are nontouching loops,

$$\Delta = 1 - (L_1 + L_2 + L_3) + (L_2 L_3)$$
$$= 1 + (G_1 G_2 G_3 G_5 H_1 + G_2 H_2 + G_4 G_5 H_1) + G_2 G_4 G_5 H_1 H_2$$

Because all three loops touch path T_1, setting $L_1 = L_2 = L_3 = 0$ in the preceding expression for Δ gives $\Delta_1 = 1$. Because loops L_1 and L_3 touch path T_2, setting $L_1 = L_3 = 0$ in the expression for Δ gives $\Delta_2 = 1 - L_2 = 1 + G_2 H_2$. Substitution of these results into Mason's gain formula yields

$$\frac{C(s)}{R(s)} = \frac{T_1 \Delta_1 + T_2 \Delta_2}{\Delta} = \frac{G_1 G_2 G_3 G_5 + (G_4 G_5)(1 + G_2 H_2)}{1 + (G_1 G_2 G_3 + G_4 + G_2 G_4 H_2)G_5 H_1 + G_2 H_2} \qquad (9.30)$$

The signal-flow graph corresponding to the block diagram of Fig. 9.3 is shown in Fig. 9.15. The symbol $D^{-1} = 1/D$ indicates integration. There are two loops, $L_1 = -3D^{-1}$ and $L_2 = -2D^{-2}$. Because there are no nontouching loops,

$$\Delta = 1 - (L_1 + L_2) = 1 + (3D^{-1} + 2D^{-2})$$

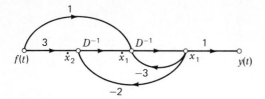

Figure 9.15 Signal-flow graph.

There are two forward paths from the input node $f(t)$ to the output node $c(t)$. The transmittances of the paths are $T_1 = D^{-1}$ and $T_2 = 3D^{-2}$ respectively. Each forward path touches both loops, so that $\Delta_1 = \Delta_2 = 1$. Application of Mason's gain formula gives

$$y(t) = \frac{T_1\Delta_1 + T_2\Delta_2}{\Delta} f(t) = \frac{D^{-1} + 3D^{-2}}{1 + 3D^{-1} + 2D^{-2}} f(t) = \frac{D + 3}{D^2 + 3D + 2} f(t) \qquad (9.1)$$

On block diagrams, the signal coming from a summing point and the system inputs and outputs are nodes. If the nodes are identified, Mason's gain formula may be applied directly to the block diagram.

Next, consider the simultaneous equations

$$x_1 = a_{11}x_1 + a_{12}x_2 + b_1 u_1$$

$$x_2 = a_{21}x_1 + a_{22}x_2 + b_2 u_2$$

The signal-flow graph for the first equation is shown in Fig. 9.16a and the graph for the second equation is shown in Fig. 9.16b. Combining these two graphs yields the complete signal-flow graph for the system, shown in Fig. 9.16c. The inputs to the system are u_1 and u_2. The system has three loops. The gain for each loop is

$$L_1 = a_{11} \qquad L_2 = a_{12} a_{21} \qquad L_3 = a_{22}$$

To obtain the equation relating u_1 and x_2, first note that there is only one forward path from u_1 to x_2. The gain of this path is

$$T_1 = b_1 a_{21}$$

(a)   (b)   (c)

Figure 9.16 Signal-flow graph for simultaneous equations.

The loops L_1 and L_3 are nontouching. Thus, the determinant for the system is

$$\Delta = 1 - (L_1 + L_2 + L_3) + (L_1 L_3) = 1 - (a_{11} + a_{12} a_{21} + a_{22}) + a_{11} a_{22}$$

Because all three loops touch the forward path, $\Delta_1 = 1$. Application of Mason's gain formula yields

$$\frac{x_2}{u_1} = \frac{T_1 \Delta_1}{\Delta} = \frac{b_1 a_{21}}{1 - (a_{11} + a_{12} a_{21} + a_{22}) + a_{11} a_{22}}$$

To obtain the equation relating u_2 and x_2, first note that there is only one forward path from u_2 to x_2. The gain of this path is $T_1 = b_2$. Because loops L_2 and L_3 touch this path, setting $L_2 = L_3 = 0$ in the equation for Δ gives

$$\Delta_1 = 1 - L_1 = 1 - a_{11}$$

The overall relationship is now found to be

$$\frac{x_2}{u_2} = \frac{T_1 \Delta_1}{\Delta} = \frac{b_2(1 - a_{11})}{1 - (a_{11} + a_{12} a_{21} + a_{22}) + a_{11} a_{22}}$$

The gain of the forward path from u_1 to x_1 is $T_1 = b_1$. Because loops L_1 and L_2 touch this path, setting $L_1 = L_2 = 0$ in the equation for Δ gives

$$\Delta_1 = 1 - L_3 = 1 - a_{22}$$

Application of Mason's gain formula shows that

$$\frac{x_1}{u_1} = \frac{T_1 \Delta_1}{\Delta} = \frac{b_1(1 - a_{22})}{1 - (a_{11} + a_{12} a_{21} + a_{22}) + a_{11} a_{22}}$$

The gain of the forward path from u_2 to x_1 is $T_1 = a_{12} b_2$. Because all loops touch this forward path, $\Delta_1 = 1$. Thus it is found that

$$\frac{x_1}{u_2} = \frac{a_{12} b_2}{1 - (a_{11} + a_{12} a_{21} + a_{22}) + a_{11} a_{22}}$$

Signal-flow graphs provide the basis for many powerful techniques employed in the design of control systems.

9.3 SOLUTION OF STATE-SPACE EQUATIONS

Consider the differential equation

$$\frac{dx}{dt} = ax \tag{9.31}$$

Integration yields

$$x(t) = x_0 + \int_0^t ax(\tau) \, d\tau \tag{9.32}$$

Replacing t by τ gives

$$x(\tau) = x_0 + \int_0^\tau ax(\tau)\, d\tau$$

Substituting this result into Eq. (9.32) shows that

$$x(t) = x_0 + \int_0^t a\left[x_0 + \int_0^\tau ax(\tau)\, d\tau \right] d\tau$$

$$= x_0 + x_0 \int_0^t a\, d\tau + \int_0^t a \int_0^\tau ax(\tau)\, d\tau\, d\tau$$

Continuing this process of replacing t by τ and substituting the result into Eq. (9.32) yields

$$x(t) = x_0\left(1 + \int_0^t a\, d\tau + \int_0^t a \int_0^\tau a\, d\tau\, d\tau + \int_0^t a \int_0^\tau a \int_0^\tau a\, d\tau\, d\tau\, d\tau + \cdots \right)$$

The result of the preceding integrations is

$$\int_0^t a\, d\tau = at$$

$$\int_0^t a \int_0^\tau a\, d\tau\, d\tau = \int_0^t a(a\tau)\, d\tau = \frac{a^2 t^2}{2!}$$

$$\int_0^t a \int_0^\tau a \int_0^\tau a\, d\tau\, d\tau\, d\tau = \int_0^t a \frac{a^2 \tau^2}{2!}\, d\tau = \frac{a^3 t^3}{3!}$$

Thus, the solution of the differential equation given by Eq. (9.31) is

$$x(t) = \left(1 + at + \frac{a^2 t^2}{2!} + \frac{a^3 t^3}{3!} + \cdots \right) x_0$$

$$= e^{at} x_0 \tag{9.33}$$

The solution of a system of differential equations is obtained in a similar manner. Thus, consider the system

$$\dot{\mathbf{x}}(t) = A\mathbf{x}(t) \qquad \mathbf{x}(0) = \mathbf{x}_0 \tag{9.34}$$

The corresponding differential equations are

$$\dot{x}_1 = a_{11}x_1 + a_{12}x_2 + \cdots + a_{1n}x_n$$

$$\dot{x}_2 = a_{21}x_1 + a_{22}x_2 + \cdots + a_{2n}x_n$$

$$\cdots\cdots\cdots\cdots\cdots\cdots\cdots\cdots\cdots\cdots$$

$$\dot{x}_n = a_{n1}x_1 + a_{n2}x_2 + \cdots + a_{nn}x_n$$

Integration of each equation gives

$$x_1 - x_1(0) = \int_0^t (a_{11}x_1 + a_{12}x_2 + \cdots + a_{1n}x_n)\, d\tau$$

$$x_2 - x_2(0) = \int_0^t (a_{21}x_1 + a_{22}x_2 + \cdots + a_{2n}x_n)\, d\tau$$

$$\cdots\cdots\cdots\cdots\cdots\cdots\cdots\cdots\cdots\cdots\cdots\cdots\cdots\cdots$$

$$x_n - x_n(0) = \int_0^t (a_{n1}x_1 + a_{n2}x_2 + \cdots + a_{nn}x_n)\, d\tau$$

The corresponding matrix representation is

$$\mathbf{x}(t) = \mathbf{x}(0) + \int_0^t A\mathbf{x}(\tau)\, d\tau$$

Replacing t by τ and substituting this result back into the preceding gives

$$\mathbf{x}(t) = \mathbf{x}(0) + \int_0^t A\left[\mathbf{x}(0) + \int_0^\tau A\mathbf{x}(\tau)\, d\tau\right] d\tau$$

$$= \mathbf{x}(0) + \mathbf{x}(0) \int_0^t A\, d\tau + \int_0^t A \int_0^\tau A\mathbf{x}(\tau)\, d\tau\, d\tau$$

Continuing this process

$$\mathbf{x}(t) = \mathbf{x}(0)\left[I + \int_0^t A\, d\tau + \int_0^t A \int_0^\tau A\, d\tau\, d\tau + \int_0^t A \int_0^\tau A \int_0^\tau A\, d\tau\, d\tau\, d\tau + \cdots \right]$$

The result of each integration is

$$\int_0^t A\, d\tau = At$$

$$\int_0^t A \int_0^\tau A\, d\tau = \int_0^t A(A\tau)\, d\tau = \frac{A^2 t^2}{2!}$$

$$\int_0^t A \int_0^\tau A \int_0^\tau A\, d\tau\, d\tau\, d\tau = \int_0^t A \frac{A^2 \tau^2}{2!}\, d\tau = \frac{A^3 t^3}{3!}$$

Thus the solution of the system of differential equations represented by Eq. (9.34) is

$$\mathbf{x}(t) = \left[I + At + \frac{A^2 t^2}{2!} + \frac{A^3 t^3}{3!} + \cdots \right]\mathbf{x}(0) = e^{At}\mathbf{x}(0)$$

$$= \mathbf{\Phi}(t)\mathbf{x}(0) \qquad (9.35)$$

where

$$\mathbf{\Phi}(t) = e^{At} = I + At + \frac{A^2 t^2}{2!} + \frac{A^3 t^3}{3!} + \cdots \qquad (9.36)$$

The function $\mathbf{\Phi}(t) = e^{At}$ is called the state-transition matrix. The state-transition matrix operates on the initial state of the system $\mathbf{x}(0)$ to yield the state of the system $\mathbf{x}(t)$ at any time t.

The exponential representation of the state-transition matrix $(e^{At} = I + At + A^2t^2/2! + \cdots)$ has properties similar to its scalar analog $(e^{at} = 1 + at + a^2t^2/2! + \cdots)$. An important property is revealed by differentiation of the preceding expression. That is,

$$\frac{d\mathbf{\Phi}(t)}{dt} = Ae^{At} = A\left[I + At + \frac{A^2t^2}{2!} + \cdots\right] \tag{9.37}$$

In using matrix methods, it is sometimes necessary to obtain the inverse of a matrix. The inverse A^{-1} of a matrix A is determined by

$$A^{-1} = \frac{\text{Adj } A}{|A|}$$

where Adj A is the adjoint of the matrix A. The adjoint is the transpose of the cofactor matrix. Consider the matrix

$$A = \begin{bmatrix} 2 & 0 & 3 \\ 4 & 2 & 1 \\ 0 & 1 & 3 \end{bmatrix}$$

The adjoint is

$$\text{Adj } A = \begin{bmatrix} \begin{vmatrix} 2 & 1 \\ 1 & 3 \end{vmatrix} & -\begin{vmatrix} 4 & 1 \\ 0 & 3 \end{vmatrix} & \begin{vmatrix} 4 & 2 \\ 0 & 1 \end{vmatrix} \\ -\begin{vmatrix} 0 & 3 \\ 1 & 3 \end{vmatrix} & \begin{vmatrix} 2 & 3 \\ 0 & 3 \end{vmatrix} & -\begin{vmatrix} 2 & 0 \\ 0 & 1 \end{vmatrix} \\ \begin{vmatrix} 0 & 3 \\ 2 & 1 \end{vmatrix} & -\begin{vmatrix} 2 & 3 \\ 4 & 1 \end{vmatrix} & \begin{vmatrix} 2 & 0 \\ 4 & 2 \end{vmatrix} \end{bmatrix}^T = \begin{bmatrix} 5 & -12 & 4 \\ 3 & 6 & -2 \\ -6 & 10 & 4 \end{bmatrix}^T$$

Thus, the inverse is

$$A^{-1} = \frac{\begin{bmatrix} 5 & 3 & -6 \\ -12 & 6 & 10 \\ 4 & -2 & 4 \end{bmatrix}}{22}$$

When the initial state of the system is known at time t_0 rather than at time $t = 0$, carrying through the preceding analysis with t_0 as the lower limit of integration yields

$$\mathbf{x}(t) = \left[I + \int_{t_0}^{t} A \, d\tau + \int_{t_0}^{t} A \int_{t_0}^{\tau} A \, d\tau \, d\tau + \cdots\right]\mathbf{x}(t_0)$$

Evaluating the integrals shows that

$$\int_{t_0}^{t} A \, d\tau = A(t - t_0)$$

$$\int_{t_0}^{t} A \int_{t_0}^{\tau} A \, d\tau \, d\tau = \int_{t_0}^{t} A[A(\tau - t_0)] \, d\tau = A^2 \frac{(t - t_0)^2}{2!}$$

Thus,

$$\mathbf{x}(t) = \left[I + A(t - t_0) + A^2 \frac{(t - t_0)^2}{2!} + \cdots \right] \mathbf{x}(t_0) = e^{A(t - t_0)} \mathbf{x}(t_0)$$

$$= \mathbf{\Phi}(t - t_0)\mathbf{x}(t_0) \tag{9.38}$$

Properties of the state-transition matrix are most readily obtained using the exponential representation. This is illustrated as follows:

$$\mathbf{\Phi}(0) = e^{A_0} = I$$

$$\mathbf{\Phi}^{-1}(t) = (e^{At})^{-1} = e^{-At} = \mathbf{\Phi}(-t)$$

$$\mathbf{\Phi}^k(t) = (e^{At})^k = e^{A(kt)} = \mathbf{\Phi}(kt) \tag{9.39}$$

$$\mathbf{\Phi}(t_1 + t_2) = e^{A(t_1 + t_2)} = e^{At_1} e^{At_2} = \mathbf{\Phi}(t_1)\mathbf{\Phi}(t_2) = \mathbf{\Phi}(t_2)\mathbf{\Phi}(t_1)$$

$$\mathbf{\Phi}(t_2 - t_1)\mathbf{\Phi}(t_1 - t_0) = e^{A(t_2 - t_1) + A(t_1 - t_0)} = \mathbf{\Phi}(t_2 - t_0)$$

Such relationships can save considerable time and computational effort.

9.4 METHODS OF COMPUTING $\mathbf{\Phi}(t)$

A major problem in solving differential equations by the state-space method is the determination of the state-transition matrix. The infinite-series representation for $\mathbf{\Phi}(t)$ involves lengthy computations and results in a very awkward form.

Laplace Transform Method

Consider the system of equations

$$\dot{x}_1 = a_{11}x_1 + a_{12}x_2 + \cdots + a_{1n}x_n$$

$$\dot{x}_2 = a_{21}x_1 + a_{22}x_2 + \cdots + a_{2n}x_n$$

$$\cdots\cdots\cdots\cdots\cdots\cdots\cdots\cdots\cdots\cdots\cdots$$

$$\dot{x}_n = a_{n1}x_1 + a_{n2}x_2 + \cdots + a_{nn}x_n$$

The Laplace transform is

$$sX_1(s) - x_1(0) = a_{11}X_1(s) + a_{12}X_2(s) + \cdots + a_{1n}X_n(s)$$

$$sX_2(s) - x_2(0) = a_{21}X_1(s) + a_{22}X_2(s) + \cdots + a_{2n}X_n(s)$$

$$\cdots\cdots\cdots\cdots\cdots\cdots\cdots\cdots\cdots\cdots\cdots\cdots\cdots\cdots\cdots$$

$$sX_n(s) - x_n(0) = a_{n1}X_1(s) + a_{n2}X_2(s) + \cdots + a_{nn}X_n(s)$$

The matrix representation for each of the preceding systems of equations is

$$\dot{\mathbf{x}}(t) = A\mathbf{x}(t)$$

and
$$sX(s) - \mathbf{x}(0) = AX(s)$$

Note that matrix differential equations transform in the same manner as ordinary differential equations. The last equation may be written in the form

$$[sI - A]X(s) = \mathbf{x}(0)$$

Premultiplying by the inverse matrix, $[sI - A]^{-1}$, yields

$$X(s) = [sI - A]^{-1}\mathbf{x}(0)$$

Inverting yields, for the desired solution,

$$\mathbf{x}(t) = \mathscr{L}^{-1}[sI - A]^{-1}\mathbf{x}(0)$$

Comparison with Eq. (9.35) reveals that

$$\mathbf{\Phi}(t) = \mathscr{L}^{-1}[sI - A]^{-1} \tag{9.40}$$

Illustrative example 9.1 Use direct programming to determine the solution of the following differential equation for the case in which $f(t) = 0$ and the initial conditions are $y_0 = 1$ and $\dot{y}_0 = 0$:

$$\ddot{y} + 5\dot{y} + 6y = f(t) \tag{9.41}$$

SOLUTION The solution when the right side is zero $[f(t) = 0]$ is called the force-free response. The next section shows that the methods used to obtain the force-free response can be readily extended to determine the response when $f(t)$ is not zero (i.e., the forced response).

For direct programming, $y = x_1$ and $\dot{y} = \dot{x}_1 = x_2$. The resulting state-space representation for the differential equation is

$$\dot{x}_1 = x_2$$
$$\dot{x}_2 = -6x_1 - 5x_2 \tag{9.42}$$

The Laplace transform is

$$sX_1(s) - x_1(0) = X_2(s)$$
$$sX_2(s) - x_2(0) = -6X_1(s) - 5X_2(s)$$

The matrix form is

$$\begin{bmatrix} s & 0 \\ 0 & s \end{bmatrix}\begin{bmatrix} X_1(s) \\ X_2(s) \end{bmatrix} = \begin{bmatrix} 0 & 1 \\ -6 & -5 \end{bmatrix}\begin{bmatrix} X_1(s) \\ X_2(s) \end{bmatrix} + \begin{bmatrix} x_1(0) \\ x_2(0) \end{bmatrix}$$

$$\left[\begin{bmatrix} s & 0 \\ 0 & s \end{bmatrix} - \begin{bmatrix} 0 & 1 \\ -6 & -5 \end{bmatrix}\right]\begin{bmatrix} X_1(s) \\ X_2(s) \end{bmatrix} = \begin{bmatrix} x_1(0) \\ x_2(0) \end{bmatrix}$$

Note that this has the general form

$$[sI - A]X(s) = x(0)$$

Premultiplying by $[sI - A]^{-1}$ gives

$$X(s) = [sI - A]^{-1}x(0)$$

The $[sI - A]$ matrix is

$$[sI - A] = \begin{bmatrix} s & -1 \\ 6 & s+5 \end{bmatrix}$$

The inverse is

$$\Phi(s) = [sI - A]^{-1} = \frac{\begin{bmatrix} s+5 & 1 \\ -6 & s \end{bmatrix}}{s^2 + 5s + 6} \tag{9.43}$$

To invert, each element is written in its partial-fraction expansion form. For the first element

$$\frac{s+5}{(s+2)(s+3)} = \frac{3}{s+2} - \frac{2}{s+3}$$

Thus

$$\Phi(s) = [sI - A]^{-1} = \begin{bmatrix} \dfrac{3}{s+2} - \dfrac{2}{s+3} & \dfrac{1}{s+2} - \dfrac{1}{s+3} \\ \dfrac{-6}{s+2} + \dfrac{6}{s+3} & \dfrac{-2}{s+2} + \dfrac{3}{s+3} \end{bmatrix}$$

Inverting yields

$$\Phi(t) = \mathscr{L}^{-1}[sI - A]^{-1} = \begin{bmatrix} 3e^{-2t} - 2e^{-3t} & e^{-2t} - e^{-3t} \\ -6e^{-2t} + 6e^{-3t} & -2e^{-2t} + 3e^{-3t} \end{bmatrix} \tag{9.44}$$

The solution $x(t) = \Phi(t)x(0)$ is

$$\begin{bmatrix} x_1 \\ x_2 \end{bmatrix} = \begin{bmatrix} 3e^{-2t} - 2e^{-3t} & e^{-2t} - e^{-3t} \\ -6e^{-2t} + 6e^{-3t} & -2e^{-2t} + 3e^{-3t} \end{bmatrix} \begin{bmatrix} x_1(0) \\ x_2(0) \end{bmatrix} \tag{9.45}$$

Because $x_1 = y$ and $x_2 = \dot{x}_1 = \dot{y}$, the initial values are $x_1(0) = y_0 = 1$ and $x_2(0) = \dot{y}_0 = 0$. Substitution of these values into the preceding matrix expression and noting that $y = x_1$ and $\dot{y} = x_2$ gives

$$y = x_1 = 3e^{-2t} - 2e^{-3t}$$

$$\dot{y} = x_2 = -6(e^{-2t} - e^{-3t}) \tag{9.46}$$

Illustrative example 9.2 Same as Illustrative example 9.1 except use series programming.

SOLUTION Comparison of Eq. (9.40) with the general form given by Eq. (9.9) shows that $a_1 = 5$, $a_2 = 6$, and $b_1 = b_2 = 0$. The resulting state-space representation is

$$\begin{bmatrix} \dot{x}_1 \\ \dot{x}_2 \end{bmatrix} = \begin{bmatrix} -5 & 1 \\ -6 & 0 \end{bmatrix} \begin{bmatrix} x_1 \\ x_2 \end{bmatrix} \tag{9.47}$$

where $x_1 = y$ and thus $\dot{x}_1 = \dot{y}$. Note from the first matrix relationship that $\dot{x}_1 = -5x_1 + x_2$ and hence for series programming $x_2 = \dot{x}_1 + 5x_1 = \dot{y} + 5y$, whereas for direct programming $x_2 = \dot{y}$. The $[sI - A]$ matrix is

$$[sI - A] = \begin{bmatrix} s+5 & -1 \\ 6 & s \end{bmatrix}$$

The inverse is

$$\Phi(s) = [sI - A]^{-1} = \frac{\begin{bmatrix} s & 1 \\ -6 & s+5 \end{bmatrix}}{s^2 + 5s + 6}$$

$$= \begin{bmatrix} \dfrac{-2}{s+2} + \dfrac{3}{s+3} & \dfrac{1}{s+2} - \dfrac{1}{s+3} \\[2ex] \dfrac{-6}{s+2} + \dfrac{6}{s+3} & \dfrac{3}{s+2} - \dfrac{2}{s+3} \end{bmatrix} \tag{9.48}$$

Inverting to obtain the state-transition matrix $\Phi(t)$ yields, for $\mathbf{x}(t) = \Phi(t)\mathbf{x}(0)$,

$$\begin{bmatrix} x_1 \\ x_2 \end{bmatrix} = \begin{bmatrix} -2e^{-2t} + 3e^{-3t} & e^{-2t} - e^{-3t} \\ -6e^{-2t} + 6e^{-3t} & 3e^{-2t} - 2e^{-3t} \end{bmatrix} \begin{bmatrix} x_1(0) \\ x_2(0) \end{bmatrix}$$

For $x_1 = y$ and $x_2 = \dot{y} + 5y$, then $x_1(0) = y_0 = 1$ and $x_2(0) = \dot{y}_0 + 5y_0 = 0 + 5 = 5$. Thus,

$$x_1 = (-2e^{-2t} + 3e^{-3t}) + 5(e^{-2t} - e^{-3t}) = 3e^{-2t} - 2e^{-3t}$$

$$x_2 = (-6e^{-2t} + 6e^{-3t}) + 5(3e^{-2t} - 2e^{-3t}) = 9e^{-2t} - 4e^{-3t} \tag{9.49}$$

For $y = x_1$ and $\dot{y} = x_2 - 5y$, then $y = 3e^{-2t} - 2e^{-3t}$ and $\dot{y} = -6e^{-2t} + 6e^{-3t}$. These are the same results for y and $\dot{y}$ as were obtained in Illustrative example 9.1.

Comparison of Illustrative examples 9.1 and 9.2 shows that the state-transition matrix for a system is not unique, but rather depends on the particular state-space representation selected to represent the system.

Illustrative example 9.3 Use direct programming to determine the solution of the following differential equation for the case in which $f(t) = 0$ and the initial conditions are $y_0 = 1$ and $\dot{y}_0 = 0$:

$$\ddot{y} + 10\dot{y} + 25y = f(t) \tag{9.50}$$

SOLUTION For direct programming $x_1 = y$ and $x_2 = \dot{x}_1 = \dot{y}$. The resulting state-space representation for this differential equation is

$$\dot{x}_1 = x_2$$

$$\dot{x}_2 = -25x_1 - 10x_2$$

The corresponding matrix form, $\dot{\mathbf{x}} = A\mathbf{x}$, is

$$\begin{bmatrix} \dot{x}_1 \\ \dot{x}_2 \end{bmatrix} = \begin{bmatrix} 0 & 1 \\ -25 & -10 \end{bmatrix}\begin{bmatrix} x_1 \\ x_2 \end{bmatrix}$$

The $[sI - A]$ matrix is

$$[sI - A] = \begin{bmatrix} s & -1 \\ 25 & s + 10 \end{bmatrix}$$

Inverting gives

$$\Phi(s) = [sI - A]^{-1} = \frac{\begin{bmatrix} s + 10 & 1 \\ -25 & s \end{bmatrix}}{s^2 + 10s + 25} \tag{9.51}$$

Obtaining the partial-fraction expansion for each element and then inverting yields

$$\Phi(t) = \begin{bmatrix} (1 + 5t)e^{-5t} & te^{-5t} \\ -25te^{-5t} & (1 - 5t)e^{-5t} \end{bmatrix} \tag{9.52}$$

The solution $\mathbf{x}(t) = \Phi(t)\mathbf{x}(0)$ is

$$\begin{bmatrix} x_1 \\ x_2 \end{bmatrix} = \begin{bmatrix} (1 + 5t)e^{-5t} & te^{-5t} \\ -25te^{-5t} & (1 - 5t)e^{-5t} \end{bmatrix}\begin{bmatrix} x_1(0) \\ x_2(0) \end{bmatrix} \tag{9.53}$$

The initial values are $x_1(0) = y_0 = 1$ and $x_2(0) = \dot{y}_0 = 0$. Thus,

$$y = x_1 = (1 + 5t)e^{-5t}$$

$$\dot{y} = x_2 = -25te^{-5t} \tag{9.54}$$

The result for $\dot{y} = x_2 = \dot{x}_1$ is verified by differentiation of the $y = x_1$ equation.

Signal-Flow-Graph Method

The use of signal-flow graphs saves considerable computational effort in determining the state-transition matrix $\Phi(t)$. The general solution for the force-free case is

$$\mathbf{x}(t) = \Phi(t)\mathbf{x}(0)$$

The Laplace transform is

$$\mathbf{X}(s) = \Phi(s)\mathbf{x}(0)$$

For a second-order system, this relationship is

$$\begin{bmatrix} X_1(s) \\ X_2(s) \end{bmatrix} = \begin{bmatrix} \phi_{11}(s) & \phi_{12}(s) \\ \phi_{21}(s) & \phi_{22}(s) \end{bmatrix} \begin{bmatrix} x_1(0) \\ x_2(0) \end{bmatrix}$$

Inversion yields

$$\begin{bmatrix} x_1(t) \\ x_2(t) \end{bmatrix} = \begin{bmatrix} \phi_{11}(t) & \phi_{12}(t) \\ \phi_{21}(t) & \phi_{22}(t) \end{bmatrix} \begin{bmatrix} x_1(0) \\ x_2(0) \end{bmatrix}$$

The matrix form for an nth-order system is

$$\begin{bmatrix} X_1(s) \\ X_2(s) \\ \vdots \\ X_n(s) \end{bmatrix} = \begin{bmatrix} \phi_{11}(s) & \phi_{12}(s) & \cdots & \phi_{1n}(s) \\ \phi_{21}(s) & \phi_{22}(0) & \cdots & \phi_{2n}(s) \\ \cdots\cdots\cdots\cdots\cdots\cdots\cdots\cdots\cdots \\ \phi_{n1}(s) & \phi_{n2}(s) & \cdots & \phi_{nn}(s) \end{bmatrix} \begin{bmatrix} x_1(0) \\ x_2(0) \\ \vdots \\ x_n(0) \end{bmatrix} \tag{9.55}$$

Inversion yields the desired result

$$\begin{bmatrix} x_1(t) \\ x_2(t) \\ \vdots \\ x_n(t) \end{bmatrix} = \begin{bmatrix} \phi_{11}(t) & \phi_{12}(t) & \cdots & \phi_{1n}(t) \\ \phi_{21}(t) & \phi_{22}(t) & \cdots & \phi_{2n}(t) \\ \cdots\cdots\cdots\cdots\cdots\cdots\cdots\cdots\cdots \\ \phi_{n1}(t) & \phi_{n2}(t) & \cdots & \phi_{nn}(t) \end{bmatrix} \begin{bmatrix} x_1(0) \\ x_2(0) \\ \vdots \\ x_n(0) \end{bmatrix} \tag{9.56}$$

Each element $\phi_{ij}(s)$ in Eq. (9.55) may be obtained directly from the signal-flow graph. To construct the signal-flow graph, first obtain the state-space form

$$\dot{\mathbf{x}}(t) = A\mathbf{x}(t)$$

Next take the Laplace transform

$$s\mathbf{X}(s) - \mathbf{x}(0) = A\mathbf{X}(s)$$

Transpose $\mathbf{x}(0)$ to the right side and then multiply through by s^{-1}. That is,

$$\mathbf{X}(s) = s^{-1}\mathbf{x}(0) + s^{-1}A\mathbf{X}(s)$$

The signal-flow graph is now constructed in which the initial values of the state variables $\mathbf{x}(0)$ are the inputs and the Laplace transform of the state variables $\mathbf{X}(s)$ are the outputs. Application of Mason's gain formula in which $x_j(0)$ is the input and $X_i(s)$ is the output yields the relationship

$$\frac{X_i(s)}{x_j(0)} = \phi_{ij}(s)$$

Thus, each element of $\Phi(s)$ in Eq. (9.55) may be obtained directly from this signal-flow graph. Inverting each $\phi_{ij}(s)$ yields the corresponding element $\phi_{ij}(t)$ in Eq. (9.56).

Illustrative example 9.4 Use the signal-flow-graph method to determine the solution of Illustrative example 9.1.

SOLUTION The matrix relationships are

$$\dot{x}_1 = x_2$$
$$\dot{x}_2 = -6x_1 - 5x_2$$

To construct the signal-flow graph, first take the Laplace transform of these matrix relationships. Thus,

$$sX_1(s) - x_1(0) = X_2(s)$$
$$sX_2(s) - x_2(0) = -6X_1(s) - 5X_2(s)$$

Taking the initial-condition terms to the right side and dividing through by s gives

$$X_1(s) = s^{-1}X_2(s) + s^{-1}x_1(0)$$
$$X_2(s) = s^{-1}[-6X_1(s) - 5X_2(s)] + s^{-1}x_2(0)$$

The signal-flow graph for the first relationship is shown in Fig. 9.17a and the graph for the second relationship is shown in Fig. 9.17b. These two graphs are combined to yield the signal-flow graph for the system, shown in Fig. 9.17c. This graph has two loops,

$$L_1 = -5s^{-1} \qquad L_2 = -6s^{-2}$$

Because there are no nontouching loops, the system determinant is

$$\Delta = 1 - (L_1 + L_2) = 1 + \frac{5}{s} + \frac{6}{s^2} = \frac{s^2 + 5s + 6}{s^2}$$

The transfer function relating the input $x_1(0)$ to the output $X_1(s)$ is $\phi_{11}(s)$. The transmittance of the forward path from $x_1(0)$ to $X_1(s)$ is $T_1 = 1/s$. Because L_2 touches T_1, setting $L_2 = 0$ in the equation for Δ gives $\Delta_1 = 1 - L_1 = 1 + 5/s = (s + 5)/s$. The transfer function relating the input $x_2(0)$ to the output $X_1(s)$ is $\phi_{12}(s)$. The transmittance of the forward path from $x_2(0)$ to $X_1(s)$ is $T_2 = 1/s^2$. Because both L_1 and L_2 touch T_2, setting $L_1 = L_2 = 0$

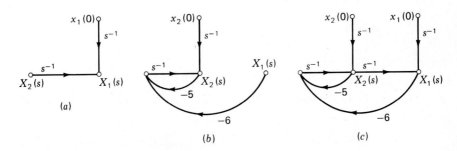

(a)

(b)

(c)

Figure 9.17 Signal-flow graph.

in the equation for Δ gives $\Delta_2 = 1$. Thus, the equation for $X_1(s)$ is

$$X_1(s) = \phi_{11}(s)x_1(0) + \phi_{12}(s)x_2(0) = \frac{T_1\Delta_1}{\Delta}x_1(0) + \frac{T_2\Delta_2}{\Delta}x_2(0)$$

$$= \frac{s+5}{s^2+5s+6}x_1(0) + \frac{1}{s^2+5s+6}x_2(0)$$

The transfer function relating $x_1(0)$ to $X_2(s)$ is $\phi_{21}(s)$. The transmittance of this path is $T_3 = -6/s^2$. Because L_1 and L_2 touch this path, setting $L_1 = L_2 = 0$ in the equation for Δ gives $\Delta_3 = 1$. The transfer function relating $x_2(0)$ to $X_2(s)$ is $\phi_{22}(s)$. The transmittance of this path is $T_4 = 1/s$. Because L_1 and L_2 touch this path, $\Delta_4 = 1$. The resulting equation for $X_2(s)$ is

$$X_2(s) = \phi_{21}(s)x_1(0) + \phi_{22}(s)x_2(0) = \frac{T_3\Delta_3}{\Delta}x_1(0) + \frac{T_4\Delta_4}{\Delta}x_2(0)$$

$$= \frac{-6}{s^2+5s+6}x_1(0) + \frac{s}{s^2+5s+6}x_2(0)$$

The matrix form of the preceding relationship is

$$\begin{bmatrix} X_1(s) \\ X_2(s) \end{bmatrix} = \frac{\begin{bmatrix} s+5 & 1 \\ -6 & s \end{bmatrix}\begin{bmatrix} x_1(0) \\ x_2(0) \end{bmatrix}}{s^2+5s+6}$$

The first term on the right is the $\Phi(s) = [sI - A]^{-1}$ matrix. This is the same result attained by the Laplace transform method [that is, Eq. (9.43)]. Note that the signal-flow-graph method eliminates the need to obtain the matrix inverse.

Illustrative example 9.5 Use the signal-flow-graph method to determine the solution of Illustrative example 9.2.

SOLUTION The matrix relationships are

$$\dot{x}_1 = -5x_1 + x_2$$
$$\dot{x}_2 = -6x_1$$

The Laplace transform of these matrix relationships is

$$sX_1(s) - x_1(0) = -5X_1(s) + X_2(s)$$
$$sX_2(s) - x_2(0) = -6X_1(s)$$

Taking the initial condition terms to the right side and dividing through by s gives

$$X_1(s) = s^{-1}[-5X_1(s) + X_2(s)] + s^{-1}x_1(0)$$
$$X_2(s) = s^{-1}[-6X_1(s)] + s^{-1}x_2(0)$$

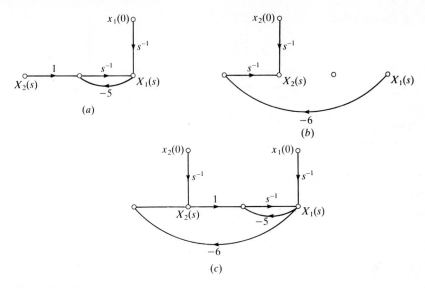

Figure 9.18 Signal-flow graph.

The signal-flow graph for the first relationship is shown in Fig. 9.18a and the graph for the second relationship is shown in Fig. 9.18b. These two graphs are combined to yield the signal-flow graph for the system shown in Fig. 9.18c. This graph has two loops,

$$L_1 = -5s^{-1} \qquad L_2 = -6s^{-2}$$

The system determinant is

$$\Delta = 1 - (L_1 + L_2) = 1 + (5s^{-1} + 6s^{-2}) = \frac{s^2 + 5s + 6}{s^2}$$

Application of Mason's gain formula yields the following equation for $X_1(s)$ and $X_2(s)$:

$$\begin{bmatrix} X_1(s) \\ X_2(s) \end{bmatrix} = \frac{\begin{bmatrix} s & 1 \\ -6 & s + 5 \end{bmatrix} \begin{bmatrix} x_1(0) \\ x_2(0) \end{bmatrix}}{s^2 + 5s + 6}$$

This is the same result for $\Phi(s)$ that was attained in Illustrative example 9.2 [that is, Eq. (9.48)].

Illustrative example 9.6 Use the signal-flow-graph method to determine the state-transition matrix for Illustrative example 9.3.

SOLUTION The matrix relationships for this illustrative example are

$$\dot{x}_1 = x_2$$

$$\dot{x}_2 = -25x_1 - 10x_2$$

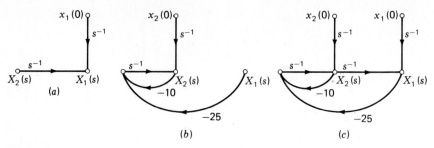

Figure 9.19 Signal-flow graph.

The Laplace transform of these relationships is

$$sX_1(s) - x_1(0) = X_2(s)$$

$$sX_2(s) - x_2(0) = -25X_1(s) - 10X_2(s)$$

Taking the initial conditions to the right side and dividing through by s gives

$$X_1(s) = s^{-1}X_2(s) + s^{-1}x_1(0)$$

$$X_2(s) = s^{-1}[-25X_1(s) - 10X_2(s)] + s^{-1}x_2(0)$$

The signal-flow graph for the first relationship is shown in Fig. 9.19*a* and the graph for the second relationship is shown in Fig. 9.19*b*. Combining these two graphs yields the signal-flow graph for the system, which is shown in Fig. 9.19*c*. This system has two loops,

$$L_1 = -10s^{-1} \qquad L_2 = -25s^{-2}$$

The determinant for the system is

$$\Delta = 1 - (L_1 + L_2) = \frac{s^2 + 10s + 25}{s^2}$$

Application of Mason's gain formula yields the following equation for $X_1(s)$ and $X_2(s)$:

$$\frac{\begin{bmatrix} X_1(s) \\ X_2(s) \end{bmatrix} = \begin{bmatrix} s + 10 & 1 \\ -25 & s \end{bmatrix} \begin{bmatrix} x_1(0) \\ x_2(0) \end{bmatrix}}{(s + 5)^2}$$

The first matrix on the right is the $\Phi(s) = [sI - A]^{-1}$ matrix. This is the same result attained by the Laplace transform method.

9.5 FORCED RESPONSE

The general representation for a system of differential equations with a forcing function is

$$\dot{\mathbf{x}}(t) = A\mathbf{x}(t) + \mathbf{b}f(t) \qquad \mathbf{x}(0) = \mathbf{x}_0 \tag{9.57}$$

To solve this system, first rewrite it in the form

$$\dot{\mathbf{x}}(t) - A\mathbf{x}(t) = \mathbf{b}f(t)$$

Multiplying through by e^{-At} shows that

$$e^{-At}[\dot{\mathbf{x}}(t) - A\mathbf{x}(t)] = \frac{d}{dt}[e^{-At}\mathbf{x}(t)] = e^{-At}\mathbf{b}f(t)$$

Integration between the limits 0 and t of the two expressions on the right side of the preceding equation gives

$$e^{-At}\mathbf{x}(t) - \mathbf{x}(0) = \int_0^t e^{-A\tau}\mathbf{b}f(\tau)\,d\tau \tag{9.58}$$

Multiplying through by e^{At} yields the desired solution:

$$\mathbf{x}(t) = e^{At}\mathbf{x}(0) + \int_0^t e^{A(t-\tau)}\mathbf{b}f(\tau)\,d\tau \tag{9.59}$$

Replacing e^{At} by $\mathbf{\Phi}(t)$ and $e^{A(t-\tau)}$ by $\mathbf{\Phi}(t-\tau) = \mathbf{\Phi}(t)\mathbf{\Phi}(-\tau)$ yields the form

$$\mathbf{x}(t) = \mathbf{\Phi}(t)\mathbf{x}(0) + \mathbf{\Phi}(t)\int_0^t \mathbf{\Phi}(-\tau)\mathbf{b}f(\tau)\,d\tau \tag{9.60}$$

The first term on the right side of Eq. (9.60) is recognized as the force-free response. The second term is the response due to the forcing function $f(t)$. Although this second term may be evaluated by the integration process indicated, the Laplace transform method usually affords considerable simplification.

The Laplace transform of Eq. (9.57) is

$$s\mathbf{X}(s) - \mathbf{x}(0) = A\mathbf{X}(s) + \mathbf{b}F(s)$$

This may be written in the form

$$[sI - A]\mathbf{X}(s) = \mathbf{x}(0) + \mathbf{b}F(s)$$

Premultiplying by the inverse $[sI - A]^{-1}$ gives

$$\mathbf{X}(s) = [sI - A]^{-1}\mathbf{x}(0) + [sI - A]^{-1}\mathbf{b}F(s) \tag{9.61}$$

Inverting yields, for the solution,

$$\mathbf{x}(t) = \mathcal{L}^{-1}[sI - A]^{-1}\mathbf{x}(0) + \mathcal{L}^{-1}[sI - A]^{-1}\mathbf{b}F(s)$$
$$= \mathcal{L}^{-1}\mathbf{\Phi}(s)\mathbf{x}(0) + \mathcal{L}^{-1}\mathbf{\Phi}(s)\mathbf{b}F(s) \tag{9.62}$$

Comparison with Eq. (9.60) shows that

$$\mathbf{\Phi}(t) = \mathcal{L}^{-1}[sI - A]^{-1} = \mathcal{L}^{-1}\mathbf{\Phi}(s) \tag{9.40}$$

and

$$\mathbf{\Phi}(t)\int_0^t \mathbf{\Phi}(-\tau)\mathbf{b}f(\tau)\,d\tau = \mathcal{L}^{-1}[sI - A]^{-1}\mathbf{b}F(s)$$
$$= \mathcal{L}^{-1}\mathbf{\Phi}(s)\mathbf{b}F(s) \tag{9.63}$$

The first result was previously established by Eq. (9.40).

Illustrative example 9.7 Use direct programming to obtain the solution of the following differential equation:

$$\ddot{y} + 5\dot{y} + 6y = f'(t) + 6f(t) \tag{9.64}$$

The forcing function is a unit step function and all initial conditions are zero.

SOLUTION The operator form for the differential equation is

$$y = \frac{D + 6}{D^2 + 5D + 6} f(t)$$

For direct programming x_1 is such that the numerator operator is unity. That is,

$$x_1 = \frac{1}{D^2 + 5D + 6} f(t) \tag{9.65}$$

Thus,

$$\ddot{x}_1 + 5\dot{x}_1 + 6x_1 = f(t)$$

For $\dot{x}_1 = x_2$, then $\ddot{x}_1 = \dot{x}_2$. The resulting state-space representation is

$$\dot{x}_1 = x_2$$
$$\dot{x}_2 = -6x_1 - 5x_2 + f(t)$$

and

$$y = 6x_1 + x_2$$

The corresponding matrix form is

$$\begin{bmatrix} \dot{x}_1 \\ \dot{x}_2 \end{bmatrix} = \begin{bmatrix} 0 & 1 \\ -6 & -5 \end{bmatrix} \begin{bmatrix} x_1 \\ x_2 \end{bmatrix} + \begin{bmatrix} 0 \\ 1 \end{bmatrix} f(t)$$

$$y = \begin{bmatrix} 6 & 1 \end{bmatrix} \begin{bmatrix} x_1 \\ x_2 \end{bmatrix}$$

Hence

$$A = \begin{bmatrix} 0 & 1 \\ -6 & -5 \end{bmatrix} \qquad b = \begin{bmatrix} 0 \\ 1 \end{bmatrix} \qquad c^T = \begin{bmatrix} 6 & 1 \end{bmatrix} \tag{9.66}$$

Because all the initial conditions are zero, $y_0 = \dot{y}_0 = f(0) = 0$. To obtain the initial conditions in terms of x, first differentiate the $y = 6x_1 + x_2$ equation. That is,

$$\dot{y} = 6\dot{x}_1 + \dot{x}_2 = 6x_2 + [-6x_1 - 5x_2 + f(t)]$$

$$= -6x_1 + x_2 + f(t)$$

Solving for x_1 and x_2 in terms of y and $\dot{y}$ and then evaluating at time $t = 0$ gives

$$x_1(0) = \frac{y_0 - \dot{y}_0 + f(0)}{12} \qquad x_2(0) = \frac{y_0 + \dot{y}_0 - f(0)}{2} \tag{9.67}$$

As would be expected when all initial conditions are zero, the initial values of the state variables are also zero.

In solving differential equations by the state-space method it is necessary to obtain the initial conditions in terms of the state variables. A general procedure for obtaining the initial conditions $x(0)$ is now described. For consistency of notation in this method, replace $\mathbf{c}^T$ by $\mathbf{c}_1^T$ so that the equation $y = \mathbf{c}^T\mathbf{x}$ becomes

$$y = \mathbf{c}_1^T\mathbf{x}$$

where $\mathbf{c}_1^T = \mathbf{c}^T$. Differentiation with respect to time yields

$$\dot{y} = \mathbf{c}_1^T\dot{\mathbf{x}} = \mathbf{c}_1^T[A\mathbf{x} + \mathbf{b}f(t)] = \mathbf{c}_1^T A\mathbf{x} + \mathbf{c}_1^T\mathbf{b}f(t)$$

$$= \mathbf{c}_2^T\mathbf{x} + d_2 f(t)$$

where $\mathbf{c}_2^T = \mathbf{c}_1^T A$ and $d_2 = \mathbf{c}_1^T\mathbf{b}$. The second derivative is

$$\ddot{y} = \mathbf{c}_2^T\dot{\mathbf{x}} + d_2 f'(t) = \mathbf{c}_2^T[A\mathbf{x} + \mathbf{b}f(t)] + d_2 f'(t)$$

$$= \mathbf{c}_3^T\mathbf{x} + d_3 f(t) + d_2 f'(t)$$

where $\mathbf{c}_3^T = \mathbf{c}_2^T A = \mathbf{c}_1^T A^2$ and $d_3 = \mathbf{c}_2^T\mathbf{b}$. The $(n-1)$ derivative is

$$y^{n-1} = \mathbf{c}_n^T\mathbf{x} + d_n f(t) + d_{n-1}f'(t) + \cdots + d_2 f^{n-1}(t)$$

where $\mathbf{c}_n^T = \mathbf{c}_{n-1}^T A = \mathbf{c}_1^T A^{n-1}$ and $d_n = \mathbf{c}_{n-1}^T\mathbf{b}$. Evaluation of the preceding equations at time $t = 0$ and transposing all given initial conditions to the left side gives

$$y_0 = \mathbf{c}_1^T\mathbf{x}(0)$$

$$\dot{y}_0 - d_2 f(0) = \mathbf{c}_2^T\mathbf{x}(0)$$

$$\cdots\cdots\cdots\cdots\cdots\cdots\cdots\cdots\cdots\cdots$$

$$y_0^{n-1} - d_n f(0) - \cdots - d_2 f^{n-1}(0) = \mathbf{c}_n^T\mathbf{x}(0)$$

These equations may be written in the matrix form

$$\begin{bmatrix} y_1(0) \\ y_2(0) \\ \vdots \\ y_n(0) \end{bmatrix} = \begin{bmatrix} c_{11} & c_{12} & \cdots & c_{1n} \\ c_{21} & c_{22} & \cdots & c_{2n} \\ \cdots\cdots\cdots\cdots\cdots\cdots \\ c_{n1} & c_{n2} & \cdots & c_{nn} \end{bmatrix} \begin{bmatrix} x_1(0) \\ x_2(0) \\ \vdots \\ x_n(0) \end{bmatrix}$$

where $y_1(0) = y_0$, $y_2(0) = \dot{y}_0 - d_2 f(0)$, $y_n(0) = y_0^{n-1} - d_n f(0) - \cdots - d_2 f^{n-1}(0)$, $\mathbf{c}_1^T = [c_{11}\, c_{12}\, \cdots\, c_{1n}]$, $\mathbf{c}_2^T = [c_{21}\, c_{22}\, \cdots\, c_{2n}]$, and $\mathbf{c}_n^T = [c_{n1}\, c_{n2}\, \cdots\, c_{nn}]$.

The preceding system may be written in the compact form

$$\mathbf{y}(0) = C\mathbf{x}(0)$$

where

$$C = \begin{bmatrix} c_{11} & c_{12} & \cdots & c_{1n} \\ c_{21} & c_{22} & \cdots & c_{2n} \\ \cdots\cdots\cdots\cdots\cdots\cdots \\ c_{n1} & c_{n2} & \cdots & c_{nn} \end{bmatrix}$$

Premultiplying by C^{-1} yields the desired result

$$\mathbf{x}(0) = C^{-1}\mathbf{y}(0)$$

For the preceding example, $n = 2$. Thus,

$$\mathbf{c}_1^T = \mathbf{c}^T = [6 \quad 1]$$

$$\mathbf{c}_2^T = \mathbf{c}_1^T A = [6 \quad 1]\begin{bmatrix} 0 & 1 \\ -6 & -5 \end{bmatrix} = [-6 \quad 1]$$

$$d_2 = \mathbf{c}_1^T \mathbf{b} = [6 \quad 1]\begin{bmatrix} 0 \\ 1 \end{bmatrix} = 1$$

Hence

$$C = \begin{bmatrix} 6 & 1 \\ -6 & 1 \end{bmatrix} \quad \text{and} \quad C^{-1} = \frac{\begin{bmatrix} 1 & -1 \\ 6 & 6 \end{bmatrix}}{12}$$

For $y_1(0) = y_0$ and $y_2(0) = \dot{y}_0 - d_2 f(0) = \dot{y}_0 - f(0)$, then

$$\begin{bmatrix} x_1(0) \\ x_2(0) \end{bmatrix} = \frac{\begin{bmatrix} 1 & -1 \\ 6 & 6 \end{bmatrix}}{12}\begin{bmatrix} y_0 \\ \dot{y}_0 - f(0) \end{bmatrix}$$

This yields the same result for $x_1(0)$ and $x_2(0)$ given by Eqs. (9.67).
The $[sI - A]$ matrix is

$$[sI - A] = \begin{bmatrix} s & -1 \\ 6 & s+5 \end{bmatrix}$$

The inverse is

$$\Phi(s) = [sI - A]^{-1} = \frac{\begin{bmatrix} s+5 & 1 \\ -6 & s \end{bmatrix}}{s^2 + 5s + 6}$$

Thus,

$$\Phi(s)\mathbf{b}F(s) = \frac{\begin{bmatrix} s+5 & 1 \\ -6 & s \end{bmatrix}}{s^2 + 5s + 6}\begin{bmatrix} 0 \\ 1 \end{bmatrix}F(s) = \frac{\begin{bmatrix} 1 \\ s \end{bmatrix}}{s^2 + 5s + 6}F(s)$$

Application of Eq. (9.62) gives

$$\begin{bmatrix} X_1(s) \\ X_2(s) \end{bmatrix} = \frac{\begin{bmatrix} s+5 & 1 \\ -6 & s \end{bmatrix}}{s^2 + 5s + 6}\begin{bmatrix} x_1(0) \\ x_2(0) \end{bmatrix} + \frac{\begin{bmatrix} 1 \\ s \end{bmatrix}}{s^2 + 5s + 6}F(s) \qquad (9.68)$$

For $\mathbf{x}(0) = \mathbf{0}$ and $F(s) = 1/s$, then

$$\begin{bmatrix} x_1(t) \\ x_2(t) \end{bmatrix} = \mathscr{L}^{-1} \begin{bmatrix} \dfrac{1}{s(s+2)(s+3)} \\ \dfrac{1}{(s+2)(s+3)} \end{bmatrix} = \begin{bmatrix} (1 - 3e^{-2t} + 2e^{-3t})/6 \\ (e^{-2t} - e^{-3t}) \end{bmatrix}$$

The desired solution is

$$y = 6x_1 + x_2 = (1 - 3e^{-2t} + 2e^{-3t}) + (e^{-2t} - e^{-3t}) = 1 - 2e^{-2t} + e^{-3t} \quad (9.69)$$

Signal-Flow-Graph Method

To apply this method, first take the Laplace transform of the state-space equation, $\dot{\mathbf{x}} = A\mathbf{x} + \mathbf{b}f(t)$. That is,

$$s\mathbf{X}(s) - \mathbf{x}(0) = A\mathbf{X}(s) + \mathbf{b}F(s)$$

The signal-flow graph is then constructed by writing this in the form

$$\mathbf{X}(s) = s^{-1}\mathbf{x}(0) + s^{-1}A\mathbf{X}(s) + s^{-1}\mathbf{b}F(s)$$

The input nodes for the resulting graph are the initial conditions $\mathbf{x}(0)$ and the transform of the input $F(s)$. The output nodes are the transform of the state variables $\mathbf{X}(s)$. Application of Mason's gain formula yields directly the following matrix:

$$\begin{bmatrix} X_1(s) \\ X_2(s) \\ \vdots \\ X_n(s) \end{bmatrix} = \begin{bmatrix} \phi_{11}(s) & \phi_{12}(s) & \cdots & \phi_{1n}(s) \\ \phi_{21}(s) & \phi_{22}(s) & \cdots & \phi_{2n}(s) \\ \cdots\cdots\cdots\cdots\cdots\cdots\cdots \\ \phi_{n1}(s) & \phi_{n2}(s) & \cdots & \phi_{nn}(s) \end{bmatrix} \begin{bmatrix} x_1(0) \\ x_2(0) \\ \vdots \\ x_n(0) \end{bmatrix} + \begin{bmatrix} \psi_1(s) \\ \psi_2(s) \\ \vdots \\ \psi_n(s) \end{bmatrix} F(s) \quad (9.70)$$

Inverting yields the desired solution. Note that each element $\phi_{ij}(s)$ and $\psi_i(s)$ is obtained individually from the signal-flow graph.

Illustrative example 9.8 Use the signal-flow-graph method to determine the solution of Illustrative example 9.7.

SOLUTION The Laplace transform of the state-space equation is

$$sX_1(s) - x_1(0) = X_2(s)$$
$$sX_2(s) - x_2(0) = -6X_1(s) - 5X_2(s) + F(s)$$

Taking the initial conditions to the right side and dividing through by s gives

$$X_1(s) = s^{-1}x_1(0) + s^{-1}X_2(s)$$
$$X_2(s) = s^{-1}x_2(0) + s^{-1}[-6X_1(s) - 5X_2(s) + F(s)]$$

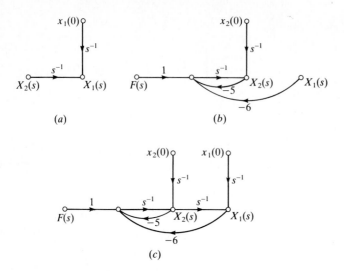

Figure 9.20 Signal-flow graph.

The signal-flow graph for the first equation is shown in Fig. 9.20a and the graph for the second equation is shown in Fig. 9.20b. These graphs are combined to yield the resultant graph for the system, shown in Fig. 9.20c. Application of Mason's gain formula to this graph yields

$$\begin{bmatrix} X_1(s) \\ X_2(s) \end{bmatrix} = \begin{bmatrix} \dfrac{s+5}{s^2+5s+6} & \dfrac{1}{s^2+5s+6} \\ \dfrac{-6}{s^2+5s+6} & \dfrac{s}{s^2+5s+6} \end{bmatrix} \begin{bmatrix} x_1(0) \\ x_2(0) \end{bmatrix} + \begin{bmatrix} \dfrac{1}{s^2+5s+6} \\ \dfrac{s}{s^2+5s+6} \end{bmatrix} F(s)$$

The first matrix on the right side is the $[sI - A]^{-1}$ matrix and the last matrix is the $[sI - A]^{-1}\mathbf{b}$ matrix. These are the same results as were obtained by the Laplace transform method [Eqs. (9.68)]. A major feature of the signal-flow-graph method is that it eliminates the need to obtain the matrix inverse.

Illustrative example 9.9 Use the signal-flow-graph method to determine the solution of the differential equation

$$\ddot{c} + 6\dot{c} + 25c = 100f(t) \tag{9.71}$$

The forcing function $f(t)$ is a unit step function and all initial conditions are zero.

SOLUTION Letting $c = x_1$ and $\dot{c} = \dot{x}_1 = x_2$, the state-space representation is

$$\dot{x}_1 = x_2$$
$$\dot{x}_2 = -25x_1 - 6x_2 + 100f(t)$$

Figure 9.21 Signal-flow graph.

Transforming gives

$$sX_1(s) - x_1(0) = X_2(s)$$

$$sX_2(s) - x_2(0) = -25X_1(s) - 6X_2(s) + 100F(s)$$

The form for constructing the signal-flow graph is

$$X_1(s) = s^{-1}x_1(0) + s^{-1}X_2(s)$$

$$X_2(s) = s^{-1}x_2(0) + s^{-1}[-25X_1(s) - 6X_2(s) + 100F(s)]$$

The resulting signal-flow graph is shown in Fig. 9.21.
Application of Mason's gain formula gives

$$\begin{bmatrix} X_1(s) \\ X_2(s) \end{bmatrix} = \frac{\begin{bmatrix} s+6 & 1 \\ -25 & s \end{bmatrix}}{s^2 + 6s + 25} \begin{bmatrix} x_1(0) \\ x_2(0) \end{bmatrix} + \begin{bmatrix} 100 \\ 100s \end{bmatrix} \frac{F(s)}{s^2 + 6s + 25} \tag{9.72}$$

For $x_1(0) = x_2(0) = 0$ and $F(s) = 1/s$, the inverse yields the desired result

$$c = x_1 = 4 + 5e^{-3t} \sin (4t - 126.9°)$$
$$\dot{c} = x_2 = 25e^{-3t} \sin 4t \tag{9.73}$$

Note that no special considerations are needed in applying this method to complex conjugate roots.

9.6 TRANSFER FUNCTIONS

The state-space representation for a system to be controlled is shown in Fig. 9.22a. The system to be controlled is referred to as the plant. In state-space notation the characteristics of the plant are given by Eqs. (9.20) and (9.21):

$$\dot{x} = Ax + bu \tag{9.20}$$

$$y = c^T x \tag{9.21}$$

The corresponding transfer function representation for the plant is shown in Fig. 9.22b. In Sec. 9.1, techniques for obtaining various state-space representations for a system were illustrated. It is now shown how the transfer function

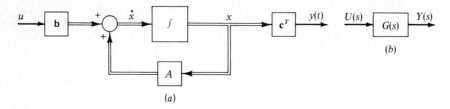

Figure 9.22 State-space representation and corresponding transfer function.

may be obtained from the state-space representation. The Laplace transform of the preceding matrix equations is

$$sX(s) - x(0) = AX(s) + bU(s)$$

$$Y(s) = c^T X(s)$$

In obtaining the transfer function, the initial conditions are ignored. Thus, the first relationship becomes

$$[sI - A]X(s) = bU(s)$$

Hence

$$X(s) = [sI - A]^{-1} bU(s) = \Phi(s)bU(s) \tag{9.74}$$

where $\Phi(s) = [sI - A]^{-1}$ is the state-transition matrix for the plant. Substitution of this result into the second transformed equation gives

$$Y(s) = c^T \Phi(s)bU(s) = G(s)U(s)$$

Hence

$$G(s) = c^T \Phi(s)b \tag{9.75}$$

Because $\Phi(s) = \text{adj } (sI - A)/|sI - A|$, the preceding may be written in the form

$$G(s) = \frac{Y(s)}{U(s)} = \frac{c^T[\text{adj } (sI - A)]b}{|sI - A|}$$

The numerator polynomial for $G(s)$ is $c^T[\text{adj } (sI - A)]b$. The denominator, which is the characteristic function, is $|sI - A|$. The roots of the characteristic equation $|sI - A| = 0$ are the eigenvalues for the system.

For a given plant $G(s)$, there are an infinite number of state-space representations. The transformation $x = Tz$ transforms Eqs. (9.20) and (9.21) into the form

$$\dot{z} = T^{-1}ATz + T^{-1}bu$$

$$y = c^T Tz$$

This new state-space representation for the plant may be written in the form

$$\dot{z} = A^*z + b^*u$$

$$y = c^{*T}z$$

where $A^* = T^{-1}AT$, $\mathbf{b}^* = T^{-1}\mathbf{b}$, and $\mathbf{c}^{*T} = \mathbf{c}^T T$ or $\mathbf{c}^* = T^T \mathbf{c}$. Note that these equations have the same form as Eqs. (9.20) and (9.21) but are different matrix representations for the same plant. In this analysis, T is any matrix which possesses an inverse T^{-1}.

The characteristic function for the plant transfer function $G(s)$ is $|sI - A|$. Because the characteristic equation is a unique property of the system and does not depend on the particular matrix representation, it follows that

$$|sI - A| = |sI - A^*|$$

To prove the preceding relationship, note that

$$|sI - A^*| = |sI - T^{-1}AT| = |sIT^{-1}T - T^{-1}AT|$$
$$= |T^{-1}(sI - A)T|$$

Because the determinant of a product is equal to the product of the determinants,

$$|sI - A^*| = |T^{-1}||sI - A||T| = |T^{-1}||T||sI - A|$$
$$= |T^{-1}T||sI - A| = |sI - A|$$

Similarly, the transfer function for the plant $G(s) = \mathbf{c}^T[sI - A]^{-1}\mathbf{b}$ is unaffected by the particular matrix representation for the plant. Thus,

$$\mathbf{c}^T[sI - A]^{-1}\mathbf{b} = \mathbf{c}^{*T}[sI - A^*]^{-1}\mathbf{b}^*$$

The preceding relationship is proved as follows:

$$\mathbf{c}^{*T}[sI - A^*]^{-1}\mathbf{b} = \mathbf{c}^T T[sI - T^{-1}AT]^{-1}T^{-1}\mathbf{b} = \mathbf{c}^T[T(sI - T^{-1}AT)T^{-1}]^{-1}\mathbf{b}$$
$$= \mathbf{c}^T[sTT^{-1} - A]^{-1}\mathbf{b} = \mathbf{c}^T[sI - A]^{-1}\mathbf{b} = G(s)$$

From Eq. (9.66), for the case of Illustrative example 9.7,

$$\mathbf{c} = \begin{bmatrix} 6 \\ 1 \end{bmatrix} \qquad A = \begin{bmatrix} 0 & 1 \\ -6 & -5 \end{bmatrix} \qquad \mathbf{b} = \begin{bmatrix} 0 \\ 1 \end{bmatrix}$$

Thus,

$$[sI - A] = \begin{bmatrix} s & -1 \\ 6 & s+5 \end{bmatrix} \qquad \text{adj } [sI - A] = \begin{bmatrix} s+5 & 1 \\ -6 & s \end{bmatrix}$$

$$|sI - A| = \begin{vmatrix} s & -1 \\ 6 & s+5 \end{vmatrix} = s^2 + 5s + 6$$

Application of Eq. (9.75) yields

$$G(s) = \frac{[6 \quad 1]\begin{bmatrix} s+5 & 1 \\ -6 & s \end{bmatrix}\begin{bmatrix} 0 \\ 1 \end{bmatrix}}{s^2 + 5s + 6} = \frac{s+6}{s^2 + 5s + 6}$$

The differential equation for this system [Eq. (9.64)] may be written in the operator form:

$$c(t) = \frac{D + 6}{D^2 + 5D + 6} f(t)$$

Replacing the operator D by s verifies the result obtained for $G(s)$.

Closed-Loop Systems

A general block-diagram representation for a closed-loop system with linear state-variable feedback is shown in Fig. 9.23a. The corresponding state-space representation is shown in Fig. 9.23b. For linear state-variable feedback, the control input u is the difference between the reference input r and a weighted sum of the state variables. That is,

$$u = K[r - (k_1 x_1 + k_2 x_2 + \cdots + k_n x_n)]$$

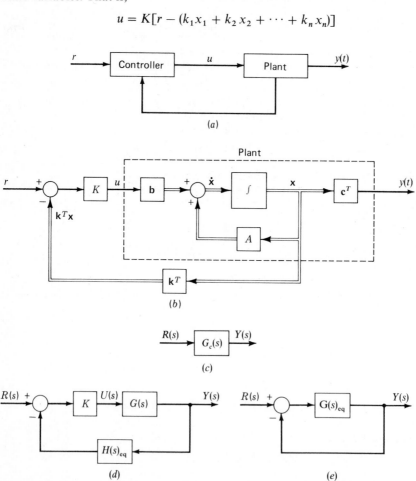

Figure 9.23 Closed-loop system with linear state-variable feedback.

In matrix notation, this relationship is

$$u = K(r - \mathbf{k}^T\mathbf{x})$$

To obtain the transfer function for the closed-loop system, first substitute u into Eq. (9.20). Thus,

$$\dot{\mathbf{x}} = (A - K\mathbf{b}\mathbf{k}^T)\mathbf{x} + K\mathbf{b}r$$

$$y = \mathbf{c}^T\mathbf{x}$$

Transforming each expression, then solving the first for $\mathbf{X}(s)$ and substituting this result into the second gives

$$Y(s) = \mathbf{c}^T\mathbf{X}(s) = K\mathbf{c}^T[sI - (A - K\mathbf{b}\mathbf{k}^T)]^{-1}\mathbf{b}R(s)$$

The closed-loop transfer function $Y(s)/R(s)$ may be written in the form

$$\frac{Y(s)}{R(s)} = K\mathbf{c}^T[sI - A_c]^{-1}\mathbf{b} = K\mathbf{c}^T\Phi_c(s)\mathbf{b} \qquad (9.76)$$

where $A_c = [A - K\mathbf{b}\mathbf{k}^T]$ and $\Phi_c(s) = [sI - A_c]^{-1}$ is the state-transition matrix for the closed-loop system. Note that the state-transition matrix for the closed-loop system is the same as that for the plant except that A is replaced by $A_c = [A - K\mathbf{b}\mathbf{k}^T]$. The transfer function representation $Y(s) = G_c(s)R(s)$ for the closed-loop system is shown in Fig. 9.23c. The closed-loop transfer function may be written in the form

$$G_c(s) = \frac{Y(s)}{R(s)} = \frac{K\mathbf{c}^T[\text{adj }(sI - A_c)]\mathbf{b}}{|sI - A_c|}$$

This shows that the denominator which is the characteristic equation for the closed-loop system is $|sI - A_c|$.

In Fig. 9.23d is shown an equivalent block-diagram representation for the system of Fig. 9.23b. The feedforward elements are the same for both cases. The feedback signal which goes to the comparator in Fig. 9.23b is $\mathbf{k}^T\mathbf{x}$. The Laplace transform for this signal is $\mathbf{k}^T\mathbf{X}(s)$. For the equivalent system the signal going to the comparator is $H(s)_{eq} Y(s)$. Thus,

$$H(s)_{eq} Y(s) = \mathbf{k}^T\mathbf{X}(s)$$

Solving for $H(s)_{eq}$ shows that

$$H(s)_{eq} = \frac{\mathbf{k}^T\mathbf{X}(s)}{Y(s)} = \frac{\mathbf{k}^T\Phi(s)\mathbf{b}}{\mathbf{c}^T\Phi(s)\mathbf{b}} \qquad (9.77)$$

where for the plant $\mathbf{X}(s) = \Phi(s)\mathbf{b}U(s)$ and $Y(s) = \mathbf{c}^T\mathbf{X}(s) = \mathbf{c}^T\Phi(s)\mathbf{b}U(s)$. In Fig. 9.23e is shown the unity-feedback system which is equivalent to Fig. 9.23d. The input-output relationships for Fig. 9.23d and e are

$$\frac{Y(s)}{R(s)} = \frac{KG(s)}{1 + KG(s)H(s)_{eq}} \qquad \frac{Y(s)}{R(s)} = \frac{G(s)_{eq}}{1 + G(s)_{eq}} \qquad (9.78)$$

Equating these expressions and solving for $G(s)_{eq}$ yields

$$G(s)_{eq} = \frac{KG(s)}{1 + KG(s)H(s)_{eq} - KG(s)} = \frac{K\mathbf{c}^T\Phi(s)\mathbf{b}}{1 + K(\mathbf{k} - \mathbf{c})^T\Phi(s)\mathbf{b}} \quad (9.79)$$

where $G(s)H(s)_{eq} = [\mathbf{c}^T\Phi(s)\mathbf{b}][\mathbf{k}^T\Phi(s)\mathbf{b}]/[\mathbf{c}^T\Phi(s)\mathbf{b}] = \mathbf{k}^T\Phi(s)\mathbf{b}$. Thus, the transfer functions $G(s)$, $G(s)_{eq}$, and $H(s)_{eq}$ may be obtained from the state-space representation for a linear state-variable feedback control system.

9.7 MULTIVARIABLE SYSTEMS

Transfer functions yield much insight into the synthesis process. In this section, it is shown how transfer functions are combined with state-space techniques to provide a powerful method for the synthesis of multivariable systems.[2-5] This method was developed by Dr. Joseph L. Peczkowski.

A multivariable system is one in which there is more than one input and more than one output. The system to be controlled is referred to as the plant. The state-space representation for the plant is given by Eqs. (9.22) and (9.23). That is,

$$\dot{\mathbf{x}} = A\mathbf{x} + B\mathbf{u} \quad (9.22)$$

$$\mathbf{y} = C\mathbf{x} + D\mathbf{u} \quad (9.23)$$

The vector $\mathbf{u}$ represents the inputs to the plant, the vector $\mathbf{x}$ represents the state variables, and the vector $\mathbf{y}$ represents the outputs from the plant. For obtaining basic information such as the transfer function for a system, initial conditions are ignored. The Laplace transform for the preceding equations is

$$s\mathbf{X}(s) = A\mathbf{X}(s) + B\mathbf{U}(s)$$

$$\mathbf{Y}(s) = C\mathbf{X}(s) + D\mathbf{U}(s)$$

The corresponding block-diagram representation is shown in Fig. 9.24a. Note that $\mathbf{U}(s)$ is the input and $\mathbf{Y}(s)$ is the output. More than one signal travels along the double-width paths. The equivalent block-diagram representation in which $\mathbf{P}(s)$ is the transfer function relating the output $\mathbf{Y}(s)$ to the input $\mathbf{U}(s)$ is shown in Fig. 9.24b. Solving the first of the preceding equations for $\mathbf{X}(s)$ and then substituting this result into the expression for $\mathbf{Y}(s)$ gives

$$\mathbf{X}(s) = [sI - A]^{-1}B\mathbf{U}(s)$$

and
$$\mathbf{Y}(s) = [C(sI - A)^{-1}B + D]\mathbf{U}(s) = \mathbf{P}(s)\mathbf{U}(s)$$

Thus,

$$\mathbf{P}(s) = C(sI - A)^{-1}B + D \quad (9.80)$$

Because there may be many different state-space representations for a given plant, the matrices A, B, C, and D are not unique. However, there is but one transfer function for a given plant. Thus, the transfer function matrix $\mathbf{P}(s)$ is unique.

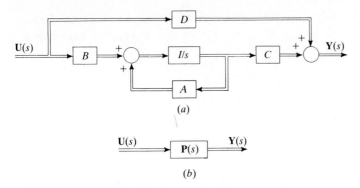

Figure 9.24 State-space representation for a plant.

In Fig. 9.25*a* is shown a multivariable system. From this block diagram, it is seen that the individual equations are

$$\mathbf{E}(s) = \mathbf{R}(s) - \mathbf{Y}(s)$$

$$\mathbf{U}(s) = \mathbf{G}(s)\mathbf{E}(s)$$

$$\mathbf{Y}(s) = \mathbf{P}(s)\mathbf{U}(s)$$

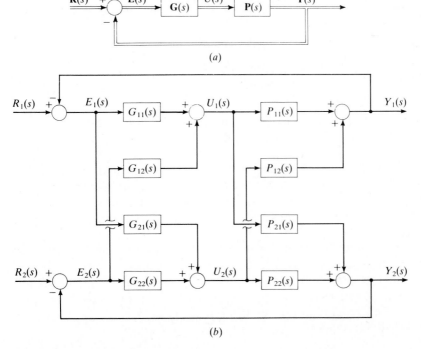

Figure 9.25 Multivariable system.

For the case in which there are two system inputs R_1 and R_2, two plant inputs U_1 and U_2, and two system outputs Y_1 and Y_2, the preceding equations become

$$\begin{bmatrix} E_1(s) \\ E_2(s) \end{bmatrix} = \begin{bmatrix} R_1(s) \\ R_2(s) \end{bmatrix} - \begin{bmatrix} Y_1(s) \\ Y_2(s) \end{bmatrix}$$

$$\begin{bmatrix} U_1(s) \\ U_2(s) \end{bmatrix} = \begin{bmatrix} G_{11}(s) & G_{12}(s) \\ G_{21}(s) & G_{22}(s) \end{bmatrix} \begin{bmatrix} E_1(s) \\ E_2(s) \end{bmatrix}$$

$$\begin{bmatrix} Y_1(s) \\ Y_2(s) \end{bmatrix} = \begin{bmatrix} P_{11}(s) & P_{12}(s) \\ P_{21}(s) & P_{22}(s) \end{bmatrix} \begin{bmatrix} U_1(s) \\ U_2(s) \end{bmatrix}$$

All of the preceding relationships are contained in the resulting block diagram shown in Fig. 9.25b. The equations for the $\mathbf{E}(s) = \mathbf{R}(s) - \mathbf{Y}(s)$ matrix are

$$E_1(s) = R_1(s) - Y_1(s)$$

$$E_2(s) = R_2(s) - Y_2(s)$$

The first summing point of the top loop yields $E_1(s)$ and the first summing point of the bottom loop yields $E_2(s)$. The equations for the $\mathbf{U}(s) = \mathbf{G}(s)\mathbf{E}(s)$ matrix are

$$U_1(s) = G_{11}(s)E_1(s) + G_{12}(s)E_2(s)$$

$$U_2(s) = G_{21}(s)E_1(s) + G_{22}(s)E_2(s)$$

The second summing point of the top loop yields $U_1(s)$ and the second summing point of the bottom loop yields $U_2(s)$. The equations for the $\mathbf{Y}(s) = \mathbf{P}(s)\mathbf{U}(s)$ matrix are

$$Y_1(s) = P_{11}(s)U_1(s) + P_{12}(s)U_2(s)$$

$$Y_2(s) = P_{21}(s)U_1(s) + P_{22}(s)U_2(s)$$

The last summing point of the top loop yields $Y_1(s)$ and the last summing point of the bottom loop yields $Y_2(s)$.

As an example of a multivariable system consider the control of a jet engine. The system inputs would be the desired value of the engine speed and the desired value of the burner exit temperature. The outputs would be the actual value of the engine speed and the actual value of the burner exit temperature. The inputs to the plant (i.e., jet engine) would be the fuel flow to the engine and the nozzle jet area. The state variables $(x_1, x_2, \ldots)$ would be various engine parameters such as engine speed, fuel flow, compressor discharge pressure, nozzle area, exit temperature, etc.

For a given system usually the plant characteristics are known. The problem then is to determine the characteristics of the controller to yield a desired relationship between the input and the output. The block-diagram representation for a closed-loop multivariable control system is shown in Fig. 9.26a. The equation for the feedforward portion of the system which relates $\mathbf{Y}(s)$ to $\mathbf{E}(s)$ is obtained as follows:

$$\mathbf{Y}(s) = \mathbf{P}(s)\mathbf{U}(s) = \mathbf{P}(s)\mathbf{G}(s)\mathbf{E}(s) = \mathbf{P}(s)\mathbf{G}(s)[\mathbf{R}(s) - \mathbf{Y}(s)]$$

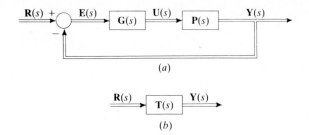

(a)

(b)

Figure 9.26 Closed-loop multivariable control system.

Solving for $\mathbf{Y}(s)$ shows that

$$[I + \mathbf{P}(s)\mathbf{G}(s)]\mathbf{Y}(s) = \mathbf{P}(s)\mathbf{G}(s)\mathbf{R}(s)$$

Thus,

$$\mathbf{Y}(s) = [I + \mathbf{P}(s)\mathbf{G}(s)]^{-1}\mathbf{P}(s)\mathbf{G}(s)\mathbf{R}(s) \tag{9.81}$$

An equivalent block-diagram representation for the control system of Fig. 9.26a is shown in Fig. 9.26b. The open-loop transfer function $\mathbf{T}(s)$ shown in Fig. 9.26b is called the transmittance. The transmittance $\mathbf{T}(s)$ relates the input to the output. That is,

$$\mathbf{Y}(s) = \mathbf{T}(s)\mathbf{R}(s) \tag{9.82}$$

Comparison of Eqs. (9.81) and (9.82) shows that

$$\mathbf{T}(s) = [I + \mathbf{P}(s)\mathbf{G}(s)]^{-1}\mathbf{P}(s)\mathbf{G}(s) \tag{9.83}$$

Equation (9.83) relates $\mathbf{G}(s)$, $\mathbf{T}(s)$, and $\mathbf{P}(s)$. To determine the equation for the controller characteristics $\mathbf{G}(s)$ in terms of the known plant characteristics $\mathbf{P}(s)$ and the desired transmittance $\mathbf{T}(s)$, first premultiply both sides of Eq. (9.83) by $[I + \mathbf{P}(s)\mathbf{G}(s)]$:

$$[I + \mathbf{P}(s)\mathbf{G}(s)]\mathbf{T}(s) = \mathbf{P}(s)\mathbf{G}(s)$$

This may be rewritten in the form

$$\mathbf{T}(s) = \mathbf{P}(s)\mathbf{G}(s) - \mathbf{P}(s)\mathbf{G}(s)\mathbf{T}(s) = \mathbf{P}(s)\mathbf{G}(s)[I - \mathbf{T}(s)]$$

Premultiplying both sides by $\mathbf{P}^{-1}(s)$ and postmultiplying both sides by $[I - \mathbf{T}(s)]^{-1}$ yields $\mathbf{G}(s)$ in terms of $\mathbf{T}(s)$ and $\mathbf{P}(s)$. That is,

$$\mathbf{G}(s) = \mathbf{P}^{-1}(s)\mathbf{T}(s)[I - \mathbf{T}(s)]^{-1} \tag{9.84}$$

This result enables the designer to determine the controller characteristics $\mathbf{G}(s)$ to yield a desired transmittance $\mathbf{T}(s)$ relating the output to the input for a given plant $\mathbf{P}(s)$.

The quantity $\mathbf{T}(s)[I - \mathbf{T}(s)]^{-1}$ in Eq. (9.84) is rather awkward and cumbersome. A more convenient form is now developed. In Fig. 9.27 is shown an equivalent representation for the multivariable control system shown in Fig. 9.26a. Comparison of Fig. 9.27 with Fig. 9.26a shows that

$$\mathbf{Y}(s) = \mathbf{Q}(s)\mathbf{E}(s) = \mathbf{P}(s)\mathbf{G}(s)\mathbf{E}(s)$$

Figure 9.27 Equivalent representation for a multivariable control system.

Thus,

$$Q(s) = P(s)G(s)$$

Premultiplying by $P^{-1}(s)$ and then solving for $G(s)$ gives

$$G(s) = P^{-1}(s)Q(s) \qquad (9.85)$$

Comparison of Eqs. (9.84) and (9.85) shows that

$$Q(s) = T(s)[I - T(s)]^{-1} \qquad (9.86)$$

The matrix $Q(s)$ is called the performance matrix. The performance matrix $Q(s)$ is thus seen to depend on the response matrix $T(s)$. When the number of inputs is equal to the number of outputs, $T(s)$ is a square matrix. Furthermore, when the system is decoupled (one input controls one output only) $T(s)$ is a diagonal matrix. That is,

$$T(s) = \begin{bmatrix} t_{11} & 0 & \cdots & 0 \\ 0 & t_{22} & \cdots & 0 \\ \multicolumn{4}{c}{\dotfill} \\ 0 & 0 & \cdots & t_{nn} \end{bmatrix}$$

The matrix $Q(s)$ has the same decoupling properties as $T(s)$. That is, if $T(s)$ is a diagonal, block diagonal, or triangular matrix, then $Q(s)$ will also be a diagonal, block diagonal, or triangular matrix.

When $T(s)$ and thus $Q(s)$ are diagonal matrices each element $q_{ii}(s)$ on the diagonal of $Q(s)$ is related to the corresponding element $t_{ii}(s)$ on the diagonal of $T(s)$ by the relationship

$$q_{ii}(s) = \frac{t_{ii}(s)}{1 - t_{ii}(s)} \qquad (9.87)$$

In Table 9.1 are shown corresponding values of elements of $Q(s)$ and $T(s)$.

Table 9.1 Unity-feedback–decoupled response

$q_{ii}(s)$	$t_{ii}(s)$
$\dfrac{1}{\tau s}$	$\dfrac{1}{1 + \tau s}$
$\dfrac{1/(\tau_1 + \tau_2)}{s\{1 + [\tau_1 \tau_2/(\tau_1 + \tau_2)]s\}}$	$\dfrac{1}{(1 + \tau_1 s)(1 + \tau_2 s)}$
$\dfrac{\omega_n/2\zeta}{s\{1 + [1/(2\zeta\omega_n)]s\}}$	$\dfrac{\omega_n^2}{s^2 + 2\zeta\omega_n s + \omega_n^2}$
$\dfrac{1/(\tau_1 + \tau_2 + \tau_3)}{s\{1 + [\tau_1 \tau_2/(\tau_1 + \tau_2 + \tau_3)]s\}(1 + \tau_3 s)}$	$\dfrac{1}{(1 + \tau_1 s)(1 + \tau_2 s)(1 + \tau_3 s)} \qquad \tau_1 \geq \tau_2 \geq \tau_3$

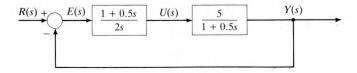

Figure 9.28 Single-input–single-output system.

Illustrative example 9.10 The transfer function for the plant of a single-input–single-output system is

$$P(s) = \frac{5}{1 + 0.5s}$$

Determine the controller characteristic $G(s)$ such that the transmittance $T(s)$ is

$$T(s) = \frac{1}{1 + 0.4s}$$

SOLUTION From Table 9.1, for $t_{ii}(s) = 1/(1 + \tau s)$ then $q_{ii}(s) = 1/\tau s$. Thus the performance matrix is

$$Q(s) = \frac{1}{0.4s} = \frac{2.5}{s}$$

The desired controller characteristic is

$$G(s) = P^{-1}(s)Q(s) = \frac{1 + 0.5s}{5} \frac{2.5}{s} = \frac{1 + 0.5s}{2s}$$

The resulting system is shown in Fig. 9.28. Notice that the numerator $(1 + 0.5s)$ of $G(s)$ cancels the denominator of $P(s)$. With this method, a considerable amount of cancellation is inherent. This tends to simplify the resulting system.

Illustrative example 9.11 The transfer function for a plant is

$$\mathbf{P}(s) = \begin{bmatrix} \dfrac{5}{1 + 2s} & \dfrac{50}{1 + 2s} \\ \dfrac{1}{1 + s} & -\dfrac{10}{1 + s} \end{bmatrix}$$

The desired transmittance is

$$\mathbf{T}(s) = \begin{bmatrix} \dfrac{1}{1 + 0.5s} & 0 \\ 0 & \dfrac{1}{1 + 0.2s} \end{bmatrix}$$

Determine the required controller $\mathbf{G}(s)$.

SOLUTION From Table 9.1, for $t_{ii}(s) = 1/(1 + \tau s)$ then $q_{ii}(s) = 1/\tau s$. Thus the performance matrix is

$$\mathbf{Q}(s) = \begin{bmatrix} 2/s & 0 \\ 0 & 5/s \end{bmatrix}$$

The transfer function $\mathbf{G}(s)$ for the required controller is

$$\mathbf{G}(s) = \mathbf{P}^{-1}(s)\mathbf{Q}(s) = \begin{bmatrix} 0.10(1 + 2s) & 0.50(1 + s) \\ 0.01(1 + 2s) & -0.05(1 + s) \end{bmatrix} \begin{bmatrix} 2/s & 0 \\ 0 & 5/s \end{bmatrix}$$

$$= \begin{bmatrix} \dfrac{0.20(1 + 2s)}{s} & \dfrac{2.50(1 + s)}{s} \\ \dfrac{0.02(1 + 2s)}{s} & -\dfrac{0.25(1 + s)}{s} \end{bmatrix}$$

Sensitivity Analysis

In practice the transfer function of the plant will change in time due to aging, wear of parts, etc. This will cause the transmittance $\mathbf{T}(s)$ to change also. For example, if $\mathbf{P}(s)$ for the system of Illustrative example 9.10 (see Fig. 9.28) were to change from $5/(1 + 0.5s)$ to $2/(1 + 0.5s)$, then the transmittance would change from $1/(1 + 0.4s)$ to $1/(1 + s)$. The response of the original system to a unit step function is shown by the solid curve in Fig. 9.29. The response for the changed value of $\mathbf{P}(s)$ is shown by the dashed curve. The effect of variations in the plant characteristics $\mathbf{P}(s)$ upon the transmittance $\mathbf{T}(s)$ may be minimized by a sensitivity analysis.

In Fig. 9.30 is shown the multiple-input–multiple-output system of Fig. 9.26a in which the feedback element $\mathbf{H}(s)$ has been added. For a change $\Delta\mathbf{P}(s)$ in the transfer function of the plant, the new transfer function will be $\mathbf{P}(s) + \Delta\mathbf{P}(s)$. Thus

Figure 9.29 Step-function response.

Figure 9.30 Multivariable system with feedback element $\mathbf{H}(s)$.

the new system output will be

$$\mathbf{Y}(s) + \Delta\mathbf{Y}(s) = [\mathbf{P}(s) + \Delta\mathbf{P}(s)]\mathbf{U}(s) = [\mathbf{P}(s) + \Delta\mathbf{P}(s)]\mathbf{G}(s)\mathbf{E}(s)$$

$$= [\mathbf{P}(s) + \Delta\mathbf{P}(s)]\mathbf{G}(s)\{\mathbf{R}(s) - \mathbf{H}(s)[\mathbf{Y}(s) + \Delta\mathbf{Y}(s)]\}$$

Algebraically rearranging shows that

$$\{I + [\mathbf{P}(s) + \Delta\mathbf{P}(s)]\mathbf{G}(s)\mathbf{H}(s)\}[\mathbf{Y}(s) + \Delta\mathbf{Y}(s)] = [\mathbf{P}(s) + \Delta\mathbf{P}(s)]\mathbf{G}(s)\mathbf{R}(s)$$

For $\Delta\mathbf{P}(s) = \mathbf{0}$, this becomes

$$[I + \mathbf{P}(s)\mathbf{G}(s)\mathbf{H}(s)]\mathbf{Y}(s) = \mathbf{P}(s)\mathbf{G}(s)\mathbf{R}(s)$$

The change $\Delta\mathbf{Y}(s)$ due to a change $\Delta\mathbf{P}(s)$ is

$$[I + \mathbf{P}(s)\mathbf{G}(s)\mathbf{H}(s)]\Delta\mathbf{Y}(s) = \Delta\mathbf{P}(s)\mathbf{G}(s)\mathbf{R}(s) \qquad (9.88)$$

This result relates the change $\Delta\mathbf{Y}(s)$ due to a change $\Delta\mathbf{P}(s)$ in the transfer function of the plant.

In Fig. 9.31 is shown the system of Fig. 9.30 in which the feedback loop has been eliminated (i.e., the open-loop system). For this open-loop system the variation in the output due to a change in the plant is

$$[\mathbf{Y}(s) + \Delta\mathbf{Y}(s)]_{\text{OL}} = [\mathbf{P}(s) + \Delta\mathbf{P}(s)]\mathbf{G}(s)\mathbf{R}(s)$$

The subscript OL represents the output if the system were an open-loop system. For $\Delta\mathbf{P}(s) = \mathbf{0}$, the output is

$$\mathbf{Y}(s)_{\text{OL}} = \mathbf{P}(s)\mathbf{G}(s)\mathbf{R}(s)$$

The change in the open-loop output due to a change in the transfer function of the plant is

$$\Delta\mathbf{Y}(s)_{\text{OL}} = \Delta\mathbf{P}(s)\mathbf{G}(s)\mathbf{R}(s) \qquad (9.89)$$

The right sides of Eqs. (9.88) and (9.89) are identical. Thus,

$$[I + \mathbf{P}(s)\mathbf{G}(s)\mathbf{H}(s)]\Delta\mathbf{Y}(s) = \Delta\mathbf{Y}(s)_{\text{OL}}$$

Premultiplying by $[I + \mathbf{P}(s)\mathbf{G}(s)\mathbf{H}(s)]^{-1}$ and postmultiplying by $\Delta\mathbf{Y}(s)_{\text{OL}}^{-1}$ shows that

$$\Delta\mathbf{Y}(s)\,\Delta\mathbf{Y}(s)_{\text{OL}}^{-1} = [I + \mathbf{P}(s)\mathbf{G}(s)\mathbf{H}(s)]^{-1}$$

R(s) → G(s) → P(s) + ΔP(s) → Y(s) + ΔY(s)

Figure 9.31 Multivariable system without feedback.

For the case of a single-input–single-output system, this becomes

$$\frac{\Delta Y(s)}{\Delta Y(s)_{OL}} = [I + P(s)G(s)H(s)]^{-1}$$

The left side is the ratio of the closed-loop response variation to the open-loop response variation. This ratio is the sensitivity **S**. That is,

$$\mathbf{S} = [I + \mathbf{P}(s)\mathbf{G}(s)\mathbf{H}(s)]^{-1} \tag{9.90}$$

It is now shown how to select **H**(s) such that the system sensitivity is considerably less than it would be for a unity-feedback system. From Fig. 9.30, it follows that

$$\mathbf{Y}(s) = \mathbf{P}(s)\mathbf{G}(s)[\mathbf{R}(s) - \mathbf{H}(s)\mathbf{Y}(s)]$$

Hence

$$[I + \mathbf{P}(s)\mathbf{G}(s)\mathbf{H}(s)]\mathbf{Y}(s) = \mathbf{P}(s)\mathbf{G}(s)\mathbf{R}(s)$$

Premultiplying by $[I + \mathbf{P}(s)\mathbf{G}(s)\mathbf{H}(s)]^{-1} = \mathbf{S}$ shows that

$$\mathbf{Y}(s) = [I + \mathbf{P}(s)\mathbf{G}(s)\mathbf{H}(s)]^{-1}\mathbf{P}(s)\mathbf{G}(s)\mathbf{R}(s) = \mathbf{S}\mathbf{P}(s)\mathbf{G}(s)\mathbf{R}(s) \tag{9.91}$$

In Fig. 9.32a and b are shown equivalent representations for the system of Fig. 9.30. From Fig. 9.32b, it follows that

$$\mathbf{Y}(s) = \mathbf{T}(s)\mathbf{R}(s) \tag{9.92}$$

Equating the right sides of Eqs. (9.91) and (9.92) shows that

$$\mathbf{S}\mathbf{P}(s)\mathbf{G}(s) = \mathbf{T}(s) \tag{9.93}$$

Premultiplying by $\mathbf{S}^{-1}$ and then $\mathbf{P}^{-1}(s)$ gives

$$\mathbf{G}(s) = \mathbf{P}^{-1}(s)\mathbf{S}^{-1}\mathbf{T}(s) \tag{9.94}$$

Comparison of Figs. 9.32a and 9.30 shows that

$$\mathbf{Q}(s) = \mathbf{P}(s)\mathbf{G}(s) \tag{9.95}$$

Substitution of **G**(s) from Eq. (9.94) into Eq. (9.95) gives

$$\mathbf{Q}(s) = \mathbf{S}^{-1}\mathbf{T}(s) \tag{9.96}$$

(a)

(b)

Figure 9.32 Equivalent representation for multivariable system with feedback element **H**(s).

Substitution of $S^{-1}T(s)$ from Eq. (9.96) into Eq. (9.94) yields

$$G(s) = P^{-1}(s)Q(s) \qquad (9.97)$$

To obtain the equation for the required feedback elements $H(s)$, first postmultiply Eq. (9.90) by $[I + P(s)G(s)H(s)]$. That is,

$$S[I + P(s)G(s)H(s)] = S + SP(s)G(s)H(s) = I$$

In accordance with Eq. (9.93), $SP(s)G(s)$ may be replaced by $T(s)$. Thus,

$$S = I - T(s)H(s) \qquad (9.98)$$

Premultiplying by $T^{-1}(s)$ and solving for $H(s)$ gives

$$H(s) = T^{-1}(s)[I - S] \qquad (9.99)$$

The usual design procedure is first to determine the sensitivity for a unity-feedback system. For $H(s) = I$, Eq. (9.98) becomes

$$S = I - T(s) \qquad (9.100)$$

By knowing the desired improvement in sensitivity, the new value of S may be determined. For a given plant, the plant characteristics $P(s)$ are known. The transmittance $T(s)$ is selected to achieve the desired performance characteristics from the input to the output. Thus, by knowing S, $P(s)$, and $T(s)$, the required controller $G(s)$ and feedback $H(s)$ may be obtained from Eqs. (9.97) and (9.99).

Illustrative example 9.12 For the system of Illustrative example 9.10, determine $G(s)$ and $H(s)$ so as to decrease the sensitivity by a factor of 10.

SOLUTION For the unity-feedback system of Illustrative example 9.10, the sensitivity is

$$S = I - T(s) = 1 - \frac{1}{1 + 0.4s} = \frac{0.4s}{1 + 0.4s}$$

To decrease the sensitivity by a factor of 10, the new value of S is

$$S = \frac{0.04s}{1 + 0.4s}$$

The required values for $Q(s)$, $G(s)$, and $H(s)$ are

$$Q(s) = S^{-1}T(s) = \frac{1 + 0.4s}{0.04s} \frac{1}{1 + 0.4s} = \frac{25}{s}$$

$$G(s) = P^{-1}(s)Q(s) = \frac{1 + 0.5s}{5} \frac{25}{s} = \frac{5(1 + 0.5s)}{s}$$

$$H(s) = T^{-1}(s)[I - S] = (1 + 0.4s)\left(1 - \frac{0.04s}{1 + 0.4s}\right) = 1 + 0.36s$$

If the plant transfer function were to change from $5/(1 + 0.5s)$ to $2/(1 + 0.5s)$ the response of the resulting system to a unit step function is shown by the alternating dashed curve of Fig. 9.29. The change in the response due to the change in the plant characteristics has been minimized considerably.

PROBLEMS

9.1 For the differential equation

$$y(t) = \frac{1}{(D + 2)(D + 3)} f(t)$$

determine the computer diagram and state-space representation by
(a) Direct programming   (b) Parallel programming
(c) Series programming   (d) General programming

9.2 For the differential equation

$$y(t) = \frac{2(D + 5)}{(D + 2)(D + 3)} f(t)$$

determine the computer diagram and state-space representation by
(a) Direct programming   (b) Parallel programming
(c) Series programming   (d) General programming

9.3 For the differential equation

$$y(t) = \frac{2(D + 5)}{(D + 2)(D + 3)(D + 4)} f(t)$$

determine the computer diagram and state-space representation by
(a) Direct programming   (b) Parallel programming
(c) Series programming   (d) General programming

9.4 For the system of Prob. 9.1, construct the signal-flow graph corresponding to the computer diagram obtained by
(a) Direct programming   (b) Parallel programming
(c) Series programming   (d) General programming
Apply Mason's gain formula to verify that the signal-flow graph represents the differential equation.

9.5 For the system of Prob. 9.2, construct the signal-flow graph corresponding to the computer diagram obtained by
(a) Direct programming   (b) Parallel programming
(c) Series programming   (d) General programming
Apply Mason's gain formula to verify that the signal-flow graph represents the differential equation.

9.6 For the system of Prob. 9.3, construct the signal-flow graph corresponding to the computer diagram obtained by
(a) Direct programming   (b) Parallel programming
(c) Series programming   (d) General programming
Apply Mason's gain formula to verify that the signal-flow graph represents the differential equation.

9.7 Determine the state-space representation for the system shown in
(a) Fig P9.7a   (b) Fig. P9.7b

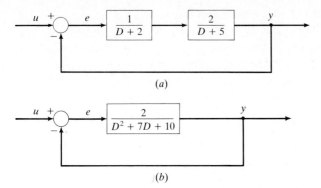

(a)

(b)

Figure P9.7

9.8 Determine the state-space representation for the system shown in
(a) Fig. P9.8a   (b) Fig. P9.8b

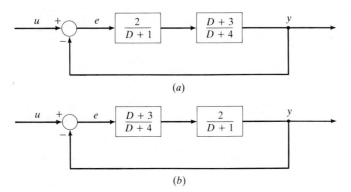

(a)

(b)

Figure P9.8

9.9 Determine the state-space representation for the system shown in
(a) Fig. P9.9a   (b) Fig. P9.9b

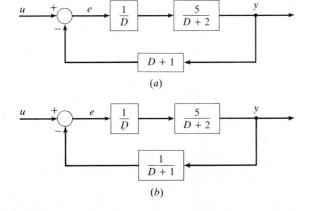

(a)

(b)   **Figure P9.9**

9.10 Determine the state-space representation for the system shown in
(a) Fig. P9.10a   (b) Fig. P9.10b

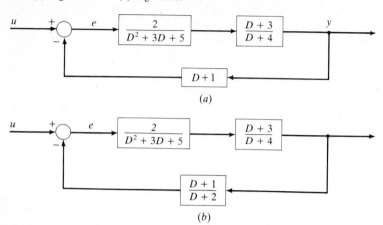

(a)

(b)

Figure P9.10

9.11 Construct the signal-flow graph for the system shown in
(a) Fig. P9.7a   (b) Fig. P9.7b
Apply Mason's gain formula to determine the differential equation relating the output $y(t)$ to the input $u(t)$.

9.12 Construct the signal-flow graph for the system shown in
(a) Fig. P9.8a   (b) Fig. P9.8b
Apply Mason's gain formula to determine the differential equation relating the output $y(t)$ to the input $u(t)$.

9.13 Construct the signal-flow graph corresponding to the block diagram shown in Fig. P9.13. Use Mason's gain formula to determine the equation relating
(a) $c(t)$ and $r(t)$ (b) $c(t)$ and $d(t)$

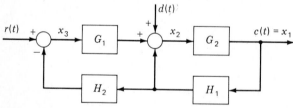

Figure P9.13

9.14 Construct the signal-flow graph corresponding to the block diagram shown in Fig. P9.14. Use Mason's gain formula to determine the equation relating $c(t)$ and $r(t)$.

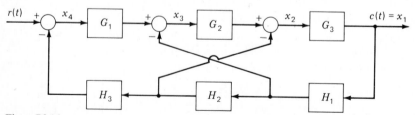

Figure P9.14

9.15 Determine the signal-flow graph corresponding to the block diagram shown in Fig. P9.15. Use Mason's gain formula to determine the equation relating $c(t)$ and $r(t)$.

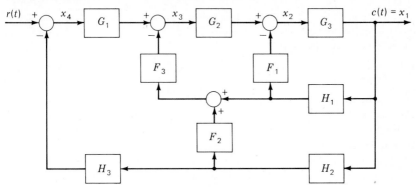

Figure P9.15

9.16 Determine the signal-flow graph corresponding to the block diagram shown in
   (a) Fig. 9.1b   (b) Fig. 9.2b   (c) Fig. 9.4
Use Mason's gain formula to determine the equation relating $y(t)$ and $f(t)$.

9.17 Obtain the signal-flow graph for the following simultaneous equations:

$$x_1 = 4x_2$$

$$x_2 = -x_1 + f(t)$$

Determine the equation relating
   (a) x_1 and $f(t)$ (b) x_2 and $f(t)$

9.18 Obtain the signal-flow graph for the following simultaneous equations:

$$x_1 = 4x_2 + u_1$$

$$x_2 = -x_1 + u_2$$

Determine the equation relating
   (a) x_1 to u_1 and u_2 (b) x_2 to u_1 and u_2

9.19 In Prob. 9.7, the state-space representation was obtained for the system shown in
   (a) Fig. P9.7a   (b) Fig. P9.7b
Use this representation and the Laplace transform method to determine the response $y(t)$ when $u(t) = 0$ and the initial conditions are $y_0 = 2$ and $\dot{y}_0 = 0$.

9.20 Consider the differential equation

$$\ddot{y} + 3\dot{y} + 2y = f(t)$$

Use direct programming and the Laplace transform method to determine the solution for the case in which $f(t) = 0$ and the initial conditions are $y_0 = 1$ and $\dot{y}_0 = 0$.

9.21 Consider the differential equation

$$\ddot{y} + 7\dot{y} + 10y = f(t)$$

Use parallel programming and the Laplace transform method to determine the solution for the case in which $f(t) = 0$ and the initial conditions are $y_0 = 3$ and $\dot{y}_0 = 0$.

9.22 Consider the differential equation

$$\ddot{y} + 6\dot{y} + 9y = f(t)$$

Use series programming and the Laplace transform method to determine the solution when $f(t) = 0$ and the initial conditions are $y_0 = 1$ and $\dot{y}_0 = 0$.

9.23 Consider the differential equation

$$\ddot{y} + 4\dot{y} + 5y = f'(t) + 5f(t)$$

Use general programming and the Laplace transform method to determine the solution for the case in which $f(t) = 0$ and the initial conditions are $y_0 = 1$ and $\dot{y}_0 = 0$.

9.24 Consider the differential equation

$$(D + 1)(D + 2)(D + 5)y(t) = f(t)$$

Use direct programming and the Laplace transform method to determine the solution for the case in which $f(t) = 0$ and the initial conditions are $y_0 = 6$ and $\dot{y}_0 = \ddot{y}_0 = 0$.

9.25 Use the signal-flow graph method to determine the solution of
    (a) Prob. 9.20   (b) Prob. 9.21   (c) Prob. 9.22
    (d) Prob. 9.23   (e) Prob. 9.24

9.26 Same as Prob. 9.20 except $f(t) = 2$ and all the initial conditions are zero.

9.27 Same as Prob. 9.21 except $f(t) = 30$ and all the initial conditions are zero.

9.28 Same as Prob. 9.22 except $f(t) = 9$ and all the initial conditions are zero.

9.29 Same as Prob. 9.23 except $f(t) = 1$ and all the initial conditions are zero.

9.30 Same as Prob. 9.24 except $f(t) = 60$ and all the initial conditions are zero.

9.31 Consider the differential equation

$$(D + 3)(D + 4)y(t) = (D + 6)f(t)$$

Use direct programming and the Laplace transform method to determine the solution when $f(t) = 2$ and the initial conditions are $y_0 = 1$, $\dot{y}_0 = 0$, and $f(0) = 0$.

9.32 Consider the differential equation

$$(D + 2)(D + 5)y(t) = (D + 10)f(t)$$

Use direct programming and the Laplace transform method to determine the solution when $f(t) = 1$ and the initial conditions are $y_0 = 1$, $\dot{y}_0 = 0$, and $f(0) = 0$.

9.33 Consider the differential equation

$$(D^2 + 9D + 20)y(t) = 2(D + 10)f(t)$$

Use series programming and the Laplace transform method to determine the response $y(t)$ when $f(t) = u(t)$ is a unit step function and the initial conditions are $y_0 = 1$ and $\dot{y}_0 = f(0) = 0$.

9.34 Consider the differential equation

$$(D + 2)(D + 5)y(t) = (D^2 + 3D + 10)f(t)$$

Use general programming and the Laplace transform method to determine the solution when $f(t) = 1$ and the initial conditions are $y_0 = 1$ and $\dot{y}_0 = f(0) = f'(0) = 0$.

9.35 Use the signal-flow graph method to determine the solution of
    (a) Prob. 9.31   (b) Prob. 9.32   (c) Prob. 9.33

9.36 For the system shown in Fig. 9.23b, $K = 2$, $\mathbf{c}^T = \mathbf{k}^T = [1 \quad 0]$,

$$A = \begin{bmatrix} 0 & 1 \\ -2 & -3 \end{bmatrix} \qquad \mathbf{b} = \begin{bmatrix} 0 \\ 1 \end{bmatrix}$$

Determine the transfer function $G(s)$ for the plant, $H(s)_{eq}$, $G(s)_{eq}$, and the closed-loop transfer function $G_c(s)$ for the system.

9.37 For the system shown in Fig. 9.23b, $K = 2$, $\mathbf{c}^T = \mathbf{k}^T = [-1 \quad 1]$,

$$A = \begin{bmatrix} -4 & 1 \\ 0 & -1 \end{bmatrix} \qquad \mathbf{b} = \begin{bmatrix} 0 \\ 1 \end{bmatrix}$$

Determine the transfer function $G(s)$ for the plant, $H(s)_{eq}$, $G(s)_{eq}$, and the closed-loop transfer function $G_c(s)$ for the system.

9.38 The state-space representation for the closed-loop control system shown in Fig. 9.23b is

$$A = \begin{bmatrix} 0 & 1 \\ 0 & -2 \end{bmatrix} \qquad b = \begin{bmatrix} 0 \\ 1 \end{bmatrix} \qquad K = 1$$

$$c^T = [5 \quad 0] \qquad k^T = [5 \quad 5]$$

Determine the transfer function $G(s)$ for the plant, $H(s)_{eq}$, $G(s)_{eq}$, and the closed-loop transfer function $G_c(s)$ for the system.

9.39 The transfer function for a plant is

$$P(s) = \frac{5}{1 + 2s}$$

Determine the controller $G(s)$ such that the transmittance will be

$$T(s) = \frac{25}{s^2 + 5s + 25}$$

9.40 The transfer function for a plant is

$$P(s) = \begin{bmatrix} \dfrac{4}{s+4} & -\dfrac{20}{s+4} \\[2mm] \dfrac{10}{s+2} & \dfrac{20}{s+2} \end{bmatrix}$$

The desired transmittance is

$$T(s) = \begin{bmatrix} \dfrac{4}{s^2 + 2s + 4} & 0 \\[2mm] 0 & \dfrac{1}{(1 + 0.2s)(1 + 0.5s)} \end{bmatrix}$$

Determine the required controller $G(s)$ for the system.

9.41 For the system of Prob. 9.39, determine $G(s)$ and $H(s)$ so as to reduce the sensitivity by a factor of 5.

9.42 For the system of Prob. 9.40, determine $\mathbf{G}(s)$ and $\mathbf{H}(s)$ so as to reduce the sensitivity by a factor of 10.

REFERENCES

1. Mason, S. J.: Feedback Theory—Some Properties of Signal-Flow Graphs, *Proc. IRE*, **41**(9), 1144–1156, September 1953.
2. Peczkowski, J. L.: Total Multivariable Synthesis with Transfer Functions, in "Proceedings Bendix Control and Control Theory Symposium," pp. 109–126, Bendix Executive Offices, April 1980.
3. Peczkowski, J. L., and Sain, M. K.: Linear Multivariable Synthesis with Transfer Functions, in "Alternatives for Multivariable Control" (Eds. M. K. Sain, J. L. Peczkowski, and J. L. Melsa), pp. 71–87, National Engineering Consortium, Chicago, 1978.
4. Sain, M. K., and Peczkowski, J. L.: Nonlinear Control by Coordinated Feedback Synthesis with Gas Turbine Applications, in "Proceedings American Control Conference," June 1985.
5. Peczkowski, J. L., and Sain, M. K.: Synthesis of System Responses: A Nonlinear Multivariable Control Design Approach, in "Proceedings American Control Conference," June 1985.

TEN

DIGITAL-CONTROL SYSTEMS

Recently, numerous systems have been devised which utilize digital computers as control elements. Such systems are generally sampled-data systems. That is, the information fed into a digital computer is the value (sample) of the corresponding signal at some instant of time. The computed output remains unchanged until new information (another sample) is fed into the digital computer.

Another well-known application of the sampled-data process is the time sharing of telemetered information from spacecraft. Instead of temperature at five different places in the vehicle and pressure at three different places being telemetered back on eight different channels, each signal uses the same channel for one second out of eight. A similar application occurs in the control of industrial processes in which system variables such as temperature, pressure, and flow are sampled periodically in order to make more effective use of the expensive control equipment, especially the digital computer. Another example is a guidance system that utilizes radar scanning in which a given sector, or region, is scanned once every revolution. Thus, in any given direction the signal is sampled at a rate equal to the scan rate of the radar. Sampling is inherent in some applications (e.g., radar scanning), it is necessary in other applications (e.g., to enter data into digital computers), and, finally, it is desirable for still other applications (e.g., time sharing).

10.1 SAMPLED-DATA SYSTEMS

A schematic representation of a sampler switch is shown in Fig. 10.1a. The switch closes every T seconds to admit the input signal. A typical input signal is represented by the continuous function $f(t)$ shown in Fig. 10.1b. The shaded pulses

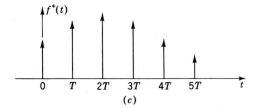

Figure 10.1 (*a*) Sampler switch, (*b*) continuous input $f(t)$ and output at switch (shaded pulses), (*c*) impulse approximation of switch output $f^*(t)$.

represent the signal at the output of the sampler. When the pulse duration is much shorter than the system time constants, the output of the sampler may be approximated by the train of impulses $f^*(t)$, as is illustrated in Fig. 10.1c. The term $f^*(t)$ is read "f star of t."

The area of each impulse is equal to the value of the input signal at the time $t = nT$ of the particular impulse. Thus, the area of the nth impulse which occurs at time $t = nT$ is $f(nT)$. The equation for the entire train of impulses is

$$f^*(t) = f(0)u_1(t) + f(T)u_1(t - T) + f(2T)u_1(t - 2T) + \cdots$$

$$= \sum_{n=0}^{\infty} f(nT)u_1(t - nT) \tag{10.1}$$

where $u_1(t)$ is a unit impulse at $t = 0$ and $u_1(t - nT)$ is a unit impulse at $t = nT$.

The Laplace transform of the sampled signal is

$$F^*(s) = \mathscr{L}[f^*(t)] = f(0) + f(T)e^{-Ts} + f(2T)e^{-2Ts} + \cdots$$

$$= \sum_{n=0}^{\infty} f(nT)e^{-nTs} \tag{10.2}$$

To illustrate the preceding concepts, consider the continuous input

$$f(t) = e^{-at}$$

The corresponding sampled signal $f*(t)$ is

$$f*(t) = u_1(t) + e^{-aT}u_1(t - T) + e^{-2aT}u_1(t - 2T) + \cdots$$

$$= \sum_{n=0}^{\infty} e^{-anT}u_1(t - nT)$$

The Laplace transform of the input $f(t)$ is

$$F(s) = \mathscr{L}[f(t)] = \frac{1}{s + a}$$

The Laplace transform of the sampled signal $f*(t)$ is

$$F*(s) = \mathscr{L}[f*(t)] = 1 + e^{-aT}e^{-sT} + e^{-2aT}e^{-2sT} + \cdots$$

$$= 1 + e^{-(s+a)T} + e^{-2(s+a)T} + \cdots$$

$$= \sum_{n=0}^{\infty} e^{-n(s+a)T}$$

The Laplace transform of a sampled signal is an infinite series. In the following, it is seen that the z transform greatly simplifies the analysis of such systems.

10.2 THE z TRANSFORM

The simple substitution

$$z = e^{Ts}$$

converts the Laplace transform to the z transform. Making this substitution in Eq. (10.2) gives

$$Z[f*(t)] = F(z) = f(0) + \frac{f(T)}{z} + \frac{f(2T)}{z^2} + \cdots$$

$$= \sum_{n=0}^{\infty} f(nT)z^{-n} \tag{10.3}$$

where $F(z)$ designates the z transform of $f*(t)$. Because only values of the signal at the sampling instants are considered, the z transform of $f(t)$ is the same as that of $f*(t)$.

Illustrative example 10.1 Determine the z transform for a unit step function. For this function $f(nT) = 1$ for $n = 0, 1, 2, \ldots$. Thus application of Eq. (10.3) gives

$$Z[u*(t)] = 1 + \frac{1}{z} + \frac{1}{z^2} + \cdots = \frac{z}{z - 1} \tag{10.4}$$

This series is convergent for $|z| > 1$. In solving problems by z transforms, the term z acts as a dummy operator. There is no need to specify the values of z over which $F(z)$ is convergent. It suffices to know that such values exist.

Illustrative example 10.2 Determine the z transform of the exponential e^{-at}.
For this function $f(nT) = e^{-anT}$. Thus,

$$Z(e^{-at}) = 1 + \frac{e^{-aT}}{z} + \left(\frac{e^{-aT}}{z}\right)^2 + \cdots = \frac{z}{z - e^{-aT}} \tag{10.5}$$

In Table 10.1 is given a partial listing of Laplace transforms and corresponding z transforms for commonly encountered functions.

When the Laplace transform of a function is known, the corresponding z transform may be obtained by the partial-fraction method.

Illustrative example 10.3 Determine the z transform for the function whose Laplace transform is

$$F(s) = \frac{1}{s(s + 1)} = \frac{1}{s} - \frac{1}{s + 1}$$

From Table 10.1, the z transform corresponding to $1/s$ is $z/(z - 1)$ and that corresponding to $1/(s + 1)$ is $z/(z - e^{-T})$. Thus,

$$F(z) = \frac{z}{z - 1} - \frac{z}{z - e^{-T}} = \frac{z(1 - e^{-T})}{(z - 1)(z - e^{-T})} \tag{10.6}$$

Table 10.1 The z transforms

Time function	Laplace transform	z transform
$u_1(t)$	1	1
$u(t)$	$\dfrac{1}{s}$	$\dfrac{z}{z - 1}$
t	$\dfrac{1}{s^2}$	$\dfrac{zT}{(z - 1)^2}$
$\dfrac{t^2}{2}$	$\dfrac{1}{s^3}$	$\dfrac{z(z + 1)T^2}{2(z - 1)^3}$
e^{-at}	$\dfrac{1}{s + a}$	$\dfrac{z}{z - e^{-aT}}$
te^{-at}	$\dfrac{1}{(s + a)^2}$	$\dfrac{zTe^{-aT}}{(z - e^{-aT})^2}$
$a^{t/T}$	$\dfrac{1}{s - (1/T) \ln a}$	$\dfrac{z}{z - a} \quad (a > 0)$
$\sin \omega t$	$\dfrac{\omega}{s^2 + \omega^2}$	$\dfrac{z \sin \omega T}{z^2 - 2z \cos \omega T + 1}$
$\cos \omega t$	$\dfrac{s}{s^2 + \omega^2}$	$\dfrac{z^2 - z \cos \omega T}{z^2 - 2z \cos \omega T + 1}$

Illustrative example 10.4 Determine the z transform of cos ωt. It is known that the Laplace transform is $s/(s^2 + \omega^2)$. Performing a partial-fraction expansion gives

$$\mathscr{L}(\cos \omega t) = \frac{s}{s^2 + \omega^2} = \frac{\frac{1}{2}}{s + j\omega} + \frac{\frac{1}{2}}{s - j\omega}$$

The corresponding z transform is

$$Z(\cos \omega t) = \frac{1}{2}\left[\frac{z}{z - e^{-j\omega T}} + \frac{z}{z - e^{j\omega T}}\right] = \frac{z^2 - z(e^{j\omega T} + e^{-j\omega T})/2}{z^2 - z(e^{j\omega T} + e^{-j\omega T}) + 1}$$

$$= \frac{z^2 - z \cos \omega T}{z^2 - 2z \cos \omega T + 1} \tag{10.7}$$

The Residue Method

This is a powerful technique for obtaining z transforms. The z transform of $f*(t)$ may be expressed in the form

$$F(z) = Z[f*(t)] = \sum \text{residues of } F(s) \frac{z}{z - e^{sT}} \text{ at poles of } F(s) \tag{10.8}$$

When the denominator of $F(s)$ contains a linear factor of the form $s - r$ such that $F(s)$ has a first-order pole at $s = r$, the corresponding residue R is

$$R = \lim_{s \to r} (s - r)\left[F(s) \frac{z}{z - e^{sT}}\right] \tag{10.9}$$

When $F(s)$ contains a repeated pole of order q, the residue is

$$R = \frac{1}{(q - 1)!} \lim_{s \to r} \frac{d^{q-1}}{ds^{q-1}}\left[(s - r)^q F(s) \frac{z}{z - e^{sT}}\right] \tag{10.10}$$

As is illustrated by the following examples, the determination of residues is similar to evaluating the constants in a partial-fraction expansion.

Illustrative example 10.5 Determine the z transform of a unit step function. For $F(s) = 1/s$, there is but one pole at $s = 0$. The corresponding residue is

$$R = \lim_{s \to 0} s\left(\frac{1}{s} \frac{z}{z - e^{sT}}\right) = \frac{z}{z - 1}$$

This verifies the result of Eq. (10.4).

Illustrative example 10.6 Determine the z transform of e^{-at}. For this function, $F(s) = 1/(s + a)$, which has but one pole at $s = -a$. Thus,

$$R = \lim_{s \to -a} (s + a)\left[\frac{1}{s + a} \frac{z}{z - e^{sT}}\right] = \frac{z}{z - e^{-aT}}$$

This verifies the result of Eq. (10.5).

Illustrative example 10.7 Determine the z transform for the function whose Laplace transform is

$$F(s) = \frac{1}{s(s + 1)}$$

The poles of $F(s)$ occur at $s = 0$ and $s = -1$. The residue due to the pole at $s = 0$ is

$$R_1 = \lim_{s \to 0} s \left[\frac{1}{s(s + 1)} \frac{z}{z - e^{sT}} \right] = \frac{z}{z - 1}$$

The residue due to the pole at $s = -1$ is

$$R_2 = \lim_{s \to -1} (s + 1) \left[\frac{1}{s(s + 1)} \frac{z}{z - e^{sT}} \right] = -\frac{z}{z - e^{-T}}$$

Adding these two residues verifies the result given by Eq. (10.6).

Illustrative example 10.8 Determine the z transform of $\cos \omega t$. The Laplace transform is

$$F(s) = \frac{s}{s^2 + \omega^2} = \frac{s}{(s - j\omega)(s + j\omega)}$$

The poles are at $s = j\omega$ and at $s = -j\omega$. Thus,

$$R_1 = \left(\frac{s}{s + j\omega} \frac{z}{z - e^{sT}} \right)_{s = j\omega} = \frac{1}{2} \frac{z}{z - e^{j\omega T}}$$

$$R_2 = \left(\frac{s}{s - j\omega} \frac{z}{z - e^{sT}} \right)_{s = -j\omega} = \frac{1}{2} \frac{z}{z - e^{-j\omega T}}$$

Adding these two residues verifies the result given by Eq. (10.7).

Illustrative example 10.9 Determine the z transform corresponding to the function $f(t) = t$. The Laplace transform is

$$F(s) = \frac{1}{s^2}$$

This has a second-order pole at $s = 0$. Application of Eq. (10.10) gives

$$R = \frac{d}{ds} \left(\frac{z}{z - e^{sT}} \right)_{s=0} = \left(\frac{zTe^{sT}}{(z - e^{sT})^2} \right)_{s=0} = \frac{Tz}{(z - 1)^2}$$

Theorems

The following basic theorems extend the usefulness of the z transform method. In addition, these theorems help one to obtain a better understanding of z transformations.

Multiplication by e^{-at} The z transform of $e^{-at}f(t)$ is

$$Z[e^{-at}f(t)] = F(ze^{aT}) \qquad (10.11)$$

Thus, replacing z in $F(z)$ by ze^{aT} yields the transform for $e^{-at}f(t)$. It is recalled that the Laplace transform of $e^{-at}f(t)$ is equal to $F(s + a)$. That is, replacing s by $s + a$ gives the effect of multiplying by e^{-at}.

To prove this theorem, note from Eq. (10.3) that

$$Z[e^{-at}f(t)] = \sum_{n=0}^{\infty} f(nT)e^{-anT}z^{-n} = \sum_{n=0}^{\infty} f(nT)(ze^{aT})^{-n}$$

The right side is $F(z)$ with z replaced by (ze^{aT}).

Multiplication by t The z transform of $tf(t)$ is

$$Z[tf(t)] = -zT \frac{d}{dz} F(z) \qquad (10.12)$$

To prove this theorem, note that

$$-zT \frac{d}{dz} F(z) = -zT \frac{d}{dz} \left[f(0) + \frac{f(T)}{z} + \frac{f(2T)}{z^2} + \frac{f(3T)}{z^3} + \cdots \right]$$

$$= T \frac{f(T)}{z} + 2T \frac{f(2T)}{z^2} + 3T \frac{f(3T)}{z^3} + \cdots$$

The right side is the z transform of $tf(t)$.

To illustrate this theorem, consider the function $f(t) = e^{-at}$, for which $F(z) = z/(z - e^{-aT})$. Application of Eq. (10.12) shows that

$$Z[te^{-at}] = -zT \frac{d}{dz} \frac{z}{z - e^{-aT}} = \frac{zTe^{-aT}}{(z - e^{-aT})^2}$$

Multiplication by $a^{t/T}$ The z transform of $a^{t/T}f(t)$ is

$$Z[a^{t/T}f(t)] = F\left(\frac{z}{a}\right) \qquad (10.13)$$

To prove this theorem, replace z by z/a in the expression for $F(z)$. Thus

$$F\left(\frac{z}{a}\right) = f(0) + a \frac{f(T)}{z} + a^2 \frac{f(2T)}{z^2} + \cdots$$

The right side is the z transform of $a^{t/T}f(t)$.

Partial differentiation This theorem states that

$$Z\left[\frac{\partial}{\partial a} [f(t, a)] \right] = \frac{\partial}{\partial a} [F(z, a)] \qquad (10.14)$$

This theorem is useful for ascertaining additional z transforms. For example, it is known that the z transform of e^{at} is $z/(z - e^{aT})$. Thus,

$$Z\left[\frac{\partial}{\partial a} e^{at}\right] = Z[te^{at}] = \frac{\partial}{\partial a} \frac{z}{z - e^{aT}} = \frac{Tze^{aT}}{(z - e^{aT})^2}$$

For the case in which $a = 0$,

$$Z[t] = \frac{Tz}{(z - 1)^2}$$

By proceeding in a similar manner, the z transforms for t^2, t^3, etc., are also obtained.

Initial-value theorem The area of the first impulse $f(0)$ of the sampled function $f^*(t)$ is

$$f(0) = \lim_{z \to \infty} F(z) \tag{10.15}$$

This theorem is verified directly by taking the limit as z approaches infinity in Eq. (10.3).

Final-value theorem The area of the impulse $f(nT)$ as n becomes infinite is

$$f(\infty) = \lim_{z \to 1} \frac{z - 1}{z} F(z) \tag{10.16}$$

To prove this theorem, consider the following sums S_n and S_{n-1}:

$$S_n = f(0) + \frac{f(T)}{z} + \cdots + \frac{f[(n - 1)T]}{z^{n-1}} + \frac{f(nT)}{z^n}$$

$$S_{n-1} = f(0) + \frac{f(T)}{z} + \cdots + \frac{f[(n - 1)T]}{z^{n-1}}$$

Dividing the second series by z and then subtracting the second from the first gives

$$S_n - \frac{1}{z} S_{n-1} = \left(1 - \frac{1}{z}\right)f(0) + \cdots + \left(1 - \frac{1}{z}\right)\frac{f[(n - 1)T]}{z^{n-1}} + \frac{f(nT)}{z^n}$$

Taking the limit as z approaches 1 gives

$$\lim_{z \to 1} \left(S_n - \frac{1}{z} S_{n-1}\right) = f(nT)$$

When n is very large, $S_{n-1} \approx S_n \approx F(z)$. Thus, the final-value theorem given by Eq. (10.16) is verified.

Real translation The solid curve shown in Fig. 10.2a is the continuous function $f(t)$. The value at time $t = 0$ is $f(0)$, the value at $t = T$ is $f(T)$, etc. When the function is shifted to the right (delayed) by a time nT, as shown in Fig. 10.2b, the

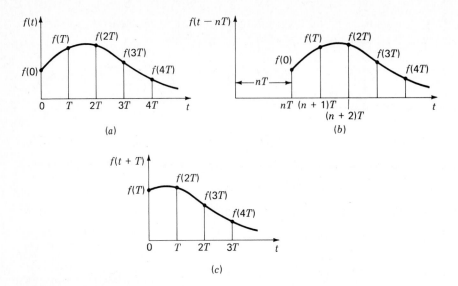

Figure 10.2 Translation of a continuous function $f(t)$.

function $f(t - nT)$ results. The value of $f(t - nT)$ at time $t = nT$ is $f(0)$, the value at $t = (n + 1)T$ is $f(T)$, etc. The z transform for $f(t)$ is

$$Z[f(t)] = f(0) + \frac{f(T)}{z} + \frac{f(2T)}{z^2} + \cdots$$

The z transform for $f(t - nT)$ is

$$Z[f(t - nT)] = \frac{f(0)}{z^n} + \frac{f(T)}{z^{n+1}} + \frac{f(2T)}{z^{n+2}} + \cdots$$

$$= \frac{1}{z^n}\left[f(0) + \frac{f(T)}{z} + \frac{f(2T)}{z^2} + \cdots \right]$$

$$= z^{-n}F(z) \tag{10.17}$$

Thus, multiplying a z transform by z^{-n} has the effect of delaying the function by a time nT. For Laplace transforms, multiplication by $e^{-t_0 s}$ has the effect of delaying the function by a time t_0.

When the function of Fig. 10.2a is shifted to the left one period T, the function $f(t + T)$ shown in Fig. 10.2c results. The value of $f(t + T)$ at time $t = 0$ is $f(T)$, the value at $t = T$ is $f(2T)$, etc. The z transform for $f(t + T)$ is

$$Z[f(t + T)] = f(T) + \frac{f(2T)}{z} + \frac{f(3T)}{z^2} + \cdots$$

Multiplying through both sides by $1/z$ and adding $f(0)$ to both sides gives

$$z^{-1}Z[f(t + T)] + f(0) = f(0) + \frac{f(T)}{z} + \frac{f(2T)}{z^2} + \cdots = F(z)$$

Thus,

$$Z[f(t + T)] = zF(z) - zf(0)$$

Similarly, it may be shown that

$$Z[f(t + 2T)] = z^2 F(z) - z^2 f(0) - zf(T)$$

and, in general,

$$Z[f(t + nT)] = z^n F(z) - z^n f(0) - z^{n-1} f(T) - \cdots - zf[(n-1)T] \quad (10.18)$$

Discrete functions The notation $f(k)$ may be used to represent $f^*(t)$. The plot of $f(k)$, shown dashed in Fig. 10.3a, is identical to that of $f^*(t)$ shown in Fig. 10.2a. The z transform for $f(k)$ is

$$Z[f(k)] = f(0) + \frac{f(1)}{z} + \frac{f(2)}{z^2} + \frac{f(3)}{z^3} + \cdots = F(z) \quad (10.19)$$

When $f(k)$ is delayed n sampling instants, the function $f(k - n)$ shown in Fig. 10.3b results. The value of $f(k - n)$ when $k = n$ is $f(0)$, the value when $k = n + 1$ is $f(1)$, etc. The z transform of $f(k - n)$ is

$$Z[f(k - n)] = \frac{f(0)}{z^n} + \frac{f(1)}{z^{n+1}} + \frac{f(2)}{z^{n+2}} + \cdots$$

$$= \frac{1}{z^n} \left[f(0) + \frac{f(1)}{z} + \frac{f(2)}{z^2} + \cdots \right]$$

$$= z^{-n} F(z) \quad (10.20)$$

Figure 10.3 Translation of a discrete function $f(k)$.

When the function $f(k)$ of Fig. 10.3a is shifted one sampling period to the left, the function $f(k + 1)$ shown in Fig. 10.3c results. The value of $f(k + 1)$ when $k = 0$ is $f(1)$, the value when $k = 1$ is $f(2)$, etc. The z transform of $f(k + 1)$ is

$$Z[f(k + 1)] = f(1) + \frac{f(2)}{z} + \frac{f(3)}{z^2} + \cdots$$

Multiplying through both sides by $1/z$ and adding $f(0)$ to both sides gives

$$z^{-1}Z[f(k + 1)] + f(0) = f(0) + \frac{f(1)}{z} + \frac{f(2)}{z^2} + \cdots = F(z)$$

Thus,

$$Z[f(k + 1)] = zF(z) - zf(0)$$

Similarly, it follows that

$$Z[f(k + 2)] = z^2F(z) - z^2f(0) - zf(1)$$

In general,

$$Z[f(k + n)] = z^nF(z) - z^nf(0) - z^{n-1}f(1) - \cdots - zf(n - 1) \qquad (10.21)$$

As shown in the following, z transform theorems for $f(k)$ are quite similar to those for $f(t)$.

Multiplication by a^k The z transform of $a^kf(k)$ is

$$Z[a^kf(k)] = F\left(\frac{z}{a}\right) \qquad (10.22)$$

This theorem is readily proved by replacing z by z/a in Eq. (10.19). Thus,

$$F\left(\frac{z}{a}\right) = f(0) + \frac{af(1)}{z} + \frac{a^2f(2)}{z^2} + \cdots$$

The right side is the z transform of $a^kf(k)$.

For a unit step function, $f(k) = 1$, and $F(z) = 1 + 1/z + 1/z^2 + \cdots = z/(z - 1)$. Application of Eq. (10.22) shows that

$$Z[a^k] = F\left(\frac{z}{a}\right) = \frac{z/a}{z/a - 1} = \frac{z}{z - a} \qquad (10.23)$$

Multiplication by k The z transform of $kf(k)$ is

$$Z[kf(k)] = -z\frac{d}{dz}F(z) \qquad (10.24)$$

Table 10.2 Properties of the z transform

$f(t)$ or $f(k)$	$Z[f(t)]$ or $Z[f(k)]$
$af(t)$ or $af(k)$	$aF(z)$
$f_1(t) + f_2(t)$ or $f_1(k) + f_2(k)$	$F_1(z) + F_2(z)$
$f(t - nT)$ or $f(k - n)$	$z^{-n}F(z)$
$f(t + T)$ or $f(k + 1)$	$zF(z) - zf(0)$
$f(t + 2T)$	$z^2F(z) - z^2f(0) - zf(T)$
$f(k + 2)$	$z^2F(z) - z^2f(0) - zf(1)$
$f(t + nT)$	$z^nF(z) - z^nf(0) - z^{n-1}f(T) - \cdots - zf[(n-1)T]$
$f(k + n)$	$z^nF(z) - z^nf(0) - z^{n-1}f(1) - \cdots - zf(n-1)$
$tf(t)$	$-zT \dfrac{d}{dz} F(z)$
$kf(k)$	$-z \dfrac{d}{dz} F(z)$
$e^{-at}f(t)$	$F(ze^{aT})$
$e^{-ak}f(k)$	$F(ze^a)$
$a^{t/T}f(t)$ or $a^kf(k)$	$F(z/a)$
$\dfrac{\partial}{\partial a} f(t, a)$	$\dfrac{\partial}{\partial a} F(z, a)$

To verify this theorem, note that

$$-z \frac{d}{dz} F(z) = -z \frac{d}{dz} \left[f(0) + \frac{f(1)}{z} + \frac{f(2)}{z^2} + \frac{f(3)}{z^3} + \cdots \right]$$

$$= \frac{f(1)}{z} + \frac{2f(2)}{z^2} + \frac{3f(3)}{z^3} + \cdots$$

This result is the z transform of $kf(k)$.

To illustrate this theorem, consider the function $f(k) = a^k$ for which $F(z) = z/(z - a)$. Application of Eq. (10.24) gives

$$Z[ka^k] = -z \frac{d}{dz} \frac{z}{z - a} = \frac{az}{(z - a)^2} \tag{10.25}$$

A listing of z transform theorems and properties is given in Table 10.2.

10.3 INVERSE z TRANSFORMS

Inspection of Table 10.1 shows that z transforms tend to be more complicated than corresponding Laplace transforms. Fortunately, there are some relatively simple techniques for obtaining inverse z transforms.

Partial-Fraction Method

To illustrate this method, let it be desired to obtain the inverse z transform for the function

$$F(z) = \frac{(1 - e^{-T})z}{(z - 1)(z - e^{-T})} \tag{10.26}$$

Performing a partial-fraction expansion gives

$$F(z) = \left[\frac{1 - e^{-T}}{(z - 1)(z - e^{-T})} \right] z = \left[\frac{K_1}{z - 1} + \frac{K_2}{z - e^{-T}} \right] z$$

$$= \frac{z}{z - 1} - \frac{z}{z - e^{-T}}$$

From Table 10.1 the corresponding time function is

$$f(t) = 1 - e^{-t}$$

Thus, the inverse $f^*(t)$ is

$$f^*(t) = \sum_{n=0}^{\infty} (1 - e^{-nT}) u_1(t - nT) \tag{10.27}$$

The z transform for a system may be written in the general form

$$C(z) = z \frac{A(z)}{B(z)} \tag{10.28}$$

For the case of distinct roots $B(z)$ may be factored as

$$B(z) = (z - r_1)(z - r_2) \cdots (z - r_n)$$

The resulting partial-fraction expansion for $C(z)$ is

$$C(z) = z \left[\frac{K_1}{z - r_1} + \frac{K_2}{z - r_2} + \cdots + \frac{K_i}{z - r_i} + \cdots + \frac{K_n}{z - r_n} \right]$$

where

$$K_i = \lim_{z \to r_i} \left[(z - r_i) \frac{A(z)}{B(z)} \right]$$

Multiplying in the z factor gives

$$C(z) = \frac{zK_1}{z - r_1} + \frac{zK_2}{z - r_2} + \cdots + \frac{zK_i}{z - r_i} + \cdots + \frac{zK_n}{z - r_n} \tag{10.29}$$

Each of these terms has the same form. The inverse z transform of this form is obtained by inverting Eq. (10.23). That is,

$$Z^{-1} \left[\frac{z}{z - a} \right] = a^k$$

Thus, the inverse z transform of Eq. (10.29) is

$$c(k) = K_1 r_1^k + K_2 r_2^k + \cdots + K_i r_i^k + \cdots + K_n r_n^k \tag{10.30}$$

Illustrative example 10.10 The z transform for a system is

$$C(z) = z \frac{(z + 2)}{(z - 1.0)(z + 0.5)(z - 0.2)}$$

Determine the inverse z transform.

SOLUTION The partial-fraction expansion is

$$C(z) = z \left[\frac{2.50}{z - 1} + \frac{1.43}{z + 0.5} - \frac{3.93}{z - 0.2} \right]$$

$$= 2.50 \frac{z}{z - 1} + 1.43 \frac{z}{z + 0.5} - 3.93 \frac{z}{z - 0.2}$$

The inverse transform is

$$c(k) = 2.50 + 1.43(-0.5)^k - 3.93(0.2)^k$$

For a root r which is repeated twice, $B(z)$ has the form

$$B(z) = (z - r)^2 (z - r_1) \cdots (z - r_{n-2})$$

The partial-fraction expansion for $C(z)$ is written as

$$C(z) = z \left(\frac{C_1 r}{(z - r)^2} + \frac{C_2}{z - r} + \frac{K_1}{z - r_1} + \cdots + \frac{K_{n-2}}{z - r_{n-2}} \right)$$

$$= \frac{z C_1 r}{(z - r)^2} + \frac{z C_2}{z - r} + \frac{z K_1}{z - r_1} + \cdots + \frac{z K_{n-2}}{z - r_{n-2}} \tag{10.31}$$

The inverse z transform of the first term is obtained by inverting Eq. (10.25). That is,

$$Z^{-1} \left[\frac{za}{(z - a)^2} \right] = ka^k$$

The inverse z transform of Eq. (10.31) is

$$c(k) = (C_1 k + C_2) r^k + K_1 r_1^k + \cdots + K_{n-2} r_{n-2}^k \tag{10.32}$$

Note that the response for repeated roots has the same form as for continuous systems.

Illustrative example 10.11 The z transform for a system is

$$C(z) = z \left[\frac{z - 2}{(z - 0.5)^2 (z - 1)} \right]$$

Determine the inverse z transform.

SOLUTION The partial-fraction expansion is

$$C(z) = z\left[\frac{3}{(z - 0.5)^2} + \frac{4}{z - 0.5} - \frac{4}{z - 1.0}\right]$$

$$= \frac{6z(0.5)}{(z - 0.5)^2} + \frac{4z}{z - 0.5} - \frac{4z}{z - 1.0}$$

Inverting yields

$$c(k) = (6k + 4)(0.5)^k - 4$$

When $B(z)$ has complex conjugate roots $a \pm jb$, then $B(z)$ has the form

$$B(z) = (z^2 - 2az + a^2 + b^2)(z - r_1) \cdots (z - r_{n-2})$$

Thus,

$$C(z) = z\frac{A(z)}{[z - (a + jb)][z - (a - jb)](z - r_1) \cdots (z - r_{n-2})}$$

The partial-fraction expansion for $C(z)$ is written as

$$C(z) = z\left[\frac{K_c}{z - (a + jb)} + \frac{K_{-c}}{z - (a - jb)} + \frac{K_1}{z - r_1} + \cdots + \frac{K_{n-2}}{z - r_{n-2}}\right] \quad (10.33)$$

The partial-fraction constant K_c is

$$K_c = \lim_{z \to a + jb} [z - (a + jb)] \frac{A(z)}{[z - (a + jb)][z - (a - jb)](z - r_1) \cdots (z - r_{n-2})}$$

$$= \frac{1}{2jb}\left[\frac{A(z)}{(z - r_1) \cdots (z - r_{n-2})}\right]_{z = a + jb} = \frac{1}{2jb}K(a + jb) \quad (10.34)$$

where

$$K(a + jb) = \left[(z^2 - 2az + a^2 + b^2)\frac{A(z)}{B(z)}\right]_{z = a + jb}$$

The quantity in brackets is $A(z)/B(z)$ in which the quadratic term $(z^2 - 2az + a^2 + b^2)$ has been canceled from $B(z)$.

The constant K_{-c} is

$$K_{-c} = \lim_{z \to a - jb} [z - (a - jb)] \frac{A(z)}{[z - (a + jb)][z - (a - b)](z - r_1) \cdots (z - r_{n-2})}$$

$$= -\frac{1}{2jb}\left[\frac{A(z)}{(z - r_1) \cdots (z - r_{n-2})}\right]_{z = a - jb} = -\frac{1}{2jb}K(a - jb) \quad (10.35)$$

where

$$K(a - jb) = \left[(z^2 - 2az + a^2 + b^2)\frac{A(z)}{B(z)}\right]_{z = a - jb}$$

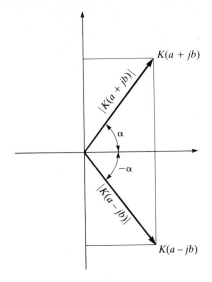

$K(a + jb)$

$|K(a + jb)|$

α

$-\alpha$

$|K(a - jb)|$

$K(a - jb)$

Figure 10.4 The vectors $K(a + jb)$ and $K(a - jb)$.

The constants $K(a + jb)$ and $K(a - jb)$ are complex conjugates. As shown in Fig. 10.4 these constants may be expressed in the form

$$K(a + jb) = |K(a + jb)|e^{j\alpha}$$
$$K(a - jb) = |K(a + jb)|e^{-j\alpha} \tag{10.36}$$

The length of both of these vectors is $|K(a + jb)| = |K(a - jb)|$. The angle of the $K(a + jb)$ vector is α and the angle of the $K(a - jb)$ vector is $-\alpha$. The inverse z transform of Eq. (10.33) is

$$c(k) = K_c(a + jb)^k + K_{-c}(a - jb)^k + K_1 r_1^k + \cdots + K_{n-2} r_{n-2}^k$$

The vectors $a + jb$ and $a - jb$ are shown in Fig. 10.5. These vectors may be expressed in the polar form

$$a + jb = Re^{j\beta}$$
$$a - jb = Re^{-j\beta}$$

The length of both of these vectors is $R = \sqrt{a^2 + b^2}$. The angle of the $a + jb$ vector is β and the angle of the $a - jb$ vector is $-\beta$. Substitution of the polar representation for $a + jb$ and $a - jb$ into Eq. (10.36) gives

$$c(k) = K_c R^k e^{jk\beta} + K_{-c} R^k e^{-jk\beta} + K_1 r_1^k + \cdots + K_{n-2} r_{n-2}^k$$

Substituting from Eqs. (10.34), (10.35), and (10.36), into the preceding yields

$$c(k) = R^k \frac{|K(a + jb)|}{b} \left(\frac{e^{j(k\beta + \alpha)} - e^{-j(k\beta + \alpha)}}{2j} \right) + K_1 r_1^k + \cdots + K_{n-2} r_{n-2}^k$$

$$= R^k \frac{|K(a + jb)|}{b} \sin(k\beta + \alpha) + K_1 r_1^k + \cdots + K_{n-2} r_{n-2}^k \tag{10.37}$$

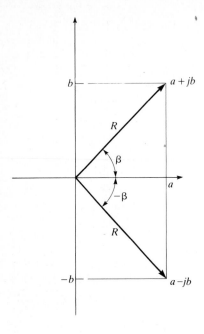

Figure 10.5 The vectors $a + jb$ and $a - jb$.

By noting that

$$|K(a + jb)|\sin(k\beta + \alpha) = |K(a + jb)|(\cos \alpha \sin k\beta + \sin \alpha \cos k\beta)$$

$$= A \sin k\beta + B \cos k\beta$$

where $A = |K(a + jb)|\cos \alpha$ is the horizontal component of the vector $K(a + jb)$ and $B = |K(a + jb)|\sin \alpha$ is the vertical component, then Eq. (10.37) may be written in the alternate form

$$c(k) = \frac{R^k}{b}(A \sin k\beta + B \cos k\beta) + K_1 r_1^k + \cdots + K_{n-2} r_{n-2}^k$$

Illustrative example 10.12 The z transform for a system is

$$C(z) = z\frac{z - 0.7}{(z^2 + 0.6z + 0.25)(z + 0.5)}$$

Determine the inverse z transform.

SOLUTION For $-2a = 0.6$ and $a^2 + b^2 = 0.25$, then $a = -0.3$ and $b = \sqrt{0.25 - 0.09} = 0.4$. The polar form for $a + jb$ is $Re^{j\beta}$. That is, $a + jb = -0.3 + j0.4 = 0.5 \underline{/126.9°}$. Thus,

$$R = 0.5 \quad \text{and} \quad \beta = 126.9°$$

Evaluation of $K(a + jb)$ yields

$$K(a + jb) = \lim_{z \to -0.3 + j0.4} (z^2 + 0.6z + 0.25) \left[\frac{z - 0.7}{(z^2 + 0.6z + 0.25)(z + 0.5)} \right]$$

$$= \left[\frac{z - 0.7}{z + 0.5} \right]_{z = -0.3 + j0.4} = \frac{-1.0 + j0.4}{0.2 + j0.4} = 2.41 \underline{\diagup 94.8^\circ}$$

Thus,

$$|K(a + jb)| = 2.41 \qquad \text{and} \qquad \alpha = 94.8^\circ$$

The partial-fraction expansion for $C(z)$ is

$$C(z) = z \left[\frac{K_c}{z + (-0.3 + j0.4)} + \frac{K_{-c}}{z - (-0.3 - j0.4)} + \frac{K_1}{z + 0.5} \right]$$

Inverting yields the general form

$$c(k) = R^k \frac{|K(a + jb)|}{b} \sin(k\beta + \alpha) + K_1(-0.5)^k$$

The constant K_1 is

$$K_1 = \lim_{z \to -0.5} (z + 0.5) \left[\frac{z - 0.7}{(z^2 + 0.6z + 0.25)(z + 0.5)} \right]$$

$$= \left[\frac{z - 0.7}{z^2 + 0.6z + 0.25} \right]_{z = -0.5}$$

$$= \frac{-1.2}{0.20} = -6.0$$

Substitution of the previously attained values into the general form yields, for the solution,

$$c(k) = (0.5)^k \frac{2.41}{0.4} \sin(126.9^\circ k + 94.8^\circ) + 6(-0.5)^k$$

Division Method

Dividing the numerator of $F(z)$ by the denominator yields a power series of the form

$$F(z) = C_0 + C_1 z^{-1} + C_2 z^{-2} + \cdots \qquad (10.38)$$

The corresponding inverse is

$$f^*(t) = \sum_{n=0}^{\infty} C_n u_1(t - nT)$$

For example, the z-transform for the function e^{-at} is $F(z) = z/(z - e^{-aT})$.

Dividing the numerator of $F(z)$ by the denominator shows that

$$z - e^{-aT})z \overline{\big)\, 1 + e^{-aT}z^{-1} + e^{-2aT}z^{-2} + \cdots}$$

$$\underline{z - e^{-aT}}$$

$$e^{-aT}$$

$$\underline{e^{-aT} - e^{-2aT}z^{-1}}$$

$$e^{-2aT}z^{-1}$$

$$\underline{e^{-2aT}z^{-1} + e^{-3aT}z^{-2}}$$

Thus,

$$F(z) = \frac{z}{z - e^{-aT}} = 1 + e^{-aT}z^{-1} + e^{-2aT}z^{-2} + \cdots$$

The constant is the value of $f(t) = e^{-at}$ at time $t = 0$, the coefficient e^{-aT} of z^{-1} is the value of $f(t)$ at $t = T$, and the coefficient e^{-2aT} is the value of $f(t)$ at $t = 2T$. Inverting $F(z)$ yields

$$f(nT) = \sum_{n=0}^{\infty} e^{anT} u_1(t - nT)$$

Residue Method

To develop this method, first write the basic relationship for $F(z)$ in the form

$$F(z) = f(0) + \frac{f(T)}{z} + \cdots + \frac{f[(n-1)T]}{z^{n-1}} + \frac{f(nT)}{z^n} + \frac{f[(n+1)T]}{z^{n+1}} + \cdots$$

Multiplication by z^{n-1} gives

$$F(z)z^{n-1} = f(0)z^{n-1} + \cdots + f[(n-1)T] + \frac{f(nT)}{z} + \frac{f[(n+1)T]}{z^2} + \cdots$$

(10.39)

From complex variable theory it is known that the coefficient $f(nT)$ of the $1/z$ term of the preceding Laurent expansion is

$$f(nT) = \frac{1}{2\pi j} \int_C F(z)z^{n-1}\, dz$$

$$= \sum \text{residues of } F(z)z^{n-1} \text{ at poles of } F(z)z^{n-1} \qquad (10.40)$$

The contour C is any closed path which encloses all the poles of $F(z)z^{n-1}$.

Replacing the function $F(s)z/(z - e^{sT})$ in Eqs. (10.9) and (10.10) by $F(z)z^{n-1}$ yields the following equations for determining the residues in Eq. (10.40). In particular, the residue due to a first-order pole at $z = r$ is

$$R = \lim_{z \to r} (z - r)[F(z)z^{n-1}] \qquad (10.41)$$

Similarly, the residue due to a repeated pole of order q is

$$R = \frac{1}{(q-1)!} \lim_{z \to r} \frac{d^{q-1}}{dz^{q-1}} [(z-r)^q F(z) z^{n-1}] \tag{10.42}$$

Application of the residue method to determine the inverse of Eq. (10.26) gives

$$R_1 = \left[\frac{(1-e^{-T})z^n}{z-e^{-T}} \right]_{z=1} = 1$$

$$R_2 = \left[\frac{(1-e^{-T})z^n}{z-1} \right]_{z=e^{-T}} = -e^{-nT}$$

Adding these residues gives $f(nT) = 1 - e^{-nT}$, which verifies the result given by Eq. (10.27). For $f(nT) = 1 - e^{-nT}$, the corresponding time function is $f(t) = 1 - e^{-t}$.

As another example, determine the inverse z transform for the function

$$F(z) = \frac{Tz}{(z-1)^2}$$

This function has a second-order pole at $z = 1$; thus,

$$R = \lim_{z \to 1} \frac{d}{dz} Tz^n = (nTz^{n-1})_{z=1} = nT$$

For $f(nT) = nT$, the corresponding time function is $f(t) = t$.

10.4 BLOCK-DIAGRAM ALGEBRA

In writing the transfer function for feedback control systems with sampling switches, one encounters some terms which are starred and some which are not. Thus, it is necessary to develop some mathematical techniques for handling such mixed terms. In Fig. 10.6a is shown a sampling switch followed by a linear

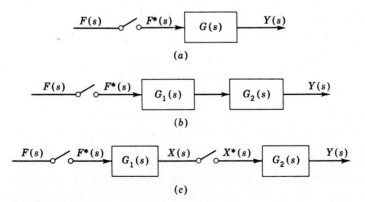

Figure 10.6 Sampler configurations.

element whose transfer function is $G(s)$. The transformed equation for the output $Y(s)$ is

$$Y(s) = F^*(s)G(s) \tag{10.43}$$

For $0 < t < T$, the response $y(t)$ is that due to the first impulse at $t = 0$ of area $f(0)$. Thus, for this interval,

$$y(t) = \mathscr{L}^{-1}[f(0)G(s)] = f(0)\mathscr{L}^{-1}[G(s)] = f(0)g(t) \tag{10.44}$$

where $g(t) = \mathscr{L}^{-1}[G(s)]$ is the response of the linear element to a unit impulse which occurs at $t = 0$.

For $T < t < 2T$, the response $y(t)$ is that due to the impulse at $t = 0$ plus that at $t = T$. For this interval, $F^*(s) = f(0) + f(T)e^{-Ts}$. Thus,

$$Y(s) = [f(0) + f(T)e^{-Ts}]G(s)$$

Inverting gives

$$y(t) = f(0)g(t) + f(T)g(t - T) \tag{10.45}$$

where $g(t - T) = \mathscr{L}^{-1}[G(s)e^{-Ts}]$ is the response of the linear element to a unit impulse which occurs at time $t = T$. The response $y(t)$ for the interval $2T < t < 3T$ is

$$y(t) = f(0)g(t) + f(T)g(t - T) + f(2T)g(t - 2T)$$

In general, the response $y(t)$ is

$$y(t) = \sum_{n=0}^{\infty} f(nT)g(t - nT) \tag{10.46}$$

When n is such that $nT > t$, then $g(t - nT)$ is zero. That is, the impulse response is zero for negative time.

Taking the limit as t approaches zero in Eq. (10.44), and the limit as t approaches T in Eq. (10.45), etc., yields the value of $y(nT)$ at the sampling instants:

$$y(0) = f(0)g(0)$$

$$y(T) = f(0)g(T) + f(T)g(0)$$

$$y(2T) = f(0)g(2T) + f(T)g(T) + f(2T)g(0) \tag{10.47}$$

$$\cdots\cdots\cdots\cdots\cdots\cdots\cdots\cdots\cdots\cdots\cdots\cdots$$

$$y(nT) = f(0)g(nT) + f(T)g[(n - 1)T] + \cdots$$

In Fig. 10.7 is shown a response function $y(t)$ which is discontinuous at the sampling instants. This occurs when the order of the denominator of $G(s)$ exceeds the order of the numerator by only one. When the order of the denominator exceeds the order of the numerator by two or more, $y(t)$ is continuous. When $y(t)$ is continuous, Eq. (10.47) yields the values at the sampling instants. When $y(t)$ is discontinuous, Eq. (10.47) yields the values immediately after the sampling instants [that is, $y(0+)$, $y(T+)$, $y(2T+)\cdots$]. Figure 10.1 shows that the values in

Figure 10.7 Response function that is discontinuous at the sampling instants.

Eq. (10.2) would also be the values immediately following the sampling instants if the function were discontinuous. Replacing f by y in Eq. (10.2) gives

$$Y^*(s) = y(0) + y(T)e^{-Ts} + y(2T)e^{-2Ts} + \cdots$$

Substitution of the values from Eq. (10.47) into the preceding expression gives

$$
\begin{aligned}
Y^*(s) = {} & f(0)[g(0) + g(T)e^{-Ts} + g(2T)e^{-2Ts} + \cdots \\
& + f(T)e^{-Ts}[g(0) + g(T)e^{-Ts} + g(2T)e^{-2Ts} + \cdots] \\
& + f(2T)e^{-2Ts}[g(0) + g(T)e^{-Ts} + g(2T)e^{-2Ts} + \cdots] \\
& + \cdots \\
= {} & [f(0) + f(T)e^{-Ts} + \cdots][g(0) + g(T)e^{-Ts} + \cdots]
\end{aligned}
$$

Thus,

$$Y^*(s) = F^*(s)G^*(s) \tag{10.48}$$

The term $G^*(s)$ is called the pulse-transfer function of the system.

Comparison of Eqs. (10.43) and (10.48) reveals a basic mathematical relationship for starring quantities. That is, starring both sides of Eq. (10.43) gives

$$[Y(s)]^* = Y^*(s)$$

$$[F^*(s)G(s)]^* = F^*(s)[G(s)]^* = F^*(s)G^*(s)$$

The z transform relationship corresponding to Eq. (10.48) is

$$Y(z) = F(z)G(z) \tag{10.49}$$

For the sampler configuration of Fig. 10.6b, the Laplace transform relationship is

$$Y(s) = F^*(s)G_1(s)G_2(s)$$

Starring gives

$$Y^*(s) = F^*(s)[G_1(s)G_2(s)]^* = F^*(s)G_1G_2^*(s)$$

where

$$G_1G_2^*(s) = [G_1(s)G_2(s)]^*$$

The corresponding z transform is

$$Y(z) = F(z)G_1G_2(z) \tag{10.50}$$

Illustrative example 10.13 Determine the z transform for Fig. 10.6b, when $G_1(s) = 1/s$ and $G_2(s) = 1/(s + 1)$. The product $G_1(s)G_2(s)$ is

$$G_1(s)G_2(s) = \frac{1}{s(s + 1)}$$

The z transform for this function is given by Eq. (10.6). That is,

$$G_1G_2(z) = \frac{z(1 - e^{-T})}{(z - 1)(z - e^{-T})} \tag{10.51}$$

The resulting z transform is

$$Y(z) = G_1 G_2(z)F(z) = \frac{z(1 - e^{-T})}{(z - 1)(z - e^{-T})} F(z)$$

For the sampler configuration shown in Fig. 10.6c, the Laplace relationships are

$$X(s) = F^*(s)G_1(s)$$

$$Y(s) = X^*(s)G_2(s)$$

Starring the first equation and then substituting this result for $X^*(s)$ into the second equation gives

$$Y(s) = F^*(s)G_1^*(s)G_2(s)$$

Starring gives

$$Y^*(s) = F^*(s)G_1^*(s)G_2^*(s)$$

The corresponding z transform is

$$Y(z) = F(z)G_1(z)G_2(z) \tag{10.52}$$

Illustrative example 10.14 Determine the z transform for Fig. 10.6c, when $G_1(s) = 1/s$ and $G_2(s) = 1/(s + 1)$. From Eqs. (10.4) and (10.5), it follows that

$$G_1(z) = \frac{z}{z - 1}$$

and

$$G_2(z) = \frac{z}{z - e^{-T}}$$

Thus,

$$G_1(z)G_2(z) = \frac{z^2}{(z - 1)(z - e^T)} \tag{10.53}$$

The resulting z transform is

$$Y(z) = G_1(z)G_2(z)F(z) = \frac{z^2}{(z - 1)(z - e^{-T})} F(z)$$

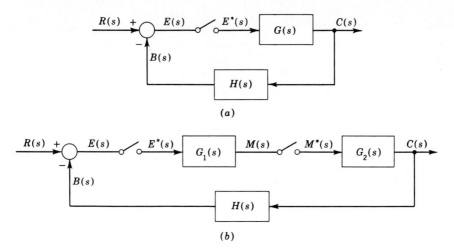

(a)

(b)

Figure 10.8 Sampled-data systems.

From the preceding two examples it is to be noted that

$$G_1 G_2(z) \neq G_1(z)G_2(z) \qquad (10.54)$$

The function $G_1 G_2(z)$ is the z transform corresponding to the product $G_1(s)G_2(s)$, whereas the function $G_1(z)G_2(z)$ is the product of the z transform for $G_1(s)$ and the z transform for $G_2(s)$.

Two sampled-data feedback control systems are shown in Fig. 10.8. The general procedure for determining the transformed equation for a sampled-data system is:

1. In addition to the actual system input R, regard all switch outputs (starred quantities) as inputs.
2. In addition to the system output C, regard all switch inputs as outputs.
3. Write equations for each output in terms of its inputs.
4. Star quantities as necessary in order to determine $C(z)$.

Application of this method to the system shown in Fig. 10.8a gives

$$C(s) = E^*(s)G(s)$$

$$E(s) = R(s) - E^*(s)G(s)H(s)$$

Starring gives

$$C^*(s) = E^*(s)G^*(s)$$

$$E^*(s) = R^*(s) - E^*(s)GH^*(s)$$

Solving the last equation for $E^*(s)$ and substituting into the first gives

$$C^*(s) = \frac{G^*(s)}{1 + GH^*(s)} R^*(s)$$

The corresponding z transform is

$$C(z) = \frac{G(z)}{1 + GH(z)} R(z) \tag{10.55}$$

The equations relating the inputs and outputs of Fig. 10.8b are

$$C(s) = M^*(s)G_2(s)$$

$$M(s) = E^*(s)G_1(s)$$

$$E(s) = R(s) - M^*(s)G_2(s)H(s)$$

Starring all equations and then solving for $C^*(s)$ gives

$$C^*(s) = \frac{G_1^*(s)G_2^*(s)}{1 + G_1^*(s)G_2 H^*(s)} R^*(s)$$

The corresponding z transform is

$$C(z) = \frac{G_1(z)G_2(z)}{1 + G_1(z)G_2 H(z)} R(z) \tag{10.56}$$

10.5 TRANSIENT RESPONSE

For continuous systems, it was found that a system is unstable if any root of the characteristic equation is in the right half of the s plane. This right half-plane may be designated by $\sigma + j\omega$ in which $\sigma > 0$. The corresponding portion of the z plane is

$$z = e^{sT} = e^{\sigma T}e^{j\omega T}$$

The magnitude is

$$|z| = e^{\sigma T}$$

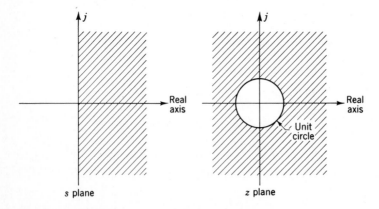

Figure 10.9 Stability regions for s plane and z plane.

Figure 10.10 Sampled-data system.

For $\sigma > 0$, then $|z| > 1$. As is illustrated in Fig. 10.9, the right half of the s plane corresponds to the outside of the unit circle of the z plane. Thus, for stability, all the roots of the z-transformed characteristic equation must lie within the unit circle.

The general form of the response for distinct roots is given by Eq. (10.30). Note that if r_i is greater than one, then r_i^k becomes infinite as k becomes infinite. Similarly, the response for the case of a repeated root is given by Eq. (10.32). Again, if r is greater than one, the response becomes infinite as k becomes infinite. The response for the case of complex conjugate roots $a \pm jb$ is given by Eq. (10.37). The term $R = \sqrt{a^2 + b^2}$ is the distance from the origin to either root. When R is greater than one, the roots lie outside the unit circle. For this case R^k becomes infinite as k becomes infinite. If any root of the characteristic equation lies outside the unit circle, the system is unstable.

Consider now the sampled-data system shown in Fig. 10.10. The z transform for the output $C(z)$ is

$$C(z) = \frac{G(z)}{1 + G(z)} R(z) = \frac{N_{G(z)}}{D_{G(z)} + N_{G(z)}} R(z) \tag{10.57}$$

where $G(z) = N_{G(z)}/D_{G(z)}$ in which $N_{G(z)}$ is the numerator of $G(z)$ and $D_{G(z)}$ is the denominator of $G(z)$. The z-transformed characteristic equation is

$$N_{G(z)} + D_{G(z)} = 0 \tag{10.58}$$

As was the case for continuous systems, the roots of the characteristic equation may be distinct, repeated, or complex conjugate. In Fig. 10.11, the root r_1 is distinct, the root r is repeated, and the roots $a \pm jb$ are complex conjugate.

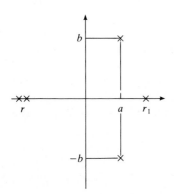

Figure 10.11 Distinct root r_1, repeated root r, and complex conjugate roots $a \pm jb$.

Figure 10.12 Sampled-data system.

For the system shown in Fig. 10.12,

$$G(s) = \frac{K}{s(s + 1)}$$

The partial-fraction expansion for $G(s)$ is

$$G(s) = K\left(\frac{1}{s} - \frac{1}{s + 1}\right)$$

The corresponding z transform is

$$G(z) = K\left(\frac{z}{z - 1} - \frac{z}{z - e^{-T}}\right) = K \frac{z(1 - e^{-T})}{(z - 1)(z - e^{-T})} \tag{10.59}$$

Thus, the characteristic equation for this system is

$$(z - 1)(z - e^{-T}) + K(1 - e^{-T})z = 0 \tag{10.60}$$

Routh's Criterion

To apply this criterion, it is necessary to transform the unit circle of the z plane to the vertical imaginary axis of the λ plane. This is accomplished by the transformation

$$\lambda = \frac{z + 1}{z - 1}$$

Solving for z gives

$$z = \frac{\lambda + 1}{\lambda - 1} \tag{10.61}$$

When the characteristic equation is expressed in terms of λ, then Routh's criterion may be applied in the same manner as for continuous systems.

Illustrative example 10.15 For a sampling period $T = 1$ s, determine the value of K such that the system shown in Fig. 10.12 becomes unstable. That is, the roots of the characteristic equation lie on the unit circle of the z plane (i.e., the imaginary axis of the λ plane).

SOLUTION For $T = 1$, Eq. (10.60) becomes

$$(z - 1)(z - 0.368) + 0.632Kz = 0 \tag{10.62}$$

Using Eq. (10.61) to transform from the z plane to the λ plane gives

$$\frac{0.632K\lambda^2 + 1.264\lambda + (2.736 - 0.632K)}{(\lambda - 1)^2} = 0$$

Routh's array is

$$
\begin{array}{ccc}
0.632K & (2.736 - 0.632K) & 0 \\
1.264 & 0 & \\
(2.736 - 0.632K) & &
\end{array}
$$

This system is unstable for $K < 0$ and for $K > 2.736/0.632 = 4.33$. The range of values of K such that this system is stable is

$$0 < K < 4.33$$

The characteristic equation for the continuous system is $s(s + 1) + K = 0$. The continuous system is stable for all values of K.

If, in the preceding example, the sampling rate is increased from 1 sample per second ($T = 1.0$) to 10 samples per second ($T = 0.1$), then the system would be unstable for $K \geq 40.1$. In general, making the sampling time shorter tends to make the system behave more like the corresponding continuous system. Usually, stability is improved as the sampling rate is increased.

Root Locus

All the techniques for constructing root-locus plots for continuous systems (in the s plane) apply equally well to sampled-data systems (in the z plane). For example, for $T = 1$ the characteristic equation given by Eq. (10.60) becomes Eq. (10.62). This characteristic equation has an $\times$ at $z = 1$ and an $\times$ at $z = 0.368$. There is a $\bigcirc$ at $z = 0$. The complete root-locus plot may now be constructed as shown in Fig. 10.13. This root-locus plot crosses the unit circle at $z = -1$. The value of the gain (0.632K) at any point on a root-locus plot is the product of the distance

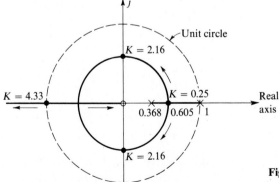

Figure 10.13 Root-locus plot for $(z - 1)$ $\times$ $(z - 0.368) + 0.632Kz = 0$.

from this point to each × divided by the distance from this point to each ○. Thus, the value of $0.632K$ at which the system becomes unstable is $0.632K = (1.368)(2)/(1)$, or

$$K = \frac{2.736}{0.632} = 4.33$$

This verifies the result ascertained by Routh's criterion.

Figure 10.14 Corresponding paths in the s plane and z plane.

Two horizontal lines of constant b are shown in the s plane of Fig. 10.14a. The corresponding paths in the z plane are radial straight lines:

$$z = e^{sT} = e^{(a \pm jb)T} = e^{aT} e^{\pm jbT}$$

The angle of inclination of these radial lines is $\theta = \pm bT$.

Two vertical lines of constant a are shown in the s plane of Fig. 10.14b. The corresponding paths in the z plane are circles of radius e^{aT}. For negative values of a the circles are inside the unit circle of the z plane. For positive values of a the circles lie outside the unit circle of the z plane.

Radial lines of constant damping ratio $\zeta = \cos \beta$ are shown in Fig. 10.14c. In polar coordinates, $s = a \pm jb = -\zeta \omega_n \pm j\omega_n \sqrt{1 - \zeta^2}$. Thus,

$$z = e^{-\zeta \omega_n T} \exp(\pm j\omega_n \sqrt{1 - \zeta^2} T)$$

The corresponding paths in the z plane are logarithmic spirals. For $\beta < 90°$ the spirals decay within the unit circle and for $\beta > 90°$ the spirals grow outside the unit circle.

Consider now how a given point, $z = re^{j\theta}$, in the z plane maps back into the s plane. For $z = re^{j\theta} = e^{sT} = e^{(a \pm jb)T}$,

$$\ln r + j\theta = aT \pm jbT \qquad -\pi < \theta < \pi$$

Equating real and imaginary parts shows that

$$\ln r = aT$$

$$\theta = \pm bT \qquad -\pi < \theta < \pi$$

This verifies the fact that a circle of constant radius r in the z plane is a vertical line of constant a in the s plane. Similarly, a ray at angle θ in the z plane is a horizontal line of constant b in the s plane.

Illustrative example 10.16 Determine the response of the system of Fig. 10.12 to a unit step function excitation for the case in which $T = 1$ and $K = 0.25$.

SOLUTION Substitution of these values into Eq. (10.59) gives

$$G(z) = \frac{0.158z}{(z - 1)(z - 0.368)}$$

Substitution into Eq. (10.57) gives

$$C(z) = \frac{G(z)}{1 + G(z)} R(z) = \frac{0.158z}{(z - 1)(z - 0.368) + 0.158z} R(z) \qquad (10.63)$$

For $K = 0.25$, the root-locus plot of Fig. 10.13 shows that the characteristic equation has a repeated root at $z = 0.605$. Thus,

$$C(z) = \frac{0.158z}{(z - 0.605)^2} R(z) = \frac{0.158z^2}{(z - 0.605)^2(z - 1)}$$

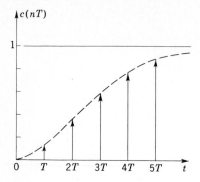

Figure 10.15 Response at sampling instants.

where $R(z) = z/(z - 1)$. Using the long-division method to determine the inverse gives

$$z^3 - 2.21z^2 + 1.58z - 0.368)\overline{0.158z^2}\qquad \dfrac{0.158z^{-1} + 0.349z^{-2} + 0.522z^{-3} + \cdots}{}$$

Because

$$C(z) = c(0) + c(T)z^{-1} + c(2T)z^{-2} + c(3T)z^{-3} + \cdots$$

then

$$c(0) = 0$$

$$c(T) = 0.158$$

$$c(2T) = 0.349 \tag{10.64}$$

$$c(3T) = 0.522$$

A plot of the response $c(nT)$ at the sampling instants is shown in Fig. 10.15.

The long-division method becomes quite cumbersome for computing $c(nT)$ for larger values of n. A more convenient procedure results from expressing the solution in the form of a difference equation.

Difference Equations

To determine the inverse z transform by this method, write the equation for $C(z)$ [Eq. (10.63)] in the form

$$C(z) = \frac{0.158z}{[(z - 1)(z - 0.368) + 0.158z]} R(z) = \frac{0.158z}{z^2 - 1.21z + 0.368} R(z)$$

Thus,

$$C(z) - 1.21z^{-1}C(z) + 0.368z^{-2}C(z) = 0.158z^{-1}R(z)$$

Application of Eq. (10.20) to invert the preceding expression yields directly the difference equation

$$c(k) = 1.21c(k - 1) - 0.368c(k - 2) + 0.158r(k - 1)$$

This difference equation gives the value $c(k)$ at the kth sampling instant in terms of values at preceding sampling instants. Application of this result to obtain the values at the sampling instants gives

$$c(0) = 0$$

$$c(1) = 0.158r(0) = 0.158$$

$$c(2) = 1.21c(1) + 0.158r(1) = 0.349$$

$$c(3) = 1.21c(2) - 0.368c(1) + 0.158r(2) = 0.522$$

Such recurrence relationships lend themselves very well to solution by a digital computer.

The response $c(k)$ at the sampling instants may also be obtained by performing a partial-fraction expansion and then inverting. Thus,

$$C(z) = z\left[\frac{0.158z}{(z-1)(z-0.605)^2}\right] = z\left[\frac{K_1}{z-1} + \frac{C_1}{(z-0.605)^2} + \frac{C_2}{z-0.605}\right]$$

The partial-fraction expansion constants are $K_1 = 1.0$, $C_1 = -0.24 = -(0.39)(0.605)$, and $C_2 = -1.0$. Thus, $C(z)$ becomes

$$C(z) = \frac{z}{z-1} - \frac{z}{z-0.605} - 0.39\frac{0.605z}{(z-0.605)^2}$$

By noting that

$$Z^{-1}\left[\frac{z}{z-1}\right] = 1 \qquad Z^{-1}\left[\frac{z}{z-a}\right] = a^k \qquad Z^{-1}\left[\frac{az}{(z-a)^2}\right] = ka^k$$

the inverse is found to be

$$c(k) = 1.0 - (1.0 + 0.39k)(0.605)^k$$

With this method, the value of $c(k)$ at any sampling instant k may be calculated directly without the need to compute the values at all the preceding instants.

10.6 FILTERS

Sampled-data systems usually incorporate a filter, as illustrated in Fig. 10.16. A perfect filter would convert the sampled signal $f^*(t)$ back to the continuous input $f(t)$. That is, the output $y(t)$ of the filter would equal $f(t)$. If such a perfect filter were possible, then the sampled-data system would behave in the same way as the continuous system.

Figure 10.16 Schematic representation of a sampler and filter.

Zero-Order Hold

The most commonly used filter is that in which the value of the last sample is retained until the next sample is taken. This type of filter is called a zero-order hold, or boxcar generator. The dashed curve in Fig. 10.17 represents the continuous function $f(t)$. The vertical arrows at the sampling instants are the impulses which represent the sampled signal $f^*(t)$. Because the zero-order hold retains the value of $f(t)$ at each sampling instant, $y(t)$ is the series of steps shown in Fig. 10.17. The equation for this series of steps (i.e., pulse functions) is

$$y(t) = f(0)[u(t) - u(t - T)] + f(T)[u(t - T) - u(t - 2T)]$$
$$+ f(2T)[u(t - 2T) - u(t - 3T)] + \cdots$$

The Laplace transform is

$$Y(s) = f(0)\frac{1 - e^{-Ts}}{s} + f(T)\frac{e^{-Ts} - e^{-2Ts}}{s} + f(2T)\frac{e^{-2Ts} - e^{-3Ts}}{s} + \cdots$$

$$= \frac{1 - e^{-Ts}}{s}[f(0) + f(T)e^{-Ts} + f(2T)e^{-2Ts} + \cdots] = \frac{1 - e^{-Ts}}{s}F^*(s)$$

This result shows that the Laplace transform for a zero-order hold is

$$\frac{1 - e^{-Ts}}{s} \tag{10.65}$$

Suppose that a zero-order hold is added to the system of Fig. 10.12. The new block diagram for this system is shown in Fig. 10.18. The transfer function for the hold is included in the overall transfer function $G(s)$. Thus,

$$G(s) = (1 - e^{-Ts})\frac{K}{s^2(s + 1)} \tag{10.66}$$

To determine $G(z)$ when $G(s)$ contains a $(1 - e^{-Ts})$ factor, first write Eq. (10.66) in the form

$$G(s) = G_1(s)G_2(s) \tag{10.67}$$

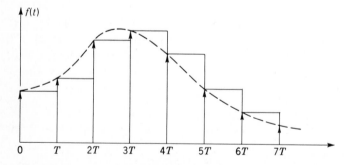

Figure 10.17 Characteristics of a zero-order hold.

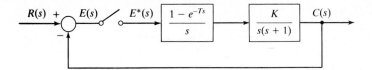

Figure 10.18 Sampled-data system with a zero-order hold.

where $G_1(s) = 1 - e^{-Ts}$ and $G_2(s)$ is the remaining portion of $G(s)$. The function $G_1(s)$ is the Laplace transform of a unit impulse at the origin and a negative unit impulse at $t = T$. The corresponding time function $g_1(t)$ is shown in Fig. 10.19. Because this time function $g_1(t)$ exists only at the sampling instants, the sampled function $g_1^*(t)$ will be the same as $g_1(t)$. Thus,

$$G_1(s) = G_1^*(s)$$

Substitution of this result into Eq. (10.67) shows that

$$G(s) = G_1^*(s)G_2(s)$$

Starring gives

$$G^*(s) = G_1^*(s)G_2^*(s)$$

The corresponding z transform is

$$G(z) = G_1(z)G_2(z) = (1 - z^{-1})G_2(z)$$

$$= \frac{z - 1}{z} G_2(z) \qquad (10.68)$$

For the case of Eq. (10.66),

$$G_2(s) = \frac{K}{s^2(s + 1)} = K\left(\frac{1}{s^2} - \frac{1}{s} + \frac{1}{s + 1}\right)$$

Thus,

$$G_2(z) = K\left[\frac{Tz}{(z - 1)^2} - \frac{z}{z - 1} + \frac{z}{z - e^{-T}}\right]$$

Substitution of this result into Eq. (10.68) gives

$$G(z) = K\left[\frac{T}{z - 1} - 1 + \frac{z - 1}{z - e^{-T}}\right] = K\left[\frac{T}{z - 1} + \frac{e^{-T} - 1}{z - e^{-T}}\right]$$

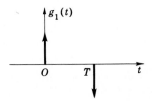

Figure 10.19 Time function $g_1(t) = \mathscr{L}^{-1}[G_1(s)] = \mathscr{L}^{-1}(1 + e^{-Ts})$.

For $T = 1$, then $G(z)$ becomes

$$G(z) = \frac{0.368K(z + 0.717)}{(z - 1)(z - 0.368)} \tag{10.69}$$

The corresponding characteristic equation for this sampled-data system is

$$D_{G(z)} + N_{G(z)} = (z - 1)(z - 0.368) + 0.368K(z + 0.717) = 0 \tag{10.70}$$

Replacing z by $(\lambda + 1)/(\lambda - 1)$ so that Routh's criterion may be applied gives

$$0.632K\lambda^2 + (1.264 - 0.528K)\lambda + (2.736 - 0.104K) = 0$$

The Routh array is

$$
\begin{array}{cc}
0.632K & (2.736 - 0.104K) \\
(1.264 - 0.528K & 0 \\
(2.736 - 0.104K) &
\end{array}
$$

This system becomes unstable for $K < 0$ and for $K > 1.264/0.528 = 2.39$. Thus, the range of values of K such that the system is stable is

$$0 < K < 2.39$$

From the characteristic equation [Eq. (10.70)] the root-locus plot may be constructed as shown in Fig. 10.20. The root-locus plot crosses the unit circle at $K = 2.39$. Repeated roots, $z = 0.65$, occur at $K = 0.2$.

Let it now be desired to determine the response of this system to a unit step function for the case in which $K = 1$. From Eq. (10.69), it follows that

$$G(z) = \frac{0.368(z + 0.717)}{(z - 1)(z - 0.368)} = \frac{0.368(z + 0.717)}{z^2 - 1.368z + 0.368}$$

The z transform for the output is

$$C(z) = \frac{G(z)}{1 + G(z)} R(z)$$

$$= \frac{N_{G(z)}}{D_{G(z)} + N_{G(z)}} R(z) = \frac{0.368(z + 0.717)R(z)}{(z^2 - 1.368z + 0.368) + 0.368(z + 0.717)}$$

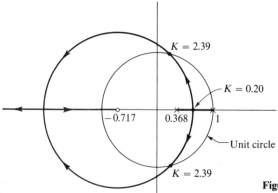

Figure 10.20 Root-locus plot for $(z - 1)$ $\times (z - 0.368) + 0.368K(z + 0.717) = 0$.

Thus,

$$C(z) - 1.0z^{-1}C(z) + 0.632z^{-2}C(z) = 0.368z^{-1}R(z) + 0.264z^{-2}R(z)$$

Application of Eq. (10.20) yields the desired computational form

$$c(k) = c(k-1) - 0.632c(k-2) + 0.368r(k-1) + 0.264r(k-2)$$

The substitution of $c(k) = r(k) = 0$ for $k < 0$ and $r(k) = 1$ for $k \geq 0$ yields the following values for $c(k)$ at the sampling instants:

$$c(0) = 0$$

$$c(1) = 0.368r(0) = 0.368$$

$$c(2) = c(1) - 0.632c(0) + 0.368r(1) + 0.264r(0) = 1.00$$

$$c(3) = c(2) - 0.632c(1) + 0.368r(2) + 0.264r(1) = 1.40$$

(10.71)

To determine the response $c(k)$ at the sampling instants by the partial-fraction expansion method, first write $C(z)$ in the form

$$C(z) = \frac{0.368(z + 0.717)R(z)}{z^2 - z + 0.632}$$

For $R(z) = z/(z - 1)$, then

$$C(z) = z \left[\frac{0.368(z + 0.717)}{(z^2 - z + 0.632)(z - 1)} \right]$$

$$= z \left[\frac{K_c}{z - (0.50 + j0.62)} + \frac{K_{-c}}{z - (0.50 - j0.62)} + \frac{K_1}{z - 1} \right]$$

Note that the quadratic term has complex conjugate roots $a \pm jb = 0.50 \pm j0.62$. Inverting yields the general form for the solution

$$c(k) = R^k \frac{|K(a + jb)|}{b} \sin(k\beta + \alpha) + K_1$$

where

$$K_1 = \left[\frac{0.368(z + 0.717)}{z^2 - z + 0.632} \right]_{z=1} = 1.0$$

$$a + jb = R\underline{/\beta} = 0.50 + j0.62 = 0.80\underline{/51.0°}$$

$$K(a + jb) = |K(a + jb)|\underline{/\alpha} = \left[\frac{0.368(z + 0.717)}{z - 1} \right]_{z=0.50 + j0.62}$$

$$= \left[\frac{0.368(1.217 + j0.62)}{-0.50 + j0.62} \right] = \frac{0.368(1.366)\underline{/27.0°}}{0.80\underline{/129.0°}}$$

$$= 0.63\underline{/-102.0°}$$

Thus,

$$c(k) = 0.80^k \left(\frac{0.63}{0.62}\right) \sin(51.0k - 102.0) + 1$$

$$= 1 + 1.02(0.80)^k \sin(51.0k - 102.0)$$

Values of $c(k)$ are the same as those given by Eq. (10.71).

The time at the kth sampling instant is $t = kT$. Replacing $c(k)$ by $c(kT)$ yields values in terms of time.

First-Order Hold

The characteristics of a first-order hold are illustrated in Fig. 10.21a. For the region $nT < t < (n + 1)T$ the output is the straight line that is the extrapolation of the two preceding sampled values. That is,

$$f(t) = f(nT) + f'(nT)(t - nT) \qquad nT < t < (n + 1)T$$

where the slope $f'(nT)$ is

$$f'(nT) = \frac{f(nT) - f[(n - 1)T]}{T}$$

The behavior of a first-order hold is illustrated in Fig. 10.21b. The dashed curve is the continuous function $f(t)$. The output of the first-order hold is the series of straight-line segments. For a first-order hold the Laplace transform is

$$\left(\frac{1}{s} + \frac{1}{Ts^2}\right)(1 - e^{-Ts})$$

The entire class of higher-order hold circuits is revealed by writing the power series approximation for $f(t)$ between the sampling instants $nT < t < (n + 1)T$. Thus,

$$f(t) = f(nT) + f'(nT)(t - nT) + \frac{f''(nT)}{2!}(t - nT)^2 + \cdots$$

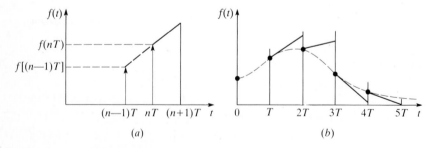

Figure 10.21 Characteristics and behavior of a first-order hold.

Figure 10.22 Characteristic of an exponential hold.

A zero-order hold approximates $f(t)$ by the first term, a first-order hold approximates $f(t)$ by the first two terms, a second-order hold approximates $f(t)$ by the first three terms, etc. Because of practical considerations, first-order and higher-order holds are seldom used. By far, the zero-order hold is the most commonly used circuit.

Exponential Hold

One of the simplest filters is the exponential hold. The transfer function for an exponential hold is

$$G(s) = \frac{1}{1 + \tau s}$$

Many components have this transfer function, and thus may be used as exponential hold filters. The output from an exponential hold is an exponential decay between the sampling periods, as is illustrated in Fig. 10.22. At each sampling instant, there is a step change or discontinuity between successive exponential decays.

10.7 DISCRETE-DATA SYSTEMS

When a system deals strictly with discrete data, the system can be described by the following set of discrete state equations:

$$\mathbf{x}[(k + 1)T] = A\mathbf{x}(kT) + \mathbf{b}f(kT) \qquad (10.72)$$

Application of Eqs. (10.19) and (10.21) yields, for the z transform,

$$z\mathbf{X}(z) - z\mathbf{x}(0) = A\mathbf{X}(z) + \mathbf{b}F(z)$$

or

$$[zI - A]\mathbf{X}(z) = z\mathbf{x}(0) - \mathbf{b}F(z)$$

Premultiplying both sides by $[zI - A]^{-1}$ gives

$$\mathbf{X}(z) = [zI - A]^{-1}z\mathbf{x}(0) + [zI - A]^{-1}\mathbf{b}F(z)$$
$$= [zI - A]^{-1}[z\mathbf{x}(0) + \mathbf{b}F(z)] \qquad (10.73)$$

Inverting yields

$$\mathbf{x}(kT) = Z^{-1}\{[zI - A]^{-1}z\}\mathbf{x}(0) + Z^{-1}\{[zI - A]^{-1}\mathbf{b}F(z)\}$$
$$= Z^{-1}\{[zI - A]^{-1}[z\mathbf{x}(0) + \mathbf{b}F(z)]\} \qquad (10.74)$$

Difference Equations

Any of the methods discussed in Sec. 9.1 may be used to obtain the state-space representation for a difference equation. For illustrative purposes, the method of general programming is described in the following. The general representation for an nth-order differential equation is given by Eq. (9.16). The corresponding state-space representation is given by Eqs. (9.17) and (9.18). Replacing $y(t)$ by $y(kT)$, $Dy(t)$ by $y[(k + 1)T]$, $D^2y(t)$ by $y[(k + 2)T]$, etc., yields the following general form for a difference equation:

$$
y[(k + n)T] + a_1 y[(k + n - 1)T] + \cdots + a_{n-1} y[(k + 1)T] + a_n y(kT)
$$
$$
= b_0 f[(k + n)T] + b_1 f[(k + n - 1)T] + \cdots
$$
$$
+ b_{n-1} f[(k + 1)T] + b_n f(kT) \tag{10.75}
$$

The corresponding state-space representation is

$$
\begin{bmatrix} x_1[(k + 1)T] \\ x_2[(k + 1)T] \\ \vdots \\ x_n[(k + 1)T] \end{bmatrix} = \begin{bmatrix} 0 & 1 & 0 & \cdots & 0 \\ 0 & 0 & 1 & \cdots & 0 \\ \hdotsfor{5} \\ -a_n & -a_{n-1} & -a_{n-2} & \cdots & -a_1 \end{bmatrix} \begin{bmatrix} x_1(kT) \\ x_2(kT) \\ \vdots \\ x_n(kT) \end{bmatrix} + \begin{bmatrix} h_1 \\ h_2 \\ \vdots \\ h_n \end{bmatrix} f(kT)
$$
$$
\tag{10.76}
$$

where

$$
x_1(kT) = y(kT) - b_0 f(kT)
$$
$$
x_2(kT) = x_1[(k + 1)T] - h_1 f(kT)
$$
$$
\cdots\cdots\cdots\cdots\cdots\cdots
$$
$$
x_n(kT) = x_{n-1}[(k + 1)T] - h_{n-1} f(kT)
$$

and

$$
h_1 = b_1 - a_1 b_0
$$
$$
h_2 = b_2 - a_2 b_0 - a_1 h_1
$$
$$
\cdots\cdots\cdots\cdots\cdots
$$
$$
h_n = b_n - a_n b_0 - a_{n-1} h_1 - \cdots - a_2 h_{n-2} - a_1 h_{n-1}
$$

The preceding equations express values in terms of time kT at the sampling instants. Replacing kT by k, $(k + 1)T$ by $(k + 1)$, $(k + n)T$ by $(k + n)$, etc., yields the corresponding equations for values at the kth sampling instant.

Illustrative example 10.17 Obtain a state-space representation for the difference equation

$$
y(k + 2) + 0.7y(k + 1) + 0.1y(k) = 2f(k + 1) + f(k)
$$

Solve for the response of this system for the case in which $y(k) = 0$ for $k \leq 0$, $f(k) = 0$ for $k < 0$ and $f(k) = 1$ for $k \geq 0$.

SOLUTION This difference equation has the form of Eq. (10.75). For $a_1 = 0.7$, $a_2 = 0.1$, $b_0 = 0$, $b_1 = 2$, and $b_2 = 1$, it follows that $h_1 = b_1 - a_1 b_0 = 2$ and $h_2 = b_2 - a_2 b_0 - a_1 h_1 = 1 - 1.4 = -0.4$. The matrix form is now obtained from Eq. (10.76). Thus,

$$\begin{bmatrix} x_1(k+1) \\ x_2(k+1) \end{bmatrix} = \begin{bmatrix} 0 & 1.0 \\ -0.1 & -0.7 \end{bmatrix} \begin{bmatrix} x_1(k) \\ x_2(k) \end{bmatrix} + \begin{bmatrix} 2.0 \\ -0.4 \end{bmatrix} f(k)$$

where $x_1(k) = y(k)$. The algebraic form of the preceding matrix equations is

$$x_1(k+1) = x_2(k) + 2f(k)$$
$$x_2(k+1) = -0.1x_1(k) - 0.7x_2(k) - 0.4f(k) \tag{10.77}$$

To prove that these relationships yield the original difference equation, solve the first expression for $x_2(k) = x_1(k+1) - 2f(k)$ and then replace k by $k+1$ to obtain $x_2(k+1) = x_1(k+2) - 2f(k+1)$. Substitution of these results for $x_2(k)$ and $x_2(k+1)$ into the second expression gives

$$x_1(k+2) - 2f(k+1) = -0.1x_1(k) - 0.7[x_1(k+1) - 2f(k)] - 0.4f(k)$$

Rearranging yields the original difference equation, with $x_1(k) = y(k)$.

Letting $k = -1$ in the original difference equation yields $y(1) + 0.7y(0) + 0.1y(-1) = 2f(0) + f(-1)$. For $y(0) = y(-1) = 0$ and $f(0) = 1$, then $y(1) = 2$. The resulting initial conditions are $x_1(0) = y(0) = 0$ and $x_2(0) = x_1(1) - 2f(0) = 2 - 2 = 0$, where $x_1(1) = y(1) = 2$. For $F(z) = z/(z-1)$,

$$[z\mathbf{x}(0) + \mathbf{b}F(z)] = \frac{z}{z-1} \begin{bmatrix} 2.0 \\ -0.4 \end{bmatrix}$$

The $[zI - A]^{-1}$ matrix is

$$[zI - A]^{-1} = \begin{bmatrix} z & -1.0 \\ 0.1 & z+0.7 \end{bmatrix}^{-1} = \frac{\begin{bmatrix} z+0.7 & 1.0 \\ -0.1 & z \end{bmatrix}}{(z+0.2)(z+0.5)}$$

From Eq. (10.73) it follows that

$$\begin{bmatrix} X_1(z) \\ X_2(z) \end{bmatrix} = z\frac{\begin{bmatrix} z+0.7 & 1.0 \\ -0.1 & z \end{bmatrix} \begin{bmatrix} 2.0 \\ -0.4 \end{bmatrix}}{(z-1)(z+0.2)(z+0.5)} = z\frac{\begin{bmatrix} 2.0 \\ -0.4 \end{bmatrix}}{(z-1)(z+0.2)}$$

The partial-fraction expansion is

$$\begin{bmatrix} X_1(z) \\ X_2(z) \end{bmatrix} = \begin{bmatrix} \dfrac{5z/3}{z-1} & -\dfrac{5z/3}{z+0.2} \\ \dfrac{-z/3}{z-1} & \dfrac{z/3}{z+0.2} \end{bmatrix}$$

Inverting gives

$$x_1(k) = \tfrac{5}{3}[1 - (-0.2)^k]$$

$$x_2(k) = -\tfrac{1}{3}[1 - (-0.2)^k]$$

where

$$y(k) = x_1(k)$$

As illustrated in this example, in the z transform method it is necessary to obtain the matrix inverse $[zI - A]^{-1}$. For higher-order systems this can become a very time-consuming process. As in the case of continuous systems, the use of the signal-flow-graph method results in the most direct solution.

Signal-Flow-Graph Method

The z transform of the discrete-data state equation, Eq. (10.72) is

$$z\mathbf{X}(z) - z\mathbf{x}(0) = A\mathbf{X}(z) + \mathbf{b}F(z)$$

To construct the signal-flow graph, express the preceding in the form

$$\mathbf{X}(z) = \mathbf{x}(0) + z^{-1}[A\mathbf{X}(z) + \mathbf{b}F(z)]$$

The inputs for this signal-flow graph are the initial conditions $\mathbf{x}(0)$ and the transform of the forcing function $F(z)$. The output nodes are the transform of the state variables $\mathbf{X}(z)$. The following matrix form is obtained directly from the signal-flow graph:

$$\begin{bmatrix} X_1(z) \\ X_2(z) \\ \vdots \\ X_n(z) \end{bmatrix} = \begin{bmatrix} A_{11}(z) & A_{12}(z) & \cdots & A_{1n}(z) \\ A_{21}(z) & A_{22}(z) & \cdots & A_{2n}(z) \\ \cdots\cdots\cdots\cdots\cdots\cdots\cdots\cdots \\ A_{n1}(z) & A_{n2}(z) & \cdots & A_{nn}(z) \end{bmatrix} \begin{bmatrix} x_1(0) \\ x_2(0) \\ \vdots \\ x_n(0) \end{bmatrix} + \begin{bmatrix} \psi_1(z) \\ \psi_2(z) \\ \vdots \\ \psi_n(z) \end{bmatrix} F(z)$$

Inverting yields the desired result. Note by comparison with Eq. (10.74) that the first matrix on the right is $[zI - A]^{-1}z$ and the last matrix is $[zI - A]^{-1}\mathbf{b}$.

Illustrative example 10.18 Use the signal-flow-graph method to determine the solution of Illustrative example 10.17.

SOLUTION The z transform of Eqs. (10.77) is

$$zX_1(z) - zx_1(0) = X_2(z) + 2F(z)$$

$$zX_2(z) - zx_2(0) = -0.1X_1(z) - 0.7X_2(z) - 0.4F(z)$$

To construct the signal-flow graph, first express these equations in the form

$$X_1(z) = x_1(0) + z^{-1}[X_2(z) + 2F(z)]$$

$$X_2(z) = x_2(0) + z^{-1}[-0.1X_1(z) - 0.7X_2(z) - 0.4F(z)]$$

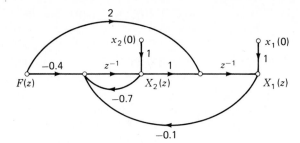

Figure 10.23 Signal-flow graph.

The resulting signal-flow graph is shown in Fig. 10.23. Application of Mason's gain formula to this signal-flow graph gives

$$\begin{bmatrix} X_1(z) \\ X_2(z) \end{bmatrix} = \frac{\begin{bmatrix} z(z+0.7) & z \\ -0.1z & z^2 \end{bmatrix} \begin{bmatrix} x_1(0) \\ x_2(0) \end{bmatrix}}{(z+0.2)(z+0.5)} + \frac{\begin{bmatrix} 2 \\ -0.4 \end{bmatrix}}{z+0.2} F(z)$$

For $x_1(0) = x_2(0) = 0$ and $F(z) = z/(z-1)$, the preceding reduces to the same transformed equation obtained in Illustrative example 10.17. Thus, the same result follows.

Discrete-Data Representation for Continuous Systems

The state-space representation for a continuous system is

$$\dot{\mathbf{x}}(t) = A\mathbf{x}(t) + \mathbf{b}f(t)$$

The solution referred to time $t = 0$ is given by Eq. (9.59). That is,

$$\mathbf{x}(t) = e^{At}\mathbf{x}(0) + \int_0^t e^{A(t-\tau)}\mathbf{b}f(\tau)\,d\tau \qquad (9.59)$$

This is the state-transition equation for a continuous system.

For the case in which the initial state is known at time $t = t_0$ rather than at time $t = 0$, the solution is obtained by first writing the state-space representation in the form

$$\dot{\mathbf{x}}(t) - A\mathbf{x}(t) = \mathbf{b}f(t)$$

Multiplying through by e^{-At} gives

$$e^{-At}[\dot{\mathbf{x}}(t) - A\mathbf{x}(t)] = \frac{d}{dt}[e^{-At}\mathbf{x}(t)] = e^{-At}\mathbf{b}f(t)$$

Integration between the limits t_0 and t of the two expressions on the right side of the preceding equation shows that

$$e^{-At}\mathbf{x}(t) - e^{-At_0}\mathbf{x}(t_0) = \int_{t_0}^t e^{-A\tau}\mathbf{b}f(\tau)\,d\tau$$

Multiplying through by e^{At} and solving for $\mathbf{x}(t)$ yields the desired solution referred to time t_0. That is,

$$\mathbf{x}(t) = e^{A(t-t_0)}\mathbf{x}(t_0) + \int_{t_0}^{t} e^{A(t-\tau)}\mathbf{b}f(\tau)\,d\tau \tag{10.78}$$

For $t_0 = kT$ and $t = (k+1)T$, the preceding becomes

$$\mathbf{x}[(k+1)T] = e^{AT}\mathbf{x}(kT) + \int_{kT}^{(k+1)T} e^{A[(k+1)T-\tau]}\mathbf{b}f(\tau)\,d\tau$$

Assuming that $f(\tau)$ remains constant over the interval $kT < t < (k+1)T$, then $f(\tau)$ may be replaced by $f(kT)$. Letting $\lambda = (k+1)T - \tau$, in which case $d\tau = -d\lambda$, then

$$\mathbf{x}[(k+1)T] = e^{AT}\mathbf{x}(kT) + \int_0^T e^{A\lambda}\,d\lambda\mathbf{b}f(kT)$$

This result may be expressed in the form

$$\mathbf{x}[(k+1)T] = \Phi(T)\mathbf{x}(kT) + G(T)f(kT) \tag{10.79}$$

where

$$\Phi(T) = e^{AT} \quad \text{and} \quad G(T) = \left(\int_0^T e^{A\lambda}\,d\lambda\right)\mathbf{b}$$

This is the discrete-data representation for the continuous system $\dot{\mathbf{x}} = A\mathbf{x} + \mathbf{b}f(t)$. Note that it has the same form as Eq. (10.72), in which $\Phi(T)$ corresponds to A and $G(T)$ corresponds to $\mathbf{b}$.

Illustrative example 10.19 Obtain the discrete-time state-space representation for the differential equation

$$\ddot{c} + 3\dot{c} + 2c = f(t)$$

Evaluate the resulting discrete-time representation for a sampling period $T = 0.2$ s.

SOLUTION Letting $c = x_1$ and $\dot{c} = \dot{x}_1 = x_2$ yields the state-space representation

$$\begin{bmatrix} \dot{x}_1 \\ \dot{x}_2 \end{bmatrix} = \begin{bmatrix} 0 & 1 \\ -2 & -3 \end{bmatrix}\begin{bmatrix} x_1 \\ x_2 \end{bmatrix} + \begin{bmatrix} 0 \\ 1 \end{bmatrix}f(t)$$

The $\Phi(t) = e^{AT}$ matrix is

$$\Phi(t) = \mathscr{L}^{-1}[sI - A]^{-1} = \mathscr{L}^{-1}\begin{bmatrix} s & -1 \\ 2 & s+3 \end{bmatrix}^{-1} = \mathscr{L}^{-1}\frac{\begin{bmatrix} s+3 & 1 \\ -2 & s \end{bmatrix}}{s^2 + 3s + 2}$$

$$= \begin{bmatrix} 2e^{-t} - e^{-2t} & e^{-t} - e^{-2t} \\ -2e^{-t} + 2e^{-2t} & -e^{-t} + 2e^{-2t} \end{bmatrix}$$

Replacing t by T yields the matrix $\Phi(T)$. The $G(T)$ matrix is obtained as follows:

$$G(T) = \begin{bmatrix} \int_0^T (2e^{-\lambda} - e^{-2\lambda})\, d\lambda & \int_0^T (e^{-\lambda} - e^{-2\lambda})\, d\lambda \\ \int_0^T (-2e^{-\lambda} + 2e^{-2\lambda})\, d\lambda & \int_0^T (-e^{-\lambda} + 2e^{-2\lambda})\, d\lambda \end{bmatrix} \begin{bmatrix} 0 \\ 1 \end{bmatrix}$$

$$= \begin{bmatrix} \dfrac{1 - 2e^{-T} + e^{-2T}}{2} \\ e^{-T} - e^{-2T} \end{bmatrix}$$

The substitution of these results into Eq. (10.79) yields

$$\begin{bmatrix} x_1[(k+1)T] \\ x_2[(k+1)T] \end{bmatrix} = \begin{bmatrix} 2e^{-T} - e^{-2T} & e^{-T} - e^{-2T} \\ -2e^{-T} + 2e^{-2T} & -e^{-T} + 2e^{-2T} \end{bmatrix} \begin{bmatrix} x_1(kT) \\ x_2(kT) \end{bmatrix}$$
$$+ \begin{bmatrix} \dfrac{1 - 2e^{-T} + e^{-2T}}{2} \\ e^{-T} - e^{-2T} \end{bmatrix} f(kT)$$

Evaluation at $T = 0.2$ gives

$$\begin{bmatrix} x_1[(k+1)T] \\ x_2[(k+1)T] \end{bmatrix} = \begin{bmatrix} 0.968 & 0.149 \\ -0.298 & 0.521 \end{bmatrix} \begin{bmatrix} x_1(kT) \\ x_2(kT) \end{bmatrix} + \begin{bmatrix} 0.016 \\ 0.149 \end{bmatrix} f(kT)$$

Note that the form of this result is identical to Eq. (10.72), in which the first matrix on the right is the A matrix and the last matrix is the **b** matrix.

Response between Sampling Instants

An important feature of the state-variable method is that it can be modified easily to determine the output between sampling instants. Thus, letting $t_0 = kT$ and $t = (k + \Delta)T$, where $0 \le \Delta < 1$, in Eq. (10.78) gives

$$\mathbf{x}[(k + \Delta)T] = e^{A\Delta T}\mathbf{x}(kT) + \int_{kT}^{(k+\Delta)T} e^{A[(k+\Delta)T - \tau]}\mathbf{b}f(\tau)\, d\tau$$

Replacing $(k + \Delta)T - \tau$ by λ and noting that $d\tau = -d\lambda$ gives

$$\mathbf{x}[(k + \Delta)T] = e^{A\Delta T}\mathbf{x}(kT) + \int_0^{\Delta T} e^{A\lambda}\, d\lambda\, \mathbf{b}f(kT)$$

$$= \Phi(\Delta T)\mathbf{x}(kT) + G(\Delta T)f(kT) \tag{10.80}$$

where

$$\Phi(\Delta T) = e^{A\Delta T} \qquad \text{and} \qquad G(\Delta T) = \left(\int_0^{\Delta T} e^{A\lambda}\, d\lambda \right)\mathbf{b}$$

The function $\Phi(\Delta T)$ is obtained by replacing T by ΔT in the function $\Phi(T)$. Similarly, $G(\Delta T)$ is obtained by replacing T by ΔT in the $G(T)$ function.

Illustrative example 10.20 For the system of Illustrative example 10.17, determine the discrete-data representation for the response halfway between the sampling instants.

SOLUTION Replacing T by ΔT in the $\Phi(T)$ and $G(T)$ matrices yields the $\Phi(\Delta T)$ and $G(\Delta T)$ matrices. Application of Eq. (10.80) yields

$$\begin{bmatrix} x_1[(k + \Delta)T] \\ x_2[(k + \Delta)T] \end{bmatrix} = \begin{bmatrix} 2e^{-\Delta T} - e^{-2\Delta T} & e^{-\Delta T} - e^{-2\Delta T} \\ -2e^{-\Delta T} + 2e^{-2\Delta T} & -e^{-\Delta T} + 2e^{-2\Delta T} \end{bmatrix} \begin{bmatrix} x_1(kT) \\ x_2(kT) \end{bmatrix}$$
$$+ \begin{bmatrix} \dfrac{1 - 2e^{-\Delta T} + e^{-2\Delta T}}{2} \\ e^{-\Delta T} - e^{-2\Delta T} \end{bmatrix} f(kT)$$

For $\Delta = 0.5$ and $T = 0.2$, then $\Delta T = 0.1$. Thus,

$$\begin{bmatrix} x_1[(k + 0.5)T] \\ x_2[(k + 0.5)T] \end{bmatrix} = \begin{bmatrix} 0.991 & 0.086 \\ -0.172 & 0.733 \end{bmatrix} \begin{bmatrix} x_1(kT) \\ x_2(kT) \end{bmatrix} + \begin{bmatrix} 0.009 \\ 0.086 \end{bmatrix} f(kT)$$

Note that to obtain the response between the sampling instants, it is necessary to use Eq. (10.79) to obtain the values $x(kT)$ used in the preceding expression.

The values $f(kT)$ are the values of the input signal at the sampling instants. Letting Δ approach one in the term $x[(k + \Delta)T]$ yields $x[(k + 1)T]$. The values $x[(k + 1)T]$ are the values for x just before the $(k + 1)$th sampling instant. Letting Δ approach zero yields the values of x just after the kth sampling instant. Inverting a z transform such as $X(z)$ yields values of x just after each sampling instant.

10.8 SAMPLED-DATA CONTROL SYSTEMS

For the system shown in Fig. 10.24a, the output of the sampler is $e^*(t)$. For the interval $kT \le t < (k + 1)T$, the output of the sampler is the unit impulse $e(kT)$. That is,

$$e^*(t) = e(kT)u_1(t - kT) \qquad kT \le t < (k + 1)T$$

The Laplace transform is

$$E^*(s) = e(kT)$$

The effect of the time delay $t_0 = kT$ will be accounted for later in the analysis by replacing t by $t - kT$ in the inverse Laplace transform. The differential equation for the continuous portion of the system is

$$c(t) = \frac{Ke^*(t)}{D(D + a)}$$

or

$$\ddot{c} + a\dot{c} = Ke^*(t)$$

(a)

(b)

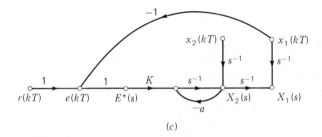

(c)

Figure 10.24 Sampled-data system and state-space representation.

Letting $c = x_1$ and $\dot{c} = \dot{x}_1 = x_2$, the state-space representation is

$$\begin{bmatrix} \dot{x}_1 \\ \dot{x}_2 \end{bmatrix} = \begin{bmatrix} 0 & 1 \\ 0 & -a \end{bmatrix} \begin{bmatrix} x_1 \\ x_2 \end{bmatrix} + \begin{bmatrix} 0 \\ K \end{bmatrix} e^*(t)$$

The Laplace transform for each of these relationships is

$$sX_1(s) - x_1(kT) = X_2(s)$$

$$sX_2(s) - x_2(kT) = -aX_2(s) + KE^*(s)$$

The signal-flow graph for the preceding relationships is shown in Fig. 10.24b. The error signal at time kT is

$$e(kT) = r(kT) - c(kT) = r(kT) - x_1(kT)$$

The signal-flow graph for the entire system may now be completed as shown in Fig. 10.24c. Application of Mason's gain formula gives

$$\begin{bmatrix} X_1(s) \\ X_2(s) \end{bmatrix} = \begin{bmatrix} \dfrac{1}{s} - \dfrac{K}{s(s+a)} & \dfrac{1}{s(s+a)} \\[2ex] \dfrac{-K}{s+a} & \dfrac{1}{s+a} \end{bmatrix} \begin{bmatrix} x_1(kT) \\ x_2(kT) \end{bmatrix} + \begin{bmatrix} \dfrac{K}{s(s+a)} \\[2ex] \dfrac{K}{s+a} \end{bmatrix} r(kT)$$

Because of the time delay $t_0 = kT$, after the inverse Laplace transform is obtained, t is replaced by $t - kT$. Thus,

$$\begin{bmatrix} x_1(t) \\ x_2(t) \end{bmatrix} = \begin{bmatrix} 1 - \dfrac{K}{a}(1 - e^{-at^*}) & \dfrac{1}{a}(1 - e^{-at^*}) \\ -Ke^{-at^*} & e^{-at^*} \end{bmatrix} \begin{bmatrix} x_1(kT) \\ x_2(kT) \end{bmatrix} + \begin{bmatrix} \dfrac{K}{a}(1 - e^{-at^*}) \\ Ke^{-at^*} \end{bmatrix} r(kT)$$

where $t^* = t - kT$. The value of the output at the sampling instants is obtained by letting $t = (k + 1)T$, in which case $t^* = t - kT = (k + 1)T - kT = T$. For $a = 1$, the preceding becomes

$$\begin{bmatrix} x_1[(k + 1)T] \\ x_2[(k + 1)T] \end{bmatrix} = \begin{bmatrix} 1 - K(1 - e^{-T}) & 1 - e^{-T} \\ -Ke^{-T} & e^{-T} \end{bmatrix} \begin{bmatrix} x_1(kT) \\ x_2(kT) \end{bmatrix} + \begin{bmatrix} K(1 - e^{-T}) \\ Ke^{-T} \end{bmatrix} r(kT)$$

For $K = 0.25$ and for a sampling period $T = 1$, these equations become

$$\begin{bmatrix} x_1[(k + 1)T] \\ x_2[(k + 1)T] \end{bmatrix} = \begin{bmatrix} 0.842 & 0.632 \\ -0.092 & 0.368 \end{bmatrix} \begin{bmatrix} x_1(kT) \\ x_2(kT) \end{bmatrix} + \begin{bmatrix} 0.158 \\ 0.092 \end{bmatrix} r(kT)$$

For $x_1(0) = c(0) = 0$, $x_2(0) = x_1(0) = \dot{c}(0) = 0$, $r(kT) = 0$ for $k < 0$, and $r(kT) = 1$ for $k \geq 0$, this problem is the same as Illustrative example 10.16. The preceding equation yields the values $x_1(0) = 0$, $x_1(T) = 0.158$, $x_1(2T) = 0.349$, and $x_1(3T) = 0.522$. These values correspond to the values $c(0)$, $c(T)$, $c(2T)$, and $c(3T)$ respectively of Eqs. (10.64). To obtain the response between sampling instants, let $t = (k + \Delta)T$, in which case $t^* = t - kT = \Delta T$. Thus,

$$\begin{bmatrix} x_1[(k + \Delta)T] \\ x_2[(k + \Delta)T] \end{bmatrix} = \begin{bmatrix} 1 - K(1 - e^{-\Delta T}) & 1 - e^{-\Delta T} \\ -Ke^{-\Delta T} & e^{-\Delta T} \end{bmatrix} \begin{bmatrix} x_1(kT) \\ x_2(kT) \end{bmatrix}$$
$$+ \begin{bmatrix} K(1 - e^{-\Delta T}) \\ Ke^{-\Delta T} \end{bmatrix} r(kT)$$

The response at the midpoint is obtained by letting $\Delta = 0.5$. Thus, for $K = 0.25$ and $T = 1$,

$$\begin{bmatrix} x_1[(k + 0.5)T] \\ x_2[(k + 0.5)T] \end{bmatrix} = \begin{bmatrix} 0.902 & 0.393 \\ -0.152 & 0.606 \end{bmatrix} \begin{bmatrix} x_1(kT) \\ x_2(kT) \end{bmatrix} + \begin{bmatrix} 0.098 \\ 0.152 \end{bmatrix} r(kT)$$

Because $c = x_1$, values midway between the sampling instants are $c(0.5T) = 0.098$, $c(1.5T) = 0.277$, $c(2.5T) = 0.458$, etc.

Zero-Order Hold

The system shown in Fig. 10.25a is the same as that of Fig. 10.24a except that a zero-order hold has been added after the sampler. The output $m(t)$ of the zero-order hold is constant between sampling instants. That is,

$$m(t) = e(kT) \qquad kT \leq t < (k + 1)T$$

The Laplace transform is

$$M(s) = \frac{e(kT)}{s} \tag{10.81}$$

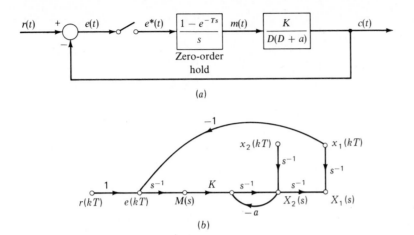

Figure 10.25 Sampled-data system with zero-order hold and state-space representation.

The effect of the time delay $t_0 = kT$ will be accounted for later by replacing t by $t - kT$. The error signal at time kT is

$$e(kT) = r(kT) - c(kT) = r(kT) - x_1(kT) \qquad (10.82)$$

The signal-flow graph for the plant is the same as that shown in Fig. 10.24b except that $E^*(s)$ is replaced by $M(s)$. Incorporation of the signal-flow graph for Eqs. (10.81) and (10.82) with that for the plant yields the signal-flow graph for the system shown in Fig. 10.25b. Application of Mason's gain formula to this signal-flow graph yields

$$
\begin{bmatrix} X_1(s) \\ X_2(s) \end{bmatrix} =
\begin{bmatrix} \dfrac{1}{s} - \dfrac{K}{s^2(s+a)} & \dfrac{1}{s(s+a)} \\[2ex] \dfrac{-K}{s(s+a)} & \dfrac{1}{s+a} \end{bmatrix}
\begin{bmatrix} x_1(kT) \\ x_2(kT) \end{bmatrix}
+
\begin{bmatrix} \dfrac{K}{s^2(s+a)} \\[2ex] \dfrac{K}{s(s+a)} \end{bmatrix} r(kT)
$$

Because of the time delay $t_0 = kT$, after the inverse Laplace transform has been obtained, t is replaced by $t - kT$. For $a = 1$, the resulting inverse is

$$
\begin{bmatrix} x_1(t) \\ x_2(t) \end{bmatrix} =
\begin{bmatrix} 1 - K(t^* - 1 + e^{-t^*}) & 1 - e^{-t^*} \\ -K(1 - e^{-t^*}) & e^{-t^*} \end{bmatrix}
\begin{bmatrix} x_1(kT) \\ x_2(kT) \end{bmatrix}
+
\begin{bmatrix} K(t^* - 1 + e^{-t^*}) \\ K(1 - e^{-t^*}) \end{bmatrix} r(kT)
$$

where $t^* = t - kT$. The value of the output at the sampling instants is obtained by letting $t = (k + 1)T$, in which case $t^* = (k + 1)T - kT = T$. Replacing t by $(k + 1)T$ and t^* by T in the preceding gives

$$
\begin{bmatrix} x_1[(k+1)T] \\ x_2[(k+1)T] \end{bmatrix} =
\begin{bmatrix} 1 - K(T - 1 + e^{-T}) & 1 - e^{-T} \\ -K(1 - e^{-T}) & e^{-T} \end{bmatrix}
\begin{bmatrix} x_1(kT) \\ x_2(kT) \end{bmatrix}
$$
$$
+
\begin{bmatrix} K(T - 1 + e^{-T}) \\ K(1 - e^{-T}) \end{bmatrix} r(kT)
$$

For $K = 1$ and $T = 1$, the response at the sampling instants is

$$\begin{bmatrix} x_1[(k+1)T] \\ x_2[(k+1)T] \end{bmatrix} = \begin{bmatrix} 0.632 & 0.632 \\ -0.632 & 0.368 \end{bmatrix} \begin{bmatrix} x_1(kT) \\ x_2(kT) \end{bmatrix} + \begin{bmatrix} 0.368 \\ 0.632 \end{bmatrix} r(kT)$$

For $x_1(0) = c_0 = 0$, $x_2(0) = \dot{x}_1(0) = \dot{c}_0 = 0$, $r(kT) = 0$ for $k < 0$, and $r(kT) = 1$ for $k \geq 0$, this problem is the same as that worked in Sec. 10.6 whose solution is given by Eqs. (10.71). The preceding matrix yields the values $x_1(0) = 0$, $x_1(T) = 0.368$, $x_1(2T) = 1.0$, and $x_1(3T) = 1.4$. These values correspond to the values $c(0) = 0$, $c(T) = 0.368$, $c(2T) = 1.0$, and $c(3T) = 1.4$ respectively of Eqs. (10.71).

The response between sampling instants is obtained by replacing t by $(k + \Delta)T$, so that $t^* = \Delta T$. The result for $K = 1$, $\Delta = 0.5$, and $T = 1$ is

$$\begin{bmatrix} x_1[(k+\Delta)T] \\ x_2[(k+\Delta)T] \end{bmatrix} = \begin{bmatrix} 0.893 & 0.393 \\ -0.393 & 0.607 \end{bmatrix} \begin{bmatrix} x_1(kT) \\ x_2(kT) \end{bmatrix} + \begin{bmatrix} 0.107 \\ 0.393 \end{bmatrix} r(kT)$$

Because $c = x_1$, then $c(0.5T) = x_1(0.5T) = 0.107$, $c(1.5T) = x_1(1.5T) = 0.684$, and $c(2.5T) = x_1(2.5T) = 1.248$. Note that in obtaining the response between sampling instants, it is necessary to know the values $x_1(kT)$ and $x_2(kT)$ at the beginning of each sampling instant. In this method, the values $x_1(kT)$, $x_2(kT)$, etc., are the values just before the sampling instants, whereas the values obtained by taking the inverse z transform are the values just after the sampling instants. If no discontinuity occurs at the sampling instants, these values are the same. In summary, letting $t^* = T$ yields the values for obtaining the response at the sampling instants, $\mathbf{x}[(k+1)T]$, and letting $t^* = \Delta T$ yields the values for obtaining the response between the sampling instants, $\mathbf{x}[(k+\Delta)T]$.

Sampled-data systems are readily programmed for solution on a digital computer. The equation relating $c(t)$ to the sampled signal $e^*(t)$ for the system shown in Fig. 10.24a is

$$c(t) = \frac{K}{D(D+a)} e^*(t)$$

For direct programming, x_1 is the differential equation that results when the numerator is replaced by one. That is,

$$x_1 = \frac{1}{D(D+a)} e^*(t)$$

For $a = 1$, the resulting state-space representation is

$$\dot{x}_1 = x_2$$

$$\dot{x}_2 = -x_2 + e^*(t)$$

$$c = Kx_1$$

The area of the pulse representing $e^*(t)$ is equal to the value of $e(t)$ at the sampling instant. By designating Δt as the integration interval (width of pulse), then

$$e^*(t)\Delta t = e(t)$$

or

$$e^*(t) = \frac{e(t)}{\Delta t}$$

The computer program for obtaining the system response $c(t)$ for the case in which $r(t) = u(t)$ is a unit step function, $K = 0.25$, and all initial conditions are zero is

```
      REAL K
      X1 = 0.0
      X2 = 0.0
      DX2 = 0.0
      DT = 0.01
      T = 0.0
      K = 0.25
      R = 1.0
      DO 25 I = 1, 5
      DO 20 J = 0, 99
      C = K*X1
      E = R−C
      ESTAR = 0.0
      IF(J.NE.0) GO TO 5
      ESTAR = E/DT
    5 IF (MOD(J,10).NE.0) GO TO 15
      WRITE (6,10) T, E, C
   10 FORMAT(3F10.2)
   15 T = T+DT
      X10 = X1
      X20 = X2
      DX20 = DX2
      DX2 = ESTAR−X2
      X2 = 0.5*DT*(DX2+DX20)+X20
      X1 = 0.5*DT*(X2+X20)+X10
   20 CONTINUE
   25 CONTINUE
      END
```

In this program X1, X2, DX2, R, C, E, and ESTAR represent x_1, x_2, $\dot{x}_2$, $r(t)$, $c(t)$, $e(t)$, and $e^*(t)$ respectively. The integration interval is DT $= 0.01$, and the starting value of time is T $= 0.0$. The statements before the first DO loop initialize the system. For the first DO loop when I $= 1$, the solution is being executed during the first sampling period. For I $= 2$, the solution is being executed during the second sampling period, etc. The second DO loop (FOR J $= 0$, 99) executes the solution 100 times during each sampling period. For DT $= 0.01$, the sampling period is $100(0.01) = 1$ s. Next, values of $c(t)$ and $e(t)$ are determined. For the first integration interval of a sampling period J $= 0$. When J $= 0$, ESTAR $= $ E/DT. For the remainder of the sampling interval J is not equal to 0, so ESTAR $= 0$. Corresponding values of T, E, and C are now printed. The T $=$ T+DT statement sets the value of time for the end of the next integration interval. The X10 $=$ X1 statement sets the present value of X1 as the starting value for the new integration interval. Similarly, the X20 $=$ X2 and the DX20 $=$ DX2 statements set the present values of X2 and DX2 as the starting values for the new integration interval. The value of DX2 at the end of this integration interval is obtained by the statement DX2 = ESTAR − X2. The new value of X2 is obtained by the statement X2 = 0.5*DT*(DX2 + DX20) + X20. The value of X20 on the right side is the value at the beginning of the integration interval. The value of the derivative at the beginning of the interval is DX20 and the value at the end is DX2; thus the term 0.5*DT*(DX2 + DX20) + X20 is the trapezoidal integration of $\dot{x}_2$ during the interval. Similarly, the new value of X1 is obtained by the statement X1 = 0.5*DT*(X2 + X20) + X10. Because $\dot{x}_1 = x_2$, the right side is the trapezoidal integration of $\dot{x}_1$ during the interval.

The system shown in Fig. 10.25a is the same as that shown in Fig. 10.24a except that a zero-order hold has been added. It is a relatively easy matter to change the computer program for the case of a zero-order hold. The input to the plant is now $m(t)$ rather than $e^*(t)$. The three consecutive statements ESTAR = 0.0, IF(J.NE.0) GO TO 5, and ESTAR = E/DT, are now replaced by the two statements

IF(J.NE.0) GO TO 5
M = E

where $m(t)$ is designated as M. At the start of the sampling interval J = 0. Thus M takes on the value E = $e(t)$ at the beginning of the sampling period. It then retains this value until the beginning of the next sampling period when J again equals 0. The statement DX2 = ESTAR − X2 is replaced by DX2 = M − X2, REAL K is replaced by REAL K,M and K = 1.0 rather than 0.25.

10.9 COMPUTER-CONTROLLED SYSTEMS

A schematic diagram of a system controlled by a digital computer is shown in Fig. 10.26. At each sampling instant, the digital controller samples the error signal $e(t)$. The controller operates on this sampled value $e^*(t)$ and previous sampled values to obtain an output $m^*(t)$. This value of $m^*(t)$ is then retained by the zero-order hold until a new value is computed at the next sampling instant.

That is, during the kth sampling interval $m(t)$ retains the value $m^*(t) = m(kT)$ at the beginning of the interval:

$$m(t) = m(kT) \qquad kT \leq t < (k+1)T$$

Figure 10.26 Digital-control system.

By designating the Laplace transform of the zero-order hold and plant as $G(s)$, it follows from Fig. 10.26 that

$$C(s) = G(s)M^*(s)$$

Starring yields

$$C^*(s) = G^*(s)M^*(s)$$

Hence

$$C(z) = G(z)M(z) \tag{10.83}$$

The z transform for the digital controller is

$$M(z) = D(z)E(z) \tag{10.84}$$

The Laplace transform for the comparator equation $[e(t) = r(t) - c(t)]$ is

$$E(s) = R(s) - C(s)$$

The corresponding z transform relationship is

$$E(z) = R(z) - C(z) \tag{10.85}$$

The z transform for the output $C(z)$ is obtained as follows:

$$C(z) = G(z)M(z) = G(z)D(z)E(z)$$

$$\doteq G(z)D(z)[R(z) - C(z)]$$

Thus,

$$C(z) = \frac{G(z)D(z)}{1 + G(z)D(z)} R(z)$$

Solving for the z transform of the controller gives

$$D(z) = \frac{C(z)}{G(z)[R(z) - C(z)]} = \frac{1}{G(z)} \frac{C(z)/R(z)}{1 - C(z)/R(z)} \tag{10.86}$$

This result shows that the controller characteristics $D(z)$ may be obtained by knowing the plant and hold characteristics $G(z)$ and the desired response $C(z)$ to a given input $R(z)$.

Illustrative example 10.21 The transfer function for a plant is $(s + 2)/s(s + 1)$. Determine the characteristics of a digital controller such that the response of the system to a unit step function will be $c(t) = 5(1 - e^{-2t})$. The sampling period is $T = 1.0$ s.

SOLUTION For the plant and zero-order hold

$$G(s) = G_1(s)G_2(s) = (1 - e^{-Ts}) \frac{s + 2}{s^2(s + 1)}$$

where

$$G_1(s) = 1 - e^{-Ts}$$

and

$$G_2(s) = \frac{s + 2}{s^2(s + 1)} = \frac{2}{s^2} - \frac{1}{s} + \frac{1}{s + 1}$$

Application of Eq. (10.68) yields

$$G(z) = \frac{z-1}{z} G_2(z) = \frac{z-1}{z} \left[\frac{2z}{(z-1)^2} - \frac{z}{z-1} + \frac{z}{z-0.368} \right]$$

$$= \frac{1.368z - 0.104}{z^2 - 1.368z + 0.368}$$

For

$$C(s) = 5\left(\frac{1}{s} - \frac{1}{s+2} \right)$$

then

$$C(z) = 5\left(\frac{z}{z-1} - \frac{z}{z-0.135} \right) = \frac{4.32z}{(z-1)(z-0.135)}$$

Thus,

$$\frac{C(z)}{R(z)} = \frac{4.32}{z-0.135} \quad \text{and} \quad 1 - \frac{C(z)}{R(z)} = \frac{z-4.45}{z-0.135}$$

Substitution of the preceding results into Eq. (10.86) gives

$$D(z) = \frac{M(z)}{E(z)} = \frac{z^2 - 1.368z + 0.368}{1.368z - 0.104} \cdot \frac{4.32}{z-4.45} = \frac{4.32z^2 - 5.91z + 1.59}{1.368z^2 - 6.202z + 0.46}$$

Dividing numerator and denominator by $1.368z^2$ and cross-multiplying gives

$$M(z) - 4.53z^{-1}M(z) + 0.34z^{-2}M(z) = 3.16E(z) - 4.32z^{-1}E(z) + 1.16z^{-2}E(z)$$

Inverting yields the desired controller characteristics

$$m(k) = 4.53m(k-1) - 0.34m(k-2) + 3.16e(k) - 4.32e(k-1) + 1.16e(k-2)$$

Discrete Equivalent of Analog Computers

One of the most versatile and widely used controllers for continuous systems is the one which provides proportional plus integral action. The equation of operation for this controller is

$$m(t) = \left(K_1 + K_2 \frac{1}{D} \right) e(t) \tag{10.87}$$

The contribution of the various actions is adjusted by varying the constants K_1 and K_2. The discrete equivalent of this controller is commonly employed as the digital controller for process control systems. Using rectangular integration to approximate the integral yields, for the discrete form,

$$m(k) = K_1 e(k) + K_2 T \sum_{n=1}^{k} e(n) + m(0)$$

Generally it is more accurate and more convenient to work with the change in m in going from one sampling period to the next. Thus replacing k by $k - 1$ in the preceding and subtracting yields

$$\Delta m(k) = m(k) - m(k - 1) = K_1[e(k) - e(k - 1)] + K_2\,Te(k)$$

Note that the preceding equation eliminates the need to know the initial value $m(0)$. The z transform of this equation is

$$M(z) - z^{-1}M(z) = K_1[E(z) - z^{-1}E(z)] + K_2\,TE(z)$$

or
$$D(z) = \frac{M(z)}{E(z)} = K_1 + K_2\,T\,\frac{z}{z - 1}$$

The Laplace transform for the continuous controller is

$$D(s) = \frac{M(s)}{E(s)} = K_1 + K_2\,T\,\frac{1}{s}$$

Note that the $z/(z - 1)$ term for the discrete controller corresponds to integration in the equivalent continuous controller (i.e., the $1/s$ term).

Optimum Response

It is now shown how the characteristics of a controller may be selected to yield an optimum response. Any one of numerous optimum-performance criterion may be used. In the following, the method developed by Kalman[1] for obtaining the controller characteristics to yield a desired transient behavior is described. In Fig. 10.26 is shown a schematic diagram of the overall system. For illustrative purposes assume that the transfer function of the plant is $ab/(s + a)(s + b)$. The transient behavior is considered optimum when the system responds to a step function in minimum time with no overshoot. In addition, there should be no steady-state error.

In Fig. 10.27, the optimum response $c(t)$ is indicated by the heavy line. The unit step input is $r(t) = 1$. During the first interval $0 \le t < T$, the input to the plant $m(t) = q_0$ is large so as to accelerate the system rapidly toward its final value. The first dotted extension in Fig. 10.27 shows the subsequent response if the plant input is not changed after the first sampling period. Thus, at time $t = T$, it is necessary to begin to decelerate the response so as to achieve the desired final value $c(\infty)$. During the second interval $T \le t < 2T$, the plant input is changed by an amount q_1, so that $m(t) = q_0 + q_1$. As drawn in Fig. 10.27, the term q_1 is a negative constant. The response $c(t)$ reaches the final value at the end of the second interval $t = 2T$. The second dotted extension shows the path of the response if the plant input is not changed after the second interval. Thus, for the third and subsequent intervals, the plant input is $q_0 + q_1 + q_2 = m(t)$.

A desirable feature of this performance criterion is that the error is reduced exactly to zero after two periods. The error then remains zero not only at the sampling points but also in between. This criterion thus eliminates the possibility of ripple (i.e., undesirable oscillations between sampling instants).

Figure 10.27 Optimum response.

Because $m(t)$ is the output of a zero-order hold, it follows that the z transform of the sampled input $m^*(t)$ is

$$M(z) = q_0 + (q_0 + q_1)z^{-1} + (q_0 + q_1 + q_2)(z^{-2} + z^{-3} + \cdots)$$

Dividing $M(z)$ by the z transform of the unit step input $[R(z) = 1 + z^{-1} + z^{-2} + \cdots]$ gives

$$\frac{M(z)}{R(z)} = q_0 + q_1 z^{-1} + q_2 z^{-2} \tag{10.88}$$

Similarly, from Fig. 10.27 the z transform of the response at the sampling instants is

$$C(z) = p_1 z^{-1} + (p_1 + p_2)(z^{-2} + z^{-3} + z^{-4} + \cdots)$$

Dividing by $R(z)$ shows that

$$\frac{C(z)}{R(z)} = p_1 z^{-1} + p_2 z^{-2} \tag{10.89}$$

The pulsed transform for the hold and plant is

$$G(z) = \frac{C(z)/R(z)}{M(z)/R(z)} = \frac{p_1 z^{-1} + p_2 z^{-2}}{q_0 + q_1 z^{-1} + q_2 z^{-2}} \tag{10.90}$$

The pulsed transform for the digital controller is

$$D(z) = \frac{M(z)}{E(z)} = \frac{M(z)}{R(z) - C(z)} = \frac{M(z)/R(z)}{1 - C(z)/R(z)} = \frac{q_0 + q_1 z^{-1} + q_2 z^{-2}}{1 - p_1 z^{-1} - p_2 z^{-2}} \tag{10.91}$$

Inverting yields

$$m(k) = p_1 m(k-1) + p_2 m(k-2) + q_0 e(k) + q_1 e(k-1) + q_2 e(k-2)$$

This is the desired relationship for the digital controller which results in an optimum response for the system. As is indicated in the preceding analysis, the controller for a digital control system is programmed to yield a desired behavior. Because of the flexibility of programming, a digital controller is more versatile than a controller for a continuous system, which is limited to some action described by a differential operator.

Application of the final-value theorem shows that the final steady-state value of $c(t)$ is

$$c(\infty) = \lim_{z \to 1} \frac{z-1}{z} C(z) = \frac{z-1}{z} [p_1 z^{-1} + (p_1 + p_2)(z^{-2} + z^{-3} + \cdots)]_{z=1}$$

$$= \frac{z-1}{z} [(p_1 z^{-1} + p_2 z^{-2})(1 + z^{-1} + z^{-2} + \cdots)]_{z=1}$$

$$= (p_1 z^{-1} + p_2 z^{-2})_{z=1} = p_1 + p_2$$

Similarly, the final steady-state value of $m(t)$ is

$$m(\infty) = \lim_{z \to 1} \frac{z-1}{z} M(z)$$

$$= \frac{z-1}{z} [q_0 + (q_0 + q_1)z^{-1} + (q_0 + q_1 + q_2)(z^{-2} + z^{-3} + z^{-4} + \cdots)]_{z=1}$$

$$= \frac{z-1}{z} [(q_0 + q_1 z^{-1} + q_2 z^{-2})(1 + z^{-1} + z^{-2} + \cdots)]_{z=1}$$

$$= (q_0 + q_1 z^{-1} + q_2 z^{-2})_{z=1} = q_0 + q_1 + q_2$$

The ratio of the steady-state value of the system output $c(\infty)$ to the system input $r(\infty)$ is the closed-loop steady-state gain K for the system:

$$K = \frac{c(\infty)}{r(\infty)} = p_1 + p_2 \tag{10.92}$$

The ratio of the steady-state value of the plant output $c(\infty)$ to the plant input is the steady-state gain K_p for the plant:

$$K_p = \frac{c(\infty)}{m(\infty)} = \frac{p_1 + p_2}{q_0 + q_1 + q_2} \tag{10.93}$$

For a second-order plant, $G(z)$ will have the form

$$G(z) = \frac{a_1 z^{-1} + a_2 z^{-2}}{b_0 + b_1 z^{-1} + b_2 z^{-2}}$$

In order that the system have the desired steady-state gain

$$K = p_1 + p_2 = k(a_1 + a_2)$$

or

$$k = \frac{K}{a_1 + a_2} \tag{10.94}$$

Multiplication of the numerator and denominator of $G(z)$ by k yields directly the form given by Eq. (10.90) in which the numerator coefficients are the p's and the denominator coefficients are the q's.

Illustrative example 10.22 Same as Illustrative example 10.21, except determine the difference equation for the digital controller to yield the optimum response with a closed-loop steady-state gain of 5.

SOLUTION In Illustrative example 10.21, it was found that the z transform for the zero-order hold and plant is

$$G(z) = \frac{1.368z - 0.104}{z^2 - 1.368z + 0.368}$$

The sum of the numerator coefficients is $a_1 + a_2 = 1.368 - 0.104 = 1.264$. To attain the desired closed-loop steady-state gain, it is necessary that $K = p_1 + p_2 = 5.0$. The scale factor k is

$$k = \frac{K}{a_1 + a_2} = \frac{5}{1.264} = 3.96$$

Thus, multiplying both the numerator and denominator of $G(z)$ by 3.96 yields

$$G(z) = \frac{5.42z - 0.42}{3.96z^2 - 5.42z + 1.46} = \frac{5.42z^{-1} - 0.42z^{-2}}{3.96 - 5.42z^{-1} + 1.46z^{-2}}$$

Comparison with Eq. (10.90) shows that $p_1 = 5.42$, $p_2 = -0.42$, $q_0 = 3.96$, $q_1 = -5.42$, and $q_2 = 1.46$. Note that $p_1 + p_2 = 5.42 - 0.42 = 5$. The resulting equation for the digital controller is

$$m(k) = 5.42m(k - 1) - 0.42m(k - 2) + 3.96e(k) - 5.42e(k - 1) + 1.46e(k - 2)$$

To obtain the computer program for this system, first write the equation relating $c(t)$ to $m(t)$. That is,

$$c(t) = \frac{D + 2}{D(D + 1)} m(t)$$

For direct programming

$$x_1 = \frac{1}{D(D + 1)} m(t)$$

The resulting state-space representation is

$$\dot{x}_1 = x_2$$
$$\dot{x}_2 = -x_2 + m(t)$$
$$c = 2x_1 + x_2$$

The computer program is

```
    REAL MK, MK1, MK2
    X1 = 0.0
    X2 = 0.0
    DX2 = 0.0
    EK = 0.0
    EK1 = 0.0
    MK = 0.0
    MK1 = 0.0
    DT = 0.01
    T = 0.0
    R = 1.0
    DO 25 I=1,5
    DO 20 J=0,99
    C = 2.0*X1 + X2
    E = R−C
    IF (J.NE.0) GO TO 5
    EK2 = EK1
    EK1 = EK
    EK = E
    MK2 = MK1
    MK1 = MK
    MK = 5.42*MK1−0.42*MK2+3.96*EK−5.42*EK1+1.46*EK2
 5  IF (MOD(J,10).NE.0) GO TO 15
    WRITE (6,10) T, E, EK, MK, C
10  FORMAT(5F10.2)
15  T = T+DT
    X10 = X1
    X20 = X2
    DX20 = DX2
    DX2 = MK−X2
    X2 = 0.5*DT*(DX2+DX20)+X20
    X1 = 0.5*DT*(X2+X20)+X10
20  CONTINUE
25  CONTINUE
    END
```

In this program EK, EK1, EK2, MK, MK1, and MK2 represent $e(k)$, $e(k-1)$, $e(k-2)$, $m(k)$, $m(k-1)$, and $m(k-2)$ respectively. The statements before the first DO loop initialize the system. For the first DO loop, when $I = 1$ the solution is being executed during the first sampling period. For $I = 2$, the solution is being executed during the second sampling period, etc. The second DO loop executes the solution 100 times during each sampling period. Values of $c(t)$ and $e(t)$ are determined for each integration interval, $DT = 0.01$. Because $J = 0$ at the beginning of the sampling period, the IF statement sets $e(k)$ equal to the value of $e(t)$ at the beginning of the sampling period. For the remainder of the sampling period J is not equal to 0 so that $e(k)$ retains this value. The value $e(k-1)$ is set equal to the value of $e(k)$ for the preceding sampling period, and $e(k-2)$ is set equal to the preceding value of $e(k-1)$. Similarly, the value $m(k-1)$ is set equal to the preceding value of $m(k)$, and the value $m(k-2)$ is set equal to the preceding value of $m(k-1)$. The present value of $m(k)$ is calculated from the equation for the digital controller. The remainder of the program uses trapezoidal integration to solve the state-space equations. Any integration method such as fourth-order Runge-Kutta can be used.

For the case of a third-order plant, $M(z)$ is

$$M(z) = q_0 + (q_0 + q_1)z^{-1} + (q_0 + q_1 + q_2)z^{-2}$$

$$+ (q_0 + q_1 + q_2 + q_3)(z^{-3} + z^{-4} + z^{-5} + \cdots)$$

Division by $R(z)$ yields

$$\frac{M(z)}{R(z)} = q_0 + q_1 z^{-1} + q_2 z^{-2} + q_3 z^{-3}$$

The z transform for the response at the sampling instants is

$$C(z) = p_1 z^{-1} + (p_1 + p_2)z^{-2} + (p_1 + p_2 + p_3)(z^{-3} + z^{-4} + z^{-5} + \cdots)$$

Dividing by $R(z)$ shows that

$$\frac{C(z)}{R(z)} = p_1 z^{-1} + p_2 z^{-2} + p_3 z^{-3}$$

The pulsed transform for the hold and plant is

$$G(z) = \frac{C(z)/R(z)}{M(z)/R(z)} = \frac{p_1 z^{-1} + p_2 z^{-2} + p_3 z^{-3}}{q_0 + q_1 z^{-1} + q_2 z^{-2} + q_3 z^{-3}} \qquad (10.95)$$

The closed-loop steady-state gain for the system is

$$K = \frac{c(\infty)}{r(\infty)} = p_1 + p_2 + p_3$$

The steady-state gain for the plant is

$$K_p = \frac{c(\infty)}{m(\infty)} = \frac{p_1 + p_2 + p_3}{q_0 + q_1 + q_2 + q_3}$$

For a third-order plant $G(z)$ has the form

$$G(z) = \frac{a_1 z^{-1} + a_2 z^{-2} + a_3 z^{-3}}{b_0 + b_1 z^{-1} + b_2 z^{-2} + b_3 z^{-3}}$$

To obtain the desired steady-state gain, then

$$K = p_1 + p_2 + p_3 = k(a_1 + a_2 + a_3) \qquad (10.96)$$

or

$$k = \frac{K}{a_1 + a_2 + a_3}$$

Multiplication of the numerator and denominator of $G(z)$ by k yields directly the form in which the numerator coefficients are the p's and the denominator coefficients are the q's. That is,

$$G(z) = \frac{C(z)/R(z)}{M(z)/R(z)} = \frac{p_1 z^{-1} + p_2 z^{-2} + p_3 z^{-3}}{q_0 + q_1 z^{-1} + q_2 z^{-2} + q_3 z^{-3}}$$

Hence

$$D(z) = \frac{M(z)}{E(z)} = \frac{q_0 + q_1 z^{-1} + q_2 z^{-2} + q_3 z^{-3}}{1 - p_1 z^{-1} - p_2 z^{-2} - p_3 z^{-3}}$$

For a plant of order n, $G(z)$ has the form

$$G(z) = \frac{a_1 z^{-1} + a_2 z^{-2} + \cdots + a_n z^{-n}}{b_0 + b_1 z^{-1} + b_2 z^{-2} + \cdots + b_n z^{-n}}$$

The steady-state gain is

$$K = p_1 + p_2 + \cdots + p_n = k(a_1 + a_2 + \cdots + a_n) \qquad (10.97)$$

Hence

$$k = \frac{K}{a_1 + a_2 + \cdots + a_n}$$

Multiplying the numerator and denominator of $G(z)$ by k yields the desired form in which the numerator coefficients are the p's and the denominator coefficients are the q's. Because

$$D(z) = \frac{M(z)}{E(z)} = \frac{q_0 + q_1 z^{-1} + \cdots + q_n z^{-n}}{1 - p_1 z^{-1} - p_2 z^{-2} - \cdots - p_n z^{-n}} \qquad (10.98)$$

then

$$m(k) = p_1 m(k-1) + p_2 m(k-2) + \cdots + p_n m(k-n)$$
$$+ q_0 e(k) + q_1 e(k-1) + \cdots + q_n e(k-n)$$

This is the required relationship for the digital controller to yield the optimum response.

PROBLEMS

10.1 For each of the following functions write

$$f^*(t) = \sum_{n=0}^{\infty} f(nT) u_1(t - nT)$$

and then take the Laplace transform to obtain $F^*(s) = \mathcal{L}[f^*(t)]$. Finally, determine the z transform $F(z)$.

  (a) $f(t) = e^{at}$ (b) $f(t) = a^{t/T}$

10.2 For each of the following functions determine the z transform $F(z)$:

(a) $f(t) = \cosh \omega t = (e^{\omega t} + e^{-\omega t})/2$ (b) $f(t) = \sin \omega t = (e^{j\omega t} - e^{-j\omega t})/2j$

10.3 Use Table 10.1 to obtain the z transform corresponding to each of the following Laplace transforms:

(a) $F(s) = \dfrac{a}{s(s + a)}$ (b) $F(s) = \dfrac{\omega}{s^2 - \omega^2}$

10.4 Determine the Laplace transform for each of the following functions and then use Table 10.1 to obtain the corresponding z transform:

(a) $f(t) = \sin \omega t$ (b) $f(t) = \cos \omega t$

10.5 Given that the z transform for a unit step function is

$$Z[u(t)] = \frac{z}{z - 1}$$

use appropriate transform theorems to determine the z transform for each of the following functions:

(a) $f(t) = t$ (b) $f(t) = t^2$ (c) $f(t) = e^{at}t^2$

10.6 Given that the z transform for a unit step function is

$$Z[u(t)] = \frac{z}{z - 1}$$

use appropriate transform theorems to determine the z transform for each of the following functions:

(a) $f(t) = e^{at}$ (b) $f(t) = te^{at}$ (c) $f(t) = t^2 e^{at}$

10.7 Use the partial-fraction method to determine the inverse z transform for each of the following:

(a) $C(z) = z\dfrac{z + 1}{(z - 0.6)(z + 0.2)}$ (b) $C(z) = z\dfrac{2.9z - 1.7}{(z - 1)(z + 0.5)(z - 0.2)}$

10.8 Use the partial-fraction method to determine the inverse z transform for each of the following:

(a) $C(z) = z\dfrac{9z - 4}{(z - 2)^2(z + 5)}$ (b) $C(z) = z\dfrac{0.5z + 0.02}{(z + 0.1)^2(z + 0.4)}$

10.9 Use the partial-fraction method to determine the inverse z transform for each of the following:

(a) $C(z) = z\dfrac{z + 0.5}{(z^2 - z + 0.5)(z - 0.5)}$ (b) $C(z) = z\dfrac{z + 0.4}{(z^2 + 0.8z + 0.25)(z + 0.7)}$

10.10 Use the residue method to determine the z transform corresponding to each of the following Laplace transforms:

(a) $F(s) = \dfrac{s + 3}{(s + 1)(s + 2)}$ (b) $F(s) = \dfrac{1}{s^2(s + 1)}$

10.11 Use the residue method to determine the z transform for each of the following functions:

(a) $f(t) = e^{at}$ (b) $f(t) = t$ (c) $f(t) = te^{at}$

Verify the resulting answers by using the method of residues to invert.

10.12 Use the method of residues to invert each of the following:

(a) $F(z) = \dfrac{z(e^T - e^{-T})}{(z - e^T)(z - e^{-T})}$ (b) $F(z) = \dfrac{zT^2(z + 1)}{(z - 1)^3}$

10.13 Use block-diagram algebra to determine $C(z)$ for each of the systems shown in Fig. P10.13.

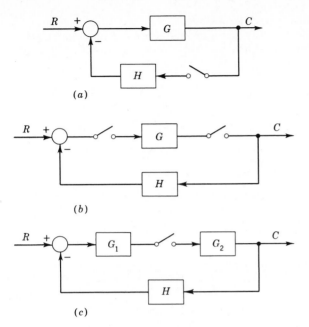

(a)

(b)

(c)   **Figure P10.13**

10.14 In Fig. P10.14 is shown a sampled-data system whose sampling rate is $T = 1$.
    (a) Apply Routh's criterion to determine whether or not the system becomes unstable.
    (b) Construct the root-locus plot for the system.

Figure P10.14

10.15 Determine the unit step function response for the system of Prob. 10.14 for the case in which $K = 1$.

10.16 For the system shown in Fig. 10.12, it is desired to use a sampling rate $T = 0.2$ s rather than $T = 1$ s. Determine the value of K at which the system becomes unstable by using Routh's criterion and also by constructing the root-locus plot.

10.17 In Fig. 10.13 is shown the root-locus plot for the system of Fig. 10.12 when $T = 1$ s. Determine the unit step function response when $K = 2.16$.

10.18 In Fig. P10.18 is shown a sampled-data system whose sampling rate is $T = 1$ s.
    (a) Apply Routh's criterion to determine the value of K at which the system becomes unstable.
    (b) Construct the root-locus plot for this system. From this plot determine the value of K at which the system becomes unstable.

Figure P10.18

10.19 Same as Prob. 10.18 except $T = 2$ s.

10.20 For the system shown in Fig. P10.20, the sampling period is $T = 1$ s. Determine the z transform for $C(z)$. Construct the root-locus plot and then determine the value of the gain K at which the system becomes unstable.

Figure P10.20

10.21 For the system shown in Fig. P10.21, the sampling period is $T = 1$ s. Determine the z transform for $C(z)$. Construct the root-locus plot and then determine the value of the gain K at which the system becomes unstable.

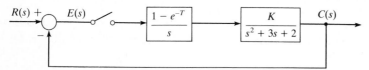

Figure P10.21

10.22 For the system shown in Fig. P10.22, the sampling period is $T = 1$ s. Determine the z transform for $C(z)$. Construct the root-locus plot and then determine the value of the gain K at which the system becomes unstable.

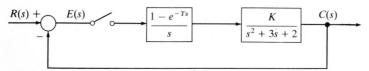

Figure P10.22

10.23 For the system shown in Fig. P10.23, the sampling period is $T = 1$ s. Determine the z transform for $C(z)$. Construct the root-locus plot and then determine the value of the gain K at which the system becomes unstable.

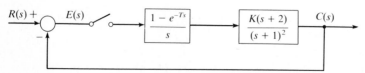

Figure P10.23

10.24 The system shown in Fig. P10.24 has an exponential hold $1/(1 + \tau s)$. For a sampling period $T = 1$ s, determine the value of the gain K at which the system becomes unstable when $\tau = 0.2$.

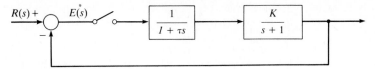

Figure P10.24

10.25 The system shown in Fig. P10.24 has an exponential hold $1/(1 + \tau s)$. For a sampling period $T = 1$ s, determine the value of the gain K at which the system becomes unstable when $\tau = 1.0$. Determine the unit step function response when $K = 4.0$ and when $K = 5.08$.

10.26 Verify the solution of Illustrative example 10.17 by (a) direct programming and (b) series programming.

10.27 Determine the discrete-time state-space representation for the response at the sampling instants and for the response midway between the sampling instants for $T = 0.2$ s for each of the following continuous systems:

    (a) $\dot{c} + c = f(t)$ (b) $\ddot{c} + 2\dot{c} + c = f(t)$

Evaluate the response at $t = 0.1, 0.2, 0.3,$ and 0.4 s, when $f(t)$ is a unit step function and all the initial conditions are zero.

10.28 For the system shown in Fig. P10.28, the input $f(t)$ is a unit step function and $T = 1$ s. Use the z transform method to determine the response at each of the first two sampling instants. Use the state-space method to determine the response just before and just after each of the first two sampling instants.

Figure P10.28

10.29 Same as Prob. 10.28 except for the system shown in Fig. P10.29.

$f(t)$ $f^*(t)$ $\dfrac{D + 4}{(D + 1)(D + 2)}$ $c(t)$

                                           Figure P10.29

10.30 Same as Prob. 10.28 except for the system shown in Fig. P10.30.

$f(t)$ $f^*(t)$ | Zero-order hold | $h(t)$ | $\dfrac{1}{D + 1}$ | $c(t)$

Figure P10.30

10.31 For the system shown in Fig. P10.31, the input $r(t)$ is a unit step function and $T = 1$ s. Use the z transform method to determine the response at each of the first two sampling instants. Use the state-space method to determine the response just before and just after each of the first two sampling instants. Then determine the value of the gain K at which the system becomes unstable.

Figure P10.31

10.32 For the system of Fig. 10.26, the transfer function for the plant is $4/(s + 1)(s + 2)$ and the sampling period is $T = 1$ s. Determine the characteristics of the digital controller such that the response of the system to a unit step function will be $c(t) = 5(1 - e^{-2t})$.

10.33 For the system of Fig. 10.26, the transfer function for the plant is $4(s + 1)/s(s + 2)$ and the sampling period is $T = 1$ s. Determine the characteristics of the digital controller such that the response of the system to a unit step function will be $c(t) = 10(1 - e^{-t})$.

10.34 For the system shown in Fig. 10.26, the transfer function for the plant is $(s + 2)/(s + 1)^2$ and the sampling period is $T = 1$ s. Determine the characteristics of the digital controller such that the response of the system to a unit step function will be $c(t) = 4(1 - e^{-t})$.

10.35 For the system of Fig. 10.26, the transfer function for the plant is $4/(s + 1)(s + 2)$ and the sampling period is $T = 1$ s. The closed-loop steady-state gain for the system is to be 5. Determine the characteristics of the digital controller to yield the optimum response.

10.36 For the system of Fig. 10.26, the transfer function for the plant is $4(s + 1)/s(s + 2)$ and the sampling period is $T = 1$ s. The closed-loop steady-state gain for the system is to be 10. Determine the characteristics of the digital controller to yield the optimum response.

10.37 For the system of Fig. 10.26, the transfer function for the plant is $(s + 2)/(s + 1)^2$ and the sampling period is $T = 1$ s. The closed-loop steady-state gain for the system is to be 4. Determine the characteristics of the digital controller to yield the optimum response.

REFERENCE

1. Kalman, R. E. (discussion of): Sampled-Data Processing Techniques for Feedback Control Systems, A. R. Bergen and J. R. Ragazzini, *Trans. AIEE*, **73** (2), 245–246, 1954.

ELEVEN

FREQUENCY-RESPONSE METHODS

Frequency-response methods provide a convenient means for investigating the dynamic behavior of control systems. By frequency response is meant the response of a system to a sinusoidal input $f = f_0 \sin \omega t$. A characteristic of linear systems is that after the effect of the initial transients has "died out," the response also becomes a sinusoid with the same angular velocity ω as the input. As is illustrated in Fig. 11.1, the response $y = y_0 \sin(\omega t + \phi)$ is displaced some phase angle ϕ from the input, and the amplitude y_0 is different from that of the input f_0. Both the phase angle ϕ and the amplitude ratio y_0/f_0 are functions of the angular velocity ω of the input signal. Graphs of ϕ versus ω and of the amplitude ratio y_0/f_0 versus ω form the basis for frequency-response methods.

11.1 FREQUENCY RESPONSE

Because frequency-response methods are based on a knowledge of ϕ versus ω and y_0/f_0 versus ω, it is now shown how these quantities may be determined directly by substitution of $j\omega$ for D in the operational form of the differential equation for the system. The general operational form of a differential equation is

$$y(t) = \frac{(a_m D^m + a_{m-1} D^{m-1} + \cdots + a_1 D + a_0) f(t)}{D^n + b_{n-1} D^{n-1} + \cdots + b_1 D + b_0} = \frac{L_m(D) f(t)}{L_n(D)} \qquad (11.1)$$

The transform of the preceding expression is

$$Y(s) = \frac{L_m(s)F(s) + I(s)}{L_n(s)} = \frac{L_m(s)N_{F(s)}}{L_n(s)D_{F(s)}} + \frac{I(s)}{L_n(s)} \qquad (11.2)$$

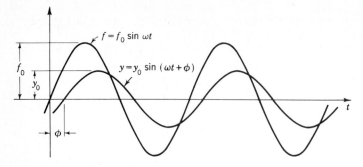

Figure 11.1 Sinusoidal response.

The transform for the input is

$$F(s) = \mathcal{L}(f_0 \sin \omega t) = \frac{\omega f_0}{s^2 + \omega^2} = \frac{N_{F(s)}}{D_{F(s)}} \tag{11.3}$$

Expanding Eq. (11.2) in a partial-fraction expansion and noting that $D_{F(s)} = s^2 + \omega^2 = (s - j\omega)(s + j\omega)$ gives

$$Y(s) = \frac{K_1}{s - r_1} + \cdots + \frac{K_n}{s - r_n} + \frac{K_C}{s - j\omega} + \frac{K_{-C}}{s + j\omega} + \frac{I_1}{s - r_1} + \cdots + \frac{I_n}{s - r_n} \tag{11.4}$$

where $r_1, r_2, \ldots, r_n$ are the zeros of $L_n(s)$; $K_1, K_2, \ldots, K_n, K_C$, and K_{-C} are the constants which arise from the partial-fraction expansion of $L_m(s)N_{F(s)}/L_n(s)D_{F(s)}$; and $I_1, I_2, \ldots, I_n$ are the constants which arise from the partial-fraction expansion of $I(s)/L_n(s)$. Inverting Eq. (11.4) gives

$$y(t) = (K_1 + I_1)e^{r_1 t} + \cdots + (K_n + I_n)e^{r_n t} + \frac{1}{b}|K(a + jb)|e^{at} \sin(bt + \phi) \tag{11.5}$$

For a stable system, $r_1, r_2, \ldots, r_n$ must have negative real parts, so that after sufficient time the effect of these terms becomes negligible. Thus, for stable systems, the steady-state sinusoidal response $y(t)_{ss}$ is determined by the last term of Eq. (11.5). For the quadratic $s^2 + \omega^2$, it follows that $a = 0$ and $b = \omega$. Thus,

$$y(t)_{ss} = \frac{1}{\omega}|K(j\omega)| \sin(\omega t + \phi) \tag{11.6}$$

The terms $|K(j\omega)|$ and $\phi = \angle K(j\omega)$ are evaluated as follows:

$$K(j\omega) = \lim_{s \to j\omega} \left[(s^2 + \omega^2) \frac{L_m(s)N_{F(s)}}{L_n(s)D_{F(s)}} \right] \tag{11.7}$$

$$= \lim_{s \to j\omega} \left[\frac{(s^2 + \omega^2)L_m(s)\omega f_0}{L_n(s)(s^2 + \omega^2)} \right]$$

$$= \frac{L_m(j\omega)\omega f_0}{L_n(j\omega)} \tag{11.8}$$

The terms $L_m(j\omega)$ and $L_n(j\omega)$ are obtained by substituting $j\omega$ for D in $L_m(D)$ and $L_n(D)$. From Eq. (11.8) it follows that

$$| K(j\omega)| = \left| \frac{L_m(j\omega)}{L_n(j\omega)} \right| \omega f_0 \tag{11.9}$$

and

$$\phi = \not{\!\!\angle} \frac{L_m(j\omega)}{L_n(j\omega)} \omega f_0 = \not{\!\!\angle} \frac{L_m(j\omega)}{L_n(j\omega)} \tag{11.10}$$

Substitution of Eq. (11.9) into Eq. (11.6) yields

$$y(t)_{ss} = \left| \frac{L_m(j\omega)}{L_n(j\omega)} \right| f_0 \sin{(\omega t + \phi)} = y_0 \sin{(\omega t + \phi)} \tag{11.11}$$

From Eq. (11.11), the amplitude y_0 of the steady-state response $y(t)_{ss}$ is seen to be

$$y_0 = \left| \frac{L_m(j\omega)}{L_n(j\omega)} \right| f_0 \tag{11.12}$$

The amplitude ratio is

$$\frac{y_0}{f_0} = \left| \frac{L_m(j\omega)}{L_n(j\omega)} \right| \tag{11.13}$$

Illustrative example 11.1 A system is described by the differential equation

$$\frac{dy}{dt} + y = f(t) \qquad \text{or} \qquad y = \frac{1}{D+1} f(t)$$

Determine the response $y(t)$ to the sinusoidal input $f(t) = f_0 \sin{\omega t}$.

SOLUTION The transform of the differential equation is

$$sY(s) - y(0) + Y(s) = F(s)$$

or

$$Y(s) = \frac{F(s)}{s+1} + \frac{y(0)}{s+1} = \frac{\omega f_0}{(s+1)(s^2 + \omega^2)} + \frac{y(0)}{s+1}$$

The partial-fraction expansion is

$$Y(s) = \frac{K_1}{s+1} + \frac{K_c}{s - j\omega} + \frac{K_{-c}}{s + j\omega} + \frac{I_1}{s+1}$$

where $I_1 = y(0)$. Inverting yields

$$y(t) = (K_1 + I_1)e^{-t} + \frac{1}{\omega} | K(j\omega)| \sin{(\omega t + \phi)}$$

where

$$K_1 = \lim_{s \to -1} \frac{(s + 1)\omega f_0}{(s + 1)(s^2 + \omega^2)} = \frac{\omega f_0}{\omega^2 + 1}$$

$$K(j\omega) = \lim_{s \to j\omega} \frac{(s^2 + \omega^2)\omega f_0}{(s + 1)(s^2 + \omega^2)} = \frac{\omega f_0}{1 + j\omega}$$

$$|K(j\omega)| = \frac{\omega f_0}{\sqrt{1 + \omega^2}}$$

$$\phi = \angle K(j\omega) = -\tan^{-1} \omega$$

The resulting transient response is

$$y(t) = \left[\frac{\omega f_0}{\omega^2 + 1} + y(0)\right] e^{-t} + \frac{f_0}{\sqrt{1 + \omega^2}} \sin(\omega t + \phi)$$

The first term is the initial transient which "dies out" because of the decaying exponential e^{-t}. The second term is the steady-state frequency response, $y(t)_{ss} = y_0 \sin(\omega t + \phi)$, where

$$y_0 = \frac{f_0}{\sqrt{1 + \omega^2}} \qquad \text{and} \qquad \phi = -\tan^{-1} \omega$$

The amplitude ratio and phase shift may be verified directly by application of Eqs. (11.10) and (11.13). From the differential equation it follows that the differential operator is $1/(D + 1)$. Replacing D by $j\omega$ shows that

$$\frac{y_0}{f_0} = \left|\frac{L_m(j\omega)}{L_n(j\omega)}\right| = \left|\frac{1}{j\omega + 1}\right| = \frac{1}{\sqrt{1 + \omega^2}}$$

$$\phi = \angle \frac{L_m(j\omega)}{L_n(j\omega)} = \angle \frac{1}{j\omega + 1} = -\tan^{-1} \omega$$

It is now shown how to obtain plots of the amplitude ratio y_0/f_0 versus ω and also phase angle ϕ versus ω. Consider the first-order linear differential equation

$$y(t) = \frac{1}{1 + \tau D} f(t) \tag{11.14}$$

The substitution of $j\omega$ for D in the differential operator gives

$$\frac{L_m(j\omega)}{L_n(j\omega)} = \frac{1}{1 + j\tau\omega} \tag{11.15}$$

The ratio of the amplitude of the output to that of the input is

$$\frac{y_0}{f_0} = \left|\frac{L_m(j\omega)}{L_n(j\omega)}\right| = \frac{1}{\sqrt{1 + (\tau\omega)^2}} \tag{11.16}$$

The phase angle ϕ is

$$\phi = \measuredangle \frac{1}{1 + j\tau\omega} \tag{11.17}$$

The angle of the numerator is $\measuredangle (1 + j0) = \tan^{-1} 0 = 0$. Thus, subtracting the angle of the denominator from the angle of the numerator gives

$$\phi = 0 - \measuredangle (1 + j\tau\omega) = -\tan^{-1} \tau\omega \tag{11.18}$$

In Fig. 11.2 is shown a graph of the amplitude ratio y_0/f_0 versus ω and a graph of the phase angle ϕ as a function of the angular velocity ω of the forcing function. When ω is small, the amplitude of the output is almost equal to that of the input and the phase angle ϕ is quite small. The output cannot keep up with the input at higher frequencies, and it begins to lag behind the input. For $\omega > 1/\tau$ this effect becomes very pronounced.

From Fig. 11.2, it is to be noticed that, for a given ω, the amplitude ratio has a given value, and also that there is a certain phase angle ϕ between the input and the output. This amplitude ratio $|L_m(j\omega)/L_n(j\omega)|$ and the phase angle $\measuredangle L_m(j\omega)/L_n(j\omega)$ determine a vector $L_m(j\omega)/L_n(j\omega)$. The path of the tip of this vector (vector loci) for values of ω from zero to ∞ is shown in Fig. 11.3. This is called a polar plot. The polar plot shown in Fig. 11.3 is seen to convey the same information as the two separate curves shown in Fig. 11.2. The polar plot for this first-order system is a semicircle, as shown in Fig. 11.3. At $\omega = 0$, the length of the vector is 1 and the phase angle is zero. At $\omega = 1/\tau$, the phase angle $\phi = -45°$ and the length of the vector is 0.707.

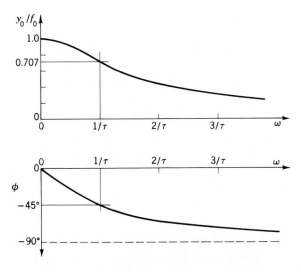

Figure 11.2 Response curves for the function $1/(1 + j\tau\omega)$.

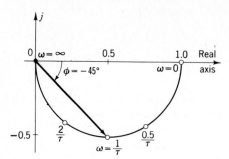

Figure 11.3 Polar plot for the function $1/(1 + j\tau\omega)$.

Consider the general first-order system

$$y(t) = \frac{K}{1 + \tau D} f(t)$$

Replacing D by $j\omega$ in the differential operator shows that

$$\frac{L_m(j\omega)}{L_n(j\omega)} = \frac{K}{1 + j\tau\omega}$$

To prove that the polar plot for this first-order system is a circle, multiply the numerator and denominator by the complex conjugate of the denominator. That is,

$$\frac{K}{1 + j\tau\omega} \frac{1 - j\tau\omega}{1 - j\tau\omega} = \frac{K}{1 + \tau^2\omega^2} - \frac{jK\tau\omega}{1 + \tau^2\omega^2} = x + jy$$

where $x = K/(1 + \tau^2\omega^2)$ is the real part and $y = -K\tau\omega/(1 + \tau^2\omega^2)$ is the imaginary part. Now, note that

$$\left(x - \frac{K}{2}\right)^2 + y^2 = \left(\frac{K}{1 + \tau^2\omega^2} - \frac{K}{2}\right)^2 + \frac{K\tau\omega}{1 + \tau^2\omega^2} = \left(\frac{K}{2}\right)^2$$

This is recognized as the equation for a circle with center at $(K/2, 0)$ and radius $R = K/2$.

The operation of the feedforward part of a control system is given by the equation

$$c(t) = G(D)e(t) \tag{11.19}$$

When the actuating signal $e(t)$ is a sinusoid $[e(t) = e_0 \sin \omega t]$, then the controlled variable $c(t)$ is also a sinusoid $[c(t) = c_0 \sin (\omega t + \phi)]$. The amplitude ratio c_0/e_0 and the phase angle ϕ are

$$\frac{c_0}{e_0} = |G(j\omega)| \tag{11.20}$$

$$\phi = ∢ G(j\omega)$$

Thus $G(j\omega)$ completely describes the frequency response of the feedforward elements. Similarly, the response of the feedback elements is determined by $H(j\omega)$.

The transfer function is the same as the differential operator except that D is replaced by s. Thus either the substitution of $j\omega$ for D in the differential operator or the substitution of $j\omega$ for s in the transfer function gives the vector equation for evaluating the frequency response.

11.2 LOGARITHMIC REPRESENTATION

Frequency-response methods are based on the response $G(j\omega)$ of the feedforward elements and $H(j\omega)$ of the feedback elements. The transfer functions for these quantities are $G(s)$ and $H(s)$ respectively. These quantities are usually obtained in factored form and are composed of multiples or ratios of one or more of the following types of terms: s, $1 + \tau s$, $(s^2 + 2\zeta\omega_n s + \omega_n^2)/\omega_n^2$. The substitution of $j\omega$ for s means that $G(j\omega)$ or $H(j\omega)$ will be composed of terms such as $j\omega$, $1 + j\tau\omega$, or $(\omega_n^2 - \omega^2 + j2\zeta\omega_n\omega)/\omega_n^2$. To obtain the resulting frequency response, the multiplication of such terms is simplified by the use of logarithms. For example, let the value of $G(s)$ be given by the equation

$$G(s) = \frac{K}{s(1 + \tau s)} \tag{11.21}$$

Substituting $j\omega$ for s yields

$$G(j\omega) = \frac{K}{j\omega(1 + j\tau\omega)} \tag{11.22}$$

The magnitude is

$$|G(j\omega)| = \frac{K}{|j\omega||1 + j\tau\omega|} = \frac{K}{\omega\sqrt{1 + \tau^2\omega^2}} \tag{11.23}$$

The logarithm of the preceding expression is

$$\log\left|\frac{G(j\omega)}{K}\right| = -\log\omega - \frac{1}{2}\log(1 + \tau^2\omega^2) \tag{11.24}$$

The angle ϕ is

$$\phi = \angle G(j\omega) = -\angle j\omega - \angle(1 + j\tau\omega)$$
$$= -90° - \tan^{-1}\tau\omega \tag{11.25}$$

The contribution due to the $1/j\omega$ term is shown in Fig. 11.4. The equation for the amplitude is $\log|1/j\omega| = -\log\omega$. When ω is 1, the value of $-\log\omega$ is 0, and when ω changes by multiples of 10, the value of $-\log\omega$ changes in increments of 1. The slope of this straight-line logarithmic response shown in Fig. 11.4 is -1 log unit/decade. A decade is the horizontal distance on the frequency scale from any value of ω to 10 times ω. Thus, the distance from $\omega = 1$ to $\omega = 10$ or from $\omega = 3$ to $\omega = 30$, etc., is a decade.

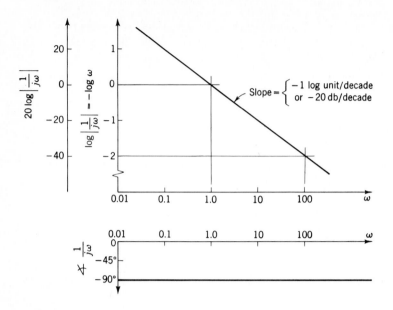

Figure 11.4 Log-magnitude plot for $1/j\omega$.

The vertical log-magnitude scale is sometimes expressed in decibel units. To convert the vertical log-magnitude scale of Fig. 11.4 to decibel units, it is necessary only to multiply by a factor of 20. Using decibel units, the slope is -20 dB/decade rather than -1 log unit/decade. In this text, the log-magnitude scale is expressed directly in logarithmic units rather than in decibels.

The contribution due to $\log |1/(1 + j\tau\omega)|$ is shown in Fig. 11.5. For small values of ω such that $\omega \ll 1/\tau$,

$$-\tfrac{1}{2} \log (1 + \tau^2\omega^2) \approx -\tfrac{1}{2} \log 1 = 0 \tag{11.26}$$

This is the equation for the low-frequency asymptote to the exact curve, as is shown in Fig. 11.5. For $\omega \gg 1/\tau$,

$$-\tfrac{1}{2} \log (1 + \tau^2\omega^2) \approx -\tfrac{1}{2} \log \tau^2\omega^2 = -\log \tau\omega \tag{11.27}$$

This is the high-frequency asymptote. When $\omega = 1/\tau$, the value of $-\log \tau\omega$ is zero, and for $\omega = 10/\tau$, the value is -1, etc. The slope of this high-frequency asymptote is thus seen to be -1 log unit/decade, or -20 dB/decade. The break frequency ($\omega = 1/\tau$) is located directly under the intersection of the two asymptotes. For most preliminary design work, the asymptotes are sufficiently close to the exact curve so that the extra effort involved in using the exact curve is generally not warranted. The maximum error occurs at the " break frequency " and is $0 - \tfrac{1}{2} \log (1 + 1) = -0.303/2 = -0.1515$. It should be noticed that Fig. 11.5 is applicable for any value of τ.

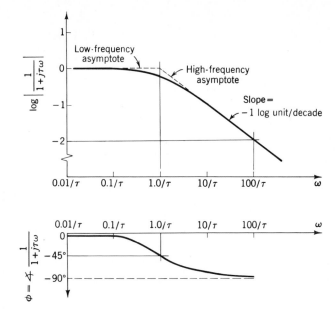

Figure 11.5 Log-magnitude plot for $1/(1 + j\tau\omega)$.

The phase-angle curve is obtained by solving the equation

$$\phi = -\tan^{-1} \tau\omega \qquad (11.28)$$

In Fig. 11.6, it is shown graphically how Figs. 11.4 and 11.5 may be added to solve Eqs. (11.24) and (11.25). For numerical purposes, it is assumed that the value of τ is 0.1 s. At any frequency ω, the resulting log is the sum of the log of each term. For example, when $\omega = 1.0$ the $\log|1/j\omega|$ is 0 and the $\log|1/(1 + 0.1j\omega)|$ is 0. The sum which is $\log|G(j\omega)/K|$ is 0. At $\omega = 100$, the $\log|1/j\omega|$ is -2 and the $\log|1/(1 + 0.1j\omega)|$ is -1. The sum which is $\log|G(j\omega)/K|$ is -3. It is also to be noted that the resulting slope is the sum of the slope of each term. For $\omega < 1/\tau = 10$, the slope of the $\log|1/j\omega|$ term is -1 and the slope of the $\log|1/(1 + 0.1j\omega)|$ term is 0. The sum is the slope of the $\log|G(j\omega)/K|$ term which is -1. For $\omega > 1/\tau = 10$, the slope of the $\log|1/j\omega|$ term is -1 and the slope of the $\log|1/(1 + 0.1j\omega)|$ term is -1. The sum is the slope of the $\log|G(j\omega)/K|$ term which is -2. Similarly, at any frequency ω, the resulting angle is the sum of the angles of each term. Such graphs of the log magnitude and angle versus log frequency are called log-magnitude or Bode diagrams. H. W. Bode[1] made many contributions to the development of frequency-response techniques.

A feature of the logarithmic method is that, if the term appears in the numerator rather than the denominator, it is necessary merely to change the sign of the amplitude and the phase-angle scales in tabulating the result. For example, if

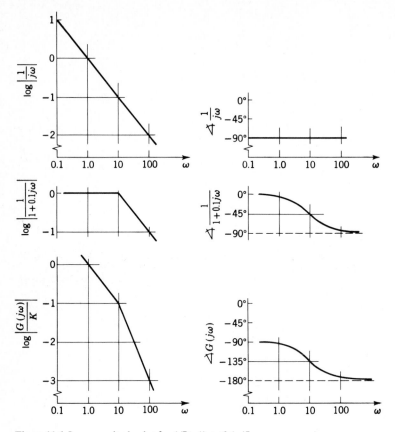

Figure 11.6 Log-magnitude plot for $1/[j\omega(1 + j0.1\omega)]$.

Eq. (11.22) were of the form

$$G(j\omega) = \frac{K(1 + j\tau\omega)}{j\omega}$$

then

$$\log \frac{|G(j\omega)|}{K} = -\log \omega + \frac{1}{2} \log (1 + \tau^2\omega^2) \qquad (11.29)$$

and

$$\angle\ G(j\omega) = -90° + \tan^{-1} \tau\omega \qquad (11.30)$$

By comparing Eqs. (11.29) and (11.30) with Eqs. (11.24) and (11.25), it is seen that only the sign of the term which went from the denominator to the numerator has changed. In Fig. 11.7 is shown the log-magnitude diagram for $G(j\omega)/K = (1 + j\tau\omega)/j\omega$, in which the value of τ is assumed to be 0.1 s. Comparison of the log-magnitude diagram for $1 + j\tau\omega$ of Fig. 11.7 with that for $1/(1 + j\tau\omega)$ of Fig. 11.6 shows that the sign of the logarithm of the amplitude ratio and the sign of the phase angle have been changed.

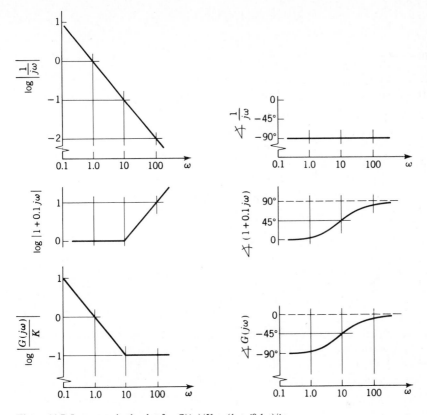

Figure 11.7 Log-magnitude plot for $G(j\omega)/K = (1 + j0.1\omega)/j\omega$.

The third type of term which occurs is of the form $(\omega_n^2 - \omega^2 + j2\zeta\omega_n\omega)/\omega_n^2$. By using a generalized graph, this term may be treated in a manner analogous to that described for evaluating terms of the form $1 + j\tau\omega$. Consider the function

$$G(j\omega) = \frac{K_1}{(1 + j\tau\omega)(\omega_n^2 - \omega^2 + j2\zeta\omega_n\omega)}$$

This may be rewritten in the form

$$G(j\omega) = \frac{K}{(1 + j\tau\omega)[(\omega_n^2 - \omega^2) + j2\zeta\omega_n\omega]/\omega_n^2} \qquad (11.31)$$

where $K = K_1\omega_n^2$.

The logarithm of the magnitude of the quadratic term is

$$\log\left|\frac{1}{[1 - (\omega/\omega_n)^2] + j2\zeta(\omega/\omega_n)}\right| = \log\frac{1}{\sqrt{[1 - (\omega/\omega_n)^2]^2 + [2\zeta(\omega/\omega_n)]^2}}$$

$$= -\frac{1}{2}\log\left\{\left[1 - \left(\frac{\omega}{\omega_n}\right)^2\right]^2 + \left(2\zeta\frac{\omega}{\omega_n}\right)^2\right\} \qquad (11.32)$$

For small values of ω/ω_n such that $\omega/\omega_n \ll 1$, Eq. (11.32) becomes

$$-\tfrac{1}{2} \log 1 = 0$$

This is the equation for the low-frequency asymptote, which is a horizontal straight line like that obtained for the $1 + j\tau\omega$ term.

The equation for the high-frequency asymptote is obtained by noting that, for $\omega/\omega_n \gg 1$, it follows that

$$\left[1 - \left(\frac{\omega}{\omega_n} \right)^2 \right]^2 \approx \left(\frac{\omega}{\omega_n} \right)^4 \gg \left(2\zeta \frac{\omega}{\omega_n} \right)^2$$

Thus, for $\omega/\omega_n \gg 1$, Eq. (11.32) becomes

$$-\frac{1}{2} \log \left(\frac{\omega}{\omega_n} \right)^4 = -2 \log \frac{\omega}{\omega_n} \tag{11.33}$$

The slope of the high-frequency asymptote is -2 log units/decade, or -40 dB/decade, and this asymptote intersects the low-frequency asymptote at $\omega/\omega_n = 1$.

The value of the phase angle is

$$\phi = -\tan^{-1} \frac{2\zeta(\omega/\omega_n)}{1 - (\omega/\omega_n)^2} \tag{11.34}$$

The nondimensional curves for the logarithm of the amplitude and the phase angle, as given by Eqs. (11.32) and (11.34) respectively, are shown in Fig. 11.8. The curves for the reciprocal of this function are obtained by merely changing the sign of the amplitude and the phase-angle scales.

The magnitude plot of Fig. 11.8 shows that the amplitude ratio attains its maximum value in the vicinity of $\omega/\omega_n = 1$. The frequency ω at which this maximum occurs is called the resonant or peak frequency ω_p. The equation for $G(j\omega)$ is

$$G(j\omega) = \frac{1}{1 - (\omega/\omega_n)^2 + j2\zeta(\omega/\omega_n)}$$

The magnitude of $G(j\omega)$ is

$$|G(j\omega)| = \frac{1}{\sqrt{[1 - (\omega/\omega_n)^2]^2 + [2\zeta(\omega/\omega_n)]^2}} \tag{11.35}$$

The maximum value of $|G(j\omega)|$ occurs when the denominator is a minimum. The quantity under the radical is a function of the frequency ratio ω/ω_n:

$$f(\omega/\omega_n) = \left[1 - \left(\frac{\omega}{\omega_n} \right)^2 \right]^2 + \left(2\zeta \frac{\omega}{\omega_n} \right)^2$$

The frequency $\omega = \omega_p$ at which this function attains its minimum value is where

$$\frac{df(\omega/\omega_n)}{d(\omega/\omega_n)} = 2 \left[1 - \left(\frac{\omega}{\omega_n} \right)^2 \right] \left(-2 \frac{\omega}{\omega_n} \right) + 2 \left(2\zeta \frac{\omega}{\omega_n} \right)(2\zeta) = 0$$

or

$$\left(\frac{\omega}{\omega_n} \right)^2 = 1 - 2\zeta^2$$

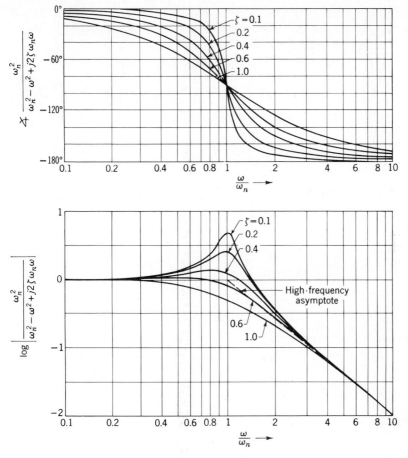

Figure 11.8 Log-magnitude plot for $1/[1 - (\omega/\omega_n)^2 + j2\zeta(\omega/\omega_n)]$.

Hence, the peak frequency $\omega = \omega_p$ is

$$\omega_p = \omega_n\sqrt{1 - 2\zeta^2} \qquad 0 \leq \zeta \leq 0.707 \qquad (11.36)$$

The corresponding amplitude ratio at this frequency is the maximum or peak value M_p. This peak value is obtained by letting $\omega = \omega_p$ in Eq. (11.35). That is,

$$M_p = |G(j\omega_p)| = \frac{1}{2\zeta\sqrt{1 - \zeta^2}} \qquad 0 \leq \zeta \leq 0.707 \qquad (11.37)$$

In summary, for any function which is composed of multiples of terms such as $j\omega$, $1 + j\tau\omega$, and $(\omega_n^2 - \omega^2 + j2\zeta\omega_n\omega)/\omega_n^2$, the value of the $\log|G(j\omega)|$ and the $\measuredangle G(j\omega)$ at any given angular velocity ω of the driving sine wave is the sum of the contribution due to each term, which is obtained from Figs. 11.4, 11.5, or 11.8. Thus, having the $|G(j\omega)|$ and the $\measuredangle G(j\omega)$, one may construct a polar plot of $G(j\omega)$, as was demonstrated by Fig. 11.3.

To write a computer program for obtaining $|G(j\omega)|$ and $\sphericalangle G(j\omega)$, first obtain the magnitude and the angle of each term in $G(j\omega)$. The product of the magnitude of each term is $|G(j\omega)|$ and the sum of the angle of each term is $\sphericalangle G(j\omega)$. To illustrate this procedure, consider the transfer function

$$G(s) = \frac{10}{s(1 + 0.25s)(s^2 + 10s + 64)/64}$$

Replacing s by $j\omega$ gives

$$G(j\omega) = \frac{10}{j\omega(1 + 0.25j\omega)(64 - \omega^2 + j10\omega)/64}$$

This may be written in the form

$$G(j\omega) = 10G_1(j\omega)G_2(j\omega)G_3(j\omega)$$

where

$$G_1(j\omega) = \frac{1}{j\omega}$$

$$G_2(j\omega) = \frac{1}{1 + 0.25j\omega}$$

$$G_3(j\omega) = \frac{64}{64 - \omega^2 + j10\omega}$$

The magnitudes and the angles are

$$|G_1(j\omega)| = \frac{1}{\omega} \qquad\qquad \sphericalangle G_1(j\omega) = -90.0° = -\pi/2$$

$$|G_2(j\omega)| = \frac{1}{\sqrt{1 + (0.25\omega)^2}} \qquad\qquad \sphericalangle G_2(j\omega) = -\tan^{-1} 0.25\omega$$

$$|G_3(j\omega)| = \frac{64}{\sqrt{(64 - \omega^2)^2 + (10\omega)^2}} \qquad\qquad \sphericalangle G_3(j\omega) = -\tan^{-1} \frac{10\omega}{64 - \omega^2}$$

The computer program is

```
W = 1.0
DO 25 K = 1,6
G1 = 1.0/W
AG1 = -3.1416/2
G2 = 1.0/SQRT(1.0+(0.25*W)**2)
AG2 = -ATAN2(0.25*W,1.0)
G3 = 64.0/SQRT((64.0-W**2)**2+(10.0*W)**2)
AG3 = -ATAN2(10.0*W,64.0-W**2)
G = 10.0*G1*G2*G3
AG = AG1+AG2+AG3
```

```
      X=G*COS(AG)
      Y=G*SIN(AG)
      AGD=180.0*AG/3.1416
      WRITE(6,10) W,G,AGD,X,Y
   10 FORMAT(5F10.2)
      W=2.0*W
   25 CONTINUE
      END
```

In this program, $W = \omega$, $G1 = |G_1(j\omega)|$, $AG1 = \measuredangle\, G_1(j\omega)$, $G2 = |G_2(j\omega)|$, $AG2 = \measuredangle\, G_2(j\omega)$, $G3 = |G_3(j\omega)|$, $AG3 = \measuredangle\, G_3(j\omega)$, $G = |G(j\omega)|$, and $AG = \measuredangle\, G(j\omega)$. The statement ATAN2(Y,X) returns the angle whose tangent is y/x. This statement puts the angle in the correct quadrant, as is determined by the sign of y and the sign of x. The real component of $G(j\omega)$ is X=G*COS(AG) and the imaginary component is Y=G*SIN(AG). The computer works with angles in radian measure. The statement AGD=180.0*AG/3.1416 changes the angle of G from radians to degrees. The first time through the DO loop, $W = \omega = 1$. Thereafter, $W = 2W$ so that corresponding values of ω, $|G(j\omega)|$, $\measuredangle\, G(j\omega)$, X, and Y are printed for values of $\omega = 1, 2, 4, 8, 16$, and 32. Many computers have plot routines that enable them to plot corresponding values of X and Y which is the polar plot.

Experimental Determination of Frequency Response

A feature of frequency-response methods is that the responses $G(j\omega)$ and $H(j\omega)$ may be determined experimentally. For example, at a given frequency ω, the value of $G(j\omega)$ is obtained by exciting the feedforward elements with a sinusoidal input of angular velocity ω and then measuring the ratio of the amplitude of the output to that of the input and also measuring the phase angle ϕ. By repeating this process for a wide range of values of ω, the frequency response is obtained. The response $H(j\omega)$ is similarly obtained by sinusoidally exciting the feedback elements.

The asymptotes of an experimentally determined log-magnitude plot are shown in Fig. 11.9. Because of the change of slope at ω_1, there is a term

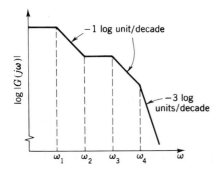

Figure 11.9 Experimentally determined log-magnitude plot.

$1/(1 + j\tau_1\omega)$, where $\tau_1 = 1/\omega_1$ in the frequency-response equation. At the angular velocity ω_2 there is a net increase of $+1$ log unit/decade, and thus the term $1 + j\tau_2\omega$, where $\tau_2 = 1/\omega_2$, appears in the numerator of the response expression. The change in slope at ω_3 is indicative of the term $1/(1 + j\tau_3\omega)$, where $\tau_3 = 1/\omega_3$. Because the slope changes by -2 log units/decade at ω_4, there is a quadratic term in the denominator. The value of the break frequency ω_4 is equal to the natural frequency ω_n for the quadratic. That is, from Fig. 11.8, it follows that at the break point for a quadratic, $\omega/\omega_n = \omega_4/\omega_n = 1$ or $\omega_n = \omega_4$. To determine the damping ratio ζ, it is necessary to compare the exact response curve for the component which causes this quadratic term in $G(j\omega)$ to the general response curves of Fig. 11.8. From the preceding, it follows that the frequency response for $G(j\omega)$ is

$$G(j\omega) = \frac{K(1 + j\tau_2\omega)}{(1 + j\tau_1\omega)(1 + j\tau_3\omega)[(\omega_4^2 - \omega^2 + j2\zeta\omega_4\omega)/\omega_4^2]} \tag{11.38}$$

In the next section, it is shown how the value of K can be determined directly from the low-frequency portion of the log-magnitude diagram.

After the equation for the frequency response has been experimentally determined, it is a simple matter to substitute s for $j\omega$ to obtain the transfer function. The substitution of s for $j\omega$ in Eq. (11.38) gives

$$G(s) = \frac{K(1 + \tau_2 s)}{(1 + \tau_1 s)(1 + \tau_3 s)[(s^2 + 2\zeta\omega_4 s + \omega_4^2)/\omega_4^2]} \tag{11.39}$$

The substitution of D for s gives the differential equation of operation.

The magnitude of $1 + j\tau\omega$ is the same as that of $1 - j\tau\omega$. However, as the phase angle for $1 + j\tau\omega$ goes from $0°$ to $+90°$, the phase angle for $1 - j\tau\omega$ goes from $0°$ to $-90°$. Most systems are minimum-phase systems; i.e., all factors are of the form $j\omega$, $1 + j\tau\omega$, or $(\omega_n^2 - \omega^2 + j2\zeta\omega_n\omega)/\omega_n^2$. For minimum-phase systems, as ω becomes infinite, the phase angle is $\phi = -90°(n - m)$, where n is the order of the denominator and m that of the numerator. Non-minimum-phase systems may be detected from the phase-angle plot, because as ω becomes infinite, $\phi \neq -90°(n - m)$. For either minimum- or non-minimum-phase systems the slope of the log-magnitude diagram at high frequencies is $-(n - m)$ log units/decade. All the frequency-response techniques to be discussed in this text are equally valid for minimum- or non-minimum-phase systems.

11.3 EVALUATING THE GAIN K

In general, a transfer function may be expressed in the form

$$G(s) = \frac{K_n\{(1 + \tau_a s) \cdots [(s^2 + 2\zeta_a\omega_{na} s + \omega_{na}^2)/\omega_{na}^2] \cdots\}}{s^n\{(1 + \tau_1 s) \cdots [(s^2 + 2\zeta_1\omega_{n1} s + \omega_{n1}^2)/\omega_{n1}^2] \cdots\}} \tag{11.40}$$

where K_n is the overall gain of the transfer function $G(s)$ and n is the power to which the s term in the denominator is raised. Usually the value of n is 0, 1, or 2.

The first time constant in the numerator is τ_a, the second τ_b, etc. The natural frequency for the first quadratic term in the numerator is ω_{n_a} and the damping ratio is ζ_a. Similarly, τ_1 is the first time constant which appears in the denominator, τ_2 is the second, etc. The natural frequency for the first quadratic term in the denominator is ω_{n_1} and its damping ratio is ζ_1.

The substitution of $j\omega$ for s in Eq. (11.40) gives

$$G(j\omega) = \frac{K_n\{(1 + j\tau_a\omega) \cdots [(\omega_{n_a}^2 - \omega^2 + j2\zeta_a\omega_{n_a}\omega)/\omega_{n_a}^2] \cdots\}}{(j\omega)^n\{(1 + j\tau_1\omega) \cdots [(\omega_{n_1}^2 - \omega^2 + j2\zeta_1\omega_{n_1}\omega)/\omega_{n_1}^2] \cdots\}} \qquad (11.41)$$

For small values of ω all the terms inside the braces of the preceding expression approach 1, so that

$$G(j\omega) = \frac{K_n}{(j\omega)^n} \qquad \omega \approx 0 \qquad (11.42)$$

As is indicated from the preceding expression, the gain K_n can be determined from the low-frequency portion of the log-magnitude plot. The techniques for evaluating K_n for $n = 0, 1,$ or 2 are described in the following.

$n = 0$ When n is zero, there is no $j\omega$ term in $G(j\omega)$. Thus, Eq. (11.42) becomes

$$G(j\omega) = K_0 \qquad \omega \approx 0 \qquad (11.43)$$

A typical log-magnitude plot for this case is shown in Fig. 11.10a. For small values of ω the low-frequency asymptote has a constant value $G(j\omega) = K_0$.

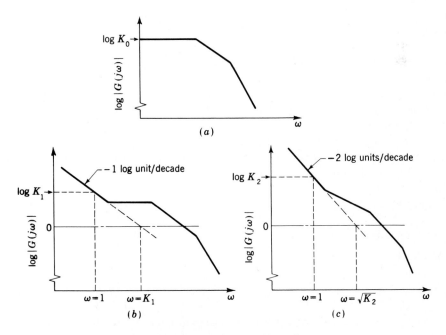

Figure 11.10 Log-magnitude plots: (a) $n = 0$, (b) $n = 1$, (c) $n = 2$.

$n = 1$ For small values of ω and for n equal to 1, it follows from Eq. (11.42) that

$$G(j\omega) = \frac{K_1}{j\omega} \qquad \omega \approx 0 \qquad (11.44)$$

The logarithm of the magnitude of $G(j\omega)$ is

$$\log |G(j\omega)| = \log K_1 - \log \omega \qquad \omega \approx 0 \qquad (11.45)$$

From Eq. (11.45), it follows that, as ω changes by a factor of 10, $\log |G(j\omega)|$ changes by -1. Thus for $\omega \approx 0$ the slope of the curve of $\log |G(j\omega)|$ versus ω is -1 log unit/decade. A typical log-magnitude diagram for the case in which $n = 1$ is shown in Fig. 11.10b. It is to be noted that the low-frequency slope of -1 log unit/decade or its extension intersects the horizontal line through $\log |G(j\omega)| = 0$ at the point where $\omega = K_1$. This fact follows directly from Eq. (11.45) by noting that, when $\log |G(j\omega)| = 0$, then $\log \omega = \log K_1$ or simply $\omega = K_1$. In addition, for ω equal to 1, Eq. (11.45) becomes $\log |G(j\omega)| = \log K_1$. Thus, as is shown in Fig. 11.10b, a vertical line through $\omega = 1$ intersects the low-frequency asymptote or its extension at the value $\log K_1$.

$n = 2$ The low-frequency equation for this case is

$$G(j\omega) = \frac{K_2}{(j\omega)^2} = -\frac{K_2}{\omega^2} \qquad (11.46)$$

The logarithm of the magnitude of $G(j\omega)$ is

$$\log |G(j\omega)| = \log K_2 - 2 \log \omega \qquad (11.47)$$

A typical log-magnitude diagram for $n = 2$ is illustrated in Fig. 11.10c. The slope at low frequencies is -2 log units/decade. From Eq. (11.47), it follows that, when $\log |G(j\omega)| = 0$, then $\log \omega = \frac{1}{2} \log K_2$ or $\omega = \sqrt{K_2}$. Thus, the low-frequency asymptote or its extension intersects the horizontal line, $\log |G(j\omega)| = 0$, at the frequency $\omega = \sqrt{K_2}$. When $\omega = 1$, then Eq. (11.47) becomes $\log |G(j\omega)| = \log K_2$. Thus, a vertical line through $\omega = 1$ intersects the low-frequency asymptote or its extension at $\log K_2$.

Replacing $j\omega$ by s in Eq. (11.42) and solving for K_n shows that

$$K_n = \lim_{s \to 0} s^n G(s)$$

Comparison of the result for $n = 0$ with Eq. (5.91) shows that $K_0 = K_p$ is the positional error constant. Similarly, comparison of the result for $n = 1$ with Eq. (5.92) shows that $K_1 = K_v$ is the velocity error constant. Finally, comparison of the result for $n = 2$ with Eq. (5.93) shows that $K_2 = K_a$ is the acceleration error constant.

11.4 EQUIVALENT UNITY-FEEDBACK SYSTEMS

Much simplification is afforded in the application of frequency-response methods to systems having unity feedback. A control system having feedback elements $H(D)$ can usually be represented by an equivalent unity-feedback system, as is illustrated in Fig. 11.11. For the case in which $H(D)$ is a constant, the equivalent unity-feedback system is readily obtained by moving the constant $H(D)$ to the input side of the main loop. The systems represented by Figs. 3.30 and 4.13b have a constant term $H(D) = C_4$ in the feedback path. Moving C_4 to the input side of the main loop yields the unity-feedback systems shown in Fig. 11.12a and b respectively.

To obtain the equivalent unity-feedback system when $H(D)$ is not a constant, first write $H(D)$ in the form $H(D) = C[1 + H_1(D)]$. The constant C may now be taken out of the feedback path, and the remaining term, $1 + H_1(D)$, may be represented as shown in Fig. 11.13 by two separate paths. The design of such systems in which there is an inner, or minor, feedback path is often facilitated by the use of inverse polar plots, as is discussed in Sec. 12.7.

Figure 11.11 Unity-feedback system.

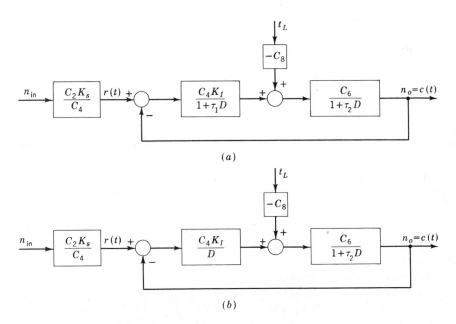

(a)

(b)

Figure 11.12 Equivalent unity-feedback systems when $H(D)$ is a constant.

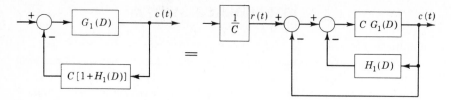

Figure 11.13 Equivalent unity-feedback system when $H(D)$ is not a constant.

In obtaining the equivalent unity-feedback system, only constant terms are to be taken outside the main loop. The fact that $r(t)$ is equal to some constant times the command signal does not affect the basic dynamic behavior of the system.

System Type

When a system is represented in its equivalent unity-feedback form, the value of n in $G(s)$ as indicated by Eq. (11.40) has a predominant effect upon the behavior of the system. When $n = 0$, the system is designated as a type 0 system. A type 1 system is one for which $n = 1$, a type 2 system is one for which $n = 2$, etc.

A type 0 system results when there is no integration, as in a proportional control. For an integral control in which there is one integrator in the feedforward elements, n is equal to 1. A type 2 system has two integrations in the feedforward elements, etc. The system shown in Fig. 11.12a is a type 0 system and the system shown in Fig. 11.12b is a type 1 system.

11.5 POLAR PLOTS

Vector loci or polar plots are better suited for the solution of certain control problems than are log-magnitude diagrams, and vice versa. As is later explained, other methods of representing frequency-response information are the log-modulus, or Nichols, plot and the inverse polar plot. A control engineer must be familiar with all these means of plotting frequency-response data in order to be able to select the method which is best suited to a particular problem.

The polar plots for a number of commonly encountered functions are shown in Fig. 11.14. Polar plots may often be roughly sketched by knowing the location at low frequencies ($\omega \to 0$) and at high frequencies ($\omega \to \infty$). For example, in Fig. 11.14a,

$$\frac{1}{1 + j\tau\omega}\bigg|_{\omega = 0+} = 1$$

$$\frac{1}{1 + j\tau\omega}\bigg|_{\omega = +\infty} \approx \frac{1}{j\tau\omega}\bigg|_{\omega = +\infty} = \frac{e^{-j90°}}{\tau\omega}\bigg|_{\omega = +\infty} = (0+)e^{-j90°}$$

$$(11.48)$$

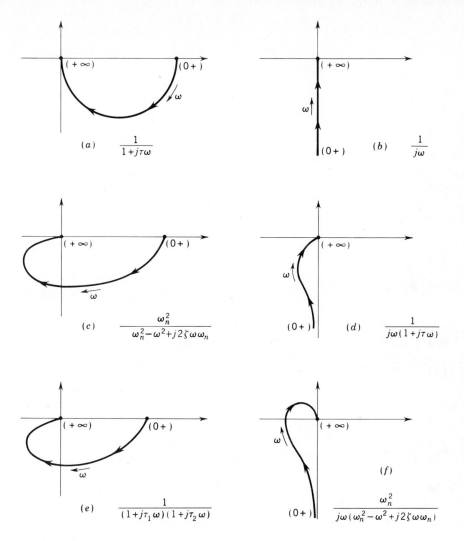

Figure 11.14 Common polar plots.

Thus, the locus of Fig. 11.14a begins at the $+1$ point on the positive real axis ($\phi = 0°$) and, as ω approaches ∞, the locus approaches 0 along the $-90°$ axis. The low- and high-frequency values for Fig. 11.14b are

$$\left.\frac{1}{j\omega}\right|_{\omega=0+} = \left.\frac{1}{\omega}e^{-j90°}\right|_{\omega=0+} = (+\infty)e^{-j90°}$$

$$\left.\frac{1}{j\omega}\right|_{\omega=+\infty} = \left.\frac{1}{\omega}e^{-j90°}\right|_{\omega=+\infty} = (0+)e^{-j90°}$$

(11.49)

For Fig. 11.14c,

$$\left.\frac{1}{1-(\omega/\omega_n)^2+j2\zeta(\omega/\omega_n)}\right|_{\omega=0+}=1$$

$$\left.\frac{1}{1-(\omega/\omega_n)^2+j2\zeta(\omega/\omega_n)}\right|_{\omega=+\infty}\approx\left.\frac{1}{-(\omega/\omega_n)^2}\right|_{\omega=+\infty} \tag{11.50}$$

$$=-\left.\left(\frac{\omega_n}{\omega}\right)^2\right|_{\omega=+\infty}=(0+)e^{-j180°}$$

The limiting value for a function composed of multiples of the preceding terms is equal to the product of the contribution of each term. For example, in Fig. 11.14d,

$$\left.\frac{1}{j\omega(1+j\tau\omega)}\right|_{\omega=0+}=\left.\frac{1}{j\omega}\right|_{\omega=0+}\left.\frac{1}{1+j\tau\omega}\right|_{\omega=0+}=(+\infty)e^{-j90°}$$

$$\left.\frac{1}{j\omega(1+j\tau\omega)}\right|_{\omega=+\infty}=\left.\frac{1}{j\omega}\right|_{\omega=+\infty}\left.\frac{1}{1+j\tau\omega}\right|_{\omega=+\infty}=(0+)e^{-j90°}(0+)e^{-j90°} \tag{11.51}$$

$$=(0+)e^{-j180°}$$

Application of this technique will verify the results shown in Fig. 11.14e and f.

The low- and high-frequency locations are summarized in Table 11.1. The reciprocal of any term in Table 11.1 is obtained by changing the sign of the phase angle and taking the reciprocal of the magnitude. At low frequencies, there is a phase shift with the $1/j\omega$ term only. Thus, for a type 0 system, the polar plot originates on the positive real axis, where ϕ is zero. Typical polar plots for type 0, 1, 2, and 3 systems are shown in Fig. 11.15. A type 1 system begins at infinity on the negative 90° axis, a type 2 system begins at infinity on the negative 180° axis, etc. For most systems, the polar plot terminates at the origin when $\omega=\infty$.

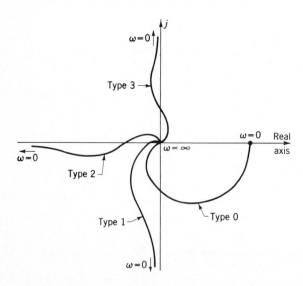

Figure 11.15 Polar plots for type 0, 1, 2, and 3 systems.

Table 11.1 Low- and high-frequency locations

ω	$\dfrac{1}{j\omega}$	$\dfrac{1}{1+j\tau\omega}$	$\dfrac{\omega_n^2}{\omega_n^2 - \omega^2 + j2\zeta\omega\omega_n}$
$0+$	$\infty\underline{/-90°}$	$1\underline{/0°}$	$1\underline{/0°}$
$+\infty$	$0\underline{/-90°}$	$0\underline{/-90°}$	$0\underline{/-180°}$

11.6 M AND α CIRCLES

Frequency-response methods make extensive use of the open-loop frequency response $G(j\omega)$. The open-loop response is the response that would be obtained if the feedback path were disconnected at the comparator (i.e., opened).

For unity-feedback systems, the closed-loop frequency response

$$\frac{C(j\omega)}{R(j\omega)}$$

is related to the open-loop response $G(j\omega)$ by the equation

$$\frac{C(j\omega)}{R(j\omega)} = \frac{G(j\omega)}{1 + G(j\omega)} \tag{11.52}$$

In Fig. 11.16 is shown a typical $G(j\omega)$ plot. The vector from the origin to a point on the curve is $G(j\omega)$ and the vector from the point $-1 + j0$ to the same point on this curve is $1 + G(j\omega)$. The ratio of these two vectors is the closed-loop frequency response for the value of ω at that point. This shows that every point on the $G(j\omega)$ plane corresponds to a certain value of $C(j\omega)/R(j\omega)$. The magnitude of the ratio of the amplitude of the output sinusoid to the input is designated by the symbol $M = |C(j\omega)/R(j\omega)|$.

In Fig. 11.17, it is to be seen that the locus of lines of constant M are circles on the $G(j\omega)$ plane. The proof of this follows. Consider any point $G(j\omega) = x + jy$ in the $G(j\omega)$ plane of Fig. 11.17. The closed-loop frequency response is

$$\frac{C(j\omega)}{R(j\omega)} = \frac{x + jy}{1 + x + jy} \tag{11.53}$$

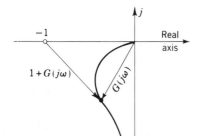

Figure 11.16 Determination of closed-loop frequency response from open-loop response.

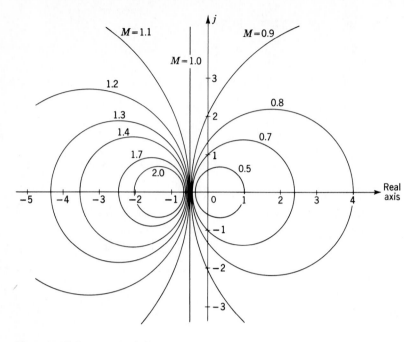

Figure 11.17 Constant M circles.

The magnitude of the preceding equation is

$$M = \left| \frac{C(j\omega)}{R(j\omega)} \right| = \left(\frac{x^2 + y^2}{1 + 2x + x^2 + y^2} \right)^{1/2} \tag{11.54}$$

Squaring and cross-multiplying gives

$$x^2(M^2 - 1) + 2xM^2 + y^2(M^2 - 1) = -M^2$$

Dividing by $M^2 - 1$ and completing the square by adding $M^4/(M^2 - 1)^2$ to both sides yields

$$x^2 + \frac{2xM^2}{M^2 - 1} + \frac{M^4}{(M^2 - 1)^2} + y^2 = \frac{M^4}{(M^2 - 1)^2} - \frac{M^2}{M^2 - 1}$$

Thus,

$$\left(x + \frac{M^2}{M^2 - 1} \right)^2 + y^2 = \frac{M^2}{(M^2 - 1)^2} \tag{11.55}$$

Equation (11.55) is the equation of a circle, as shown in Fig. 11.18, with center at

$$x = \frac{-M^2}{M^2 - 1} \qquad y = 0 \tag{11.56}$$

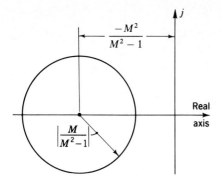

Figure 11.18 Typical M circle.

and radius

$$r = \left| \frac{M}{M^2 - 1} \right| \tag{11.57}$$

The closed-loop frequency response may be expressed in the form

$$\frac{C(j\omega)}{R(j\omega)} = Me^{j\alpha} \tag{11.58}$$

where $M = |C(j\omega)/R(j\omega)|$ and $\alpha = \measuredangle C(j\omega)/R(j\omega)$.

The loci of lines of constant phase angle α for the closed-loop response are also circles. The circles of constant α are shown in Fig. 11.19. From Eq. (11.53),

$$\alpha = \measuredangle \frac{x + jy}{1 + x + jy}$$

Multiplying numerator and denominator by the complex conjugate of the denominator gives

$$\alpha = \measuredangle \frac{x^2 + x + y^2 + jy}{(1 + x)^2 + y^2}$$

Letting $N = \tan \alpha$, then

$$N = \tan \alpha = \frac{y}{x^2 + x + y^2}$$

This may be written in the form

$$x^2 + x + y^2 - \frac{y}{N} = 0$$

Completing the square gives

$$\left(x + \frac{1}{2} \right)^2 + \left(y - \frac{1}{2N} \right)^2 = \frac{N^2 + 1}{4N^2}$$

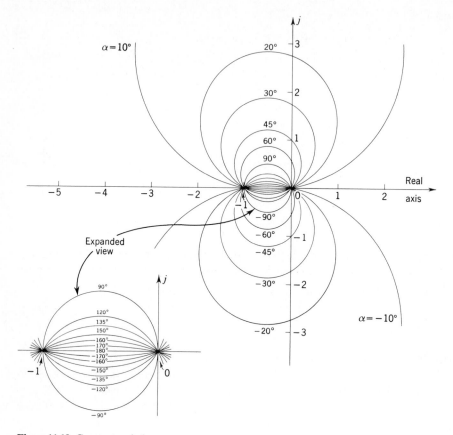

Figure 11.19 Constant α circles.

The centers of these circles are located at

$$x = -\frac{1}{2} \quad \text{and} \quad y = \frac{1}{2N} \tag{11.59}$$

The radius of each circle is

$$r = \frac{1}{2N} \sqrt{N^2 + 1} \tag{11.60}$$

11.7 CORRELATION BETWEEN TRANSIENT AND FREQUENCY RESPONSE

The transient response of a system can be ascertained from the frequency response. The correlation for a type 1 system is developed first.

Type 1 Systems

In Fig. 11.20 is shown the block-diagram representation for a second-order type 1 system. As mentioned in Sec. 11.4, a type 1 system has a lone s term in the denominator of $G(s)$. The closed-loop transfer function $C(s)/R(s)$ is

$$\frac{C(s)}{R(s)} = \frac{K_1/\tau}{s^2 + (1/\tau)s + K_1/\tau} = \frac{\omega_n^2}{s^2 + 2\zeta\omega_n s + \omega_n^2} \qquad (11.61)$$

The closed-loop response is described by a second-order differential equation in which $\omega_n^2 = K_1/\tau$ and $2\zeta\omega_n = 1/\tau$ or $\zeta = 1/(2\sqrt{K_1\tau})$. Thus ζ and ω_n completely describe the transient behavior. To correlate the transient behavior with the frequency response, note that

$$M = \left|\frac{C(j\omega)}{R(j\omega)}\right| = \frac{\omega_n^2}{|(\omega_n^2 - \omega^2) + j2\zeta\omega\omega_n|} = \frac{1}{\sqrt{[1 - (\omega/\omega_n)^2]^2 + (2\zeta\omega/\omega_n)^2}} \qquad (11.62)$$

The value of ω at which M is a maximum is obtained by differentiating Eq. (11.62) with respect to ω and then setting the resulting expression equal to zero. Thus,

$$\frac{\omega_p}{\omega_n} = \sqrt{1 - 2\zeta^2} \qquad 0 \le \zeta \le 0.707 \qquad (11.63)$$

where ω_p is the value of ω at which M attains its peak or maximum value. Substituting ω_p/ω_n for ω/ω_n in Eq. (11.62) gives the peak or maximum value of M, which is designated M_p. That is,

$$M_p = \frac{1}{2\zeta\sqrt{1 - \zeta^2}} \qquad 0 \le \zeta \le 0.707, \ M_p \ge 1 \qquad (11.64)$$

The preceding result has significance only for $0 \le \zeta \le 0.707$, in which case $M_p \ge 1$. In Fig. 11.21 is shown a plot of M_p versus ζ.

Oftentimes, it is desired to know the damping ratio ζ that corresponds to a certain value of M_p. Solving Eq. (11.64) for the damping ratio ζ shows that

$$\zeta = [\tfrac{1}{2}(1 - \sqrt{1 - 1/M_p^2})]^{1/2} \qquad (11.65)$$

As previously discussed, the zeros of the characteristic function which are located nearest the imaginary axis have a predominant effect upon the transient behavior of higher-order systems. Thus, the transient behavior of a higher-order type 1 system for which $M_p \ge 1$ (the value of M_p may be obtained from a polar plot for the system) may be approximated by a second-order system whose damping ratio ζ as obtained from Eq. (11.65) corresponds to the value of M_p for the system.

Figure 11.20 Second-order type 1 system.

Figure 11.21 M_p versus ζ for a second-order system.

Damping ratio ζ

Illustrative example 11.2 In Fig. 11.22 is shown the polar plot for a type 1 system for which $M_p = 1.6$ and $\omega_p = 3$. Determine the value of ζ and ω_n to be used for approximating the transient behavior.

SOLUTION From Eq. (11.65), it follows that for $M_p = 1.6$ the corresponding value of ζ is 0.33. The value of ω_n may now be ascertained from Eq. (11.63). That is,

$$\omega_n = \frac{\omega_p}{\sqrt{1 - 2\zeta^2}} = \frac{3}{\sqrt{1 - 2(0.33)^2}} = 3.4$$

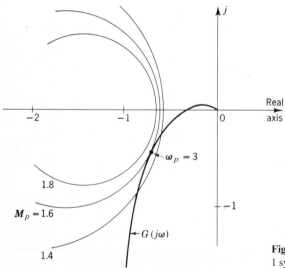

Figure 11.22 Polar plot for a type 1 system.

Figure 11.23 First-order type 1 system.

It is to be noticed from Fig. 6.9 that for $\zeta > 0.707$ ($M_p < 1$) there is no over-shoot of the response to a step change in the input. This is similar to the type of response that is obtained from a first-order system. Thus when M_p is less than 1, the transient response is approximated by an equivalent first-order system.

In Fig. 11.23 is shown a first-order type 1 system. The correlation between the frequency and the transient response for this system is obtained by first writing the equation for the closed-loop response:

$$\frac{C(s)}{R(s)} = \frac{K_1}{K_1 + s} = \frac{1}{1 + (1/K_1)s} = \frac{1}{1 + \tau_c s}$$

where $\tau_c = 1/K_1$ is the closed-loop time constant.

In Fig. 11.24 is shown a plot of the open-loop frequency response

$$G(j\omega) = \frac{K_1}{j\omega}$$

When the magnitude of $G(j\omega)$ is 1, then

$$|G(j\omega)| = \frac{K_1}{|j\omega_c|} = \frac{K_1}{\omega_c} = 1$$

where ω_c designates the value of ω at which $|G(j\omega)| = 1$. That is, ω_c is the value of ω at which the $G(j\omega)$ plot crosses the unit circle, as is illustrated in Fig. 11.24. Because $K_1 = \omega_c$ and $\tau_c = 1/K_1$, then

$$\tau_c = \frac{1}{K_1} = \frac{1}{\omega_c}$$

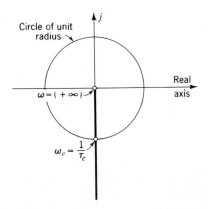

Figure 11.24 Polar plot for $G(j\omega) = K_1/j\omega$.

Table 11.2 Correlation criteria for type 1 systems

$M_p < 1$	$M_p \geq 1$
$\tau_c = \dfrac{1}{\omega_c}$	$M_p = \dfrac{1}{2\zeta\sqrt{1-\zeta^2}}$
	$\dfrac{\omega_p}{\omega_n} = \sqrt{1 - 2\zeta^2}$

Thus, for type 1 systems in which $M_p < 1$, an indication of the equivalent time constant τ_c is obtained by taking the reciprocal of the angular frequency ω_c at which the $G(j\omega)$ plot crosses the unit circle.

The preceding results correlating transient and frequency response for type 1 systems are summarized in Table 11.2.

For $M_p < 1$, the response is approximated by a first-order system in which τ_c describes the transient behavior. For $M_p \geq 1$, the response is approximated by a second-order system in which ζ and ω_n describe the transient behavior.

Although the preceding correlation criteria are developed for type 1 systems, these criteria yield good approximations for type 2 systems, type 3 systems, etc. In the next section, it is shown that the preceding correlation criteria should be modified somewhat for the case of type 0 systems.

Type 0 Systems

In Fig. 11.25 is shown the block-diagram representation for a second-order type 0 system. The closed-loop transfer function $C(s)/R(s)$ is

$$\frac{C(s)}{R(s)} = \frac{K_0/b}{s^2 + (a/b)s + (1 + K_0)/b} = \frac{[K_0/(1 + K_0)]\omega_n^2}{s^2 + 2\zeta\omega_n s + \omega_n^2}$$

where $\omega_n^2 = (1 + K_0)/b$ and $2\zeta\omega_n = a/b$.

To correlate the frequency response with the transient response, note that

$$M = \left| \frac{C(j\omega)}{R(j\omega)} \right| = \frac{K_0/(1 + K_0)}{|1 - (\omega/\omega_n)^2 + 2j\zeta(\omega/\omega_n)|}$$

$$= \frac{K_0/(1 + K_0)}{\sqrt{[1 - (\omega/\omega_n)^2]^2 + (2\zeta\omega/\omega_n)^2}} \tag{11.66}$$

Figure 11.25 Second-order type 0 system.

Differentiating to determine the value of ω at which M is a maximum gives the same result as for the corresponding type 1 system, that is,

$$\omega_p = \omega_n \sqrt{1 - 2\zeta^2} \qquad 0 \le \zeta \le 0.707 \qquad (11.63)$$

Substituting ω_p for ω in Eq. (11.66) gives

$$M_p = \frac{K_0/(1 + K_0)}{2\zeta\sqrt{1 - \zeta^2}} \qquad 0 \le \zeta \le 0.707$$

or

$$\frac{M_p}{K_0/(1 + K_0)} = \frac{1}{2\zeta\sqrt{1 - \zeta^2}} \qquad \frac{M_p}{K_0/(1 + K_0)} \ge 1 \qquad (11.67)$$

Thus, for type 0 systems, the ordinate $M_p/[K_0/(1 + K_0)]$ of Fig. 11.21 is to be used. It is to be noted that for large values of K_0, the preceding criterion becomes the same as that for a type 1 system.

Solving Eq. (11.67) for the damping ratio ζ that corresponds to a certain value of M_p and K_0 shows that

$$\zeta = \{\tfrac{1}{2}[1 - \sqrt{1 - (K/M_p)^2}]\}^{1/2} \qquad (11.68)$$

where $K = K_0/(1 + K_0)$.

Illustrative example 11.3 In Fig. 11.26 is shown the polar plot for a type 0 system in which $M_p = 1.6$, $\omega_p = 3$, and $K_0 = 4$. Determine the value of ζ and ω_n to be used in approximating the transient behavior of this system.

SOLUTION From Eq. (11.68), it follows that for $K/M_p = [K_0/(1 + K_0)]/M_p = 0.8/1.6 = 0.5$, the corresponding value of ζ is 0.26. The value of ω_n is now ascertained from Eq. (11.63). Thus,

$$\omega_n = \frac{\omega_p}{\sqrt{1 - 2\zeta^2}} = \frac{3}{\sqrt{1 - 2(0.26)^2}} = 3.22$$

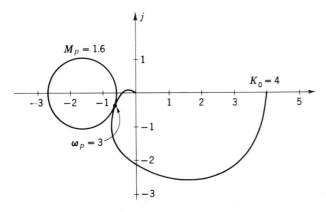

Figure 11.26 Polar plot for a type 0 system.

Figure 11.27 First-order type 0 system.

For a type 0 system in which $M_p/[K_0/(1 + K_0)] < 1$, the transient behavior is best described by a first-order system. In Fig. 11.27 is shown a first-order type 0 system. The closed-loop transfer function is

$$\frac{C(s)}{R(s)} = \frac{K_0}{1 + K_0 + \tau s} = \frac{K_0/(1 + K_0)}{1 + \tau_c s}$$

where $\tau_c = \tau/(1 + K_0)$ is the time constant for the closed-loop response. In Fig. 11.28 is shown a plot of the open-loop frequency response $G(j\omega)$. That is,

$$G(j\omega) = \frac{K_0}{1 + j\tau\omega}$$

This plot crosses a circle of radius r at $\omega = \omega_c$. That is,

$$|G(j\omega)| = \frac{K_0}{\sqrt{1 + (\tau\omega_c)^2}} = r \qquad (11.69)$$

The value of r such that ω_c is the reciprocal of the closed-loop time constant is obtained by substituting $\omega_c = 1/\tau_c = (1 + K_0)/\tau$ into Eq. (11.69). This gives

$$K_0^2 = r^2[1 + (1 + K_0)^2]$$

or

$$r = \frac{K_0}{\sqrt{1 + (1 + K_0)^2}}$$

For large values of K_0, the radius r approaches the unit circle and the preceding criterion becomes the same as that for a type 1 system.

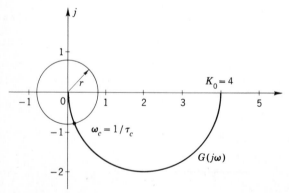

Figure 11.28 Polar plot for $G(j\omega) = K_0/(1 + j\omega)$.

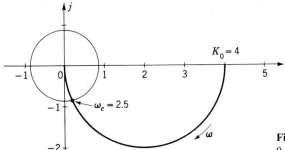

Figure 11.29 Polar plot for a type 0 system.

Illustrative example 11.4 In Fig. 11.29 is shown the polar plot of a type 0 system. Determine the value of τ_c to be used in approximating the transient behavior of this system.

SOLUTION From the polar plot, it is seen that $K_0 = 4$. The value of r is

$$r = \frac{4}{\sqrt{1 + (5)^2}} = \frac{4}{\sqrt{26}} = 0.785$$

The polar plot shows that $\omega_c = 2.5$. Thus,

$$\tau_c = \frac{1}{\omega_c} = 0.4$$

The preceding correlation criteria for type 0 systems are summarized in Table 11.3.

Bandwidth

The closed-loop transfer function for the system of Fig. 11.27 is

$$\frac{C(s)}{R(s)} = \frac{K_0}{(1 + K_0) + \tau s} = \frac{K_0/(1 + K_0)}{1 + [\tau/(1 + K_0)]s} = \frac{K}{1 + \tau_c s}$$

Table 11.3 Correlation criteria for type 0 systems

$\dfrac{M_p}{K_0(1 + K_0)} < 1$	$\dfrac{M_p}{K_0/(1 + K_0)} \geq 1$
$\tau_c = \dfrac{1}{\omega_c}$	$\dfrac{M_p}{K_0/(1 + K_0)} = \dfrac{1}{2\zeta\sqrt{1 - \zeta^2}}$
$r = \dfrac{K_0}{\sqrt{1 + (1 + K_0)^2}}$	$\dfrac{\omega_p}{\omega_n} = \sqrt{1 - 2\zeta^2}$

where $K = K_0/(1 + K_0)$ is the steady-state gain and $\tau_c = \tau/(1 + K_0)$ is the closed-loop time constant. Replacing s by $j\omega$ yields for the closed-loop frequency response

$$\frac{C(j\omega)}{R(j\omega)} = \frac{K}{1 + j\tau_c \omega}$$

The amplitude ratio M of the closed-loop frequency response is

$$M(j\omega) = \left| \frac{C(j\omega)}{R(j\omega)} \right| = \frac{K}{\sqrt{1 + (\tau_c \omega)^2}}$$

A plot of the amplitude ratio is shown in Fig. 11.30a. For small values of ω, the amplitude ratio is

$$M(j0) = \left| \frac{C(j\omega)}{R(j\omega)} \right|_{\omega = 0} = K$$

The frequency $\omega = \omega_b$ at which the amplitude ratio decreases to $\sqrt{2}/2 = 0.707$ times its value for low frequencies $[M(j\omega) = 0.707M(j0)]$ is

$$M(j\omega) = \frac{K}{\sqrt{1 + (\tau_c \omega_b)^2}} = 0.707K$$

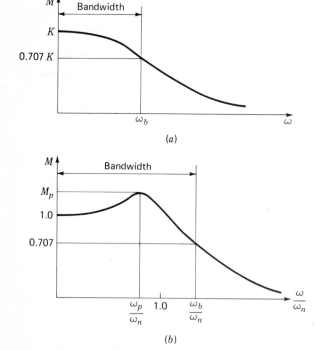

(a)

(b)

Figure 11.30 Closed-loop frequency response for (a) a type 0 system and (b) a type 1 system.

Squaring both sides and solving for ω_b gives

$$\omega_b = \frac{1}{\tau_c}$$

The frequency ω_b is called the bandwidth. For $\tau_c = 1$, the bandwidth is $\omega_b = 1$ rad/s. For $\tau_c = 0.1$, the bandwidth is $\omega_b = 1/0.1 = 10$ rad/s. The bandwidth is a measure of the frequency range $(0 < \omega < \omega_b)$ over which the closed-loop amplitude ratio is greater than $\sqrt{2}/2$ times its value for low frequencies. The response follows the input reasonably well for frequencies within the bandwidth. Note from Fig. 11.30a that the amplitude ratio decreases rapidly for frequencies greater than ω_b.

The bandwidth provides a measure of the speed of response of a system. In general, the larger the bandwidth, the faster is the speed of response. For the example under consideration, the larger bandwidth $\omega_b = 10$ rad/s corresponds to the faster-responding system $\tau_c = 0.1$.

For the system of Fig. 11.20, the closed-loop transfer function is given by Eq. (11.61) and the amplitude ratio M of the closed-loop frequency response is given by Eq. (11.62). That is,

$$\frac{C(s)}{R(s)} = \frac{K_1/\tau}{s^2 + (1/\tau)s + K_1/\tau} = \frac{\omega_n^2}{s^2 + 2\zeta\omega_n s + \omega_n^2} \tag{11.61}$$

$$M(j\omega) = \left| \frac{C(j\omega)}{R(j\omega)} \right| = \frac{1}{\sqrt{[1 - (\omega/\omega_n)^2]^2 + (2\zeta\omega/\omega_n)^2}} \tag{11.62}$$

where $\omega_n^2 = K_1/\tau$ and $\zeta = 1/(2\sqrt{K_1\tau})$. A plot of this closed-loop amplitude ratio is shown in Fig. 11.30b. The peak value M_p is determined by Eq. (11.64) and the frequency at which the peak value occurs is determined by Eq. (11.63). For numerical purposes, if $\zeta = 0.4$, then $M_p = 1.36$ and $\omega_p/\omega_n = 0.82$.

For small values of ω, the amplitude ratio is

$$M(j0) = M(j\omega)\Big|_{\omega=0} = \left| \frac{C(j\omega)}{R(j\omega)} \right|_{\omega=0} = 1$$

The frequency $\omega = \omega_b$ at which the amplitude ratio decreases to 0.707 times its value for low frequencies is

$$M(j\omega) = \frac{1}{\sqrt{[1 - (\omega_b/\omega_n)^2]^2 + (2\zeta\omega_b/\omega_n)^2}} = 0.707 M(j0) = 0.707$$

Squaring and solving for $(\omega_b/\omega_n)^2$ gives

$$\left(\frac{\omega_b}{\omega_n} \right)^2 = (1 - 2\zeta^2) + \sqrt{2 - 4\zeta^2(1 - \zeta^2)}$$

For $\zeta = 0, 0.4$, and 0.707, the corresponding values of ω_b/ω_n are 1.55, 1.36, and 1.0.

11.8 DETERMINING THE GAIN K TO YIELD A DESIRED M_p

In Fig. 11.31 is shown a typical polar plot of $G(j\omega)$. If the gain K of the original function is doubled, the value of $G(j\omega)$ is doubled at every point. As shown in Fig. 11.31, it is not necessary to change the shape of the polar plot, but merely to change the scale by multiplying the old scale by the factor 2. Values of this new scale are shown in parentheses. It is now shown how the gain K can be adjusted so that the polar plot $G(j\omega)$ will be tangent to any desired $M_p > 1$ circle. This, in effect, is determining the gain K so that the system will have a desired M_p.

An M circle is shown in Fig. 11.32. From Eqs. (11.56) and (11.57), it follows that the center is located on the real axis at $-M^2/(M^2 - 1)$ and the radius is $M/(M^2 - 1)$. The line drawn from the origin, tangent to the M circle at the point P, has an included angle of ψ. The value of $\sin \psi$ is

$$\sin \psi = \frac{1}{M} \tag{11.70}$$

A characteristic feature of the point of tangency P is that a line drawn from the point P perpendicular to the negative real axis intersects this axis at the -1 point. This characteristic may be proved from the geometry of Fig. 11.32.

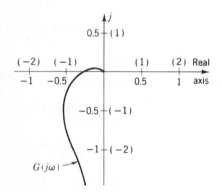

Figure 11.31 Typical polar plot.

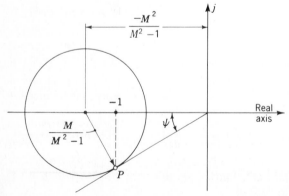

Figure 11.32 Tangent to an M circle.

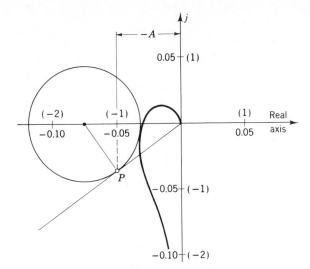

Figure 11.33 Determination of K to yield a desired M_p.

The procedure for determining the gain K so that $G(j\omega)$ will have a desired value of M_p is as follows:

1. Draw the polar plot for $G(j\omega)/K$.
2. Draw the tangent line to the desired M_p circle [Eq. (11.70)].
3. Draw the circle with its center on the negative real axis that is tangent to both the $G(j\omega)/K$ plot and the tangent line, as is shown in Fig. 11.33.
4. Erect the perpendicular to the negative real axis from point P, the point of tangency of this circle and the tangent line. This perpendicular intersects the negative real axis at a value $-A + j0 = -A$.
5. In order that the circle drawn in step 3 corresponds to the desired M_p circle, this point should be $-1 + j0 = -1$ rather than $-A$. The desired gain is that value of K which changes the scale so that this does become the -1 point; thus $K(-A) = -1$ or $K = 1/A$.

As is illustrated in Fig. 11.33, the perpendicular drawn from point P to the negative real axis intersects the negative real axis at a value of -0.05. However, this value should be -1. Multiplication of the scale by a factor of 20 (that is, $-0.05 \times 20 = -1$), as is shown in Fig. 11.33 by the numbers in parentheses, converts this point to the -1 point. Thus, the original function should have a gain of 20 in order that the circle drawn will be the desired M_p circle.

Illustrative example 11.5 Let it be desired to determine the gain K such that the unity-feedback system for which $G(s) = K/[s(1 + 0.1s)]$ will have a peak value $M_p = 1.4$. What values of ζ and ω_n should be used in approximating the transient behavior of this system?

SOLUTION First construct the polar plot for

$$\frac{G(j\omega)}{K} = \frac{1}{j\omega(1 + 0.1j\omega)}$$

as is shown in Fig. 11.34. The value of ψ is obtained from Eq. (11.70):

$$\psi = \sin^{-1}\frac{1}{M_p} = \sin^{-1}\frac{1}{1.4} = 45.6° \tag{11.71}$$

By trial and error, the circle which is tangent to both the $G(j\omega)/K$ plot and also to the line drawn at the angle $\psi = 45.6°$ is determined. The perpendicular drawn from point P intersects the negative real axis at the point $-A = -0.06$, or $A = 0.06$. Thus, the gain is $K = 1/0.06 = 16.7$.

From Eq. (11.65), for $M_p = 1.4$ the corresponding value of ζ is 0.387. From Fig. 11.34, $\omega_p = 10.8$; hence from Eq. (11.63) the value of ω_n is found to be

$$\omega_n = \frac{10.8}{\sqrt{1 - 2(0.387)^2}} = 12.9$$

Because the transfer function for this second-order system is known, the results of the preceding correlation between the frequency response and the transient response may be checked analytically. The open-loop transfer function is

$$G(s) = \frac{K}{s(1 + 0.1s)} = \frac{10K}{s(s + 10)}$$

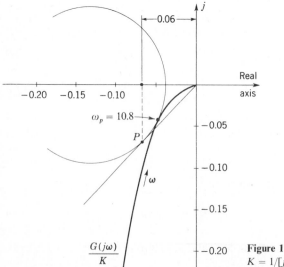

Figure 11.34 Polar plot for $G(j\omega)/K = 1/[j\omega(1 + j0.1\omega)]$.

The corresponding closed-loop transfer function for this unity-feedback system is

$$\frac{C(s)}{R(s)} = \frac{G(s)}{1 + G(s)} = \frac{10K}{s^2 + 10s + 10K}$$

From the characteristic function $(s^2 + 10s + 10K)$, it follows that $\omega_n^2 = 10K$ and $2\zeta\omega_n = 10$. Thus,

$$\omega_n = \sqrt{10K} \qquad \text{and} \qquad \zeta = \frac{10}{2\omega_n} = \frac{1}{2}\sqrt{\frac{10}{K}}$$

Replacing s by $j\omega$ in the expression for the closed-loop transfer function gives

$$\frac{C(j\omega)}{R(j\omega)} = \frac{10K}{10K - \omega^2 + j10\omega}$$

The amplitude ratio M is

$$M = \left|\frac{C(j\omega)}{R(j\omega)}\right| = \frac{10K}{\sqrt{(10K - \omega^2)^2 + (10\omega)^2}}$$

Differentiating the denominator with respect to ω and setting it equal to zero yields the frequency ω_p at which M is a maximum. Thus,

$$2(10K - \omega^2)(-2\omega) + 2(10\omega)(10) = 0$$

or

$$\omega = \omega_p = \sqrt{10K - 50}$$

Substitution of this value of $\omega = \omega_p$ into the expression for M yields, for the peak value of the amplitude ratio M,

$$M_p = \frac{10K}{\sqrt{(50)^2 + 100(10K - 50)}} = \frac{10K}{\sqrt{1000K - 2500}}$$

The value of K such that $M_p = 1.4$ is $K = 100/6 = 16.7$. Thus,

$$\omega_n = \sqrt{10K} = \sqrt{167} = 12.9$$

$$\zeta = \frac{1}{2}\sqrt{\frac{10}{K}} = \frac{\sqrt{0.6}}{2} = 0.387$$

In Fig. 11.35a is shown the log-magnitude plot of the open-loop frequency response

$$G(j\omega) = \frac{16.7}{j\omega(1 + 0.1j\omega)}$$

The low-frequency asymptote is

$$|G(j\omega)| \approx \frac{16.7}{\omega} \qquad \omega \approx 0$$

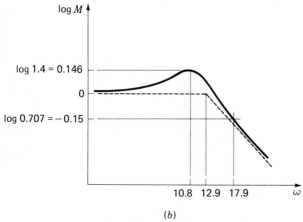

Figure 11.35 Log-magnitude plot of (a) open-loop frequency response and (b) closed-loop frequency response.

The point on this asymptote at $\omega = 1$ is located by the coordinate, $\log |G(j\omega)| = \log 16.7 = 1.22$. The slope changes from -1 to -2 at the break frequency $1/\tau = 1/0.1 = 10 \, \text{rad/s}$. The velocity error constant is $K_v = 16.7$. In Fig. 11.35b is shown the amplitude ratio M for the closed-loop frequency response:

$$M = \left| \frac{C(j\omega)}{R(j\omega)} \right| = \frac{16.7}{|j\omega(1 + 0.1j\omega) + 16.7|}$$

$$= \frac{16.7}{\sqrt{(16.7 - 0.1\omega^2)^2 + \omega^2}}$$

For small values of ω, the amplitude ratio M approaches unity ($\log 1 = 0$). The peak value $M_p = 1.4$ ($\log 1.4 = 0.146$) occurs at the frequency $\omega_p = 10.8$. The low-frequency asymptote (slope $= 0$) and the high-frequency asymptote (slope $= -2$) intersect at the natural frequency $\omega_n = 12.9$. The bandwidth is the value of ω such that M is 0.707 times its low-frequency value of 1.0. Thus, letting $M = 0.707$ in the preceding equation and solving for ω yields $\omega = \omega_b = 17.9$ rad/s.

Illustrative example 11.6 For the system shown in Fig. 11.36a, the transfer function $G(s)$ is

$$G(s) = \frac{100}{(1 + 0.1s)(s^2 + 8s + 25)} = \frac{4}{(1 + 0.1s)(s^2 + 8s + 25)/25}$$

The corresponding frequency response is

$$G(j\omega) = \frac{4}{(1 + 0.1j\omega)[1 - \omega^2/25 + j(8\omega/25)]}$$

(a) Determine the factor K_c by which the gain of the system should be changed so that the resulting system will have an M_p of 1.4.

(b) What values of ζ and ω_n should be used to approximate the transient behavior of this resulting system?

SOLUTION (a) For an M_p of 1.4, the tangent line is constructed at an angle of $\psi = 45.6°$. As shown in Fig. 11.36b, the perpendicular from the point P to the real axis intersects the real axis at the point -1.85. In order that this be the -1 point, the original scale must be changed by a factor of $1/1.85 = 0.54$.

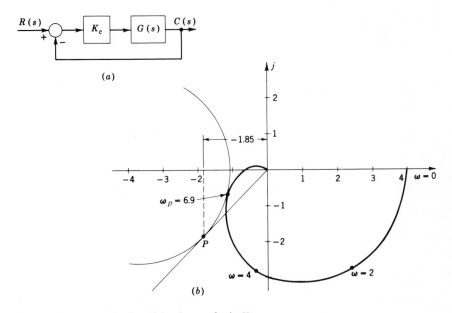

Figure 11.36 Determination of the change of gain K_c.

This is the factor K_c by which the gain of the system should be changed. Thus,

$$\text{New gain} = (K_c)(\text{original gain}) = (0.54)(4) = 2.16$$

(b) For this type 0 system,

$$\frac{M_p}{K_0/(1 + K_0)} = \frac{1.4}{2.16/3.16} = 2.05$$

From Eq. (11.68), it follows that for $K/M_p = [K_0/(1 + K_0)]/M_p = 1/2.05 = 0.488$ the corresponding value of ζ is 0.25. The polar plot shows that $\omega_p = 6.9$. Hence, from Eq. (11.63),

$$\omega_n = \frac{6.9}{\sqrt{1 - 2(0.25)^2}} = 7.37$$

This last example shows that if the $G(j\omega)$ plot is employed rather than the $G(j\omega)/K$ plot, then $1/A$ is equal to the factor K_c by which the gain should be changed in order to obtain a desired M_p.

Log-Modulus Plots

In addition to log-magnitude and polar plots, another method of representing frequency-response information is log-modulus, or Nichols, plots.[2] The log-modulus curve is a plot of $\log |G(j\omega)|$ versus $\angle G(j\omega)$ for various values of angular velocity ω. The log-modulus curve for $G(j\omega)/K = 1/[j\omega(1 + 0.1j\omega)]$ is constructed in Fig. 11.37. From the equation for $G(j\omega)/K$ it is seen that, for a given angular velocity ω, the corresponding phase angle and amplitude ratio can be obtained. These values determine one point on the log-modulus plot. By repeating this process for other values of ω, the log-modulus plot of $G(j\omega)/K$ shown in Fig. 11.37 is obtained. Every point on the log-modulus plot corresponds to a certain value of $G(j\omega)$. Lines of constant M and constant α which are circles on the polar plot (Figs. 11.17 and 11.19) become contours when drawn on

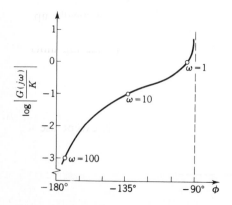

Figure 11.37 Log-modulus plot for $G(j\omega)/K = 1/[j\omega(1 + j0.1\omega)]$.

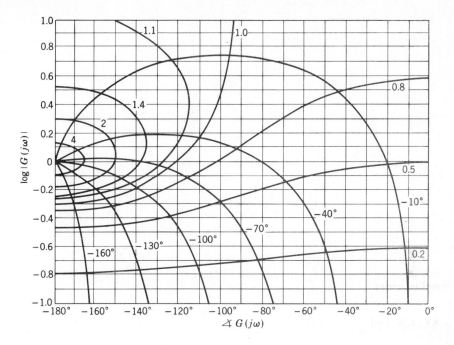

Figure 11.38 Log-modulus representation for lines of constant M and lines of constant α.

the log-modulus plot. These M and α contours are shown in Fig. 11.38. It is now shown how the log-modulus techniques may be used to determine the gain K so that a system will have a desired value of M_p.

Illustrative example 11.7 Same as Illustrative example 11.5, except use log-modulus techniques.

SOLUTION In Fig. 11.39 is shown the log-modulus plot of $G(j\omega)/K$ corresponding to the polar plot of Fig. 11.34. Changing the gain K does not affect the phase angle, but merely moves the log-modulus curve vertically up for $K > 1$ and down for $K < 1$. In Fig. 11.39 it is to be noticed that the original function represented by the solid line must be moved up 1.22 log units so that it will be tangent to the desired M_p contour. Because $\log K = 1.22$, it follows that the required gain is $K = 16.7$. The values of ζ and ω_n are found as before.

Illustrative example 11.8 Same as Illustrative example 11.6, except use log-modulus techniques.

SOLUTION In Fig. 11.40 is shown the log-modulus plot of $G(j\omega)$ corresponding to the polar plot of Fig. 11.36b. Because $\log K_c = -0.27$, then $K_c = 0.54$.

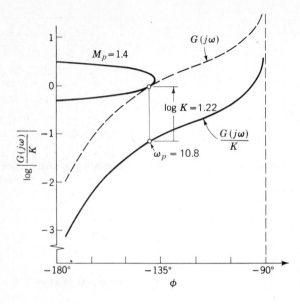

Figure 11.39 Determination of K for a desired M_p on the log-modulus plot.

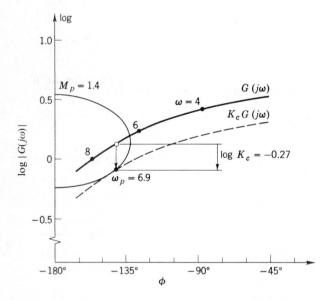

Figure 11.40 Determination of K_c for a desired M_p on the log-modulus plot.

PROBLEMS

11.1 Each of the mechanical systems shown in Fig. P11.1 is excited sinusoidally by a force $f = f_0 \sin \omega t$. For each system, determine

    (a) The equation for the amplitude ratio y_0/f_0

    (b) The equation for the phase shift ϕ

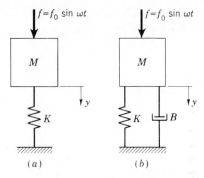

Figure P11.1

11.2 Each of the mechanical systems shown in Fig. P11.2 is excited sinusoidally by a motion of the support $x = x_0 \sin \omega t$. For each system, determine

(a) The equation for the amplitude ratio y_0/x_0

(b) The equation for the phase shift ϕ

Figure P11.2

11.3 In Fig. P11.3 is shown a seismic instrument. The motion of the pen relative to the base is $z = y - x$. Show that the differential equation relating z to the motion of the base is

$$(MD^2 + BD + K)z = -MD^2x$$

(a) Determine the equation for the amplitude ratio z_0/x_0 for the case in which the motion of the base is $x = x_0 \sin \omega t$.

(b) Show that for very large values of ω/ω_n, $z_0/x_0 = -1$. Such an instrument which may be used to measure displacements is called a vibrometer.

(c) Show that for small values of ω/ω_n,

$$\frac{z_0}{a_0} = \frac{1}{\omega_n^2}$$

where

$$\ddot{x} = -x_0 \omega^2 \sin \omega t = a_0 \sin \omega t$$

Such an instrument which measures acceleration is called an accelerometer.

Base

Figure P11.3

11.4 For each of the functions given below, evaluate $|G(j\omega)|$ and $\phi = \measuredangle G(j\omega)$ for $\omega = 4(1/0.25) = 16$, $\omega = 2(1/0.25) = 8$, $\omega = 1/0.25 = 4$, $\omega = (1/2)(1/0.25) = 2$, and $\omega = (1/4)(1/0.25) = 1$. [Note that values of ω in the vicinity of the break frequency $(1/0.25)$ yield the most significant information.] Construct the exact log-magnitude diagram for each function and sketch in the asymptotes.

(a) $G(j\omega) = \dfrac{10}{1 + 0.25j\omega}$

(b) $G(j\omega) = \dfrac{10}{j\omega(1 + 0.25j\omega)}$

(c) $G(j\omega) = \dfrac{10}{(1 + 0.25j\omega)^2}$

11.5 For the quadratic given below, evaluate $|G(j\omega)|$ and $\phi = \measuredangle G(j\omega)$ for values of $\omega/\omega_n = 4$, $\omega/\omega_n = 2$, $\omega/\omega_n = 1$, $\omega/\omega_n = \frac{1}{2}$, and $\omega/\omega_n = \frac{1}{4}$ and for $\zeta = 0.4$. Repeat for the case in which $\zeta = 0.1$. Construct the exact log-magnitude diagram and sketch in the asymptotes.

$$G(j\omega) = \frac{10}{1 - (\omega/\omega_n)^2 + j2\zeta(\omega/\omega_n)}$$

11.6 Construct the polar plot for each of the $G(j\omega)$ functions given in Probs. 11.4 and 11.5.

11.7 For each of the following functions:

$$G(s) = \frac{10}{(1 + s)(1 + 0.1s)}$$

$$G(s) = \frac{10(1 + s)}{s(s^2 + 8s + 100)/100}$$

(a) Draw the asymptotes for the log-magnitude diagram.
(b) Sketch the polar plot.

11.8 Sketch the asymptotes of the log-magnitude diagram and sketch the polar plot for each of the following transfer functions:

(a) $G(s) = \dfrac{10(1 + s)}{(1 + 10s)(1 + 0.1s)}$

(b) $G(s) = \dfrac{10(1 + s)}{s(1 + 0.1s)^2}$

11.9 For each of the transfer functions given in Prob. 11.7, write a computer program to determine $|G(j\omega)|$, $\measuredangle G(j\omega)$, $\log \omega$, $\log |G(j\omega)|$, X, and Y when $\omega = 0.25, 0.50, 1.0, 2.0, 4.0, 8.0, 16.0,$ and 32.0.

11.10 For each of the transfer functions given in Prob. 11.8, write a computer program to determine $|G(j\omega)|$, $\measuredangle G(j\omega)$, $\log \omega$, $\log |G(j\omega)|$, X, and Y when $\omega = 0.125, 0.25, 0.5, 1.0, 2.0, 4.0, 8.0, 16.0,$ and 32.0.

11.11 The asymptotes of the log-magnitude diagram for two $G(j\omega)$ functions are shown in Fig. P11.11. For each case, the value of ϕ is $-270°$ at very high frequencies. Determine the equation for $G(j\omega)$ and evaluate the gain K for each case.

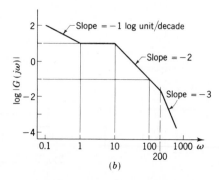

Figure P11.11

11.12 The asymptotes of the log-magnitude diagram for two $G(j\omega)$ functions are shown in Fig. P11.12. Determine the equation for $G(j\omega)$ and evaluate the gain K for each case.

Figure P11.12

11.13 For the system shown in Fig. P11.13, the frequency-response curves for $G_1(j\omega)$ and $G_2(j\omega)$ were determined experimentally. For both $G_1(j\omega)$ and $G_2(j\omega)$ the phase angle at very high frequencies is $\phi_1 = \phi_2 = -90°$. Construct the log-magnitude plot for $G(j\omega) = G_1(j\omega)G_2(j\omega)$. Determine the equation for $G(j\omega)$ and evaluate the gain K.

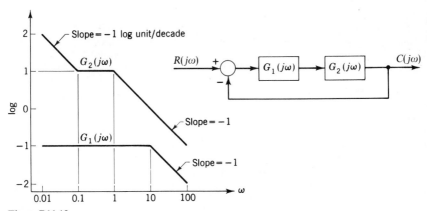

Figure P11.13

11.14 Convert each of the systems shown in Fig. P11.14 to an equivalent unity-feedback system.

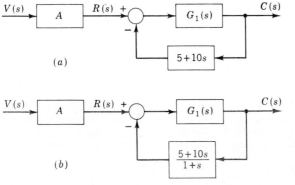

Figure P11.14

11.15 From the polar plot of a type 1 system, it is found that $M_p = 1.4$ and $\omega_p = 5$. What are the approximate values of the damping ratio ζ and natural frequency ω_n?

11.16 From the polar plot of a type 0 system, it is found that $K_0 = 4$, $M_p = 1.4$, and $\omega_p = 5$. What are the approximate values of the damping ratio ζ and natural frequency ω_n?

11.17 The polar plots of $G(j\omega)$ for two unity-feedback systems (system A and system B) are shown in Fig. P11.17. For each system, determine whether the response would be better approximated by a first- or a second-order system and also the corresponding value of τ or ζ and ω_n to be used.

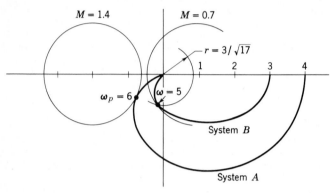

Figure P11.17

11.18 For the system shown in Fig. P11.18, write the characteristic equation and then determine the value of K such that $\omega_n = 1$. What is the corresponding value of the damping ratio ζ? Sketch the asymptotes of the log-magnitude diagram for both the open-loop and the closed-loop frequency response. Obtain the equation for M and differentiate to obtain the value $\omega = \omega_p$ at which $M = M_p$ has its peak value. Determine the peak value M_p. Will Eqs. (11.63) and (11.65) yield exact or approximate values for ω_n and ζ for this case?

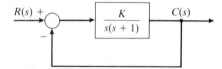

Figure P11.18

11.19 For the system shown in Fig. P11.19, write the characteristic equation and then determine the value of K such that $\omega_n = 3$. What is the corresponding value of the damping ratio ζ? Sketch the asymptotes of the log-magnitude diagram for both the open-loop and closed-loop frequency response. Determine ω_p and M_p. Will Eqs. (11.63) and (11.68) yield exact or approximate values for ω_n and ζ for this case?

Figure P11.19

11.20 (a) It is desired to have an M_p of 1.4 for a unity-feedback type 0 control system in which $K_0 = 5$. From the polar plot of $G(j\omega)$, it is found that $-A = -\frac{1}{3}$ and $\omega_p = 4$. Determine the factor by which the gain should be changed to yield the desired value of M_p. What is the approximate damping ratio and natural frequency for the resulting system?

(b) Same as part (a) except for a type 1 system.

11.21 For the system shown in Fig. P11.21,

$$G(s) = \frac{3}{(s^2 + 2\zeta\omega_n s + \omega_n^2)/\omega_n^2} = \frac{3}{(s^2 + 8s + 4)/4}$$

(a) Determine the closed-loop transfer function $C(s)/R(s)$. What is the corresponding value of ζ and ω_n for this closed-loop transfer function?

(b) Construct the polar plot for $G(j\omega)$ and then determine the factor K_c by which the gain should be changed in order to yield an M_p of $2/\sqrt{3}$. By using the correlation criterion between the frequency response and transient response, determine the approximate value of ζ and ω_n for the resulting closed-loop system.

$R(s)$ +
$C(s)$

$$\frac{3}{(s^2 + 8s + 4)/4}$$

Figure P11.21

11.22 By use of polar plots, determine the value of K to yield an M_p of 1.4 for each of the following unity-feedback systems:

(a) $G(s) = \dfrac{K}{s(1 + 0.25s)}$

(b) $G(s) = \dfrac{K}{s(0.25s^2 + 0.40s + 1)}$

(c) $G(s) = \dfrac{K}{(1 + 0.25s)(0.25s^2 + 0.40s + 1)}$

11.23 Same as Prob. 11.22 except use log-modulus plots rather than polar plots.

11.24 For each system of Prob. 11.22, let $K = 4$, and then determine the factor K_c by which the gain should be changed in order to obtain an M_p of 1.4.

11.25 Same as Prob. 11.24 except use log-modulus plots rather than polar plots.

11.26 The open-loop transfer function for a unity-feedback system is

$$G(s) = \frac{K}{s(s + 4)}$$

From the polar plot determine the gain K such that $M_p = 1.365$. What are the corresponding damping ratio ζ, natural frequency ω_n, and bandwidth ω_b?

11.27 Same as Prob. 11.26 except $M_p = 1.155$.

11.28 The open-loop transfer function for a unity-feedback system is

$$G(s) = \frac{K}{(s + 4)^2}$$

From the polar plot determine the gain K such that $M_p = 1.365$. What are the corresponding values of the damping ratio ζ, natural frequency ω_n, and bandwidth ω_b?

11.29 Same as Prob. 11.28 except $M_p = 1.155$.

REFERENCES

1. Bode, H. W.: "Network Analysis and Feedback Amplifier Design," D. Van Nostrand Company, Inc., Princeton, N.J., 1945.
2. James, H. M., Nichols, N. B., and Phillips, R. S.: "Theory of Servomechanisms," McGraw-Hill Book Company, New York, 1947.

TWELVE

SYSTEM COMPENSATION

Additional insight into the correlation between the shape of a polar plot and the dynamic behavior of a system is obtained by the Nyquist stability criterion.[1] For many design problems, it is not only necessary to change the gain K as discussed in the preceding chapter but it is also necessary to reshape the polar plot. In this chapter, the significance of the Nyquist stability criterion is first presented and then it is shown how system performance may be improved by reshaping the polar plot.

12.1 NYQUIST STABILITY CRITERION

Consider the function

$$F(s) = s - r$$

The vector from the origin to the point r and the vector from the origin to the point s are indicated in Fig. 12.1a. The term $s - r$ is represented by the vector from r to s. The point s travels around the closed path indicated as the path of values of s. That is, s goes along the path to point s_1, then to s_2, and then back to its original position. The point r is a fixed point located inside the path of values of s. The dotted vector from r to s_1 indicates the location of the $s - r$ vector when s is located at the point s_1. Similarly, the dotted vector from r to s_2 represents $s - r$ when s is at the point s_2. Figure 12.1a shows that when r is located inside the path of values of s, the vector $s - r$ rotates through one complete revolution as s traverses the path from s to s_1 to s_2 and then back to s. Thus, the angle of the function $F(s) = s - r$ increases by 2π radians (one revolution) as s traverses the path of values of s.

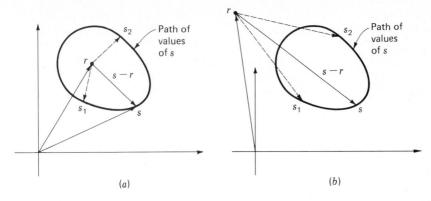

Figure 12.1 Rotation of $s - r$ vector when point r is (a) inside path of values of s, (b) outside path of values of s.

For the reciprocal, $1/(s - r)$, the angle is the negative of the angle for the term $s - r$. Thus, the angle of the reciprocal decreases by 2π radians (one revolution) as s traverses the path.

The case in which r is located outside the path is shown in Fig. 12.1b. The vector $s - r$ goes from the point r to s. The angle of $s - r$ changes slightly as s travels from its initial position to the point s_1. In traveling from s_1 to the point s_2, the angle of the function returns to its initial value and then changes slightly in the opposite direction. In returning to its starting position the angle of the vector $s - r$ returns to its initial value. Thus, as s traverses the path of values, the vector $s - r$ eventually returns to its initial position, so that there is no net rotation of the vector.

Consider the function

$$F(s) = \frac{(s - z_1)(s - z_2) \cdots (s - z_m)}{(s - p_1)(s - p_2) \cdots (s - p_n)}$$

where $z_1, z_2, \ldots, z_m$ are the zeros of the function and $p_1, p_2, \ldots, p_n$ are the poles of the function. For each zero of $F(s)$ located within the path of values of s, the angle of the function $F(s)$ increases by 2π radians (one revolution) as s traverses the path. For each pole of $F(s)$ located within the path of values, the angle decreases by 2π radians as s traverses the path. Poles and zeros outside the path do not affect the angle.

The following general equation may be formulated:

$$N = Z - P \tag{12.1}$$

where P = number of poles of $F(s)$ located inside path of values for s
Z = number of zeros of $F(s)$ located inside path of values for s
N = net number of encirclements of origin of $F(s)$ plane

When the net number of encirclements N is in the same sense as motion around the path of the s contour, an excess of zeros Z is indicated. The opposite sense

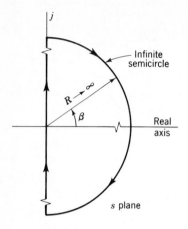

Figure 12.2 Path of values of s that encloses the entire right half-plane.

signifies an excess of poles P. From Eq. (12.1), it follows that when N is in the same sense, N is a positive number. Similarly, when N is in the opposite sense, N is a negative number.† Generally, the path of values for s is traversed in a clockwise direction. Thus, for a net number of clockwise encirclements N is positive and there is an excess of zeros. For a net number of counterclockwise encirclements N is negative and there is an excess of poles.

For control work, the path of values for s is usually taken as shown in Fig. 12.2. This contour is seen to proceed from the origin up the imaginary axis to infinity; then an infinite semicircle ($R \to \infty$) sweeps around to the bottom of the imaginary axis; whence it returns to the origin. This contour in effect encloses the entire right half-plane.

As was discussed in Chap. 6, a system is basically unstable if any zeros of the characteristic function are located in the right half-plane. Note that

$$1 + G(s)H(s) = 1 + \frac{N_{G(s)}N_{H(s)}}{D_{G(s)}D_{H(s)}} = \frac{D_{G(s)}D_{H(s)} + N_{G(s)}N_{H(s)}}{D_{G(s)}D_{H(s)}} \qquad (12.2)$$

It is apparent that the zeros of $1 + G(s)H(s)$ are the zeros of the characteristic function $(D_{G(s)}D_{H(s)} + N_{G(s)}N_{H(s)})$. Similarly, the poles of $1 + G(s)H(s)$ are the zeros of $D_{G(s)}D_{H(s)}$. By letting s assume the values indicated along the contour of Fig. 12.2 and by letting $F(s) = 1 + G(s)H(s)$, Eq. (12.1) becomes

$$Z = N + P \qquad (12.3)$$

where Z = number of zeros of characteristic function [i.e., zeros of $1 + G(s)H(s)$] in right half-plane
  P = number of zeros of $D_{G(s)}D_{H(s)}$ [i.e., poles of $1 + G(s)H(s)$] in right half-plane
  N = net number of encirclements of origin of $1 + G(s)H(s)$ map

---

† Often Eq. (12.1) is written in the form $N = P - Z$. When this form is used, N is negative for a net number of encirclements in the same sense and is positive for the opposite sense.

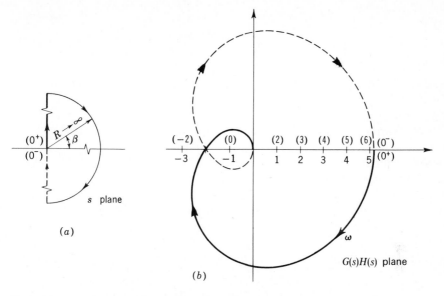

Figure 12.3 (a) Path of values of s, (b) map of $G(s)H(s)$ for a type 0 system.

Because $G(s)H(s)$ is usually obtained in factored form, it is more convenient to construct the map for $G(s)H(s)$ rather than $1 + G(s)H(s)$. The effect of adding $+1$ to each point of the $G(s)H(s)$ map to obtain the $1 + G(s)H(s)$ map is accomplished simply by adding $+1$ to the scale of the real axis, as shown by the numbers in parentheses in Fig. 12.3b. It is seen that the -1 point of the $G(s)H(s)$ map corresponds to the origin of the $1 + G(s)H(s)$ map. Thus, N is equal to the net number of encirclements of the -1 point of the $G(s)H(s)$ plot. The application of the Nyquist stability criterion to type 0, type 1, and type 2 systems is now illustrated.

Type 0 System

Consider the following type 0 system:

$$G(s)H(s) = \frac{52}{(s + 2)(s^2 + 2s + 5)} = \frac{52/10}{(0.5s + 1)(0.2s^2 + 0.4s + 1)} \quad (12.4)$$

In Fig. 12.3a is shown the path of values of s that encloses the entire right half-plane. Along the positive imaginary axis (indicated by the heavy line), s equals $j\omega$. Thus, for this region Eq. (12.4) becomes

$$G(j\omega)H(j\omega) = \frac{5.2}{(1 + 0.5j\omega)(1 - 0.2\omega^2 + 0.4j\omega)}$$

This is the frequency response for the system. It is indicated by the heavy line in Fig. 12.3b. For values of s along the infinite semicircle of Fig. 12.3a, Eq. (12.4) becomes

$$G(s)H(s) = \lim_{R \to \infty} \frac{52}{(Re^{j\beta} + 2)(R^2 e^{2j\beta} + 2Re^{j\beta} + 5)} = \lim_{R \to \infty} \frac{52}{R^3 e^{3j\beta}} = 0$$

Thus, the $G(s)H(s)$ plot runs into the origin for values of s on the infinite semi-circle.

For values of s on the negative imaginary axis (indicated by the dotted line), $s = -j\omega$. Because $G(-j\omega)H(-j\omega)$ is the complex conjugate of $G(j\omega)H(j\omega)$, the plot for $G(-j\omega)H(-j\omega)$ is obtained by reflecting the frequency response $G(j\omega)H(j\omega)$ about the real (horizontal) axis. The resulting $G(-j\omega)H(-j\omega)$ plot is shown dotted in Fig. 12.3b.

The poles of $1 + G(s)H(s)$ are the poles of $G(s)H(s)$. Because there are no poles of Eq. (12.4) in the right half-plane (i.e., within the path of values of s), then $P = 0$. There are two clockwise encirclements of the -1 point of Fig. 12.3b; hence $N = 2$. Note that the -1 point of the $G(s)H(s)$ plot is the origin of the $1 + G(s)H(s)$ plot. Application of the Nyquist stability criterion to determine Z gives

$$Z = N + P = 2$$

Because there are two zeros of the characteristic function located in the right half-plane, this system is basically unstable.

In general, for any type 0 system the frequency response $G(j\omega)H(j\omega)$ begins on the real axis for $\omega = 0$ and terminates at the origin for $\omega = \infty$. The $G(s)H(s)$ plot remains at the origin as the path of values of s traverses the infinite semi-circle. For values of s along the negative imaginary axis (that is, $s = -j\omega$), then $G(-j\omega)H(-j\omega)$ is the complex conjugate of the frequency response $G(j\omega)H(j\omega)$.

At the intersection of a polar plot and the real axis the imaginary part of $G(j\omega)H(j\omega)$ is zero. The frequency at which a polar plot crosses the real axis is obtained by setting the imaginary part of $G(j\omega)H(j\omega)$ equal to zero. Substitution of this value of ω into the expression for $G(j\omega)H(j\omega)$ yields the location at which the polar plot crosses the real axis. Similarly, the location and frequency at which a polar plot crosses the imaginary axis are obtained by setting the real part of $G(j\omega)H(j\omega)$ equal to zero.

To illustrate this procedure, consider the system whose polar plot is shown in Fig. 12.3. The frequency response $G(j\omega)H(j\omega)$ may be written in the form

$$G(j\omega)H(j\omega) = \frac{5.2}{(1 + 0.5j\omega)(1 - 0.2\omega^2 + 0.4j\omega)} = \frac{5.2}{(1 - 0.4\omega^2) + j\omega(0.9 - 0.1\omega^2)}$$

Multiplying numerator and denominator by the complex conjugate of the denominator gives

$$G(j\omega)H(j\omega) = 5.2 \frac{(1 - 0.4\omega^2) - j\omega(0.9 - 0.1\omega^2)}{(1 - 0.4\omega^2)^2 + \omega^2(0.9 - 0.1\omega^2)^2}$$

The imaginary part is zero when $0.9 - 0.1\omega^2 = 0$ or $\omega = 3$. Substitution of this value of ω into the expression for $G(j\omega)H(j\omega)$ yields the value at which

$G(j\omega)H(j\omega)$ crosses the real axis. That is,

$$G(j3)H(j3) = \frac{5.2}{[1 - 0.4(3)^2] + j(0)} = \frac{5.2}{1 - 3.6} = -2.0$$

The real part is zero when $1 - 0.4\omega^2 = 0$ or $\omega = \sqrt{2.5} = 1.58$. This is the value of ω at which the polar plot crosses the imaginary axis. That is,

$$G(j1.58)H(j1.58) = \frac{5.2}{0 + j(1.58)(0.9 + 0.25)} = -\frac{j5.2}{1.027} = -j5.06$$

Type 1 System

For a type 1 system, there is an s term in the denominator of $G(s)H(s)$. For example, consider the system

$$G(s)H(s) = \frac{40}{s(s + 1)(s + 4)} = \frac{10}{s(1 + s)(1 + 0.25s)} \tag{12.5}$$

In Fig. 12.4a it is to be noted that the assumed path of values of s excludes the origin. The reason for this will be made apparent in the following development.

The portion of the plot for $s = j\omega$ as ω takes on values from $(0+)$ to $(+\infty)$ is indicated by the heavy line. This is the frequency response for the system

$$G(j\omega)H(j\omega) = \frac{10}{j\omega(1 + j\omega)(1 + 0.25j\omega)}$$

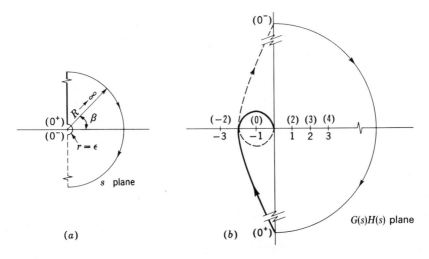

(a)   (b) (0^+)

Figure 12.4 (a) Path of values of s that excludes pole at origin, (b) map of $G(s)H(s)$ for a type 1 system.

For values of s on the infinite semicircle, the $G(s)H(s)$ plot runs into the origin, that is,

$$G(s)H(s) = \lim_{R \to \infty} \frac{40}{Re^{j\beta}(Re^{j\beta} + 1)(Re^{j\beta} + 4)} \approx \lim_{R \to \infty} \frac{40}{R^3 e^{3j\beta}} = 0$$

The dotted portion of the $G(s)H(s)$ plot corresponds to $s = -j\omega$, where $G(-j\omega)H(-j\omega)$ is the complex conjugate of $G(j\omega)H(j\omega)$. To complete the Nyquist plot it is necessary to connect the ends corresponding to $\omega = (0-)$ and $\omega = (0+)$. If the path of values of s were to run through the origin, then for $s = 0$, the function $G(s)H(s)$ in Eq. (12.5) would be infinite and there would be no indication of how to join the $\omega = (0-)$ and $\omega = (0+)$ ends of the $G(s)H(s)$ plot in Fig. 12.4b. Hence a small semicircle of radius ε is constructed about the origin, as illustrated in Fig. 12.4a. The equation for $G(s)H(s)$ is obtained by the substitution of $s = \varepsilon e^{j\beta}$, where β varies from $-90°$ to $0°$ to $+90°$ as s traverses the small semicircle. Hence

$$G(s)H(s) = \lim_{\varepsilon \to 0} \frac{40}{\varepsilon e^{j\beta}(\varepsilon e^{j\beta} + 1)(\varepsilon e^{j\beta} + 4)} \approx \frac{40}{\varepsilon e^{j\beta}(1)(4)} \approx \frac{10}{\varepsilon} e^{-j\beta} \approx \infty e^{-j\beta} \quad (12.6)$$

Thus $G(s)H(s)$ is an infinite semicircle. As s traverses the small semicircle from $0-$ to $0+$, then β goes from $-90°$ to $0°$ to $+90°$. The corresponding infinite semicircle goes from $+90°$ to $0°$ to $-90°$ (i.e., the angle of the infinite semicircle is $-\beta$). The complete Nyquist plot is shown in Fig. 12.4b.

Because the origin has been excluded from the path of values of s, then from Eq. (12.5) it follows that there are no poles of $G(s)H(s)$ within the path of values of s, and thus $P = 0$. As indicated in Fig. 12.4b, there are two clockwise encirclements of the -1 point, so that $N = 2$. Application of the Nyquist stability criterion gives

$$Z = N + P = 2$$

Thus, there are two roots of the characteristic equation in the right half-plane. Consequently the system is unstable.

The general procedure for constructing the complete Nyquist plot for a type 1 system is summarized as follows. A type 1 system always has a pole at the origin, so that it is necessary to exclude the origin from the path of values of s, as indicated in Fig. 12.4a. The frequency response $G(j\omega)H(j\omega)$ corresponds to values of $s = j\omega$ as the path of values of s traverses the positive imaginary axis from $0+$ to $+\infty$. For values of s on the infinite semicircle, the $G(s)H(s)$ plot runs into the origin. As s traverses the negative imaginary axis then $G(s)H(s) = G(-j\omega)H(-j\omega)$ is the complex conjugate of the frequency response.

This complex conjugate of the frequency response is obtained by "flipping" the frequency-response plot about the horizontal axis. To complete the Nyquist plot, the ends at $\omega = (0-)$ and $\omega = (0+)$ are joined by assuming values of s on the small semicircle of radius ε. Equation (12.6) holds for any type 1 system, that is,

$$G(s)H(s) = \infty e^{-j\beta}$$

This is an infinite semicircle.

In the application of the Nyquist stability criterion it would make no differ-ence if the origin were included in the path of values of s as shown in Fig. 12.5a or if it were excluded as shown in Fig. 12.4a. For the case in which the origin is included, β varies from $-90°$ to $-180°$ to $-270°$ as s traverses the small semi-circle. Thus, from Eq. (12.6), the infinite semicircle goes from $90°$ to $180°$ to $270°$. The resultant $G(s)H(s)$ plot is shown in Fig. 12.5b. For this case, $N = +1$. Because $s = 0$ is a pole of $G(s)H(s)$ which is now included in the path of values of s, $P = 1$. Hence

$$Z = N + P = 1 + 1 = 2$$

Thus, the same result $(Z = 2)$ is obtained whether the pole at the origin is included in or excluded from the path of values of s. When a pole of $G(s)H(s)$ occurs at the origin $(s = 0)$, it is customary to exclude the origin from the path of values of s as indicated in Fig. 12.4a.

A direct method for determining the net encirclements N of the -1 point is to draw a radial line from the -1 point out through the plot, as illustrated in Fig. 12.5b. Note the direction of the arrow on the loci at each point where the radial line crosses the loci. In Fig. 12.5b, the radial line crosses the loci three times. At two crossings the direction of the arrow is such as to rotate the radial line in a clockwise direction. At the third crossing the direction is such as to rotate the radial line in a counterclockwise direction. The value of N is the number of crossings at which the arrow on the loci tends to rotate the radial line clockwise minus the number of loci crossings at which the arrow tends to rotate the radial line counterclockwise. Thus, for this case $N = 2 - 1 = 1$. The radial line may be drawn at any angle from the -1 point.

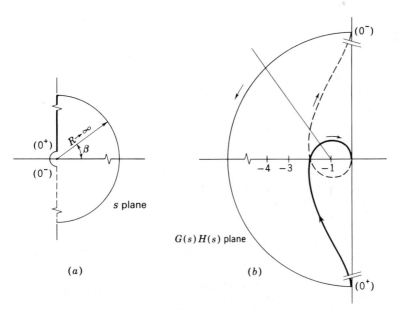

Figure 12.5 (a) Path of values of s that includes pole at origin, (b) map of $G(s)H(s)$ for a type 1 system.

Type 2 System

A type 2 system is characterized by the fact that there is a double pole at the origin. For example, consider the system

$$G(s)H(s) = \frac{4(1 + s)}{s^2(1 + 0.1s)} \tag{12.7}$$

As for a type 1 system the path of values of s is taken to exclude the origin, as shown in Fig. 12.6a. The frequency-response plot for this system is indicated by the heavy line in Fig. 12.6b. The $G(s)H(s)$ plot runs into the origin as s traverses the infinite semicircle. The $G(-j\omega)H(-j\omega)$ plot is merely the complex conjugate of the frequency response.

For values of s on the small semicircle which excludes the origin,

$$G(s)H(s) = \lim_{\varepsilon \to 0} \frac{4(1 + \varepsilon e^{j\beta})}{\varepsilon^2 e^{2j\beta}(1 + 0.1\varepsilon e^{j\beta})} \approx \frac{4}{\varepsilon^2} e^{-j2\beta} \approx \infty e^{-j2\beta} \tag{12.8}$$

Equation (12.8) is valid for any type 2 system. As s traverses the small semicircle from $0-$ to $0+$, then β goes from $-90°$ to $0°$ to $+90°$. The corresponding infinite semicircle changes from $+180°$ to $0°$ to $-180°$ (i.e., the angle of the infinite semicircle is -2β).

For the $G(s)H(s)$ plot described by Eq. (12.7), there are no poles within the path of values of s, so that P is zero. Because there are no encirclements of the -1 point of Fig. 12.6b, N is also zero. Thus, $Z = N + P = 0$, so that the system is basically stable.

When the polar plot for $G(s)H(s)$ goes through the -1 point, the system is on the borderline between being stable and unstable. For this case the characteristic

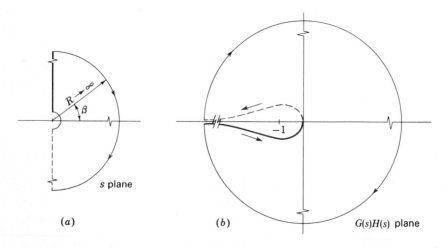

Figure 12.6 (a) Path of values of s, (b) map of $G(s)H(s)$ for a type 2 system.

equation has roots on the imaginary axis. For example, from Fig. 12.3*b*, it is to be noticed that the polar plot crosses the negative real axis at a value of -2. If the gain were halved, the polar plot would go through the -1 point. For this case, the characteristic equation becomes

$$(s + 2)(s^2 + 2s + 5) + 26 = s^3 + 4s^2 + 9s + 36$$

$$= (s + 4)(s^2 + 9) \qquad (12.9)$$

Two roots, $s = \pm j3$, are seen to lie on the imaginary axis.

Similarly, from Fig. 12.4*b* it is to be noticed that the polar plot also crosses the negative real axis at -2. Halving the gain yields, for the characteristic equation,

$$s(s + 1)(s + 4) + 20 = s^3 + 5s^2 + 4s + 20$$

$$= (s + 5)(s^2 + 4) \qquad (12.10)$$

Two roots, $s = \pm j2$, are seen to lie on the imaginary axis.

For most systems, the open loop is stable. For this case, all the zeros of $D_G D_H$ lie in the left half-plane ($P = 0$). Thus, the Nyquist stability criterion reduces to

$$Z = N$$

For a stable system ($Z = 0$), there are no net encirclements N of the -1 point.

In applying frequency-response methods, one always works with the open-loop response. It is well to note that the closed loop may be stable even though the open loop is unstable, and vice versa.

12.2 GAIN MARGIN AND PHASE MARGIN

From the preceding discussion, the -1 point of the $G(s)H(s)$ map was seen to have great significance with regard to the stability of a system. In Fig. 12.7 is shown a typical $G(j\omega)H(j\omega)$ plot in the vicinity of the -1 point. If the gain were multiplied by an amount K_M, called the gain margin, the $G(j\omega)H(j\omega)$ plot would go through the -1 point. Thus, the gain margin is an indication of how much the gain can be increased before the curve goes through the critical point.

The angle γ in Fig. 12.7 is the angle measured from the negative real axis to the radial line through the point where the polar plot crosses the unit circle. If the angle γ were zero, the polar plot would go through the -1 point. The angle γ, called the phase margin, is thus seen to be another indication or measure of the closeness of the polar plot to the critical point. It is to be noted that a positive phase margin indicates a stable system, as does a gain margin greater than 1.

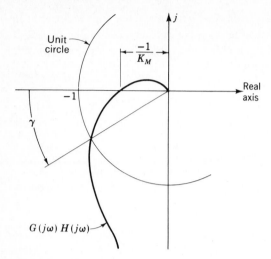

Figure 12.7 Gain margin and phase margin on the polar plot.

Log-Modulus Plots

In Fig. 12.8 is shown the log-modulus plot corresponding to Fig. 12.7. At the -1 point, $|G(j\omega)H(j\omega)| = 1$ and $\phi = \angle G(j\omega)H(j\omega) = -180°$. Thus, on the log-modulus plot the -1 point is located at the intersection of the ordinate $\log |G(j\omega)H(j\omega)| = \log 1 = 0$ and the abscissa $\phi = -180°$. From Fig. 12.8, the vertical distance that the $G(j\omega)H(j\omega)$ plot may be raised before it goes through the -1 point is $\log K_M$.

The value of the phase margin may be obtained from a log-modulus plot as follows. The horizontal line in Fig. 12.8 of $\log |G(j\omega)H(j\omega)| = \log 1 = 0$ corresponds to the unit circle of Fig. 12.7. The angle γ between the point where this horizontal line intersects the $G(j\omega)H(j\omega)$ plot and the value $\phi = -180°$ is the phase margin.

Figure 12.8 Gain margin and phase margin on the log-modulus plot.

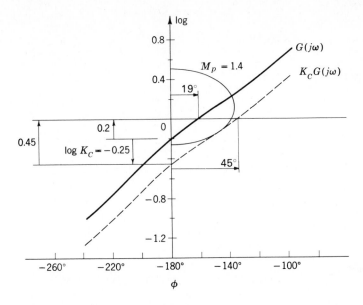

Figure 12.9 Log-modulus plot for a system.

Illustrative example 12.1 The log-modulus plot of $G(j\omega)$ for a unity-feedback system is shown in Fig. 12.9. What is the value of the gain margin and the phase margin for this system? By what factor K_c should the gain of the system be changed so that M_p will be 1.4? What is the new value for the gain margin and the phase margin?

SOLUTION From the $G(j\omega)$ plot for the original system, it follows that log $K_M = 0.2$, or $K_M = 1.59$, and $\gamma = 19°$. The factor K_c is obtained by moving the $G(j\omega)$ locus straight down until the new locus $K_c G(j\omega)$ is tangent to the desired $M_p = 1.4$ contour. From Fig. 12.9, it follows that

$$\log K_c = -0.25 \qquad \text{or} \qquad K_c = 0.563$$

The new gain margin is log $K_M = 0.45$ or $K_M = 2.81$, and the new phase margin is $\gamma = 45°$.

Bode Diagrams

In Fig. 12.10 is shown a Bode diagram. By entering the phase angle plot at $\phi = -180°$, it is noted that the corresponding value of log $|G(j\omega)H(j\omega)| = -0.3$. Hence, the gain of the system may be increased by $+0.3$ log units before the $G(j\omega)H(j\omega)$ curve goes through the -1 point. Thus, the value of log K_M may be represented graphically, as illustrated in Fig. 12.10.

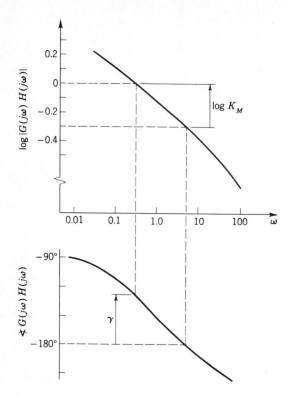

Figure 12.10 Gain margin and phase margin on the Bode diagram.

The unit circle of the polar plot corresponds to the horizontal line through the log-magnitude plot at $\log |G(j\omega)H(j\omega)| = \log 1 = 0$. The phase margin γ is the difference between the corresponding angle and the angle $\phi = -180°$, as illustrated in the angle diagram of Fig. 12.10.

Illustrative example 12.2 In Fig. 12.11 is shown the Bode diagram for a control system. By what factor K_c should the gain of the system be changed so that the resulting phase margin is 45°?

SOLUTION Entering the angle plot at $\phi = -180° + \gamma = -135°$, the corresponding value of $\log |G(j\omega)H(j\omega)|$ is 0.3. To have a phase margin of 45°, the magnitude diagram must be lowered by 0.3 log units. Thus,

$$\log K_c = -0.3$$

or $$K_c = 0.5$$

In Sec. 11.7 the transient-response parameters ζ and ω_n were related to the frequency-response parameters M_p and ω_p. This correlation is limited to use with polar or log-modulus plots. That is, M circles may be drawn on polar plots and M contours on log-modulus plots, but it is impossible to construct M lines on

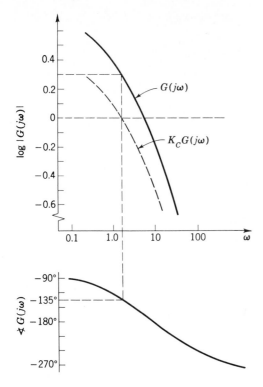

Figure 12.11 Bode diagram for a system.

Bode diagrams because the magnitude and phase angle are on separate plots. When designing with Bode diagrams, the transient and frequency responses are correlated by relating ζ and ω_n to the phase margin.

Correlation between ζ, ω_n, and Phase Margin

It is now shown that the transient-response parameters ζ and ω_n may also be related to the phase margin γ. Although the following analysis is for a second-order type 1 system, the results yield good approximations for higher-order systems and for systems of other type numbers, such as a type 0 or a type 2 system.

The closed-loop transfer function for the system of Fig. 11.20 is

$$\frac{C(s)}{R(s)} = \frac{K/\tau}{s^2 + (1/\tau)s + K/\tau} = \frac{\omega_n^2}{s^2 + 2\zeta\omega_n s + \omega_n^2} \tag{12.11}$$

where $\omega_n^2 = K/\tau$

$2\zeta\omega_n = 1/\tau$

The open-loop frequency response is

$$G(j\omega) = \frac{K}{j\omega(1 + j\tau\omega)}$$

By letting ω_c be the value of ω when $|G(j\omega)| = 1$, then

$$|G(j\omega)| = \frac{K}{\omega_c \sqrt{1 + \tau^2 \omega_c^2}} = 1$$

Squaring gives

$$\tau^2 \omega_c^4 + \omega_c^2 - K^2 = 0$$

or

$$\left(\frac{\tau}{K}\right)^2 \omega_c^4 + \frac{1}{K^2} \omega_c^2 - 1 = 0$$

Because $\tau/K = 1/\omega_n^2$ and $K = \omega_n^2 \tau = \omega_n^2/2\zeta\omega_n = \omega_n/2\zeta$, then

$$\left(\frac{\omega_c}{\omega_n}\right)^4 + 4\zeta^2 \left(\frac{\omega_c}{\omega_n}\right)^2 - 1 = 0$$

Application of the quadratic equation yields

$$\left(\frac{\omega_c}{\omega_n}\right)^2 = \sqrt{4\zeta^4 + 1} - 2\zeta^2 \tag{12.12}$$

The preceding expression relates ζ and the ratio ω_c/ω_n. A plot of this relationship is shown in Fig. 12.12a.

The phase angle is

$$\gamma = 180° + \sphericalangle G(j\omega_c)$$

where $\sphericalangle G(j\omega_c)$ is the angle of $G(j\omega)$ at which the polar plot crosses the unit circle. From Eq. (12.12), it follows that

$$\sphericalangle G(j\omega_c) = 0° - 90° - \tan^{-1} \tau\omega_c$$

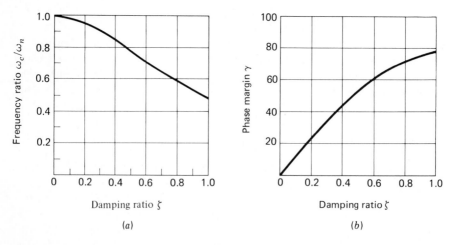

Figure 12.12 Correlation between ζ, ω_n, and phase margin γ: (a) plot of (ω_c/ω_n) versus ζ; (b) plot of γ versus ζ.

Hence

$$\gamma = 90° - \tan^{-1} \tau\omega_c = \tan^{-1} \frac{1}{\tau\omega_c} = \tan^{-1} \frac{2\zeta}{\omega_c/\omega_n}$$

or
$$\gamma = \tan^{-1} \frac{2\zeta}{\sqrt{-2\zeta^2 + \sqrt{4\zeta^4 + 1}}} \qquad (12.13)$$

Solving Eq. (12.13) for ζ in terms of γ gives

$$\zeta = \frac{\tan \gamma \sqrt{\cos \gamma}}{2} \qquad (12.14)$$

In Fig. 12.12b is shown a plot of γ versus ζ. If γ is known, then the value of ζ may be determined from this plot. The corresponding value of ω_c/ω_n may then be ascertained from Fig. 12.12a.

Illustrative example 12.3 Determine the value of ζ and ω_n for the system shown by the dotted line in Fig. 12.11.

SOLUTION For this system, the phase margin γ is 45°. From Eq. (12.14) or from Fig. 12.12b, the corresponding value of ζ is 0.42. Next, from Eq. (12.12) or from Fig. 12.12a, the ratio ω_c/ω_n is found to be 0.84. The frequency ω_c is the value of ω at which $\log |G(j\omega)H(j\omega)| = \log 1 = 0$. From Fig. 12.11, it is found that $\omega_c = 1.7$ and thus $\omega_n = 2.03$.

12.3 LEAD COMPENSATION

In Sec. 11.8, it is shown how the gain K is selected in order to obtain a desired value of M_p. A change in the gain K in effect changes the scale factor of the polar plot but does not change the basic shape of the plot. In the design of control systems, it is often necessary to change the shape of the polar plot in order to achieve the desired dynamic performance. A common means of doing this is to insert elements in series with the feedforward portion of the control. This method of compensating the performance of the control system is called series compensation.

In general, the frequency-response characteristics of a component which is used to provide series compensation are such that the output of the component either lags or leads the input. In some cases, it is advantageous to use a component in which the output lags the input for a certain range of frequencies and then the output leads the input for other frequencies. This is known as lag-lead series compensation. A component which is used to provide series compensation is sometimes referred to as a series equalizer.

In this section, the design of series lead compensators is discussed. The next two sections consider series lag compensators and series lag-lead compensators respectively.

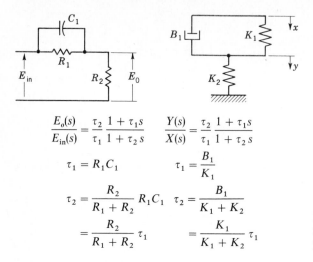

$$\frac{E_o(s)}{E_{in}(s)} = \frac{\tau_2}{\tau_1} \frac{1 + \tau_1 s}{1 + \tau_2 s} \qquad \frac{Y(s)}{X(s)} = \frac{\tau_2}{\tau_1} \frac{1 + \tau_1 s}{1 + \tau_2 s}$$

$$\tau_1 = R_1 C_1 \qquad \qquad \tau_1 = \frac{B_1}{K_1}$$

$$\tau_2 = \frac{R_2}{R_1 + R_2} R_1 C_1 \qquad \tau_2 = \frac{B_1}{K_1 + K_2}$$

$$= \frac{R_2}{R_1 + R_2} \tau_1 \qquad \qquad = \frac{K_1}{K_1 + K_2} \tau_1$$

Figure 12.13 An electrical and a mechanical circuit used to obtain phase lead.

In Fig. 12.13 is shown both an electrical and a mechanical circuit which have the general phase-lead characteristics given by

$$\frac{E_o(s)}{E_{in}(s)} = \frac{Y(s)}{X(s)} = \frac{\tau_2}{\tau_1} \frac{1 + \tau_1 s}{1 + \tau_2 s} \qquad \tau_1 > \tau_2$$

The frequency response for the preceding transfer function is

$$\frac{E_o(j\omega)}{E_{in}(j\omega)} = \frac{Y(j\omega)}{X(j\omega)} = \frac{\tau_2}{\tau_1} \frac{1 + j\tau_1 \omega}{1 + j\tau_2 \omega} \qquad \tau_1 > \tau_2 \qquad (12.15)$$

Because the steady-state gain ($\omega \approx 0$) is τ_2/τ_1, additional amplification equal to τ_1/τ_2 must be provided to maintain the original system gain.

The construction of the log-magnitude diagram for Eq. (12.15) in which $\tau_2 = \tau_1/10$ is illustrated in Fig. 12.14. For the numerator the break frequency occurs at $\omega = 1/\tau_1$ and for the denominator the break frequency occurs at $\omega = 1/\tau_2 = 10/\tau_1$. Adding the diagram for the numerator to that for the denominator yields the resultant diagram for phase lead.

The use of a phase-lead compensator placed in series with the feedforward portion of a control system to improve stability is now illustrated. In Fig. 12.15 the dashed curve is the frequency response $G(j\omega)H(j\omega)$ of the uncompensated control system. This control is one that would inherently be unstable. The addition of the lead compensator $G_c(j\omega)$ to reshape the high-frequency portion of the polar plot is shown by the solid-line curve of Fig. 12.15. It is to be noted that lead compensation rotates a typical vector such as that for $\omega = 2$ in a counterclockwise direction away from the -1 point. It is also to be noted that the length of this vector is increased. Because of the counterclockwise rotation of a typical vector, the effect of lead compensation is to increase ω_p. This tends to increase the speed of response of the system.

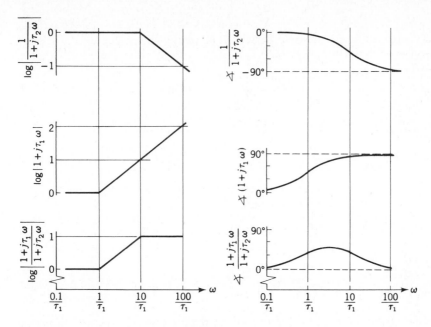

Figure 12.14 Construction of the log-magnitude diagram for $(1 + j\tau_1\omega)/(1 + j\tau_2\omega)$ when $\tau_2 = \tau_1/10$.

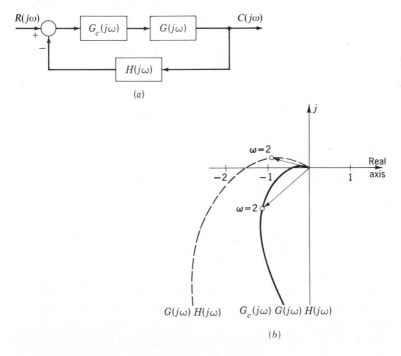

Figure 12.15 Use of phase lead to reshape a polar plot.

To select a lead compensator, it is necessary to be familiar with some of the general properties which are now derived. The phase shift ϕ due to a lead compensator is

$$\phi = \sphericalangle \frac{1 + j\tau_1 \omega}{1 + j\tau_2 \omega} = \tan^{-1} \tau_1 \omega - \tan^{-1} \tau_2 \omega \qquad (12.16)$$

Differentiating with respect to ω gives

$$\frac{d\phi}{d\omega} = \frac{\tau_1}{1 + (\tau_1 \omega)^2} - \frac{\tau_2}{1 + (\tau_2 \omega)^2}$$

Setting this derivative equal to zero and solving for ω yields the frequency $\omega = \omega_m$ at which the phase shift is a maximum. That is,

$$\tau_1 [1 + (\tau_2 \omega_m)^2] - \tau_2 [1 + (\tau_1 \omega_m)^2] = 0$$

$$(\tau_1 - \tau_2)(1 - \tau_1 \tau_2 \omega_m^2) = 0$$

$$\omega_m^2 = \frac{1}{\tau_1 \tau_2}$$

In Fig. 12.16a is shown the log-magnitude diagram for a lead compensator. It is to be noted that the maximum phase shift ϕ_m occurs at the frequency $\omega_m = 1/\sqrt{\tau_1 \tau_2}$. This frequency at which the maximum phase shift occurs is located at the midpoint between the break frequencies $\omega_1 = 1/\tau_1$ and $\omega_2 = 1/\tau_2$. Because frequency is plotted on a logarithmic scale, the midpoint is

$$\log \omega_m = \frac{1}{2} (\log \omega_1 + \log \omega_2) = \frac{1}{2} \log \omega_1 \omega_2 = \log \frac{1}{\sqrt{\tau_1 \tau_2}}$$

Hence

$$\omega_m = \frac{1}{\sqrt{\tau_1 \tau_2}} \qquad (12.17)$$

The value of the maximum phase shift ϕ_m is obtained by substituting $\omega = \omega_m = 1/\sqrt{\tau_1 \tau_2}$ into Eq. (12.16). That is,

$$\phi_m = \sphericalangle \frac{1 + j\sqrt{\tau_1/\tau_2}}{1 + j\sqrt{\tau_2/\tau_1}} \frac{1 - j\sqrt{\tau_2/\tau_1}}{1 - j\sqrt{\tau_2/\tau_1}} = \sphericalangle \frac{1 + 1 + j(\sqrt{\tau_1/\tau_2} - \sqrt{\tau_2/\tau_1})}{1 + \tau_2/\tau_1}$$

$$= \tan^{-1} \frac{\sqrt{\tau_1/\tau_2} - \sqrt{\tau_2/\tau_1}}{2} = \tan^{-1} \frac{\tau_1/\tau_2 - 1}{2\sqrt{\tau_1/\tau_2}} \qquad (12.18)$$

This result shows that ϕ_m is a function of the ratio τ_1/τ_2. In Fig. 12.16b is shown a plot of ϕ_m versus $\log \tau_1/\tau_2$. The double scale for the horizontal axis displays both $\log \tau_1/\tau_2$ and τ_1/τ_2.

In accordance with Eq. (12.18), ϕ_m may be regarded as the angle of a right triangle whose opposite side is $\tau_1/\tau_2 - 1$ and whose adjacent side is $2\sqrt{\tau_1/\tau_2}$.

Lead

(a)

Figure 12.16 Lead compensator characteristics: (a) log-magnitude diagram, (b) maximum phase shift ϕ_m versus $\log \tau_1/\tau_2$.

From this triangle, it follows that

$$\sin \phi_m = \frac{\tau_1/\tau_2 - 1}{\tau_1/\tau_2 + 1}$$

Solving for the ratio τ_1/τ_2 shows that

$$\frac{\tau_1}{\tau_2} = \frac{1 + \sin \phi_m}{1 - \sin \phi_m} \tag{12.19}$$

To select a lead compensator, it is necessary to specify the values of both τ_1 and τ_2. Because of the two unknowns, τ_1 and τ_2, the selection of a lead compensator to achieve a desired design specification is basically a trial-and-error process. However, a systematic procedure which rapidly converges the trial-and-error process is described in the following example, in which the procedure for obtaining a desired phase margin is described.

Illustrative example 12.4 The log-modulus plot of $G(j\omega)$ for a unity-feedback system is shown in Fig. 12.17a. Determine the values of τ_1 and τ_2 for a series lead compensator such that the compensated system will have a phase margin of $45° \pm 3°$. What are the approximate values of ζ and ω_n for the resulting compensated system?

SOLUTION From Fig. 12.16a, it is to be noted that when the phase shift is a maximum, the lead compensator increases the gain by a factor of $\frac{1}{2} \log \tau_1/\tau_2$. The first step is to assume a value of τ_1/τ_2. For example, for $\tau_1/\tau_2 = 4$, Fig. 12.16b shows that $\log \tau_1/\tau_2 = 0.6$ and $\phi_m = 36°$. At the frequency $\omega_m = 1/\sqrt{\tau_1\tau_2}$, the lead compensator increases the gain by a factor $\frac{1}{2} \log \tau_1/\tau_2 = 0.3$ and the phase angle is increased by $\phi_m = 36°$. The $\times$ in Fig. 12.17a is located at the point through which the log-modulus plot should pass to yield a phase margin of $45°$. Thus, by measuring down -0.3 log units from the $\times$ as shown in Fig. 12.17a, when the lead compensator is added, the point indicated by the $\bigcirc$ which corresponds to ω_m will be raised vertically 0.3 log units to the zero horizontal axis and shifted to the right $36°$. The new location is indicated by the solid dot. To have a phase margin of $45°$, the solid dot should be located at the $\times$. Thus, it is necessary to assume a larger value of τ_1/τ_2. The result for $\tau_1/\tau_2 = 10$ is illustrated in Fig. 12.17b. It is to be noted that this yields a phase margin of $43°$, which is satisfactory. Because of the rapid convergence of this process, usually two or three trials suffice for obtaining the desired ratio τ_1/τ_2. The desired values of τ_1 and τ_2 are now computed from the relationships $\tau_1/\tau_2 = 10$ and $\omega_c = \omega_m = 1/\sqrt{\tau_1\tau_2} = 2.6$. Hence,

$$\tau_1 = 1.2 \quad \text{and} \quad \tau_2 = 0.12$$

The complete log-modulus plot for the resulting $G_c(j\omega)G(j\omega)$ system may now be constructed as illustrated in Fig. 12.17c.

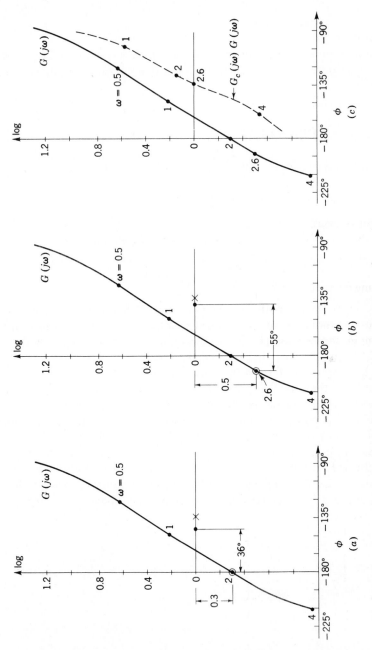

Figure 12.17 Use of series lead compensation to obtain a desired phase margin.

For a phase margin of 43°, Eq. (12.14) or Fig. 12.12b shows that the approximate value of ζ is 0.40. From Eq. (12.12) or from Fig. 12.12a, it is found that $\omega_c/\omega_n = 0.85$. Thus, the approximate value of the natural frequency ω_n is found to be 3.1.

By putting an $\times$ at the point through which the resulting polar plot should pass, this procedure may be used to obtain a desired gain margin or a desired value of M_p.

12.4 LAG COMPENSATION

The output lags the input for any component which has a transfer function of the form

$$\frac{E_o(s)}{E_{in}(s)} = \frac{Y(s)}{X(s)} = \frac{1 + \tau_2 s}{1 + \tau_1 s} \qquad \tau_1 > \tau_2 \tag{12.20}$$

The frequency response for the preceding transfer function is

$$\frac{E_o(j\omega)}{E_{in}(j\omega)} = \frac{Y(j\omega)}{X(j\omega)} = \frac{1 + j\tau_2 \omega}{1 + j\tau_1 \omega} \qquad \tau_1 > \tau_2 \tag{12.21}$$

In Fig. 12.18 is shown both an electrical and a mechanical circuit in which the output lags the input, as is described by Eq. (12.21). The construction of the log-magnitude plot for Eq. (12.21) for the case in which $\tau_2 = \tau_1/10$ is illustrated in Fig. 12.19. For the term $1/(1 + j\tau_1\omega)$ the break frequency occurs at $\omega = 1/\tau_1$ and for the numerator $1 + j\tau_2\omega$ the break frequency occurs at $\omega = 1/\tau_2 = 10/\tau_1$. The addition of the log-magnitude diagram for the numerator to that for the denominator yields the resulting diagram for $(1 + j\tau_2\omega)/(1 + j\tau_1\omega)$.

The use of a phase-lag component placed in series with the feedforward portion of a control system to improve stability is now illustrated. In Fig. 12.20 the dashed curve is the frequency response $G(j\omega)H(j\omega)$ of the uncompensated control system. This control is one which would inherently be unstable. The addition of series compensation $G_c(j\omega)$ reshapes the polar plot, as shown by the solid-line curve of Fig. 12.20. The resultant system has good dynamic response. It is to

Figure 12.18 An electrical and a mechanical circuit used to provide lag compensation.

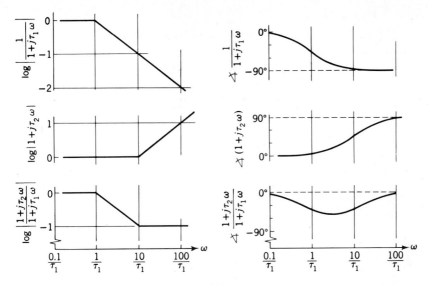

Figure 12.19 Construction of the log-magnitude diagram for $(1 + j\tau_2\omega)/(1 + j\tau_1\omega)$ when $\tau_2 = \tau_1/10$.

be noted that the effect of lag compensation is to shorten a typical vector such as that for $\omega = 0.5$ and also to rotate it in a clockwise direction. The shortening is due to the attenuation. By attenuation is meant multiplication by a factor less than 1. The attenuation caused by use of lag compensation can be seen from Fig. 12.21a. The greater the spread in time constants τ_1 and τ_2, the more pronounced is the attenuation which occurs at higher frequencies. Series lag compensation has little effect on the low-frequency portion of the curve. By reshaping

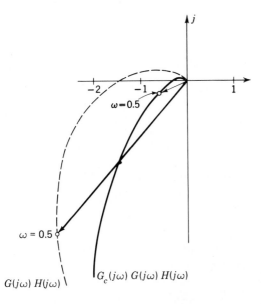

Figure 12.20 Use of phase lag to reshape a polar plot.

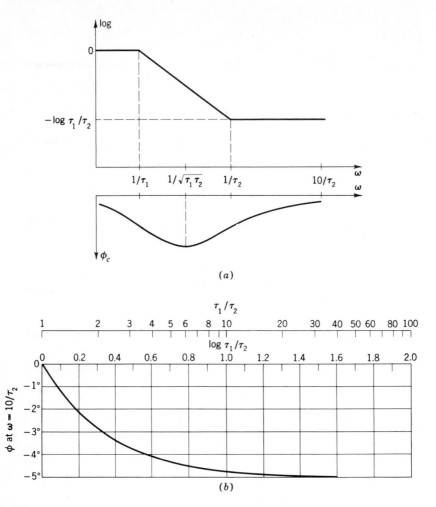

Figure 12.21 Lag compensator characteristics: (a) log-magnitude diagram, (b) phase shift ϕ at $\omega = 10/\tau_2$.

the polar plot, it has been possible to achieve good dynamic performance without changing the value of the gain K. Although it would have been possible to make this system stable by decreasing the gain K only, errors caused by friction, hysteresis, backlash, etc., tend to predominate as the gain is decreased; thus, in general, the higher the value of K, the more accurate will be the control system.

In Fig. 12.21a is shown the log-magnitude diagram for a lag compensator. Comparison of Figs. 12.16a and 12.21a shows that the phase-shift characteristics of a lag compensator are the same as for a lead compensator except that the sign of the phase angle is negative. Similarly, the log-magnitude characteristics for a lag compensator are the same as for a lead compensator except that the sign is reversed.

The negative phase shift associated with lag compensation is usually undesirable. The effectiveness of lag compensation is attributed to the attenuation which occurs at higher frequencies. In Fig. 12.21a, it is to be noted that when $\omega = 10/\tau_2$, then the negative phase shift is very small. The solid-line curve in Fig. 12.21b is a plot of the value of the negative phase shift at $\omega = 10/\tau_2$. A significant difference between selecting a lag compensator and selecting a lead compensator is that for the lag compensator the region of the small phase shift ($\omega = 10/\tau_2$) is located in the vicinity of the point of interest, whereas for a lead compensator the region of maximum phase shift ($\omega = \omega_m$) is located in the vicinity of the point of interest.

Illustrative example 12.5 Same as Illustrative example 12.4 except use a lag rather than a lead compensator. The log-modulus diagram of $G(j\omega)$ is shown in Fig. 12.22a.

SOLUTION For a phase margin of 45°, the resultant $G_c(j\omega)G(j\omega)$ plot must pass through the point indicated by the × in Fig. 12.22a. From Fig. 12.21b, it is to be noted that for values of $\tau_1/\tau_2 > 4$, the phase shift introduced by a lag compensator at $\omega = 10/\tau_2$ is about $-5°$. Thus, a vertical line is erected 5° to the right of the ×, as shown in Fig. 12.22a. The intersection of this vertical line with the $G(j\omega)$ plot is indicated by a ◯. In order that the compensator move this point to the ×, $-\log \tau_1/\tau_2 = -0.6$, and thus $\tau_1/\tau_2 = 4.0$.

Figure 12.22 Use of series lag compensation to obtain a desired phase margin.

Figure 12.22a shows that the frequency $\omega = 10/\tau_2$ is 0.5. The particular values of τ_1 and τ_2 are now computed from the relationships $\tau_1/\tau_2 = 4$ and $10/\tau_2 = 0.5$. Thus,

$$\tau_1 = 80 \quad \text{and} \quad \tau_2 = 20$$

With these values of τ_1 and τ_2, the log-modulus diagram for the resultant compensated system $G_c(j\omega)G(j\omega)$ may now be constructed as shown in Fig. 12.22b.

For a phase margin of 45°, it follows from Eq. (12.14) or from Fig. 12.12b that the approximate value of ζ is 0.42. Similarly, from Eq. (12.12) or from Fig. 12.12a the ratio ω_c/ω_n is found to be 0.84. Because the crossover frequency ω_c is 0.5, then the approximate natural frequency ω_n is found to be 0.6.

It is to be noted that the values of τ_1 and τ_2 for this lag compensator are larger than the values of τ_1 and τ_2 for the corresponding lead compensator. Thus, the system with lag compensation is slower than the corresponding system with lead compensation. This fact is also substantiated by noting that the approximate natural frequency for the system with lag compensation is considerably smaller ($\omega_n = 0.6$) than the approximate natural frequency for the system with lead compensation ($\omega_n = 3.1$).

By proceeding in a similar manner, it is possible to select lag compensators to yield a desired M_p or a desired gain margin. When selecting a compensator to yield a desired phase margin or a desired gain margin, one may use either Bode diagrams or log-modulus diagrams. Because it is not possible to construct M contours on Bode diagrams, it is necessary to use log-modulus diagrams for obtaining a compensator to yield a desired value of M_p.

12.5 LAG-LEAD COMPENSATION

A lag-lead compensator is a series combination of a lag and a lead network. The general transfer function for a lag-lead compensator is

$$\frac{E_o(s)}{E_{in}(s)} = \frac{Y(s)}{X(s)} = \frac{1 + c\tau_2 s}{1 + c\tau_1 s} \frac{1 + \tau_1 s}{1 + \tau_2 s} \qquad \tau_1 > \tau_2 \tag{12.22}$$

where $c > 1$ is a constant. The substitution of $j\omega$ for s gives

$$\frac{E_o(j\omega)}{E_{in}(j\omega)} = \frac{Y(j\omega)}{X(j\omega)} = \frac{1 + jc\tau_2\omega}{1 + jc\tau_1\omega} \frac{1 + j\tau_1\omega}{1 + j\tau_2\omega} \tag{12.23}$$

Rather than using a lag and a lead compensator in series, it is possible to use a single compensator, as is shown in Fig. 12.23.

The log-magnitude diagram for a typical lag-lead compensator is shown in Fig. 12.24. Because $c\tau_1 > c\tau_2 > \tau_1 > \tau_2$, the first break frequency occurs at $1/c\tau_1$. This break frequency belongs to a denominator term, so that the magnitude plot

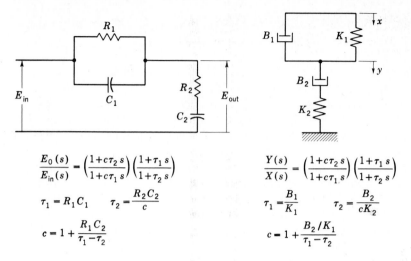

$$\frac{E_0(s)}{E_{in}(s)} = \left(\frac{1+c\tau_2 s}{1+c\tau_1 s}\right)\left(\frac{1+\tau_1 s}{1+\tau_2 s}\right)$$

$$\tau_1 = R_1 C_1 \qquad \tau_2 = \frac{R_2 C_2}{c}$$

$$c = 1 + \frac{R_1 C_2}{\tau_1 - \tau_2}$$

$$\frac{Y(s)}{X(s)} = \left(\frac{1+c\tau_2 s}{1+c\tau_1 s}\right)\left(\frac{1+\tau_1 s}{1+\tau_2 s}\right)$$

$$\tau_1 = \frac{B_1}{K_1} \qquad \tau_2 = \frac{B_2}{cK_2}$$

$$c = 1 + \frac{B_2/K_1}{\tau_1 - \tau_2}$$

Figure 12.23 An electrical and a mechanical circuit used to provide lag-lead compensation.

has a slope of -1 log unit/decade between $1/c\tau_1$ and $1/c\tau_2$. The second break frequency, $1/c\tau_2$, is associated with a numerator term, and thus the magnitude plot again becomes horizontal. The third break frequency, $1/\tau_1$, also occurs in the numerator. This results in a slope of $+1$ log unit/decade in the region from $1/\tau_1$ to $1/\tau_2$. Finally, the break frequency $1/\tau_2$, which occurs in the denominator, causes the magnitude curve to become horizontal again. It is to be noted that the maximum phase shift ϕ_m occurs at $\omega = 1/\sqrt{\tau_1\tau_2}$ and the corresponding attenuation is $-\frac{1}{2}\log \tau_1/\tau_2$. This is the same as for a lead compensator only (see Fig. 12.16a) except that the sign of the log of the magnitude is negative. This feature makes the lag-lead compensator considerably more effective than the lead compensator only.

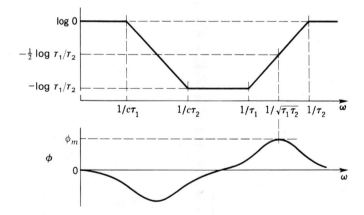

Figure 12.24 Log-magnitude diagram for a lag-lead compensator.

Illustrative example 12.6 Same as Illustrative example 12.4 except use a lag-lead compensator rather than a lead compensator.

SOLUTION In Fig. 12.25*a* is shown the log-modulus plot for the uncompensated system $G(j\omega)$. As a first trial, it is assumed that $\tau_1/\tau_2 = 2$; thus $\phi_m = 20°$ and $-\frac{1}{2}\log \tau_1/\tau_2 = -0.15$ log units. The circle in Fig. 12.25*a* will be shifted downward -0.15 log units and it will be moved to the right 20°. The resulting location is indicated by the ×. The corresponding phase margin is $\gamma = 45°$.

The particular values of τ_1 and τ_2 for this lag-lead compensator are determined from the relationships $\tau_1/\tau_2 = 2$ and $\omega_m = 1/\sqrt{\tau_1\tau_2} = 1.1$. Thus, it is found that $\tau_1 = 1.30$ and $\tau_2 = 0.65$.

A factor of 5 provides a reasonable separation between the second and third break frequencies (see Fig. 12.24), that is, $1/\tau_1 = 5(1/c\tau_2)$. Solving this relationship for c gives

$$c = 5\frac{\tau_1}{\tau_2} = 10$$

With the compensator specified, the log-modulus diagram for the resulting compensated system may now be constructed as shown in Fig. 12.25*b*. As is indicated by Fig. 12.24, lag-lead compensation does not affect the low- or high-frequency regions but rather the mid-frequency region.

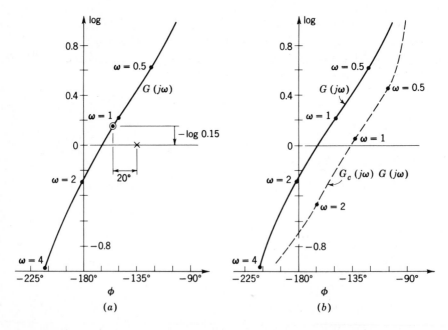

Figure 12.25 Use of series lag-lead compensation to obtain a desired phase margin.

12.6 INTERNAL FEEDBACK

Another method commonly used to alter frequency-response characteristics is that of providing a separate internal-feedback path about certain components. In employing log-magnitude diagrams to investigate the effect of internal feedback, the use of a few approximations affords much simplification. This approximate analysis in effect puts the designer in the right "ball park." In the latter design stages, it may be desirable to make an exact analysis. The approximations which are used to evaluate the effect of placing a feedback element $H_1(s)$ around a component $G_1(s)$ as shown in Fig. 12.26 are that, when $|G_1(j\omega)H_1(j\omega)| \ll 1$,

$$G(j\omega) = \frac{G_1(j\omega)}{1 + G_1(j\omega)H_1(j\omega)} \approx G_1(j\omega) \tag{12.24}$$

For the case when $|G_1(j\omega)H_1(j\omega)| \gg 1$,

$$G(j\omega) = \frac{G_1(j\omega)}{1 + G_1(j\omega)H_1(j\omega)} \approx \frac{G_1(j\omega)}{G_1(j\omega)H_1(j\omega)} = \frac{1}{H_1(j\omega)} \tag{12.25}$$

Note that when $|G_1(j\omega)H_1(j\omega)| \ll 1$, then $|G_1(j\omega)| \ll |1/H_1(j\omega)|$. Thus, Eq. (12.24) indicates that $G(j\omega)$ is to be approximated by $G_1(j\omega)$ when $|G_1(j\omega)| \ll |1/H_1(j\omega)|$. Similarly, when $|G_1(j\omega)H_1(j\omega)| \gg 1$, then $|1/H_1(j\omega)| \ll |G_1(j\omega)|$. Thus, Eq. (12.25) indicates that $G(j\omega)$ is to be approximated by $1/H_1(j\omega)$ when $|1/H_1(j\omega)| \ll |G_1(j\omega)|$. In summary, $G(j\omega)$ is always approximated by the smaller of $G_1(j\omega)$ or $1/H_1(j\omega)$.

When $|G_1(j\omega)H_1(j\omega)| = 1$, then $|G_1(j\omega)| = |1/H_1(j\omega)|$. Thus, the intersection of the log-magnitude diagram for $|G_1(j\omega)|$ and that for $|1/H_1(j\omega)|$ determines the point at which $|G_1(j\omega)H_1(j\omega)| = 1$. Usually at high frequencies $|G_1(j\omega)H_1(j\omega)| \ll 1$, in which case $|G_1(j\omega)| \ll |1/H_1(j\omega)|$. Thus, the high-frequency response is approximated by $G_1(j\omega)$ and the low-frequency response is approximated by $1/H_1(j\omega)$.

In summary, this approximation converts an internal-feedback path to an approximate open-loop element for which a standard analysis can be made. With the use of inverse polar plots, as is next described, no approximations are employed.

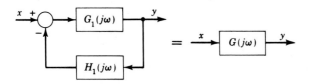

Figure 12.26 Internal feedback $H_1(s)$ placed about an element $G_1(s)$.

12.7 INVERSE POLAR PLOTS

A plot of the function $G^{-1}(j\omega) = 1/G(j\omega)$ is called an inverse polar plot. In Fig. 12.27 is shown a typical inverse polar plot of the function $G^{-1}(j\omega)$. At any frequency ω, the vector from the origin to a point on the plot defines the vector $G^{-1}(j\omega)$ for that frequency. The length of the vector is $|G^{-1}(j\omega)| = |1/G(j\omega)|$, and the angle is

$$\angle G^{-1}(j\omega) = \angle \frac{1}{G(j\omega)} = -\angle G(j\omega)$$

A plot of M circles and α lines for inverse polar plots is accomplished by first taking the reciprocal of Eq. (11.52), that is,

$$\frac{R(j\omega)}{C(j\omega)} = \frac{1 + G(j\omega)}{G(j\omega)} = G^{-1}(j\omega) + 1 \tag{12.26}$$

A typical vector $G^{-1}(j\omega)$ as shown in Fig. 12.27 may be written in the general form $G^{-1}(j\omega) = x + jy$. Substitution of this general representation for $G^{-1}(j\omega)$ into Eq. (12.26) gives

$$\frac{R(j\omega)}{C(j\omega)} = x + 1 + jy \tag{12.27}$$

Because $M = |C(j\omega)/R(j\omega)|$, from the magnitude of Eq. (12.27) it follows that

$$\left|\frac{R(j\omega)}{C(j\omega)}\right| = \frac{1}{M} = \sqrt{(x + 1)^2 + y^2}$$

Squaring this result gives

$$(x + 1)^2 + y^2 = \frac{1}{M^2} \tag{12.28}$$

Thus, on the inverse plane, lines of constant M are circles of radius $1/M$. The center of these concentric M circles is at the point $x = -1$ and $y = 0$, that is, the -1 point. A plot of these M circles on the inverse plane is shown in Fig. 12.28. Because the reciprocal of -1 is still -1, this point has the same significance for an inverse polar plot as for a direct polar plot. Polar plots are referred to as direct polar plots when it is necessary to distinguish them from inverse polar plots.

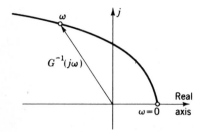

Figure 12.27 Typical inverse polar plot $G^{-1}(j\omega)$.

Figure 12.28 M circles and α rays on the inverse plane.

The lines of constant $\alpha = \sphericalangle [C(j\omega)/R(j\omega)] = - \sphericalangle[R(j\omega)/C(j\omega)]$ are determined from Eq. (12.27) as follows:

$$\alpha = - \sphericalangle \frac{R(j\omega)}{C(j\omega)} = -\tan^{-1} \frac{y}{x+1} \qquad (12.29)$$

When plotted on the inverse plane, as is shown in Fig. 12.28, constant α contours are radial straight lines (rays) which pass through the $-1 + j0$ point.

As is illustrated in Fig. 12.29, the angle ψ of a radial line drawn from the origin tangent to any M circle is

$$\sin \psi = \frac{1}{M} \qquad (12.30)$$

The use of the inverse polar plot for determining the gain K to yield a desired value of M_p is similar to that for the direct polar plot. Consider the same function $G(j\omega) = K/[j\omega(1 + 0.1j\omega)]$, discussed in Sec. 11.8. The plot of the inverse function

$$\frac{K}{G(j\omega)} = KG^{-1}(j\omega) = j\omega(1 + 0.1j\omega) \qquad (12.31)$$

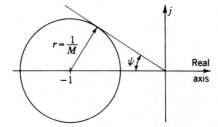

Figure 12.29 Tangent line to an M circle.

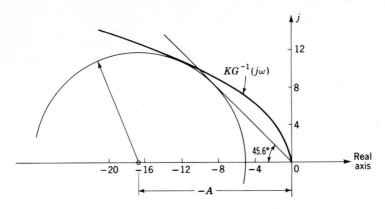

Figure 12.30 Inverse polar plot $KG^{-1}(j\omega) = j\omega(1 + j0.1\omega)$.

is shown in Fig. 12.30. For a desired $M_p = 1.4$, the angle ψ of the tangent line is

$$\psi = \sin^{-1}\frac{1}{M_p} = \sin^{-1}\frac{1}{1.4} = 45.6° \qquad (12.32)$$

Next construct by trial and error the circle which is tangent to both the $KG^{-1}(j\omega)$ plot and the tangent line. In order for this circle to be the desired M_p circle, its center must be at the -1 point. From Fig. 12.30, it is to be seen that the center is at $-A = -16.7$. To convert this to the -1 point, the scale factor must be multiplied by $1/A = 1/16.7 = 0.06$. The resulting function $G^{-1}(j\omega)$, which is tangent to the desired M circle is

$$G^{-1}(j\omega) = \frac{1}{A}[KG^{-1}(j\omega)] = 0.06j\omega(1 + 0.1j\omega)$$

or $$G(j\omega) = \frac{16.7}{j\omega(1 + 0.1j\omega)}$$

Thus, the value of A yields directly the required gain K.

The general procedure for obtaining K by use of the inverse polar plot is:

1. Plot the inverse function $KG^{-1}(j\omega)$.
2. Construct the tangent line in accordance with Eq. (12.30).
3. By trial and error, determine the circle which is tangent to both the $KG^{-1}(j\omega)$ plot and the tangent line.
4. The desired gain is $K = A$.

When the function $G^{-1}(j\omega)$ is plotted rather than $KG^{-1}(j\omega)$, then A is equal to the factor K_c by which the gain should be changed to yield the desired M_p.

Illustrative example 12.7 Consider the function

$$G(j\omega) = \frac{5}{j\omega(1 + 0.1j\omega)}$$

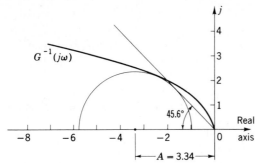

Figure 12.31 Inverse polar plot of $G^{-1}(j\omega) = j\omega(1 + j0.1\omega)/5$.

The inverse polar plot $G^{-1}(j\omega)$ is shown in Fig. 12.31. Determine the factor K_c by which the gain should be changed so that the resulting plot is tangent to the $M_p = 1.4$ circle.

SOLUTION In order that the point $-A$ in Fig. 12.31 be the -1 point, the scale factor must be multiplied by $1/A$. The resulting function is

$$\frac{1}{A} G^{-1}(j\omega) = \frac{1}{AG(j\omega)} = \frac{1}{K_c\, G(j\omega)}$$

or

$$K_c\, G(j\omega) = \frac{(3.34)(5)}{j\omega(1 + 0.1j\omega)} = \frac{16.7}{j\omega(1 + 0.1j\omega)}$$

The preceding analysis shows that $K_c = A$.

The major advantage of using the inverse plane is realized for systems with internal feedback. The reciprocal of Eq. (12.24) is

$$G^{-1}(j\omega) = \frac{1 + G_1(j\omega)H_1(j\omega)}{G_1(j\omega)} = G_1^{-1}(j\omega) + H_1(j\omega) \tag{12.33}$$

As is illustrated in Fig. 12.32, the vectors $G_1^{-1}(j\omega)$ and $H_1(j\omega)$ may be added as vector quantities to yield $G^{-1}(j\omega)$.

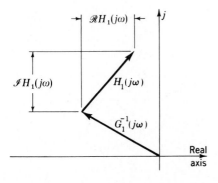

Figure 12.32 Vector addition of $G_1^{-1}(j\omega)$ and $H_1(j\omega)$ to yield $G^{-1}(j\omega)$.

Illustrative example 12.8 For the system shown in Fig. 12.33a, the value of K_1 is 10 and K_2 is 2.5. From the inverse polar plot $1/[K_2 G_1(j\omega)]$ of Fig. 12.33b, the value of M_p is found to be 5. It is desired to obtain an M_p of 1.4 by the use of an internal feedback loop as shown in Fig. 12.34a. What feedback element $H_1(j\omega)$ should be used?

SOLUTION The equation for the inverse polar plot with internal feedback is

$$\frac{1}{G(j\omega)} = \frac{1}{K_2} \frac{1 + G_1(j\omega)H_1(j\omega)}{G_1(j\omega)} = \frac{1}{K_2 G_1(j\omega)} + \frac{H_1(j\omega)}{K_2} \qquad (12.34)$$

The quantity $H_1(j\omega)/K_2$ must be such that when it is added to $1/[K_2 G_1(j\omega)]$, the resulting plot will be tangent to the $M = 1.4$ circle. To obtain this, it is necessary only to raise the original plot vertically. It is to be seen from Fig. 12.32 that the horizontal component of $H_1(j\omega)$ is its real part $\mathscr{R}H_1(j\omega)$ and the vertical component is its imaginary part $\mathscr{I}H_1(j\omega)$. It is thus necessary only that $H_1(j\omega)$ be purely imaginary, i.e., of the form $H_1(s) = \beta s$ or $H_1(j\omega) = j\beta\omega$. In Fig. 12.34b, it is to be noted that, at $\omega = 5$, the addition of $H_1(j5)/K_2 = 0.5j$ to the $1/[K_2 G_1(j5)]$ vector causes the resulting curve to pass through the top of the $M = 1.4$ circle. For this case, the value of β is $j\beta5/2.5 = 0.5j$, or $\beta = 0.25$. The resulting curve may now be constructed as indicated by the dashed line in Fig. 12.34b. Because this curve is not tangent to the $M = 1.4$ circle, another trial value must be taken. From the dashed loci of Fig. 12.34b, it now appears that the point of tangency is more likely to occur in the neighborhood of $\omega = 4.5$. The addition of $H_1(j4.5)/K_2 = 0.47j$ to $1/[K_2 G_1(j4.5)]$ causes the new resulting curve to be tangent to the desired M circle. In this case $j\beta4.5/2.5 = 0.47j$, or $\beta = 0.26$. Thus, the desired result is $H_1(j\omega) = 0.26j\omega$ or $H_1(s) = 0.26s$.

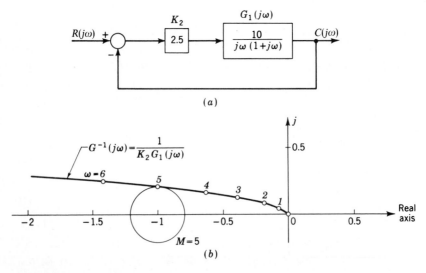

(a)

(b)

Figure 12.33 System without internal feedback.

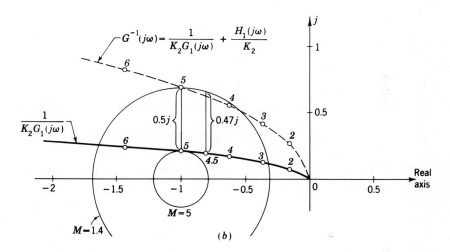

Figure 12.34 System of Fig. 12.33 with internal feedback.

When $H_1(j\omega)$ is a constant, $H_1(j\omega)$ is entirely real. As may be seen from Fig. 12.32, the effect of a constant $H_1(j\omega)$ is to shift the inverse plot horizontally to the right. Suppose in the preceding problem that it is desired to increase the speed of response by having ω_p equal to 6 rather than 4.5. After assuming a few trial values for $H_1(j6)/K_2$, it is found that $H_1(j6)/K_2 = 0.6 + 0.4j$ makes the resultant plot tangent to the $M = 1.4$ circle at $\omega = 6$, as is illustrated in Fig. 12.35. Because $H_1(j6)/K_2 = 0.6 + 0.4j = 0.6 + j\beta6$, then $\beta = 0.4/6 = 0.067$; whence the required $H_1(j\omega)$ is $K_2(0.6 + j\beta\omega) = 1.5 + j0.167\omega$. Thus,

$$H_1(s) = 1.5 + 0.167s \tag{12.35}$$

The general procedure followed in this illustrative example was to assume a value of $H_1(j\omega)$ which makes a point lie on the desired M circle. From this assumed value, the general equation for the resulting plot was obtained. When the assumed point is not the point of tangency, then another trial point must be selected.

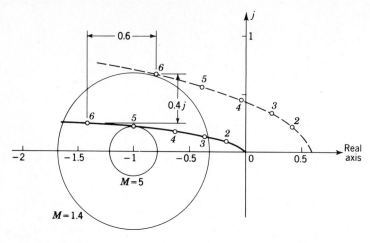

Figure 12.35 Use of inverse polar plot to increase speed of response.

12.8 STABILITY CRITERIA IN THE INVERSE PLANE

The inverse function $1/[G(s)H(s)]$ may be expressed in terms of numerator and denominator terms as follows:

$$\frac{1}{G(s)H(s)} = \frac{D_G D_H}{N_G N_H}$$

In Fig. 12.36 is shown the complete inverse polar plot for the function

$$G(s)H(s) = \frac{10(s + 2)}{s(s + 1)(s + 4)}$$

or

$$\frac{1}{G(s)H(S)} = \frac{s(s + 1)(s + 4)}{10(s + 2)}$$

As s proceeds up the vertical (imaginary) axis of Fig. 12.36a, then the corresponding inverse plot in Fig. 12.36b is indicated by the heavy line.

For values of s on the infinite semicircle $s = Re^{j\beta}$, then

$$\frac{1}{G(s)H(s)} \approx \frac{R^3 e^{3j\beta}}{Re^{j\beta}} = R^2 e^{2j\beta}$$

In general, if n is the highest order of s in the denominator of $G(s)H(s)$ and m is the highest order of s in the numerator of $G(s)H(s)$, then for values of s on the infinite semicircle,

$$\frac{1}{G(s)H(s)} = R^{n-m} e^{j(n-m)\beta}$$

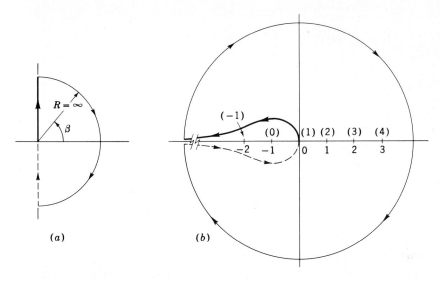

Figure 12.36 Inverse polar plot for $1/[G(s)H(s)] = s(s + 1)(s + 4)/[10(s + 2)]$.

Because β changes by an angle of $-\pi$, the corresponding inverse plot transcribes an infinite semicircle which goes through the angle $-(n - m)\pi$. The negative imaginary axis $s = -j\omega$ is the complex conjugate of the positive imaginary axis $s = j\omega$; thus the portion shown dotted is the mirror image of the heavy-line portion.

The effect of adding $(+1)$ to the $1/G(s)H(s)$ plot is obtained by the new scale shown in parentheses above the real axis of Fig. 12.36*b*. Thus, the -1 point of the $1/G(s)H(s)$ plot is seen to be the origin of the $[1/G(s)H(s)] + 1$ plot. With respect to the numerator and denominator terms, this new plot is

$$\frac{1}{G(s)H(s)} + 1 = \frac{D_G D_H}{N_G N_H} + 1 = \frac{N_G N_H + D_G D_H}{N_G N_H}$$

Application of the Nyquist stability criterion to the inverse plot shows that

$$Z = N + P$$

where Z = number of zeros of the characteristic function $(N_G N_H + D_G D_H)$ in the right half-plane

P = number of zeros of $N_G N_H$ in the right half-plane

N = net number of encirclements of the origin of the $1 + 1/G(s)H(s)$ plot (that is, -1 point of the $1/G(s)H(s)$ plot)

It is to be noted that N and Z have the same meaning for the inverse plots as for direct plots. However, P is the number of zeros of $N_G N_H$ when using inverse polar plots, whereas P is the number of zeros of $D_G D_H$ when using direct polar plots. For the system of Fig. 12.36, $N = 0$ and $P = 0$. Application of the Nyquist stability criterion shows that $Z = N + P = 0$. Hence, the system is stable.

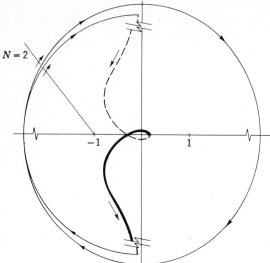

Figure 12.37 Inverse polar plot for $1/[G(s)H(s)] = (s + 2)(s^2 + 2s + 5)/52$.

Illustrative example 12.9 Determine the inverse polar plot and the value of Z for the type 0 system whose direct polar plot is shown in Fig. 12.3.

SOLUTION For a type 0 system the inverse polar plot begins on the positive real axis at the value $1/K$. The heavy-line portion of Fig. 12.37 is the portion of the inverse plot for $s = j\omega$. The dotted line is the portion for $s = -j\omega$. From Eq. (12.4), it is to be noted that $n - m = 3$. Thus, the inverse plot transcribes an infinite semicircle which goes through the angle $-(n - m)\pi = -3\pi$. For the system of Fig. 12.37, $N = 2$ and $P = 0$. Application of the Nyquist stability criterion gives $Z = N + P = 2$. Thus, the system is basically unstable, as was previously determined using the direct polar plot.

On direct polar plots instability is indicated when the frequency response crosses the negative real axis outside the -1 point (e.g., see Fig. 12.3). On inverse polar plots instability is indicated when the frequency response crosses the negative real axis inside the -1 point (e.g., see Fig. 12.37).

PROBLEMS

12.1 For each of the $G(s)H(s)$ plots shown in Fig. P12.1, the path of values for s is the same as that shown in Fig. 12.2. For each plot, determine the number of roots of the characteristic equation which are located in the right half-plane ($P = 0$ in all cases).

For each stable system, determine the factor by which the gain should be changed so that the system will just become unstable.

For each unstable system, determine the factor by which the gain should be changed so that the system will just become stable.

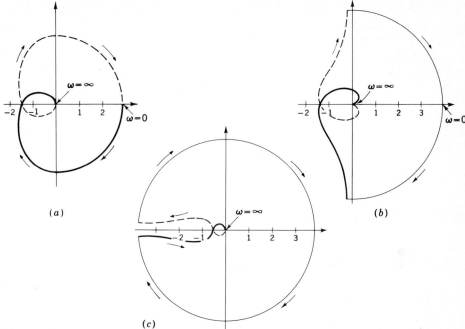

Figure P12.1

12.2 For each of the following unity-feedback systems, sketch the complete $G(s)$ plot and determine the number of roots of the characteristic equation that lie in the right half-plane:

(a) $G(s) = \dfrac{10}{(1 + 0.25s)(0.25s^2 + 0.40s + 1)}$

(b) $G(s) = \dfrac{10}{s(1 + 0.25s)}$

(c) $G(s) = \dfrac{10}{s^2(1 + 0.25s)}$

(d) $G(s) = \dfrac{10(1 + s)}{s^2(1 + 0.25s)}$

12.3 Plot the log-magnitude diagram for the system given in Prob. 12.2(a). For this system, determine

(a) The gain margin and the phase margin

(b) The factor by which the gain should be changed to yield a gain margin of 5

(c) The factor by which the gain should be changed to yield a phase margin of 40°

12.4 The open-loop transfer function for a unity-feedback system is

$$G(s) = \frac{2}{s(1 + s)(1 + 2s)}$$

The Bode diagram for this system is shown in Fig. P12.4.

(a) Determine the factor K_c by which the gain should be changed so that the resulting system will have a phase margin of 45°.

(b) Determine the factor K_c by which the gain should be changed so that the resulting system will have a gain margin of 5.

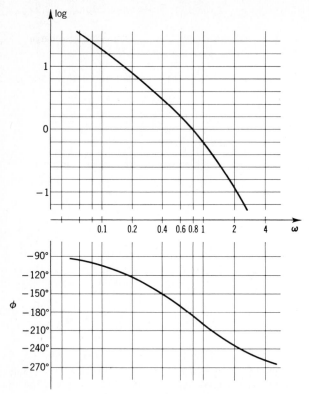

Figure P12.4

12.5 The Nichols plot of $G(j\omega)$ for a unity-feedback system is shown in Fig. P12.5.

    (a) Determine the gain margin and the phase margin for this system.

    (b) By what factor K_c should the gain K of the system be changed so that the system will have a phase margin of 45°?

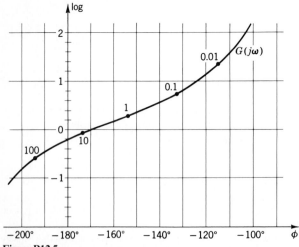

Figure P12.5

12.6 A plot of $G(j\omega)$ for a unity-feedback system is shown in Fig. P12.6. What is the value of the gain margin and the phase margin for this system? By what factor K_c should the gain of the system be changed so that M_p will be 1.4? What are the new values of the gain margin and the phase margin?

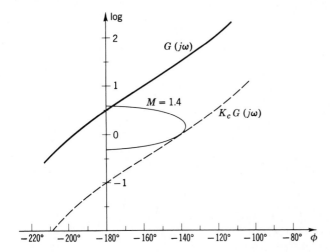

Figure P12.6

12.7 From the polar plot of a type 0 system ($K_0 = 4$), it is noticed that $G(j5)H(j5) = 1.0 \underline{/-145°}$. What value of ζ and ω_n should be used to approximate the transient response of this system?

12.8 From the polar plot of a type 0 system ($K_0 = 4$) it is found that $M_p = 1.4$ and $\omega_p = 3$. What are the approximate values of the damping ratio ζ, natural frequency ω_n, phase margin γ, and crossover frequency ω_c?

12.9 The open-loop transfer function for a control system is

$$G(s)H(s) = \frac{10}{(1 + 0.5s)(s^2 + 6s + 16)/16}$$

Corresponding values of ω, $|G(j\omega)H(j\omega)|$, and $\sphericalangle G(j\omega)H(j\omega)$ are

ω	$|G(j\omega)H(j\omega)|$	$\sphericalangle G(j\omega)H(j\omega)$
0	10.00	0.0
2.0	6.67	−90.0
3.0	4.60	−125.1
3.5	3.72	−140.1
4.0	2.98	−153.4
4.5	2.38	−165.0
5.0	1.90	−174.9
∞	0.00	−270.0

Determine the factor K_c by which the gain should be changed to yield a phase margin of 45°. What is the approximate value of ζ and ω_n for the resulting system?

12.10 The open-loop transfer function for a control system is

$$G(s)H(s) = \frac{10}{s(1 + 0.25s)(1 + 0.5s)}$$

Corresponding values of ω, $|G(j\omega)H(j\omega)|$, and $\not< G(j\omega)H(j\omega)$ are

ω	$\|G(j\omega)H(j\omega)\|$	$\not< G(j\omega)H(j\omega)$
0.6	15.79	-115.2
1.0	8.68	-130.6
1.4	5.52	-144.3
1.8	3.77	-156.2
2.2	2.68	-166.5
2.6	1.97	-175.5
3.0	1.48	-183.2
∞	0.00	-270.0

Determine the value of K to yield a phase margin of $45°$. What is the approximate value of ζ and ω_n for the resulting system?

12.11 The open-loop transfer function for a control system is

$$G(s)H(s) = \frac{K(s + 3)}{s^2 - 2s + 10}$$

Determine the values of $G(j\omega)H(j\omega)$ for $\omega = 0, 1, 2, 3, 4, 5,$ and 6. Construct the entire Nyquist plot and then determine

(a) The range of values of K such that the system is stable
(b) The value of K to yield a phase margin of $45°$

12.12 The open-loop transfer function for a control system is

$$G(s)H(s) = \frac{K(s + 2)}{(s - 1)(s + 1)}$$

Determine the values of $G(j\omega)H(j\omega)$ for $\omega = 0, 1, 2, 3,$ and 4. Construct the entire Nyquist plot and then determine

(a) The range of values of K such that the system is stable
(b) The value of K to yield a phase margin of $45°$

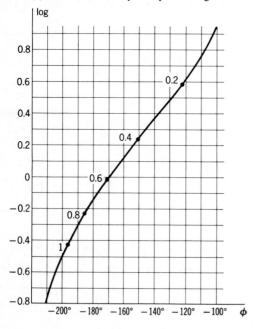

Figure P12.14

12.13 The open-loop transfer function for a control system is

$$G(s)H(s) = \frac{K(s + 2)}{s(s - 1)}$$

Evaluate $G(j\omega)H(j\omega)/K$ for $\omega = 0, 1, 2, 3, 4,$ and 5. Sketch the complete polar plot and then determine
(a) The range of values of K for which the system is stable
(b) The value of K to yield a phase margin of $45°$

12.14 The log-modulus diagram for a control system is shown in Fig. P12.14. Determine the values of τ_1 and τ_2 for a series lead compensator such that the resulting system will have a phase margin of $35° \pm 3°$.

12.15 Same as Prob. 12.14 except use a lag compensator.

12.16 Same as Prob. 12.14 except use a lag-lead compensator.

12.17 The log-magnitude diagram shown in Fig. P12.17 is for the unity-feedback system

$$G(s) = \frac{4}{(s + 1)[(s^2 + 8s + 25)/25]}$$

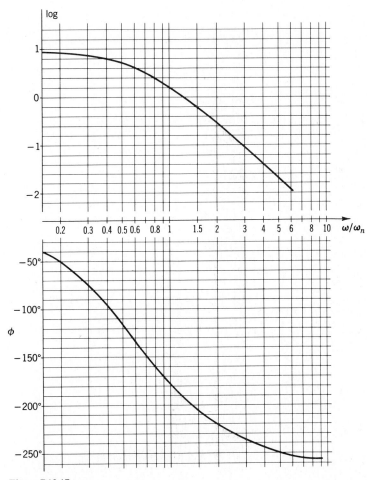

Figure P12.17

(a) Find the values of τ_1 and τ_2 for a series lag-lead compensator such that the resulting system will have a phase margin of $45° \pm 3°$.

(b) Use Fig. P12.17 to construct the log-modulus plot for this system. Using this plot, verify the results of part (a).

12.18 In Fig. P12.18 is shown the log-magnitude diagram for the unity-feedback system

$$G(s) = \frac{2.5}{s(1 + 0.25s)(1 + s)}$$

Determine the values of τ_1 and τ_2 for a series lead compensator $(1 + \tau_1 s)/(1 + \tau_2 s)$ such that the resulting system will have a phase margin of $45° \pm 5°$.

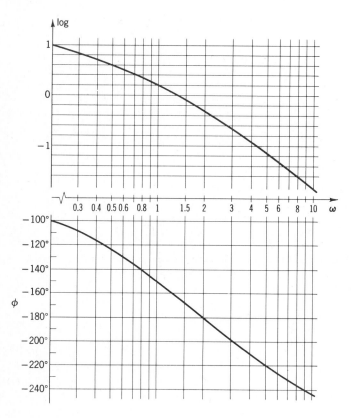

Figure P12.18

12.19 Same as Prob. 12.18 except use a lag compensator.

12.20 Same as Prob. 12.19 except use a lag-lead compensator.

12.21 Construct the approximate log-magnitude plots for each of the two systems shown in Fig. P12.21. For each system, write the equation for the open-loop transfer function corresponding to the asymptotes. To obtain the exact transfer function for each system, use block-diagram algebra to eliminate the minor feedback loops. Compare the exact and approximate transfer functions.

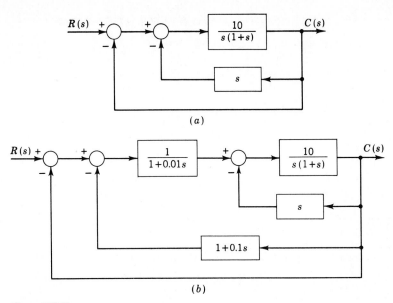

(a)

(b)

Figure P12.21

12.22 Same as Prob. 11.22, but use inverse polar plots to determine the value of the gain K to yield an M_p of 1.4.

12.23 The inverse polar plot for the unity-feedback system

$$G_1(s) = \frac{2.5}{s(1 + 0.25s)(1 + s)} = \frac{10}{s(s + 1)(s + 4)}$$

is shown in Fig. P12.23. For internal-feedback compensation of the form $H_1(s) = \alpha + \beta s$, determine the values of α and β such that the resulting system will have an M_p of 1.4 and ω_p occurs at $\omega = 2$.

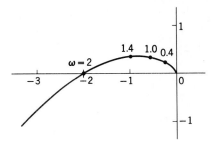

Figure P12.23

12.24 Use inverse polar plots to determine the value of β such that the system of Fig. P12.24 will have an M_p of 1.4.

Figure P12.24

12.25 For the system shown in Fig. P12.25, determine the value of K_1 to yield an M_p of 1.4.

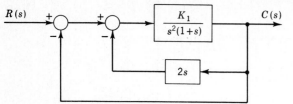

Figure P12.25

12.26 Same as Prob. 12.2 except use the inverse plane.

12.27 Same as Prob. 12.3 except use the inverse plane rather than the log-magnitude diagram.

REFERENCE

1. Nyquist, H.: Regeneration Theory, *Bell System Tech. J.*, **11**, 126–147, 1932.

EQUILIBRIUM FLOW

Many pneumatic components use two orifices in series to obtain a controlled pressure in the chamber between the orifices. If the component is available, the chamber pressure may be experimentally determined for various operating conditions. However, in the initial design stages, before any parts have been manufactured, it is desirable to be able to predict the value of the chamber pressure. This may be accomplished by use of the nondimensional family of curves shown in Fig. A.1a. As is illustrated by the insert above Fig. A.1a, the pressures P_1, P_2, and P_3 represent the inlet pressure, chamber pressure, and discharge pressure respectively. The symbol A_1 is the area of the first orifice times the coefficient of discharge and A_2 is the area of the second orifice times the coefficient of discharge.

Usually, the overall pressure ratio P_1/P_3 is known, and also the ratio A_2/A_1 is known, so that the ratio P_2/P_1 can be found from Fig. A.1a. The value of the chamber pressure P_2 is then computed as the product of the ratio P_2/P_1 and the value of the inlet pressure P_1. In using Fig. A.1a, it is necessary to use absolute pressure. Because these are nondimensional curves, any consistent set of units may be used.

The derivation of this nondimensional family of curves is accomplished as follows. Assuming that the fluid is a perfect gas and that the kinetic energy at the inlet is negligible compared with other terms in the energy equation, the mass rate of flow through the first orifice is

$$M_1 = \frac{A_1 P_1}{\sqrt{T_1}} \left\{ 2g_c \frac{k}{k-1} \frac{1}{R} \left[\left(\frac{P_{1_t}}{P_1}\right)^{2/k} - \left(\frac{P_{1_t}}{P_1}\right)^{(k+1)/k} \right] \right\}^{1/2} \qquad (A.1)$$

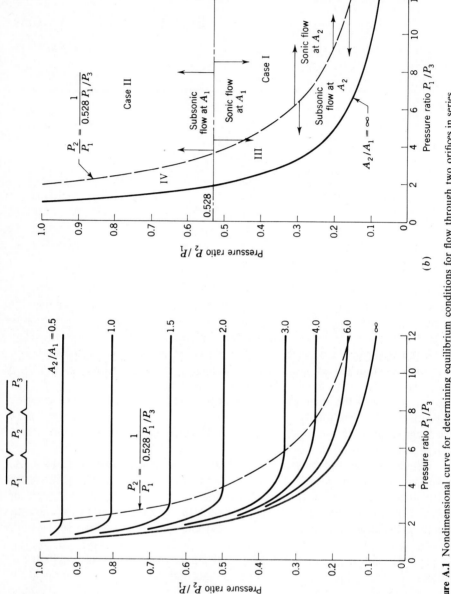

Figure A.1 Nondimensional curve for determining equilibrium conditions for flow through two orifices in series.

where T_1 = stagnation temperature at inlet

g_c = mass conversion constant

k = ratio of specific heat at constant pressure to that at constant volume

P_{1_t} = throat pressure at first orifice

R = gas constant

By replacing the subscript 1 in Eq. (A.1) by 2 and the subscript 1_t by 2_t, the equation for the mass rate of flow through the second orifice is obtained. For equilibrium to exist, the mass rate of flow in must equal that out, so that

$$\frac{A_1 P_1}{\sqrt{T_1}} B_1^{1/2} = \frac{A_2 P_2}{\sqrt{T_2}} B_2^{1/2} \tag{A.2}$$

where

$$B_1 = \left(\frac{P_{1_t}}{P_1}\right)^{2/k} - \left(\frac{P_{1_t}}{P_1}\right)^{(k+1)/k}$$

$$B_2 = \left(\frac{P_{2_t}}{P_2}\right)^{2/k} - \left(\frac{P_{2_t}}{P_2}\right)^{(k+1)/k}$$

Because there is little time for heat transfer to take place, the flow may be considered to be adiabatic, so that $T_1 = T_2$. Thus Eq. (A.2) becomes

$$A_1 P_1 B_1^{1/2} = A_2 P_2 B_2^{1/2} \tag{A.3}$$

In the following analysis, it is assumed that the fluid is air, for which $k = 1.4$, and the critical pressure ratio is $P_2/P_1 = P_3/P_2 = 0.528$. By using the appropriate value of k and the critical ratio, this analysis is applicable for any gas.

When sonic flow exists at the first orifice, $B_1^{1/2} = 0.259$, and similarly for sonic flow at the second orifice $B_2^{1/2} = 0.259$. Thus, for sonic flow at both orifices Eq. (A.3) reduces to

$$\frac{P_2}{P_1} = \frac{1}{A_2/A_1} \tag{A.4}$$

When sonic flow exists at the first orifice,

$$\frac{P_2}{P_1} \leq 0.528$$

Above the line $P_2/P_1 = 0.528$ shown in Fig. A.1b subsonic flow exists at A_1, and below this line sonic flow exists at A_1.

The equation for the line of separation between subsonic and sonic flow at A_2 is obtained by noting that the critical ratio is $P_3/P_2 = 0.528$; thus,

$$\frac{P_2}{P_1} = \frac{1}{(P_3/P_2)(P_1/P_3)} = \frac{1}{0.528(P_1/P_3)} \tag{A.5}$$

The curve defined by the preceding expression is shown in Fig. A.1b. To the right of this curve sonic flow exists at A_2 and to the left of this curve subsonic flow exists at A_2. The regions in which each of the four possible combinations of sonic or subsonic flow exist at the first and second orifices are shown in Fig. A.1b, that is,

Case I.   Sonic flow at both orifices
Case II. Subsonic flow at the first orifice and sonic at the second
Case III. Sonic flow at the first orifice and subsonic at the second
Case IV. Subsonic flow at both orifices

For case I, it follows from Eq. (A.4) that the lines of constant values of A_2/A_1 are horizontal straight lines as shown in Fig. A.1a.

For case II, $P_2 = P_{1_t}$ and $B_2^{1/2} = 0.259$, so that Eq. (A.3) reduces to

$$\frac{P_2}{P_1} = \frac{1}{B_1^{1/2}} = \frac{1}{0.259(A_2/A_1)} \tag{A.6}$$

For a given area ratio A_2/A_1, there is but one value of P_2/P_1 which makes the left side of Eq. (A.6) equal to the right side. Thus, for case II, lines of constant A_2/A_1 are also horizontal.

By applying these techniques to case III and case IV, the complete family of curves shown in Fig. A.1a is obtained. This method of analysis may be extended to determine equilibrium flow conditions for three or more orifices in series.

Equation (A.1) is an awkward form for computing partial derivatives. However, for the case of sonic flow, this expression reduces to

$$M_1 = \frac{0.53}{\sqrt{T_1}} A_1 P_1 \tag{A.7}$$

where the preceding expression has units of pounds-mass (lb_m), pounds-force (lb_f), feet, degrees Rankine, and seconds. For the usual design case in which the stagnation temperature of the inlet air is 60°F or 520°R, then $M_1 = 0.0074 A_1 P_1$.

For the case of subsonic flow, $P_{1_t} = P_2$, Eq. (A.1) may be approximated by noting that a plot of the function B_1 versus $(P_1 - P_2)P_2/P_1^2$ is very nearly a straight line. The slope of this line is such that

$$B_1 \approx \frac{0.261(P_1 - P_2)P_2}{P_1^2} \tag{A.8}$$

Substitution of the preceding approximation into Eq. (A.1) gives

$$M_1 \approx \frac{2.06}{\sqrt{T_1}} A_1 \sqrt{0.261(P_1 - P_2)P_2}$$

$$= \frac{1.05}{\sqrt{T_1}} A_1 \sqrt{(P_1 - P_2)P_2} \tag{A.9}$$

where $\sqrt{(2g_c/R)[k/(k-1)]} = 2.06$. For a stagnation temperature of 520°R, Eq. (A.9) is

$$M_1 = 0.046 A_1 \sqrt{(P_1 - P_2)P_2} \tag{A.10}$$

The length units cancel in Eqs. (A.7) to (A.10); hence any unit of length may be used.

Because the curves shown in Fig. A.1 are nondimensional, they are valid for SI units as well as British gravitational units. When using SI units (kg, m, N, s, K), the g_c term does not appear in Eq. (A.1). Equations (A.2) to (A.6) are valid for SI units as well as British gravitational units. For SI units the constant in Eq. (A.7) should be 0.04 rather than 0.53, the constant 0.261 in Eq. (A.8) remains the same, the constant 2.06 in Eq. (A.9) becomes 0.156, and the constant 1.05 becomes 0.08. Finally, the constant 0.046 in Eq. (A.10) becomes 0.0047 for SI units.

B

FOURIER SERIES, FOURIER INTEGRAL, AND LAPLACE TRANSFORM

A greater understanding of the Laplace transform $F(s)$ of a time function $f(t)$ may be obtained by examining the similarities which exist between Laplace transforms and the more familiar Fourier series and Fourier integral.

B.1 FOURIER SERIES

A periodic function as shown in Fig. B.1 may be represented by the series

$$f(t) = K + \sum_{n=1}^{\infty} (A_n \cos n\omega_0 t + B_n \sin n\omega_0 t) \tag{B.1}$$

where $\omega_0 = 2\pi/T$, in which T is the period. The constant K is evaluated as follows. Integration of each term in Eq. (B.1) over a complete period causes each term in the summation on the right side to vanish. Thus,

$$\int_{-T/2}^{T/2} f(t)\, dt = K \int_{-T/2}^{T/2} dt + 0 = KT$$

or

$$K = \frac{1}{T} \int_{-T/2}^{T/2} f(t)\, dt \tag{B.2}$$

The value of K is seen to be equal to the average value of the function over a period.

Figure B.1 Periodic function.

To evaluate A_n each term of Eq. (B.1) is multiplied by $\cos m\omega_0 t$ and then integrated over a period. Because

$$\int_{-T/2}^{T/2} \cos n\omega_0 t \cos m\omega_0 t \; dt = \begin{cases} 0 & \text{for } m \ne n \\ T/2 & \text{for } m = n \end{cases} \qquad (B.3)$$

and

$$\int_{-T/2}^{T/2} \sin n\omega_0 t \cos m\omega_0 t \; dt = 0 \qquad (B.4)$$

it follows that

$$A_n = \frac{2}{T} \int_{-T/2}^{T/2} f(t) \cos n\omega_0 t \; dt \qquad (B.5)$$

Similarly, multiplication of each term of Eq. (B.1) by $\sin m\omega_0 t$ and integration over the period yields the following result for B_n:

$$B_n = \frac{2}{T} \int_{-T/2}^{T/2} f(t) \sin n\omega_0 t \; dt \qquad (B.6)$$

Equation (B.1) may be telescoped into a more convenient form by using Eqs. (5.32) and (5.33) to express the cosine and sine in exponential form:

$$A_n \cos n\omega_0 t = \frac{A_n}{2} (e^{jn\omega_0 t} + e^{-jn\omega_0 t})$$

$$B_n \sin n\omega_0 t = -j \frac{B_n}{2} (e^{jn\omega_0 t} - e^{-jn\omega_0 t})$$

Thus,

$$f(t) = K + \frac{1}{2} \sum_{n=1}^{\infty} (A_n - jB_n)e^{jn\omega_0 t} + (A_n + jB_n)e^{-jn\omega_0 t} \qquad (B.7)$$

By also writing Eqs. (B.5) and (B.6) in exponential form, it can be seen that

$$A_n - jB_n = \frac{2}{T} \int_{-T/2}^{T/2} f(t)(\cos n\omega_0 t - j \sin n\omega_0 t) \; dt$$

$$= \frac{2}{T} \int_{-T/2}^{T/2} f(t)e^{-jn\omega_0 t} \; dt \qquad (B.8)$$

and
$$A_n + jB_n = \frac{2}{T} \int_{-T/2}^{T/2} f(t)(\cos n\omega_0 t + j \sin n\omega_0 t) \, dt$$

$$= \frac{2}{T} \int_{-T/2}^{T/2} f(t) e^{jn\omega_0 t} \, dt \tag{B.9}$$

Substitution of the preceding results into Eq. (B.7) gives

$$f(t) = K + \frac{1}{T} \sum_{n=1}^{\infty} e^{jn\omega_0 t} \int_{-T/2}^{T/2} f(t) e^{-jn\omega_0 t} \, dt$$

$$+ \frac{1}{T} \sum_{n=1}^{\infty} e^{-jn\omega_0 t} \int_{-T/2}^{T/2} f(t) e^{jn\omega_0 t} \, dt \tag{B.10}$$

By noting that the last summation is unaltered by changing the sign of n, the Fourier series becomes

$$f(t) = K + \sum_{n=1}^{\infty} \frac{e^{jn\omega_0 t}}{T} \int_{-T/2}^{T/2} f(t) e^{-jn\omega_0 t} \, dt$$

$$+ \sum_{n=-1}^{-\infty} \frac{e^{jn\omega_0 t}}{T} \int_{-T/2}^{T/2} f(t) e^{-jn\omega_0 t} \, dt \tag{B.11}$$

Because the value of the summation for $n = 0$ is K,

$$f(t) = \sum_{n=-\infty}^{\infty} \frac{e^{jn\omega_0 t}}{T} \int_{-T/2}^{T/2} f(t) e^{-jn\omega_0 t} \, dt \tag{B.12}$$

Equation (B.12) is frequently written in the form

$$f(t) = \sum_{n=-\infty}^{\infty} C_n e^{jn\omega_0 t} \tag{B.13}$$

where

$$C_n = \frac{1}{T} \int_{-T/2}^{T/2} f(t) e^{-jn\omega_0 t} \, dt$$

B.2 FOURIER INTEGRAL

As the period T becomes infinite, the Fourier series expression given by Eq. (B.12) is

$$f(t) = \lim_{T \to \infty} \left[\sum_{n=-\infty}^{\infty} \frac{e^{jn\omega_0 t}}{T} \int_{-T/2}^{T/2} f(t) e^{-jn\omega_0 t} \, dt \right] \tag{B.14}$$

For large values of T it is more appropriate to use the following notation:

$$\lim_{T \to \infty} \omega_0 = \lim_{T \to \infty} \frac{2\pi}{T} = \Delta\omega$$

and
$$\lim_{T \to \infty} n\omega_0 = n \, \Delta\omega = \omega$$

Thus, Eq. (B.14) becomes

$$f(t) = \lim_{\substack{\Delta\omega \to 0 \\ T \to \infty}} \left[\frac{1}{2\pi} \sum_{n=-\infty}^{\infty} e^{j\omega t} \Delta\omega \int_{-T/2}^{T/2} f(t) e^{-j\omega t} \, dt \right] \tag{B.15}$$

The limit of Eq. (B.15) is the Fourier integral.

$$f(t) = \frac{1}{2\pi} \int_{-\infty}^{\infty} e^{j\omega t} \left[\int_{-\infty}^{\infty} f(t) e^{-j\omega t} \, dt \right] d\omega \tag{B.16}$$

The Fourier integral is frequently expressed by the Fourier transform pair

$$f(t) = \frac{1}{2\pi} \int_{-\infty}^{\infty} F(j\omega) e^{j\omega t} \, d\omega \tag{B.17}$$

$$F(j\omega) = \int_{-\infty}^{\infty} f(t) e^{-j\omega t} \, dt \tag{B.18}$$

Equation (B.18) is referred to as the direct Fourier transform and Eq. (B.17) is the inverse Fourier transform.

For most physical problems, it is desired to know the solution for $t > 0$. Thus, if the initial conditions are known, then the lower limit of integration in Eq. (B.18) may be taken as zero.

To illustrate the use of the Fourier transform, consider the function

$$f(t) = e^{at} \qquad t \geq 0 \tag{B.19}$$

Application of the direct Fourier transform gives

$$F(j\omega) = \int_0^{\infty} e^{at} e^{-j\omega t} \, dt = \frac{e^{at} e^{-j\omega t}}{a - j\omega} \bigg|_0^{\infty} \tag{B.20}$$

If the exponent a is less than zero, the preceding expression becomes

$$F(j\omega) = 0 - \frac{1}{a - j\omega} = \frac{1}{j\omega - a} \qquad a < 0 \tag{B.21}$$

However, if the exponent a is positive, e^{at} becomes infinite when evaluated at $t = \infty$ and thus $F(j\omega)$ diverges.

B.3 LAPLACE TRANSFORM

To extend the usefulness of the Fourier transform so that it is applicable to divergent functions, a converging factor $e^{-\sigma t}$ is introduced. Thus, the general transform equation is

$$F(\sigma + j\omega) = \int_0^{\infty} f(t) e^{-\sigma t} e^{-j\omega t} \, dt$$

$$= \int_0^{\infty} f(t) e^{-(\sigma + j\omega)t} \, dt \tag{B.22}$$

The transform for the time function given by Eq. (B.19) is

$$F(\sigma + j\omega) = \frac{e^{(a - \sigma - j\omega)t}}{a - \sigma - j\omega}\bigg|_0^\infty$$

$$= \frac{1}{(\sigma + j\omega) - a} \qquad \sigma > a \qquad (B.23)$$

The preceding expression is seen to converge when σ is greater than a. To ensure convergence of the Fourier transform, it was necessary that $\int_0^\infty |f(t)| \, dt < \infty$. However, the transform which is indicated by Eq. (B.22) converges when $\int_0^\infty | f(t)|e^{-\sigma t} \, dt < \infty$ for some finite σ.

The substitution of $s = \sigma + j\omega$ and $F(s) = F(\sigma + j\omega)$ into Eq. (B.22) yields the Laplace transform equation

$$F(s) = \int_0^\infty f(t)e^{-st} \, dt \qquad (B.24)$$

It is only necessary that some finite value of σ exists such that $\int_0^\infty | f(t)|e^{-\sigma t} \, dt < \infty$ in order to verify the existence of the transform indicated by Eq. (B.24). For most functions $f(t)$ encountered in engineering work, the transform $F(s)$ is convergent. It should also be noted that, to solve differential equations by Laplace transforms, it is not necessary to determine the value or values of σ over which $F(s)$ is convergent. It suffices to know that such a value or values of σ exist.

In effect, Eq. (B.24) is the result of substituting s for $j\omega$ and $F(s)$ for $F(j\omega)$ in Eq. (B.18). The use of these same substitutions in Eq. (B.17) yields the inverse transform, that is,

$$f(t) = \frac{1}{2\pi j} \int_{\sigma - j\infty}^{\sigma + j\infty} F(s)e^{st} \, ds \qquad (B.25)$$

The new limits of integration are obtained by noting that, when $\omega = \pm\infty$, then $s = \sigma \pm j\omega = \sigma \pm j\infty$. Equation (B.25) is a line integral for which the path of integration is a vertical line which is displaced a distance σ from the imaginary axis, as is shown in Fig. B.2. For convergence, it is necessary that σ be such that all the values of s which make $F(s)$ infinite [i.e., poles of $F(s)$] lie to the left of the vertical line shown in Fig. B.2.

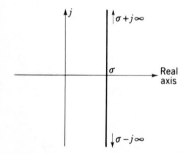

Figure B.2 Vertical line $s = \sigma + j\omega$.

INDEX